Advances in Science, Technology & Innovation

IEREK Interdisciplinary Series for Sustainable Development

Advances in Science, Technology & Innovation (ASTI) is a series of peer-reviewed books based on important emerging research that redefines the current disciplinary boundaries in science, technology and innovation (STI) in order to develop integrated concepts for sustainable development. It not only discusses the progress made towards securing more resources, allocating smarter solutions, and rebalancing the relationship between nature and people, but also provides in-depth insights from comprehensive research that addresses the **17 sustainable development goals (SDGs)** as set out by the UN for 2030.

The series draws on the best research papers from various IEREK and other international conferences to promote the creation and development of viable solutions for a **sustainable future and a positive societal** transformation with the help of integrated and innovative science-based approaches. Including interdisciplinary contributions, it presents innovative approaches and highlights how they can best support both economic and sustainable development, through better use of data, more effective institutions, and global, local and individual action, for the welfare of all societies.

The series particularly features conceptual and empirical contributions from various interrelated fields of science, technology and innovation, with an emphasis on digital transformation, that focus on providing practical solutions to **ensure food, water and energy security to achieve the SDGs.** It also presents new case studies offering concrete examples of how to resolve sustainable urbanization and environmental issues in different regions of the world.

The series is intended for professionals in research and teaching, consultancies and industry, and government and international organizations. Published in collaboration with IEREK, the Springer ASTI series will acquaint readers with essential new studies in STI for sustainable development.

ASTI series has now been accepted for Scopus (September 2020). All content published in this series will start appearing on the Scopus site in early 2021.

Pir Mohammad · Shahfahad ·
Jesús Rodrigo Comino · Abraham R. Matamanda
Editors

Geographic Approaches to Climate Change and Mitigation: Urban and Rural Perspectives (Volume 1)

A Culmination of Selected Research Papers from the International Conference on Geographic Perspectives on Climate Change Mitigation in Urban and Rural Environment–1st Edition—University of Salento, Italy

Springer

Editors
Pir Mohammad
Department of Land Surveying
and Geo-Informatics
The Hong Kong Polytechnic University
Hung Hom, Hong Kong

Jesús Rodrigo Comino
Department of Regional Geographic Analysis
and Physical Geography
University of Granada
Granada, Spain

Shahfahad
Department of Geography, Institute of Science
Banaras Hindu University
Varanasi, Uttar Pradesh, India

Abraham R. Matamanda
Department of Urban and Regional Planning
University of the Free State
Bloemfontein, South Africa

ISSN 2522-8714 ISSN 2522-8722 (electronic)
Advances in Science, Technology & Innovation
IEREK Interdisciplinary Series for Sustainable Development
ISBN 978-3-031-92118-6 ISBN 978-3-031-92119-3 (eBook)
https://doi.org/10.1007/978-3-031-92119-3

This Springer imprint is published by the registered company Springer Nature Switzerland AG
The registered company address is: Gewerbestrasse 11, 6330 Cham, Switzerland

Scientific Committee

Jose Maria cabeza Lainez, University of Seville, Spain
Putu Indra Chirstiawan, Ganesha University of Education, Indonesia
Bonin Mahdavi Estalkhsari, Shahid Beheshti University, Iran
Darya Gachkar, Universidad De Sevilla, Spain
Bijay Halder, Universiti Kebangsaan Malaysia, Malaysia
Asadollah Khoorani, University of Hormozgan, Iran
Gomal Amin, The Hong Kong Polytechnic University, China
Saraah Imran, Indian Institute of Technology Roorkee, India
Aqil Tariq, Mississippi State University, USA
Casandra Muñoz Gómez, Universidad Autónoma de México, Mexico
Rabin Chakrabortty, University of Burdwan, India
Malay Prananik, Asian Institute of Technology, Thailand
Surendra Tiwari, Asian Institute of Technology, Thailand
María Teresa González, University of Granada, Spain
Rinku Mahanta, Indian Institute of Technology, Roorkee, India
Deepak Rawat, Indian Institute of Technology, Roorkee, India
Mahin Kalehoui, Tarbiat Modares University, Iran
Antonio Pena Garcia, University of Granada, Spain
Priyaka Negi, Indian Institute of Technology, Roorkee, India
Suvamoy Pramanik, University of Delhi, New Delhi, India
Arijit Das, Technical University Dortmund, Germany
Mohana Sundaram Shanmugam, Asian Institute of Technology, Thailand
Saurav K.C., Asian Institute of Technology, Thailand
Priyaka Negi, Indian Institute of Technology, Roorkee, India
Swapan Talukdar, Asutosh College, Kolkata, India
Takuji W. Tsusaka, Asian Institute of Technology, Thailand

The Editors warmly thank all the Reviewers who have contributed their authority to the double-blind review process, to ensure the quality of this publication.

Preface

As we travel through the intertwined narratives of climate change mitigation in urban and rural environments, it is essential to recognize the urgency and complexity of the challenge we face. Climate change is not merely an environmental issue; it is a multifaceted crisis that touches every aspect of human life, economy, and society. From the bustling streets of metropolitan areas to the serene landscapes of rural communities, its effects are felt universally, albeit in diverse ways.

This book, *Geographic Approaches to Climate Change and Mitigation: Urban and Rural Perspectives (Volume 1)*, is an outcome of the selected papers presented at the 1st International Conference on Geographic Perspectives on Climate Change Mitigation in Urban and Rural Environments (GCUE 2024) held from 25 to 27, June 2024, at the University of Salento, Italy. This conference is a shared commitment to exploring these varied impacts and to highlight innovative strategies that can be employed across different settings. Our aim is to close the gap between urban and rural approaches, encouraging an understanding of how climate solutions can be tailored to the unique characteristics of each environment. This edited book is divided into two volumes. This volume will be divided into six parts, consisting of 33 chapters. Each chapter is devoted to distinct yet interconnected aspects of the subject matter.

This book is not merely an assortment of academic papers; it represents a collective endeavor to understand and address the complexities of urban and rural environments. Each contributor brings a wealth of knowledge and insight, sharing case studies and research that illuminate successful initiatives, and the lessons learned along the way. This collaborative effort is a testament to the power of interdisciplinary work and the importance of engaging multiple perspectives in addressing global challenges.

As you go through the chapters that follow, I encourage you to reflect on your own role in climate conversation. Whether you are a researcher, a student, a policymaker, or a community advocate, each of us has a part to play in shaping the future of our planet. The insights and strategies presented here are intended not only to inform but also to inspire action.

In closing, I would like to express my gratitude to all the contributors, reviewers, and supporters who have made this book possible. Together, we can forge a path toward a more sustainable and resilient future, one that honors the unique contributions of both urban and rural environments in the fight against climate change.

Hung Hom, Hong Kong — Pir Mohammad

Acknowledgments

The completion of this book, *Geographic Approaches to Climate Change and Mitigation: Urban and Rural Perspectives (Volume 1)*, represents the collective effort of many individuals and organizations whose contributions have been invaluable. First and foremost, we would like to express our deepest gratitude to the contributing authors, whose expertise and insights have enriched this volume. Your dedication to advancing the understanding of climate change mitigation in diverse contexts is evident in every chapter. We appreciate the time, effort, and passion you invested in your research and writing.

We extend our heartfelt thanks to the reviewers, whose constructive feedback and critical evaluations have greatly enhanced the quality of this book. Your attention to detail and commitment to excellence have ensured that the content is both rigorous and accessible.

We acknowledge and appreciate the efforts of IEREK in organizing the conference in collaboration with University of Salento, Italy and giving a platform of various researches worldwide to come together and discuss the issues of urban and rural environments in regard to climate change and its mitigation. We also appreciate and acknowledge the hosts and co-hosts of the conference for organizing it. We are extremely grateful to all our supporting organizations of the conference.

A special thank you goes to our editorial team and publishing staff, whose professionalism and guidance have been instrumental in bringing this book to fruition. Your expertise in navigating the publishing process has been greatly appreciated.

To our families and friends, thank you for your understanding and encouragement during the long hours spent working on this project. Your support has been a source of strength and motivation.

Finally, we would like to acknowledge the communities, both urban and rural, who are at the forefront of climate change impacts. Your resilience, creativity, and commitment to sustainability inspire us all. This book is dedicated to you and to those who strive to create a more equitable and sustainable future for our planet.

Introduction

Climate change is one of the most pressing challenges of our time, affecting ecosystems, economies, and communities across the globe. From rising sea levels and increased frequency of extreme weather events to shifts in agricultural productivity and public health crises, the ramifications of climate change are profound and far-reaching. As the urgency to address these issues escalates, the need for effective and innovative climate change mitigation strategies becomes necessary.

This book, *Geographic Approaches to Climate Change and Mitigation: Urban and Rural Perspectives (Volume 1)*, aims to explore the diverse approaches to climate change mitigation through the lens of geography, emphasizing the distinct contexts of urban and rural areas.

Urban environments are at the forefront of the climate crisis. They are characterized by high population density and significant infrastructure, facing unique challenges in mitigating climate change. The challenges faced by urban centers are multifaceted: from managing waste and transportation emissions to ensuring equitable access to green spaces and sustainable resources. Cities are often hotspots for greenhouse gas emissions, but they also serve as centers for innovation, policymaking, and community mobilization. They possess the infrastructure, economic resources, and human capital necessary to implement transformative change. By harnessing technology, fostering public engagement, and promoting sustainable practices, urban areas can lead the way in mitigating climate change impacts.

In addition, rural areas are often overlooked in climate discussions even though they play a critical role in both climate change mitigation and adaptation. Though they contribute less to global emissions, rural regions are uniquely vulnerable to the effects of climate change, including altered precipitation patterns, extreme heat, and threats to food security. These communities are often dependent on natural resources for their livelihoods, making them particularly sensitive to environmental changes. Understanding the challenges faced by rural areas is essential for creating strategies that not only reduce emissions but also enhance resilience and support sustainable development.

This book seeks to close the gap between urban and rural perspectives, recognizing that both environments must work together to achieve meaningful climate action. The interplay between urban and rural areas is complex; for instance, urban demand for resources can drive unsustainable practices in rural regions, while rural ecological health can directly impact urban quality of life. By encouraging collaboration between urban and rural stakeholders, we can develop integrated strategies that use the strengths of both environments.

Throughout this volume, readers will find a diverse range of perspectives from scholars, practitioners, and policymakers who are actively engaged in climate change mitigation. This volume explores six broad topic areas: Applications of Climatology in Urbanism, The Role of Geography in Climate Change Mitigation, Meteorology and Disaster Prevention in Urban and Rural Environments, Urban Planning and Climate Change Mitigation, Rural Land Use and Climate Change Mitigation, and Geographic Mapping and Innovative Systems for Rural Fragmentation Protection. Each chapter addresses key questions that are vital to our collective understanding: How can urban areas leverage their resources and infrastructure to spearhead climate mitigation efforts? What innovative practices are rural communities employing to enhance resilience while minimizing their carbon footprints? How can policies be designed to promote synergy between urban and rural stakeholders, ensuring a cohesive approach to climate action?

The contributions in this book are collected in the form of case studies and empirical research, showcasing successful initiatives and lessons learned from various geographic contexts. From urban green infrastructure projects aimed at reducing heat island effects to rural agricultural practices that enhance soil health and carbon sequestration, the chapters illustrate the breadth of strategies available. Additionally, theoretical frameworks are presented to help contextualize these practices within broader climate change narratives.

Ultimately, this book aims to work as a resource for researchers, students, and decision-makers committed to combating climate change. By integrating geographic perspectives and emphasizing the unique challenges and opportunities within urban and rural environments, we hope to inspire new ideas and collaborative efforts. Our goal is to promote resilient communities that are not only equipped to face the challenges of a changing climate but also actively contribute to a sustainable and equitable future for all.

Pir Mohammad

Contents

Applications of Climatology in Urbanism

Soaked Cities: Climatic Moisture and Urbanization Patterns in India from 1971 to 2011

Eva Klaus

Abstract

Rapid urbanization and climate change will be immensely transformative processes in the decades to come, particularly in many developing countries. It is essential to understand how their interaction will compound growing urban pressures. More specifically, how does climate change affect urbanization trends and city growth? This study considers the experience of India, employing a 40-year panel dataset from 1971 to 2011 covering 485 districts across the country to assess how changes in climatic moisture affect urbanization, urban population growth, and rural population growth at the district level. The findings suggest that drying leads to more rapid growth in the urban population share, which is consistent with the literature on climate change acting as a "push" factor for urbanization. However, this paper also finds that changes in climatic moisture had a statistically insignificant effect on a district's total urban population growth; instead, drying contributes to slower rural population growth. This suggests that increases in urban share growth in drying districts may be attributable, in part, to stagnating rural growth, potentially driven by rural out-migration. Second, this study employs another 40-year panel dataset covering 2222 urban centers in India to assess the impact of changes in climatic moisture on city population growth. The findings suggest that cities located in districts that experienced an increase in moisture saw more rapid growth than cities located in districts that experienced drying. Taken together, these results suggest that drying districts may indeed urbanize more rapidly, but districts experiencing increases in climatic moisture may face more intense population pressures.

E. Klaus (✉)
Charter Cities Institute, Washington, D.C., USA
e-mail: eva@cci.city

Keywords

Urbanization · Moisture · Climate change · India · Population growth

1 Introduction and Motivation

Over the coming decades, urbanization and climate change will profoundly transform our world. However, these two powerful phenomena are projected to have a particularly large impact on developing countries. Cities are currently home to over half of the world's population, but the UN Framework Convention on Climate Change projects they will accommodate over 70% of the total population by 2050, with the overwhelming majority of urban growth concentrated in countries across Africa, Asia, and South America (UNFCC 2017). Rapidly burgeoning urban centers may pose major challenges for national and municipal governments struggling to keep up with the pace of expansion and growing demand for public services. Historically, unplanned urban growth has resulted in the "proliferation of slums, congestion, pollution, lack of affordable housing, poor access to sanitation and waste management, and vulnerability to natural hazards" (World Bank 2021).

As a result, rapid and unsustainable urbanization "could make the world's society and economy increasingly vulnerable to the impacts of climate change…, [particularly] urban centers… in developing countries" (UNFCC 2017). Indeed, the United Nations warns that the "effects of urbanization and climate change are converging in dangerous ways" (UN Habitat n.d.). For example, hasty, unsupervised, and unsound construction may increase the likelihood of damage to infrastructure and dwelling units during severe storms, which are projected to become increasingly intense and unpredictable as climate change continues (IPCC 2022, 938). Similarly, land degradation resulting from urban sprawl

P. Mohammad et al. (eds.), *Geographic Approaches to Climate Change and Mitigation: Urban and Rural Perspectives (Volume 1)*,
Advances in Science, Technology & Innovation, https://doi.org/10.1007/978-3-031-92119-3_1

may heighten susceptibility to devastating landslides during periods of heavy rainfall (Ozturk et al. 2022). Rising temperatures also threaten to turn cities into urban heat islands. Recent work by Tuholske et al. (2021) finds that between 1982 and 2016, global exposure to extreme heat in urban areas "increased nearly 200%, affecting 1.7 billion people." Remarkably, 52.4% of the increase in exposure to extreme heat in urban areas was concentrated in India alone, due to both the country's rapid urban population growth and its vulnerabilities to climate change (Tuholske et al. 2021).

On the other hand, sustainable urbanization has the potential to foster human flourishing at all levels and support climate adaptation efforts.[1] In fact, there is "a near-perfect correlation between urbanization and prosperity across nations. On average, as the share of a country's population that is urban rises by 10%, the country's per capita output increases by 30%" (Glaeser 2012, 7). Cities, at their best, are epicenters of opportunity, innovation, and economic growth. As such, they play an important role in the fight against climate change. For one, individuals and households whose livelihoods stand to be negatively impacted as climate change "reshapes the comparative advantage of regions" will increasingly rely on urban centers for alternative employment opportunities and stability (Adger et al. 2020). For example, work by Rigaud et al. (2018) at the World Bank finds that climatic variability, drought, and other hazards caused by climate change will negatively impact agricultural livelihoods in South Asia—a region projected to see 35.6 million internal climate migrants by 2050—leading to urban migration as people seek alternative sources of income (90). Furthermore, efficient and informed urban planning can also enable households to better withstand the impacts of climate change through high-quality public service provision—such as water and sanitation services, electricity, and waste management—and sound infrastructure and dwelling construction. Finally, the knowledge sharing and innovation that arise from urban agglomerations have the potential to spur the development of new green technologies that support climate resilience in both urban and rural areas.

India is one of many countries across the Global South which is expected to experience significant urban growth and expansion over the coming decades. In fact, by 2050, India is expected to add "404 million inhabitants" to its cities (Abhishek et al. 2017, 48). Furthermore, the subcontinent is "predicted to acquire 18.6 million more hectares [of urban land] by 2030" (Harari 2020, 1). Cities such as Hyderabad, which are expected to grow by over 10 million residents before 2041, will likely undergo substantial physical expansion (Alam and Markandey 2020; Hatab et al. 2021). Managing such urbanization and adequately preparing cities for rapid growth will allow citizens to harness the economic opportunities that come with urban development and help improve overall climate resilience. Christiaensen et al. (2013) point out that "how the world urbanizes may well be as important as urbanization itself."

As such, it is essential to understand how urbanization and climate change will interact to compound growing population pressures on urban areas. Existing literature suggests that climate change may influence urbanization dynamics and city growth in a variety of ways.[2] For one, regions experiencing higher temperatures, drier conditions, or more extreme weather patterns may see declines in agricultural productivity, driving rural-to-urban migration or rural out-migration as residents engaged in agriculture search for new employment opportunities (Barrios et al. 2006; Henderson et al. 2017). Alternatively, regions experiencing milder conditions may see agricultural productivity gains and rural growth, leading to population and savings spillovers into nearby cities, thus contributing to greater urban population growth (Asher et al. 2022; Bustos et al. 2019). Recent research suggests that these two processes may, in fact, occur simultaneously and in close proximity to one another (Tuholske et al. 2019).

This study makes an original contribution to this debate by exploiting district-level heterogeneity in climatic moisture to estimate the impact of climate change on urbanization and city growth in India between the years of 1971 and 2011 (see Fig. 1).[3] The central contention of this study is that changes in climatic moisture are exogenous and directly impact agricultural productivity, which subsequently influences urbanization patterns by either pushing rural residents into urban areas or supporting nearby city growth through spillovers. The guiding research question is thus divided into two parts regarding how climate change affects: (i) district-level urbanization patterns and (ii) city-level population growth across India.[4] To answer these questions, two 40-year datasets are constructed from census data, climatic and meteorological data, and industry-related carbon emissions data. The first dataset aggregates population statistics to the district level,

[1] Cities also have an important role to play in climate change mitigation efforts; however, the focus of this section is climate adaptation, so discussion of mitigation efforts is left to other studies.

[2] There is an important distinction to be made between urbanization and urban growth. As Tacoli et al. (2015) explain, urbanization is defined as "the proportion of the total national population living in areas classed as urban." This is also called the "national urban population share." Urban growth, on the other hand, refers to "the absolute number of people living in areas classed as urban." These transitions are driven by different processes, so it is important to investigate them both separately and together.

[3] Construction and justification for the use of climatic moisture are provided in later sections.

[4] India's governing apparatus is divided into 28 states and 8 territories. According to the 2011 census, states were further partitioned into 640 districts; this number has increased to 741 as of 2020. As outlined in the "Methods" section, this paper employs the district names and boundaries from the 2011 census to allow for panel comparison.

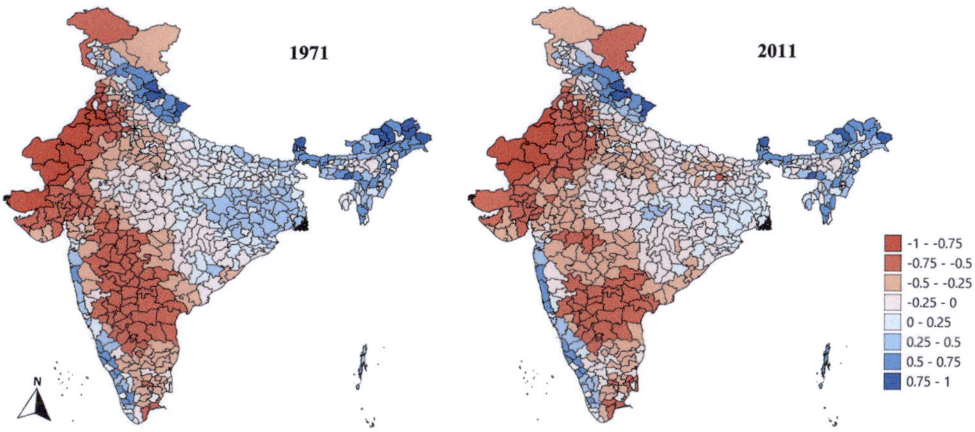

Fig. 1 Moisture map of India, 1971–2011. *Source* Author's calculations using data from Willmott and Feddema's Moisture Index and the 2011 Census of India. This map shows the average Moisture Index score for each district. Districts that are dark red have lower moisture scores, representing drier conditions. Districts that are dark blue have higher moisture scores, representing wetter conditions. The 1971 map is displayed on the left and the 2011 map on the right. Districts in gray have no gridded observations in Willmott and Feddema's data archive

resulting in a balanced panel of 485 districts. This is used to estimate the effect of climatic moisture changes on the share of the population living in urban centers, urban population growth, and rural population growth at the district level. The second dataset is constructed at the city level, resulting in a balanced panel of 2222 urban centers throughout India; this is used to estimate the effect of climatic moisture changes on city population growth.[5] An important difference between these two datasets is that while the first accounts for the classification of new urban centers over time, the second includes only cities that were classified in the 1971 census.[6]

This study finds evidence that declines in climatic moisture—alternatively referred to as "drying"—lead to greater growth in urbanization across districts in India. However, the effect of climatic moisture decline on a district's total urban population growth is statistically insignificant. Instead, evidence suggests that districts experiencing drying have slower rural population growth compared to districts with increased climatic moisture. As such, the observed increase in the rate of urbanization in drying districts may be partially due to rural out-migration to districts with land more suitable to agriculture, potentially contributing to the observed increase in rural population growth in moister districts. This is consistent with the literature and the Indian context, where substantial frictions exist in sectoral labor reallocation and rural-to-urban migration is slow (Liu et al. 2020). Moreover, this study also finds strong, statistically significant evidence that established cities have grown most rapidly in districts experiencing increased moisture. This is largely consistent with the findings of Asher et al. (2022). Taken together, these findings suggest that districts in India experiencing drying may become increasingly urbanized, but major population pressures will likely arise in urban centers and rural regions located in districts experiencing moisture growth.

This study contributes to the literature in three important ways. For one, most scholarship at the nexus of urbanization and climate change employs cross-country regression methods. This study exploits within-country heterogeneity in climatic moisture changes to investigate internal urbanization patterns. Focusing on a single country also has the advantage of removing unobserved cross-country heterogeneity. Second, this study considers the effect of changes in moisture on urbanization, as opposed to precipitation or temperature, which are the most common proxies for climate change. This is an important contribution, as moisture is arguably a more pertinent measure for estimating the effect of climate change on atmospheric agricultural potential and thus agricultural productivity. Finally, this study goes beyond the consideration of urban share growth to investigate rural and city-level dynamics. These findings have important implications for understanding how climate change may alter or compound population pressures on urban centers.

Nonetheless, this study also suffers from several shortcomings. For one, it does not directly observe district-level agricultural output. Although there are limited data available on agricultural yields for recent years, consistent and accurate time-series data on yields over the period examined are not readily available. Instead, this study finds its theoretical underpinnings in the abundant climatology literature, which provides evidence for the close relationship

[5] The data identification and analysis methods follow closely in the precedent set by Hendersen et al. (2017).

[6] Cities originally classified in the 1971 census are henceforth referred to as "established cities." In subsequent Indian censuses, new towns were identified and classed as urban areas; these emergent "census towns" are, on average, much smaller than established cities. The reclassification of new towns is captured in the urban share and total urban population statistics.

between climatic moisture and agricultural potential (Abiy et al. 2019; Gornall et al. 2010; Kumar and Gautam 2014). Additionally, this study does not directly measure internal migration patterns, nor does it consider seasonality in migration, primarily focusing on permanent, district-level local migration instead. Although it is true that rural-to-urban migration patterns vary across regions, states, districts, and cities—for example, some destinations attract migrants from across the country, while others pull primarily from nearby populations—the goal of this paper is to understand how climatic changes impact local urbanization dynamics and city growth. Finally, the use of annual averages for moisture, temperature, and precipitation may obscure seasonal fluctuations and extreme weather events, which can devastate crop yields and directly influence internal migration patterns. Nonetheless, this study is concerned primarily with long-run, slow-onset climatic changes, so annual averages in climatic moisture are the more pertinent measure. Future studies may seek to overcome these limitations by employing data on crop yields, directly observing internal migration and demographic dynamics, constructing average measures for climatic moisture during the wet and dry seasons, and considering exposure to extreme weather events.

The rest of this paper is structured as follows: Section 2 surveys the existing literature on the relationship between climate change and urbanization; Section 3 provides an overview of the empirical strategy, including the relevance and construction of the climatic moisture index, data sources, and empirical specification; Section 4 outlines the main findings of this study; Section 5 discusses the results and proposes an agenda for future research; Section 6 concludes.

2 Literature

A growing body of research suggests there are clear yet nuanced relationships between climate change and urbanization patterns. For one, regions negatively impacted by climate change may experience higher rates of urbanization, with declining climatic and environmental conditions acting as a "push" factor driving rural residents into nearby cities in search of alternative off-farm employment opportunities. Several recent studies support this argument. Henderson et al. (2017) present evidence that there are "strong, but differentiated links between climate and urbanization." More specifically, these authors find that districts in countries throughout Sub-Saharan Africa that experienced greater loss in moisture but had a more robust industrial presence, experienced greater urbanization, measured as growth in urban population share, leading them to argue that local "urban migration provides an 'escape' from negative agricultural moisture shocks." Furthermore, Barrios et al. (2006) find that decreases in rainfall increase urbanization rates in countries across Sub-Saharan Africa. Using global data, Peri and Sasahara (2019) argue that increases in temperature may have differential effects on urban migration depending on a country's socioeconomic status. They find evidence that higher temperatures increase internal migration to urban areas in middle-income countries but decrease rural-to-urban migration in countries situated in the bottom quartile of the income distribution. Bohra-Mishra et al. (2014) find that higher temperatures have contributed to permanent out-migration from warming provinces in Indonesia.

Nonetheless, not all regions have been, or are projected to be, equally impacted by climate change. While some areas may be exposed to increasingly adverse climatic conditions, others may benefit from milder conditions. As such, regions experiencing more favorable climatic and meteorological conditions may experience improved agricultural productivity, leading to local economic growth and potentially subsequent urbanization. In this way, changes such as increased climatic moisture may act as a "land-augmenting technological change" that "leads to an increase in the marginal product of labor in agriculture" (Bustos et al. 2016). This may lead to increases in rural population growth, as more amenable land attracts rural migrants from areas where land has become less suitable to agriculture. Recent work by Asher et al. (2022) finds that villages in India with better access to canals—a land-augmenting technology that increases agricultural productivity—have "higher population density [which is evidence of] increased demand for [agricultural] labor." The authors present compelling evidence that there are "concentrated population gains in proximate urban areas…[despite] net population gains [that seem to be] considerably higher in rural areas than in urban areas."

The burgeoning rural population may migrate into local cities, but high agricultural yields may also contribute indirectly to the growth of nearby urban centers. In a setting such as India, where "rural–urban mobility is low and structural transformation, particularly the movement from agriculture to manufacturing is slow and 'stunted,'" agricultural gains may also contribute to urban growth through the flow of increased rural savings and subsequent investment (Liu et al. 2020). Bustos et al. (2019) find evidence that "agricultural productivity growth can lead to structural transformation through its impact on capital accumulation," as more savings are made available to financially integrated urban centers. In this way, climate change-induced agricultural productivity gains may lead to structural transformation in urban centers as a byproduct of rural growth.

Taken together, these two strands of literature provide evidence that urbanization patterns may be impacted by climate change in two important ways: regions experiencing increasingly adverse climatic and meteorological conditions may see a higher proportion of the population concentrated in urban areas as rural residents are "pushed" to find off-farm employment; and regions experiencing climatic changes more amenable to agricultural productivity may see urban population gains in nearby cities as the entire region grows and thrives. Recent work by Tuholske et al. (2019) finds evidence of such a complex relationship between climatic conditions, city size, and urban growth patterns. Using gridded population data and a robust method of urban identification, the authors present evidence that "small and medium-sized urban settlements in arid regions are growing faster compared to larger urban settlements across Africa's arid regions…By comparison, in semi-arid and humid regions of Africa, larger cities are absorbing a greater share of urban population growth." In essence, secondary and tertiary cities have emerged in drier regions throughout Africa, while city populations have exploded in wetter regions. The authors' conclusions provide a neat bridge between the primary strands of work on climate change and urban growth patterns, suggesting that climatic conditions may give rise to two distinct forms of urbanization, which can occur simultaneously and in close proximity to one another. In concert with the study conducted by Henderson et al. (2017), these findings also provide support for the relevance of climatic moisture as a driver of urbanization.

This study contributes to the growing body of literature that employs climatic moisture as a proxy measure for climate change and investigates its impact on urbanization and urban growth. However, in contrast to the studies conducted by Henderson et al. (2017) and Tuholske et al. (2019), this study focuses solely on within-country heterogeneity in moisture growth as opposed to cross-country differences. Furthermore, it considers the simultaneous yet differentiated processes of urbanization and urban growth, as opposed to previous studies, which often consider only one or the other.

3 Empirical Approach

3.1 Relevance of Climatic Moisture

A majority of the literature investigating the impacts of climate change on urbanization patterns measures climatic conditions using data on precipitation or temperature; however, there are three distinct advantages of estimating climatic variability and change using moisture: (i) it is highly reflective of long-term climatic changes; (ii) it is closely related to agricultural potential; and (iii) it is a good source of exogenous heterogeneity, particularly across districts in India.

Thornthwaite was one of the first climatologists to recognize the importance of atmospheric water balance in the assessment and classification of climatic conditions, arguing that "the combined evaporation from the soil surface and transpiration from plants, called evapotranspiration, is a critical and often-overlooked part of the hydrological process (Thornthwaite 1948). Furthermore, Thornthwaite argued that it is not only necessary to track the actual movement of water "from the Earth to the atmosphere" (ibid.), but also the "evapotranspiration that would occur if the vegetation experiences no water stress" (Willmott and Feddema 1992). Thus, "potential evapotranspiration"—according to Thornthwaite's construction—may also be understood as the water required by a hypothetical, full-cover reference plant, given typical meteorological conditions for a region, such that no water is returned to the atmosphere through evapotranspiration. Comparing potential evapotranspiration to precipitation levels, Thornthwaite argued, would provide valuable information on the dryness or wetness of the climate: "As water deficiency becomes larger with respect to potential evapotranspiration, the climate becomes arid; as water surplus becomes larger, the climate becomes more humid" (Thornthwaite 1948).

Following from this observation, Thornthwaite put forth a method for estimating a moisture index based on the ratio between precipitation and potential evapotranspiration—which may also be interpreted as an "aridity index"—to measure long-run climatic tendencies toward water deficiency or surplus. Willmott and Feddema (1992) subsequently proposed a modified version of Thornthwaite's index that lends itself to clearer interpretation.[7] Their Moisture Index is defined as follows:

$$I_m = \begin{cases} \left(\frac{r}{E^\circ} - 1\right), & r < E^\circ \\ \left(1 - \frac{E^\circ}{r}\right), & r \geq E^\circ \end{cases}, \tag{1}$$

where r is the average monthly precipitation rate and E° is the average monthly potential evapotranspiration. Estimating potential evapotranspiration is a complex process for which multiple methods have been proposed. In this model, potential evapotranspiration is calculated as a function of spatially interpolated meteorological data—specifically, temperature, humidity, windiness, and solar radiation—using the method proposed by Shuttleworth and Wallace (1985), as it has been found to perform well across regions in different climatic zones (Abeysiriwardana et al. 2022; Stannard 1993).[8] This

[7] Willmott and Feddema's construction yields a clearly bounded measure, as opposed to Thornthwaite's unbounded measure, thus allowing for clearer and more consistent interpretation.

[8] See Shuttleworth and Wallace (1985) for a full description of the methods used to calculate potential evapotranspiration (PET). This method is considered to be stronger than others because it combines separate estimates for soil evapotranspiration and canopy transpiration. Stannard (1993) and Abeysiriwardana et al. (2022) also provide straightforward explanations of the methods to calculate PET, along with a comparison between the Shuttleworth–Wallace method, Penman–Monteith method (a simpler, more commonly used calculation), and other accepted PET-calculation methods.

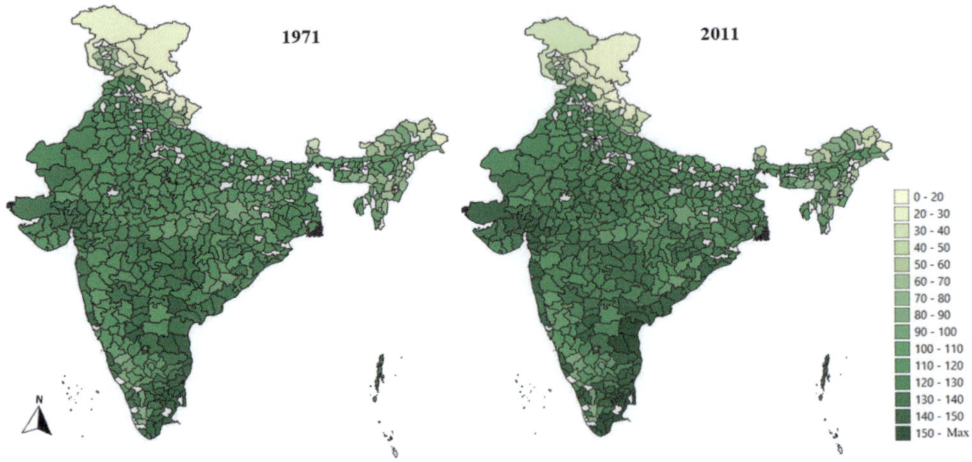

Fig. 2 Annual mean potential evapotranspiration, 1971–2011. *Source* Author's calculations using data from Willmott and Feddema's Moisture Index and the 2011 Census of India. This map shows the average annual potential evapotranspiration for each district in millimeters per year. Districts that are dark green have higher levels of potential evapotranspiration, representing higher water demand. Districts that are light green have lower levels of potential evapotranspiration, representing lower water demand. The 1971 map is displayed on the left and the 2011 map on the right. Districts in gray have no gridded observations in Willmott and Feddema's data archive

calculation yields a symmetric index ranging from −1 to 1, such that negative scores correspond to a deficiency of moisture relative to atmospheric demand and positive scores correspond to a moisture surplus relative to atmospheric demand.[9]

Using Willmott and Feddema's Moisture Index to proxy climate change has a variety of advantages over more commonly used measures such as temperature and precipitation. For one, climatic moisture provides a stable measure of slow-onset climatic changes. Neither potential evapotranspiration nor mean precipitation are highly volatile measures; as such, they reflect long-run, enduring changes in "mean climate state" as opposed to extreme events (Gornall et al. 2010). As climate change continues, scientists predict that for every additional degree Celsius of warming, the atmosphere will be capable of holding up to 7% more water, resulting in increased evaporation of surface moisture (Buis 2022; Kumar and Gautam 2014). In other words, evapotranspiration and potential evapotranspiration are expected to increase globally (see Fig. 2). In fact, the Intergovernmental Panel on Climate Change states with high confidence that "evapotranspiration increased by between approximately 0.5 and 1.5 mm yr^2 between the 1980s and early 2010s due to warming-induced increased atmospheric demand worldwide" (IPCC 2022, 568). Increases in evapotranspiration are expected to "accelerate the hydrologic cycle, altering rainfall, magnitude and timing of run-off…, [as well as] affect the soil moisture, groundwater recharge, and frequency of flood or drought, and finally groundwater level in different areas" (Kumar and Guatam 2014). Furthermore, as evapotranspiration rises, if increasing water demand is not met by higher average levels of precipitation, then there will likely be a "greater soil [moisture] deficit in summer" which is not only expected to "exacerbate impacts of heat waves as well as drought stress," but also to negatively impact agricultural productivity (IPCC 2022, 225). In regions expected to experience a decrease in mean precipitation levels, increased evapotranspiration will further exacerbate drying (see Figs. 2 and 3). In essence, changes in climatic moisture reflect the impacts of climate change on evapotranspiration levels and mean precipitation patterns, which will interact to make "wet regions wetter and dry regions drier" (Buis 2022). This is expected to have profound effects on human activities and livelihoods, particularly in the agricultural sector.

Indeed, agricultural productivity is closely tied to climatic and meteorological conditions (Abiy et al. 2019; Gornall et al. 2010; Henderson et al. 2017; Kumar and Gautam 2014). For example, in their 2017 study, Henderson, Storeygard, and Deichmann use a moisture index to proxy climatic conditions, arguing that because "plant growth is also a function of temperature, dividing precipitation by potential evapotranspiration, which is the appropriate non-linear function for temperature…, creates a better measure of climatic agricultural potential." It follows from the relevance of moisture for agricultural potential that changes in climatic moisture levels, resulting from climate change, will have a direct impact on changes in agricultural productivity. Essentially, as global warming continues, rising temperatures will increase evapotranspiration, impacting hydrological processes and increasing water demand for agriculture, thereby diminishing agricultural productivity if adequate adaptation measures are

[9] Willmott and Feddema's index is recentered around 1 to allow for calculation of the annualized growth rate.

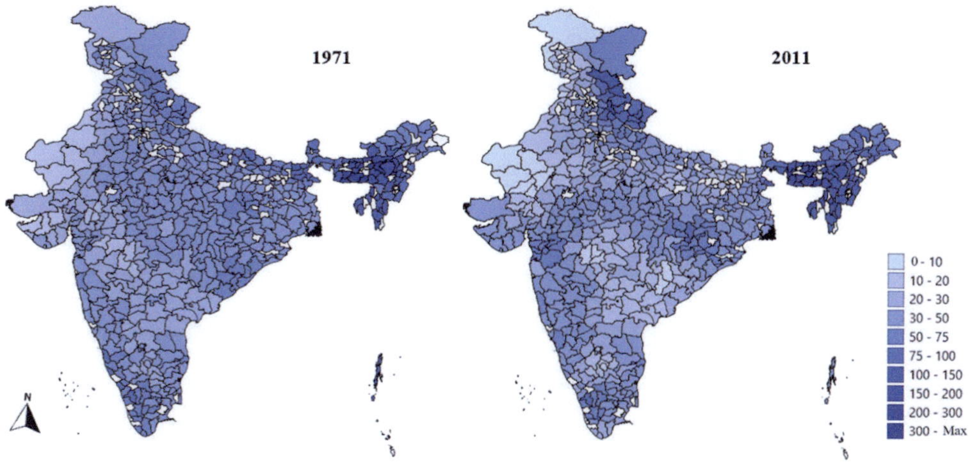

Fig. 3 Annual mean precipitation, 1971–2011. *Source* Author's calculations using data from Willmott and Feddema's Moisture Index and the 2011 Census of India. This map shows the average annual precipitation in millimeters for each district. Districts that are dark blue received more precipitation; those that are white received less precipitation. The 1971 map is displayed on the left and the 2011 map on the right. Districts in gray have no gridded observations in Willmott and Feddema's data archive

not implemented. Furthermore, Abiy et al. (2019) find that an "atmospheric moisture deficit at critical times of crop development causes stunting of crop development attributed to soil deficit." Climate change will not only affect agricultural productivity through its impact on growing conditions, but also its impact on the efficacy of farming systems, such as "established infrastructure, local farming practice, and individual experience" (Gornall et al. 2010). In this way, changes in mean climatic conditions, as measured by moisture, have profound effects on agricultural productivity.

Finally, because climatic moisture is calculated as an annual average using spatially interpolated meteorological data, the statistic is largely unaffected by agricultural practices and urbanization, limiting endogeneity concerns. Willmott and Feddema's climatic moisture measure is estimated from meteorological data using a data-interpolation strategy based on "the spherical version of Shepard's algorithm, which employs an enhanced distance-weighting method (Shepard 1968; Willmott et al. 1985)" (Willmott and Matsuura 2018b). Furthermore, "the number of nearby [weather] stations that influenced a grid-node estimate was increased to an average of 20, from an average of 7…, [resulting in] smaller cross-validation errors" (ibid.).[10] Willmott and Feddema's Moisture Index uses these interpolations to estimate potential evapotranspiration, rather than directly observing actual evapotranspiration. The advantage of this approach is that the moisture measure is not impacted by local agricultural practices or urbanization, mitigating issues caused by reverse causality. For example, in theory, regions with low agricultural productivity may lack the capacity to invest in irrigation systems, resulting in decreased evapotranspiration and lower near-surface atmospheric moisture (Zhang et al. 2019); similar processes may occur in urban areas, which are vulnerable to the urban "heat island" effect. However, these sources of endogeneity are mitigated through robust estimation of potential evapotranspiration as opposed to direct observation of evapotranspiration. Furthermore, substantial variation in moisture changes, particularly across the Indian subcontinent, provides a good source of heterogeneity (see Figs. 1 and 4). Overall, as measured by Willmott and Feddema's Moisture Index, moisture provides a pertinent, exogenous, and heterogeneous proxy for slow-onset climatic changes.

3.2 Data

To estimate the relationships between changes in climatic moisture, urbanization, and urban growth, this study combines data from three sources: the 2011 Census of India,[11] the Willmott and Feddema Moisture Index Gridded Dataset, and the European Commission Emissions Database for Global Atmospheric Research.

As part of the 2011 Census of India, the Office of the Registrar General and Census Commissioner published statistics on the "adjusted population figures for the twelve censuses from 1901 to 2011 for India, States, Union Territories, and Districts according to [2011] jurisdictions" (Census of India 2011). Accordingly, the 2011 census includes population statistics

[10] Willmott and Feddema estimate moisture at spatial resolution of 0.5 × 0.5 degrees. A full overview of the methods used to calculate gridded data on moisture, temperature, and precipitation can be found in Willmott and Matsuura (1999).

[11] The 2011 Census of India is the most recent census. Due to the Covid-19 pandemic, the 2021 Census of India has been continually postponed.

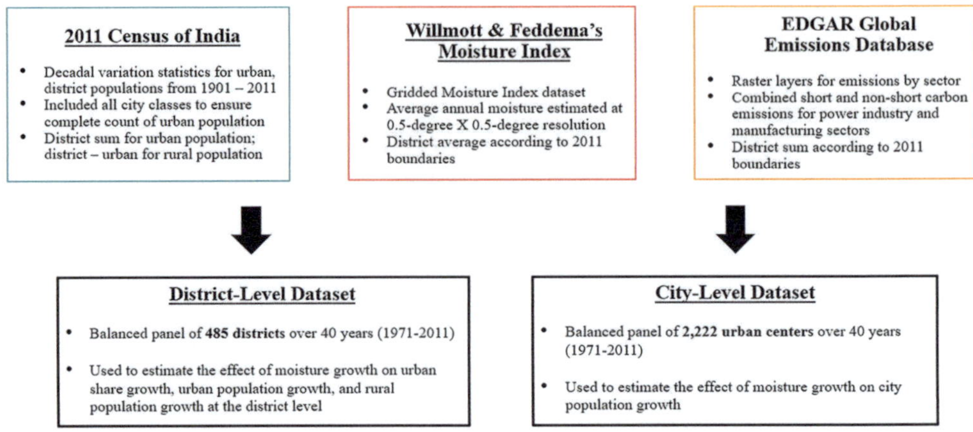

Source: Author. Flowchart displays the data sources compiled for this study.

Fig. 4 Data sources. *Source* Author. Flowchart displays the data sources compiled for this study

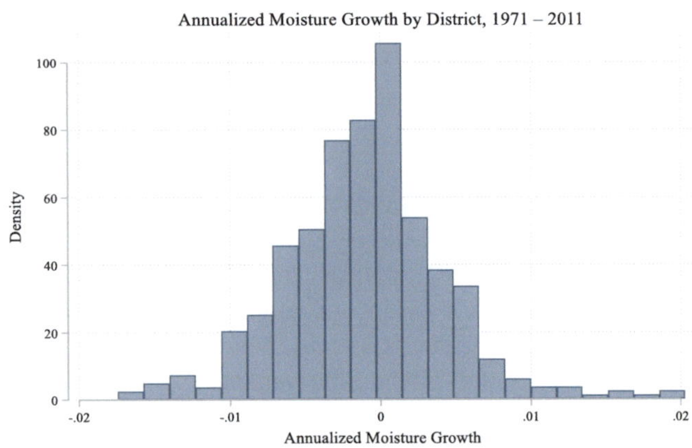

Source: Author's calculations using Willmott and Feddema's Moisture Index data and the 2011 Census of India. This graph shows the distribution of districts experiencing moisture growth and decline from 1971–2011.

Fig. 5 Distribution of annualized moisture growth. *Source* Author's calculations using Willmott and Feddema's Moisture Index data and the 2011 Census of India. This graph shows the distribution of districts experiencing moisture growth and decline from 1971–2011

for towns and urban agglomerations dating back to 1901, as classified by 2011 standards. In theory, this would allow for the creation of a balanced panel dataset containing twelve observations at 10-year intervals between the years of 1901 and 2011 on the population of every urban area in India according to 2011 classifications. However, population statistics before 1961 are often interpolated or unavailable due to failure or inability to conduct the census in various states during early rounds. As such, this study uses population statistics from 1971 and 2011 on Class I, II, III, IV, V, and VI cities (i.e., all urban centers identified in the 2011 census) to identify urbanization patterns and city growth.[12] For cases in which cities spill into multiple districts, large urban agglomerations are broken down into subsets to ensure that the urban population of a district is not over- or under-estimated. District-level urban populations are a sum of the population of all urban centers located in the district, including urban centers that were not classified as urban areas in 1971 but were by 2011. District-level population statistics with decadal variation to 1901 are also taken from the 2011 Census of India. Because the aggregation of Class I–VI cities provides a complete measure of the urban population, rural populations are calculated by subtracting the district-level urban population from the total district population. In the city-level analysis, each urban subset is treated as a separate observation. Cleaning and combining data on all cities in India that were classified as an urban area in both 1971 and 2011 yields a balanced panel dataset of 2222 urban areas. These statistics are used to calculate annual city growth.

Data on moisture are taken from the Willmott and Feddema Moisture Index Gridded Dataset. The methods used to calculate Willmott and Feddema's Moisture Index are outlined in Sect. 3.1 of this paper. For districts containing multiple gridded observations, the average moisture index score is calculated using 2011 shapefiles (Willmott and Matsuura 2018c). Data on precipitation and temperature are also taken from Willmott and Feddema's gridded datasets and averaged at the district level using similar methods (Willmott and Matsuura 2018a, b). Precipitation is expressed in millimeters and temperature in degrees Celsius. District-level annualized growth rates for precipitation and temperature are calculated using the same methods employed for moisture. These measures are included in the regression analysis to completely isolate the effect of moisture and ensure robustness (Henderson et al. 2017). For some districts, there are no gridded observations available; these districts are dropped from the analysis. Because the estimates are calculated at a constant spatial interval, it may be argued that districts are missing moisture, temperature, and precipitation observations at random. A valid counterargument is that the

[12] In order, the population thresholds for Class I–VI cities are as follows: 100,000 and above; 50,000–99,999; 20,000–49,999; 10,000–19,999; 5000–9999; and less than 5000.

smaller districts which often comprise cities are systematically more likely to be missing observations compared to large rural districts. This is a limitation that should be rectified in future studies. One solution may be to impute district-level annual means by taking the average from surrounding districts; however, this estimation method is unlikely to yield a robust measure such as those calculated using Willmott and Feddema's spatial-interpolation algorithm.

Industrial presence is calculated at the district level using emissions data from the European Commission Emissions Database for Global Atmospheric Research (EDGAR) (Monforti-Ferrario et al. 2019, 2021; Olivier et al. 2021). EDGAR publishes raster layers containing annual sector-specific statistics on non-short and short-cycle carbon emissions, expressed in tons and calculated at the 0.1×0.1-degree level.[13] Non-short and short-cycle carbon emissions caused by power industry and manufacturing are combined to estimate total industrial presence by district for 1971 and 2011. Emissions are summed on the district level using 2011 shapefiles. Emissions data are then matched to urban centers by district. This has a variety of drawbacks. For one, industrial presence cannot be directly tied to a single urban area within the district; rather it is an aggregate measure of total district-level emissions. This may lead to over-estimation of industrial presence for separate urban areas located within the same district. However, for cases in which there are multiple observations within the same district, the individual urban subsets often belong to the same larger urban agglomeration. As such, calculating total emissions yields a more accurate measure for local industrial presence than the district-level mean. Ideally, polygons for each urban area in the 2011 census would be used to calculate emissions within the city boundaries; however, such shapefiles are not available. This should be pursued in future studies.

3.3 Empirical Specification

Using two panel datasets, this study estimates the effect of moisture growth on four outcomes of interest: annualized district-level urban share growth, annualized district-level urban population growth, annualized district-level rural population growth, and annualized city population growth. The base specification follows:

$$Y_i = \beta_0 + \beta_1 \text{MOISTURE\_GROWTH}_i + \beta_2 X_i + \beta_3 n_i + e_i, \quad (2)$$

where Y_i is one of the four outcomes of interest; MOISTURE_GROWTH$_i$ is the annualized moisture growth rate, the primary variable of interest; X_i is a vector of controls, including baseline urban share or logged baseline urban/rural/city population, baseline moisture, and a dummy for whether the district is coastal; and ν_i is a vector of state fixed effects. Models that estimate the impact of moisture growth on city population growth are also clustered at the district level. Logged baseline industry emissions and annualized industry emissions growth are included as controls in the extended version of the model. As discussed in the previous section, two additional models for each dependent variable are tested, which control for baseline average annual temperature, annualized temperature growth, baseline average annual precipitation, and annualized precipitation growth in the primary and extended models. Following the precedent set by Henderson et al. (2017), this is intended to serve as a robustness check for the impact of moisture on urbanization.

All dependent variables are winsorized at three standard deviations above the mean. Results for the winsorized estimates are discussed in the following section; non-winsorized estimates are recorded in Tables A.3–A.6 in the Appendix. Results are largely consistent. Table A.1 in the Appendix provides summary statistics for each of the variables used in the regressions for urban share, urban population, and rural population. Table A.2 in the appendix contains summary statistics for the city population growth regressions. Figure A.1 in the Appendix shows the distribution of urban share growth; Figure A.2 the distribution of total district urban population growth; Figure A.3 the distribution of rural population growth; and Figure A.4 the distribution of city population growth. Figure 5 shows the distribution of annualized moisture growth by district.

4 Results

4.1 District-Level Results

The results of the regression analysis with annualized growth of urban share as the dependent variable are recorded in Table 1. They suggest that increased drying may contribute to increased growth in urban share. In the primary specification, the coefficient on moisture growth is significant at the 5% level and suggests that a 1 percentage point decrease in annualized moisture growth increases annualized urban share growth by 0.210 percentage points, all else equal. In other words, a decrease of one standard deviation in the annualized moisture growth rate increases urban share growth by 0.105 percentage points. This is consistent with the literature which finds that drying increases urbanization (Henderson et al. 2017). The magnitude of the coefficient on moisture growth diminishes slightly as subsequent controls are

[13] Non-short cycle emissions data includes "all fossil CO_2 sources, such as fossil fuel combustion, non-metallic mineral processes, metal (ferrous and non-ferrous) production processes, urea production, agricultural liming, and solvents use" (Montforti-Ferrario et al. 2021). EDGAR primarily uses data from the International Energy Agency World Energy Balances.

Table 1 Regression results: urban share growth, winsorized

Independent variable	(1)	(2)	(3)	(4)	(5)	(6)
Baseline urban share	−0.029*** (0.003)	−0.031*** (0.003)	−0.032*** (0.003)	−0.032*** (0.003)	−0.032*** (0.003)	−0.033*** (0.003)
Baseline moisture	0.003*** (0.001)	0.002	0.002	0.002	0.002	0.002
Coastal	0.006*** (0.001)	0.003** (0.001)	0.003** (0.001)	0.003** (0.001)	0.003** (0.001)	0.003** (0.001)
Annualized moisture growth	−0.035 (0.086)	−0.210** (0.103)	−0.220** (0.103)	−0.202** (0.103)	−0.217* (0.103)	−0.211** (0.104)
Baseline industry emissions	–	–	0.0006 (0.0005)	0.0006 (0.0006)	–	0.0005 (0.0006)
Annualized industry emissions growth	–	–	–	0.0008 (0.0023)	–	0.014 (0.023)
Baseline precipitation	–	–	–	–	0.00001 (0.00001)	0.00001 (0.00001)
Annualized precipitation growth	–	–	–	–	−0.024 (0.037)	−0.026 (0.037)
Baseline air temperature	–	–	–	–	0.001 (0.001)	0.001 (0.001)
Annualized air temperature growth	–	–	–	–	0.160*** (0.048)	0.159*** (0.048)
State fixed effects	NO	YES	YES	YES	YES	YES
Constant	0.015*** (0.001)	0.016*** (0.002)	0.027*** (0.009)	0.026** (0.010)	0.016*** (0.002)	0.024*** (0.010)
Adjusted R-Squared	0.200	0.336	0.337	0.336	0.350	0.350

*p < 0.010, **p < 0.05, ***p < 0.01. *Notes* This table regresses moisture growth and various additional controls (listed in the leftmost column) on annualized urban share growth. The dependent variable is winsorized to three standard deviations outside the mean. Regressions 2–6 employ state fixed effects. Urban share is calculated by dividing urban population over total district population

added to the regression but remains significant at the 10% level. The regression results also suggest that coastal districts saw greater urban share growth between 1971 and 2011. In every regression specification, the results suggest that coastal districts experienced urban share growth that was 0.3 percentage points greater than non-coastal districts, significant at the 5% level, all else equal. This is consistent with the literature, which argues that human settlements tend to concentrate along coastlines (Neumann et al. 2015; Small and Nicholls 2003).

These findings are robust to the inclusion of controls for baseline average annual precipitation, baseline average annual temperature, precipitation growth, and temperature growth. The point estimates on moisture growth are largely unchanged and remain statistically significant at the 5% level. Baseline precipitation, precipitation growth, and baseline temperature are all statistically insignificant in robustness checks on the primary specification and the extended specification; however, temperature growth emerges as statistically significant at the 1% level. The results suggest that a 1 percentage point increase in annualized average temperature growth increases urban share growth by 0.160 percentage points, all else equal.

Looking a bit deeper into the dynamics underpinning this increase in urban share growth yields intriguing results. As shown in Table 2, the point estimate on moisture growth is statistically insignificant in every urban population growth regression, except for the model that does not include state fixed effects. Instead, what emerges as statistically significant for urban population growth is coastal location, high baseline industrial presence, industrial growth, and positive temperature growth. The most marked difference between the findings on urban share growth and total urban population growth pertains to the importance of baseline industrial presence and growth. The findings suggest that higher baseline industrial presence and industrial growth increase urban population growth, all else equal. Coastal location and, in robustness checks, temperature growth have a similar effect on total urban population growth when compared to urbanization growth.

Table 2 Regression results: total urban population growth, winsorized

Independent variable	(1)	(2)	(3)	(4)	(5)	(6)
Log(Baseline urban population)	-0.004^{***} (0.0004)	-0.004^{***} (0.0004)	-0.005^{***} (0.0005)	-0.005^{***} (0.0005)	-0.004^{***} (0.0004)	-0.005^{***} (0.0005)
Baseline moisture	−0.0004 (0.001)	−0.0008 (0.002)	−0.001 (0.002)	−0.002 (0.002)	−0.0008 (0.002)	−0.002 (0.002)
Coastal	0.004^{**} (0.001)	0.004^{**} (0.002)	0.004^{***} (0.002)	0.004^{**} (0.002)	0.004^{***} (0.002)	0.004^{**} (0.002)
Annualized moisture growth	-0.178^{*} (0.097)	−0.105 (0.120)	−0.070 (0.117)	−0.073 (0.117)	−0.110 (0.121)	−0.084 (0.118)
Baseline industry emissions	–	–	0.004^{***} (0.0007)	0.003^{***} (0.0007)	–	0.003^{***} (0.0007)
Annualized industry emissions growth	–	–	–	0.043^{*} (0.026)	–	0.047^{*} (0.026)
Baseline precipitation	–	–	–	–	0.00001 (0.00001)	0.00002 (0.00002)
Annualized precipitation growth	–	–	–	–	−0.003 (0.043)	−0.004 (0.042)
Baseline air temperature	–	–	–	–	0.001 (0.001)	0.001 (0.001)
Annualized air temperature growth	–	–	–	–	0.168^{***} (0.056)	0.154^{***} (0.054)
State fixed effects	NO	YES	YES	YES	YES	YES
Constant	0.074^{***} (0.005)	0.075^{***} (0.006)	0.152^{***} (0.016)	0.144^{***} (0.017)	0.075^{***} (0.006)	0.140^{***} (0.017)
Adjusted R-Squared	0.175	0.275	0.313	0.315	0.282	0.321

$^{*}p < 0.010$, $^{**}p < 0.05$, $^{***}p < 0.01$. *Notes* This table regresses moisture growth and various additional controls (listed in the leftmost column) on annualized total district-level urban population growth. The dependent variable is winsorized to three standard deviations outside the mean. Regressions 2–6 employ state fixed effects. Urban population is calculated as the sum of the population of Class I–VI cities located within a district

The results of the regression analysis using rural population as the dependent variable are more interesting, as shown in Table 3. The point estimate on moisture growth is positive and statistically significant in all regression specifications. In the primary specification, moisture growth is significant at the 5% level, and the results suggest that a 1 percentage point increase in moisture growth increases rural population growth by 0.130 percentage points, all else equal. In other words, a one standard deviation decrease in moisture growth decreases rural population growth by 0.065 percentage points. The point estimate on moisture is significant at the 1% level after controlling for temperature and precipitation, and the magnitude is slightly larger. The other variables that emerge as consistently statistically significant for rural population growth are inland location, lower baseline moisture—perhaps because such regions are now experiencing increases in moisture—low baseline industrial emissions, and industrial growth. Taken together, these results offer weak evidence that drier conditions increase total urban population growth but strong evidence that drying decreases rural population growth. Instead, districts with increased moisture experience faster rural growth. This may underlie the apparent increase in urban share in drying districts.

4.2 City-Level Results

The results of the regression analysis using annualized city population growth as the dependent variable are recorded in Table 4. The point estimate for moisture growth is significant at the 1% level in all specifications that include state fixed effects. In the primary specification, the point estimate on moisture growth suggests that a 1 percentage point increase in moisture growth leads to a 0.373 percentage point increase in city population growth, all else equal. In other words, a one standard deviation increase in moisture growth in the surrounding district leads to a 0.187 percentage point increase in city population growth. The point estimate declines slightly as subsequent controls are added and more sharply after the inclusion of temperature and precipitation controls, down to 0.329; however, the coefficient remains statistically significant at the 1% level. These results suggest that established cities (i.e., cities that existed at baseline) grew more quickly in districts that experienced an increase in moisture. This finding is in line with the literature on agricultural productivity improvements contributing to urban growth. Baseline industry emissions, annualized

Table 3 Regression results: rural population growth, winsorized

Independent variable	(1)	(2)	(3)	(4)	(5)	(6)
Baseline rural population	-0.002^{***} (0.0003)	-0.0009^{**} (0.0004)	−0.0003 (0.0005)	0.0003 (0.0005)	-0.0009^{**} (0.0004)	0.0003 (0.0005)
Baseline moisture	-0.001^{*} (0.0008)	-0.002^{***} (0.0009)	-0.002^{***} (0.0009)	-0.002^{***} (0.0009)	-0.002^{**} (0.0009)	-0.002^{**} (0.0009)
Coastal	-0.007^{***} (0.0008)	-0.002^{***} (0.0007)	-0.002^{***} (0.0007)	-0.002^{***} (0.0007)	-0.002^{***} (0.0007)	-0.002^{***} (0.0007)
Annualized moisture growth	-0.155^{***} (0.057)	0.130^{**} (0.051)	0.130^{**} (0.051)	0.131^{**} (0.050)	0.143^{***} (0.051)	0.143^{***} (0.050)
Log(Baseline industry emissions)	–	–	-0.0006^{*} (0.0003)	-0.001^{***} (0.0003)	–	-0.001^{***} (0.0003)
Annualized industry emissions growth	–	–	–	0.043^{***} (0.011)	–	0.042^{***} (0.011)
Baseline precipitation	–	–	–	–	−9.54e−06 (6.84e−06)	−9.48e−06 (6.73e−06)
Annualized precipitation growth	–	–	–	–	0.005 (0.018)	−0.0006 (0.018)
Baseline air temperature	–	–	–	–	−0.0006 (0.0005)	−0.0004 (0.0005)
Annualized air temperature growth	–	–	–	–	-0.049^{**} (0.023)	−0.036 (0.023)
State fixed effects	NO	YES	YES	YES	YES	YES
Constant	0.042^{***} (0.005)	0.031^{**} (0.006)	0.013 (0.011)	−0.005 (0.012)	0.031^{***} (0.006)	−0.004 (0.012)
Adjusted R-Squared	0.2141	0.609	0.611	0.623	0.615	0.627

$^{*}p < 0.010$, $^{**}p < 0.05$, $^{***}p < 0.01$. *Notes* This table regresses moisture growth and various additional controls on annualized total district-level rural population growth. The dependent variable is winsorized to three standard deviations outside the mean. Regressions 2–6 employ state fixed effects. Rural population is calculated as the difference between total population and the sum of the population of Class I–VI cities located within a district

industry emissions growth, baseline temperature, and annualized temperature growth also emerge as statistically significant for city population growth. Of these variables, industry growth has the strongest statistical significance. The magnitude of the point estimate suggests that a 1% point increase in district industry growth leads to a 0.058 increase in city growth.

5 Discussion

This study finds that drying increases urban share growth in districts across India. This is broadly in line with the literature on climate change acting as a "push" factor contributing to increased urbanization (Henderson et al. 2017; Barrios et al. 2006). However, the results of this study also suggest that moisture growth does not have a statistically significant effect on district-level urban population growth. Instead, drying contributes to slower rural population growth; conversely, moisture increase contributes to more rapid rural population growth. This has interesting implications for the findings on urban share growth, suggesting that the observed increase in urbanization may be driven by stifled rural growth and/or rural out-migration rather than by urban population growth. Rural residents may simply be relocating to areas with more favorable farming conditions rather than to nearby urban areas. This may also contribute to the observed increase in rural population growth in districts experiencing moisture increase.

Indeed, a recent report by the World Bank finds that the "southern Indian highlands… will be climate in-migration hotspots" while the "northern part of the Gangetic Plain, and the corridor from Delhi to Lahore" and the "irrigated areas and rice-growing areas are likely to see population dampening as a result of out-migration" (Rigaud et al. 2018, 89). A rough comparison between the moisture map displayed in Fig. 1 and a map from the National Aeronautics and Space Administration's Socioeconomic Data and Applications Center (SEDAC) database showing internal migration flows across India provides support for the argument that individuals may be moving out of drying districts and into districts more conducive to agriculture (de Sherbinin et al. 2015). In this way, drying districts may be urbanizing more rapidly

Table 4 Regression results: city population growth, winsorized

Independent Variable	(1)	(2)	(3)	(4)	(5)	(6)
Log(Baseline urban population)	0.0005 (0.0005)	0.0006 (0.0005)	0.0006 (0.0005)	0.0006 (0.0005)	0.0006 (0.0005)	0.0006 (0.0005)
Baseline moisture	0.002 (0.001)	0.0004 (0.002)	0.0003 (0.002)	0.0003 (0.002)	0.0002 (0.002)	−0.0001 (0.002)
Coastal	−0.002 (0.002)	0.002 (0.002)	0.002 (0.002)	0.001 (0.002)	0.002 (0.002)	0.002 (0.002)
Annualized moisture growth	0.082 (0.095)	0.373*** (0.131)	0.371*** (0.132)	0.370*** (0.134)	0.334*** (0.133)	0.329** (0.136)
Baseline industry emissions	–	–	−0.0007 (0.0005)	−0.001****** (0.0005)	–	−0.001* (0.0005)
Annualized industry emissions growth	–	–	–	0.058*** (0.022)	–	0.059*** (0.022)
Baseline precipitation	–	–	–	–	−5.17e−06 (0.00001)	−1.10e−06 (0.00001)
Annualized precipitation growth	–	–	–	–	0.028 (0.029)	0.023 (0.029)
Baseline air temperature	–	–	–	–	0.001** (0.0007)	0.002** (0.0007)
Annualized air temperature growth	–	–	–	–	0.070 (0.045)	0.081* (0.046)
State fixed effects	NO	YES	YES	YES	YES	YES
Constant	0.018*** (0.005)	0.017*** (0.006)	0.005** (0.011)	−0.006 (0.011)	0.017*** (0.006)	−0.007 (0.011)
Adjusted R-Squared	0.007	0.108	0.110	0.116	0.109	0.118

*p < 0.010, **p < 0.05, ***p < 0.01. *Notes* This table regresses moisture growth and various additional controls on annualized city population growth. The dependent variable is winsorized to three standard deviations outside the mean. Regressions 2–6 employ state fixed effects. All regressions cluster standard errors at the district level

than districts experiencing moisture gains, though rural out-migration likely plays a particularly important role in this transition.

This study also finds evidence that cities in districts experiencing increases in climatic moisture tend to grow faster than cities located in drying regions. Together with the results on rural population dynamics, these findings suggest that increased moisture—and thus improvements to agricultural productivity—may contribute to rural growth, which subsequently results in gains to nearby urban centers. This is in line with the literature which suggests urban growth may be augmented by local gains to agricultural productivity (Asher et al. 2022; Bustos et al. 2019). However, it is important to reconcile this finding with the district-level regression analysis, which finds that moisture has no impact on total urban population growth. As previously noted, the city-level analysis employs a panel dataset of urban centers and does not account for cities that emerged or were reclassified between 1971 and 2011.[14] While "established" urban centers experienced more rapid growth in moister districts, it may be that rural villages coalesced into small census towns in drying districts as individuals sought off-farm employment to escape declining agricultural conditions.[15]

The emergence of new towns may explain why population growth in "established" cities does not correspond with significantly higher total urban population growth in moister districts as compared to drier districts. In essence, climate change may hasten two types of urbanization: (i) the rise of small census towns, which result from increased demand for off-farm employment but ultimately stagnate due to rural out-migration; and (ii) rapid, sustained urban center growth driven by improvements to agricultural productivity. Additional analysis is necessary to confirm or refute this theory, though scholars have become increasingly interested in India's small and dispersed "emergent urban areas" (van Duijne and Nijman 2019). The role of reclassification is

[14] "Reclassification" refers to "the shift of a settlement from one category (rural) to another (urban) when its boundaries expand, it is annexed to an adjacent settlement, or it passes specific criteria such as a population density threshold" (Menashe-Oren and Bocquier 2021).

[15] "Census towns" refers to areas that are classified as urban for the purposes of the census based on the following criteria: "population exceeds 500; at least 75% of main male working population is employed outside the agricultural sector; minimum population density of 400 persons per km^2" (Mitra and Tripathi 2021).

also an understudied and important phenomenon, particularly in India, where it is estimated that reclassification may have contributed to almost 9.6% of measured urban growth (Menashe-Oren and Bocquier 2021).

Additionally, future studies should further explore the nuanced relationships between climate change, migration, and urbanization. One of the major drawbacks of this study is that migration and the motivations for migration are not directly observed. Furthermore, this study focuses primarily on local, intra-district migration as opposed to cross-district, cross-state, and international migration; most Indian districts are small geographic units, so changes in climatic conditions are likely to induce cross-district or cross-state migration. Future studies should also explore how moisture conditions, urbanization levels, and industrial presence in neighboring districts affect intra-district dynamics. Similarly, fertility and mortality trends are not directly observed in this study. Both migration and natural demographic trends play an important role in urbanization and urban population growth (Menashe-Oren and Bocquier 2021); because these processes are intricately intertwined, this study does not explicitly distinguish between the two but considers overall population and demographic changes. However, because there are important and distinct policy implications for addressing different types of migration, as well as natural demographic trends, future studies should seek to disentangle the dynamics of these phenomena and their impacts on urbanization and urban growth.

Another drawback of this study is that agricultural output is not directly measured. As more accurate, systematic data on crop yields across India become available, future studies should seek to establish more robust causal relationships between climatic changes, agricultural yields, migration, and urbanization. However, future studies should not only consider the impact of long-term, slow-onset climatic changes, but also shifts in the seasonal variation of rainfall patterns and extreme weather events, as both directly impact agricultural yields. Climatic and meteorological conditions immediately before and during the growing season may have particularly poignant effects on output and thus more directly influence decisions to migrate. Similarly, extreme weather events—such as intense rains, floods, heatwaves, and droughts—not only have the potential to destroy crop yields, but also to directly induce migration as households flee devastation. Nonetheless, future studies should be aware of endogeneity concerns, particularly if governments introduce targeted legislation to improve conditions for households in climate-affected areas, to stem the flow of migration, or to induce urbanization patterns.[16]

6 Conclusion

Overall, this study finds that residents in drying districts may become increasingly concentrated in urban areas, but the highest population pressures may be found in districts experiencing an increase in climatic moisture. This finding has important implications for urban planning and climate resilience efforts throughout India. For one, residents in burgeoning towns must have access to the infrastructure and services necessary to help parlay the growth of off-farm employment into structural transformation.[17] Promoting the growth of smaller urban areas will not only lead to economic growth in rural districts, but also help support those who stand to be most severely impacted by climate change. Furthermore, creating more local opportunities for off-farm and industrial employment through the sustainable growth of towns may help towns absorb rural residents who would otherwise move to areas with better agricultural land, thus helping relieve pressure on in-demand regions (Tripathi 2020). Mitra and Tripathi (2021) explain: "The proper management of new census towns will play a pivotal role in the context of higher and balanced urbanization…, [which] will help reduce pressure on large cities and the impact of the other diseconomies." Harnessing the potential of growing towns and small cities will be a crucial challenge for both India's urban development and climate change adaptation efforts.

Additionally, districts in areas experiencing moisture growth need to be ready to accommodate significant population increases. Rapid urban expansion in major cities across India has already put significant strain on municipal authorities. In a World Bank report, Ellis and Roberts (2016) find that "South Asia's cities are not fulfilling their development potential as characterized by either prosperity or livability because of the congestion pressures that growing populations are exerting on infrastructure, basic services, land, housing, and the environment" (77). Lofty policy solutions such as strengthening urban governance, increasing infrastructure investment, or implementing climate-conscious urban expansion plans need to be tailored to address local needs and capacities. However, in many contexts, particularly in India and across the Global South, implementing land use reform, improving land use administration system, and pre-emptively laying arterial roads and

[16] One policy initiative that may be cause for endogeneity concerns in this study is the introduction of the Mahatma Gandhi National Rural Employment Guarantee Act (MGNREGA) program. In response to poor agricultural performance, this employment guarantee program was introduced in 2005, first to the most backward districts, then expanded in 2008 to cover all districts in India. MGNREGA ties guaranteed employment to rural residence to help reduce rural poverty and unemployment. However, one externality of this program may be slower migration to urban areas. Although all districts now receive the program, differences in longevity and implementation quality may be correlated with both agricultural productivity and urbanization patterns, causing bias. Future studies should further explore this issue.

[17] Poor infrastructure and inadequate service provision are only two constraints among many affecting the pace of structural transformation in India.

sanitation infrastructure will contribute to building sustainable cities from the ground up. Policy makers should also consider the "anatomy of density" in the cities they govern (Angel et al. 2021). For example, focusing on "a policy to increase dwelling packing without maintaining the occupancy rate and dwelling unit occupancy (which would only be possible with an increase in the population and number of households) would produce a neutral result at great cost" (Angel et al. 2021, 278). Indeed, policies with shape the city's built environment have an outsized role to play in the realization of Sustainable Development Goal 11 (Harari 2020). Climate change and urbanization pose highly localized challenges that require highly localized solutions. As such, future research should consider how these processes will manifest in specific, meso-level settings.

Understanding how climate change and urbanization converge will help inform more effective, integrated policy responses to address two of the most formidable challenges facing our world in the decades to come. The evidence presented in this study suggests that changing climatic conditions will have a significant impact on urbanization patterns, whereby populations in drying regions become increasingly ossified into small towns while population pressures in both cities and rural areas increase in regions experiencing greater climatic moisture. Policies to support urbanization efforts should take these dynamics into account when planning for future development; otherwise, the challenges posed by urban growth and climate change may compound one another and prove too heavy for local, national, and international governments to bear. On the other hand, sustainable urbanization has the potential to support human flourishing at all levels and insulate against the impacts of climate change. Indeed, as climate change continues to transform our environment and impact livelihoods, safe and sustainable urban development will be an integral part of creating a prosperous world for all.

Acknowledgements Thank you to Professor Matthew McCartney, Professor Sam Asher, and Professor Giovanna M. D. Dore for their insightful feedback and comments on early drafts of this paper. All errors remain my own.

Appendix

See Tables A.1, A.2 and Figs. A.1, A.2, A.3, A.4.

Supplementary Materials

See Tables A.3, A.4, A.5 and A.6.

Table A.1 District-level descriptive statistics

Independent variable	Mean	SD	Minimum	Maximum
Urban share (2011)	0.246	0.175	0.035	0.981
Urban share (1971)	0.153	0.129	0.008	0.906
Annualized urban share growth	0.014	0.010	−0.007	0.044
Urban population (2011)	609,538	905,518	4,644	8,760,511
Urban population (1971)	167,302	229,865	331	2,074,354
Annualized urban population growth	0.034	0.011	0.008	0.068
Rural population (2011)	1,523,144	1,004,746	23,610	6,074,188
Rural population (1971)	804,629	520,695	12,683	3,362,126
Annualized rural population growth	0.016	0.006	−0.004	0.037
Baseline moisture	0.913	0.405	0.069	1.9000
Annualized moisture growth	−0.001	0.005	−0.018	0.020
Baseline industry emissions	1.53e−07	2.16e−07	7.54e−10	1.93e−06
Annualized industry emissions growth	0.036	0.018	−0.032	0.108
Annualized district population growth	0.020	0.006	0.003	0.057
Baseline average annual precipitation	48.497	45.439	11.868	716.508
Annualized precipitation growth	−0.006	0.012	−0.044	0.025
Baseline average annual temperature	0.944	0.729	0.304	7.725
Annualized temperature growth	−0.007	0.011	−0.048	0.022

Notes This table displays the summary statistics for variables used in the urban share growth, total district urban population growth, and rural population growth regressions. The variable name is displayed in the leftmost column. Column 2 displays the mean; Column 3 displays the standard deviation; Column 4 displays the minimum value the variable takes; and Column 5 displays the maximum value. Annualized urban share, urban population, and rural population growth rates reflect winsorization

Table A.2 City-level descriptive statistics

Independent variable	Mean	SD	Minimum	Maximum
Urban population (2011)	114,110	355,764	1,204	8,495,492
Urban population (1971)	36,730	91,043	96	1,750,134
Annualized urban population growth	0.023	0.011	−0.013	0.060
Baseline moisture	0.808	0.386	0.069	1.9
Annualized moisture growth	−0.0001	0.005	−0.017	0.020
Baseline industry emissions	2.04e−07	2.38e−07	7.54e−10	1.93e−06
Annualized industry emissions growth	0.038	0.019	−0.032	0.108
Baseline average annual precipitation	42.222	30.408	11.868	716.508
Annualized precipitation growth	−0.004	0.012	−0.044	0.025
Baseline average annual temperature	0.868	0.573	0.304	7.725
Annualized temperature growth	−0.006	0.011	−0.048	0.022

Notes This table displays the summary statistics for variables used in the city population growth regressions. The variable name is displayed in the leftmost column. Column 2 displays the mean; Column 3 displays the standard deviation; Column 4 displays the minimum value the variable takes; and Column 5 displays the maximum value. City population growth rates reflect winsorization

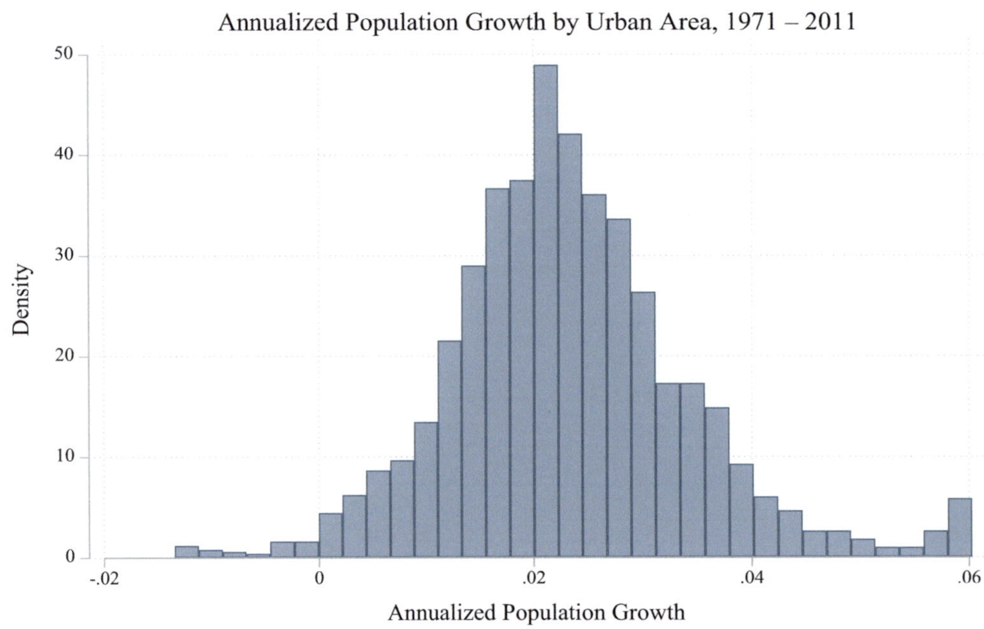

Fig. A.1 Distribution of annualized urban share growth. *Source* Author's calculations using data from the 2011 Census of India. Graph shows distribution of annualized city population growth. Annualized city growth is winsorized to three standard deviations from the mean

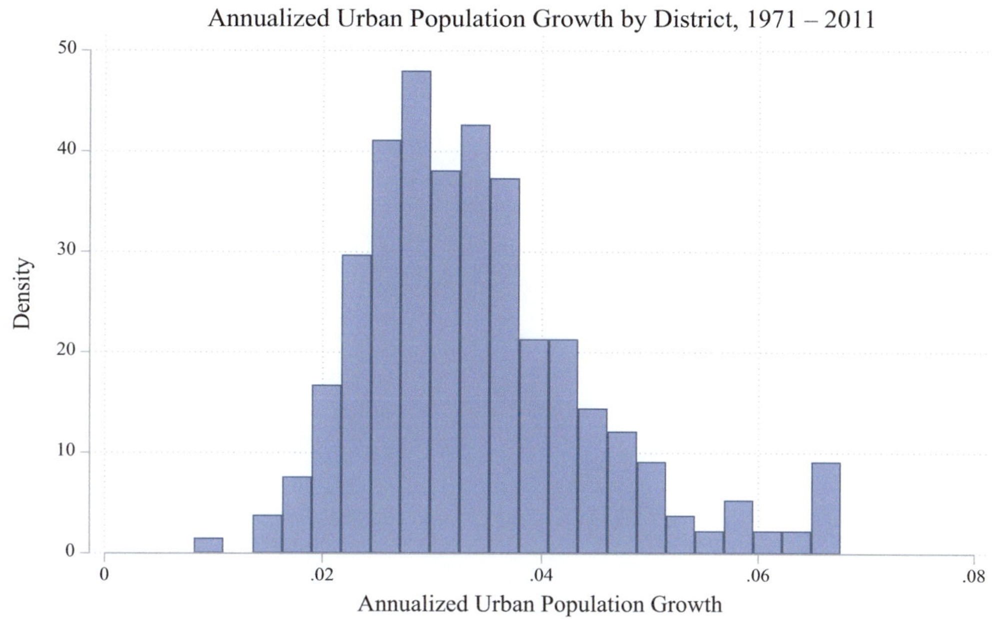

Fig. A.2 Distribution of annualized urban population growth. *Source* Author's calculations using data from the 2011 India Census. This graph shows the distribution of annualized growth in rural population. Annualized rural growth is winsorized to three standard deviations from the mean

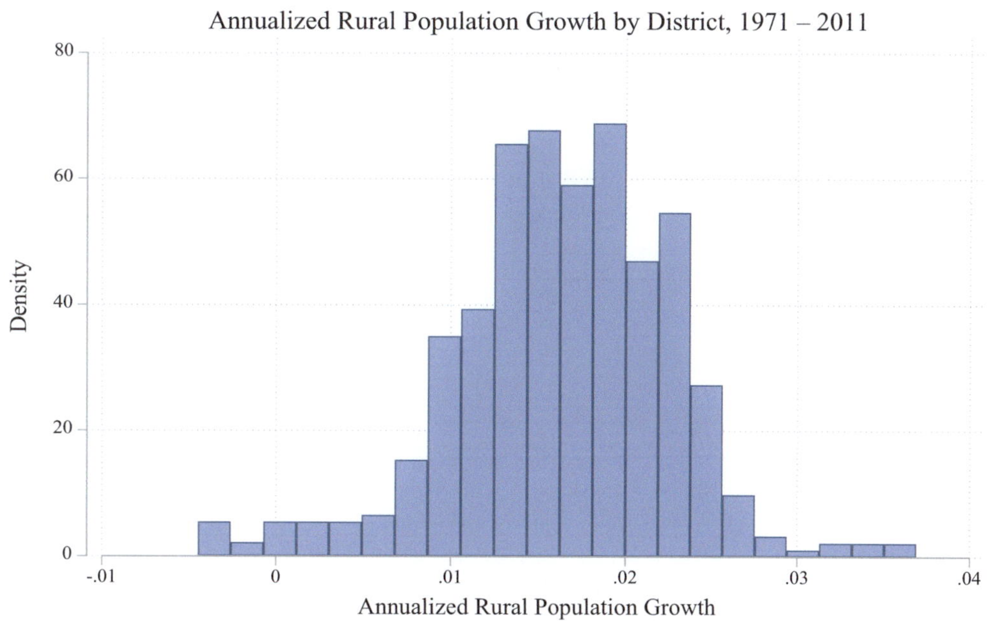

Fig. A.3 Distribution of annualized rural population growth. *Source* Author's calculations using data from the 2011 India Census. This graph shows the distribution of annualized growth in rural population. Annualized rural growth is winsorized to three standard deviations from the mean

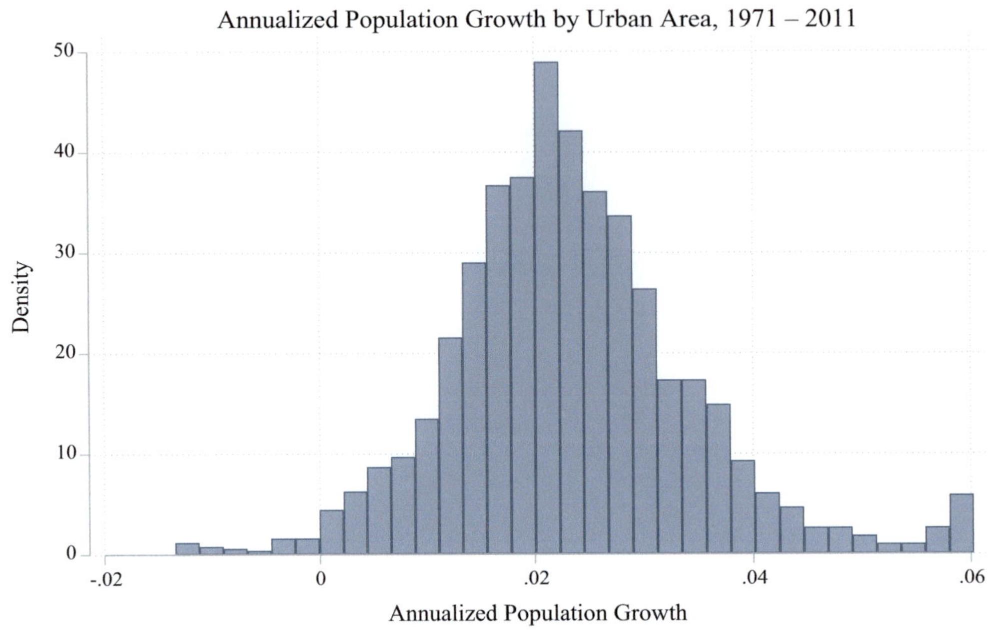

Fig. A.4 Distribution of annualized city population growth. *Source* Author's calculations using data from the 2011 Census of India. Graph shows distribution of annualized city population growth. Annualized city growth is winsorized to three standard deviations from the mean

Table A.3 Regression results: urban share growth, non-winsorized

Independent variable	(1)	(2)	(3)	(4)	(5)	(6)
Baseline urban share	-0.030^{***} (0.003)	-0.032^{***} (0.003)	−0.033 (0.004)	-0.033^{***} (0.004)	-0.033^{***} (0.003)	-0.034^{***} (0.004)
Baseline moisture	0.003^{**} (0.001)	0.002 (0.002)	0.002 (0.002)	0.002 (0.002)	0.002 (0.002)	0.001 (0.002)
Coastal	0.006^{***} (0.001)	0.004^{**} (0.002)	0.003^{**} (0.002)	0.003^{**} (0.002)	0.004^{***} (0.002)	0.004^{**} (0.002)
Annualized moisture growth	−0.040 (0.090)	-0.212^{**} (0.109)	-0.204^{*} (0.109)	-0.205^{*} (0.109)	-0.220^{**} (0.120)	-0.214^{**} (0.120)
Log(Baseline industry emissions)	–	–	0.0006 (0.0006)	0.0006 (0.0006)	–	0.0005 (0.0006)
Annualized industry emissions growth	–	–	–	0.006 (0.024)	–	0.012 (0.024)
Baseline precipitation	–	–	–	–	0.00001 (0.00001)	0.00001 (0.00001)
Annualized precipitation growth	–	–	–	–	−0.022 (0.039)	−0.024 (0.039)
Baseline air temperature	–	–	–	–	0.001 (0.001)	0.001 (0.001)
Annualized air temperature growth	–	–	–	–	0.162^{***} (0.050)	0.162^{***} (0.051)
State fixed effects	NO	YES	YES	YES	YES	YES
Constant	0.015^{***} (0.001)	0.017^{***} (0.002)	0.027^{**} (0.010)	0.026^{***} (0.010)	0.016^{***} (0.002)	0.024^{**} (0.010)
Adjusted R-Squared	0.195	0.320	0.320	0.319	0.333	0.331

$^{*}p < 0.010$, $^{**}p < 0.05$, $^{***}p < 0.01$. *Notes* This table regresses moisture growth and various additional controls (listed in the leftmost column) on annualized urban share growth. Regressions 2–6 employ state fixed effects. Urban share is calculated by dividing urban population over total district population

Table A.4 Regression results: urban population growth, non-winsorized

Independent Variable	(1)	(2)	(3)	(4)	(5)	(6)
Log(Baseline urban share)	−0.004*** (0.0004)	−0.004*** (0.0005)	−0.005*** (0.0005)	−0.005*** (0.0005)	−0.004*** (0.0005)	−0.005*** (0.0005)
Baseline moisture	−0.0007 (0.001)	−0.001 (0.002)	−0.002 (0.002)	−0.002 (0.002)	−0.001 (0.002)	−0.002 (0.002)
Coastal	0.004** (0.002)	0.004** (0.002)	0.004** (0.002)	0.004** (0.002)	0.005*** (0.002)	0.004** (0.002)
Annualized moisture growth	−0.190* (0.100)	−0.108** (0.124)	−0.072 (0.121)	−0.076* (0.121)	−0.114** (0.125)	−0.086 (0.122)
Log(Baseline industry emissions)	–	–	0.004*** (0.0007)	0.003*** (0.0008)	–	0.003*** (0.0008)
Annualized industry emissions growth	–	–	–	0.041 (0.027)	–	0.046* (0.026)
Baseline precipitation	–	–	–	–	0.00001 (0.00001)	0.00002 (0.00002)
Annualized precipitation growth	–	–	–	–	−0.004 (0.044)	−0.006 (0.043)
Baseline air temperature	–	–	–	–	0.001 (0.001)	0.001 (0.001)
Annualized air temperature growth	–	–	–	–	0.176*** (0.058)	0.161*** (0.056)
State fixed effects	NO	YES	YES	YES	YES	YES
Constant	0.076*** (0.005)	0.078*** (0.006)	0.157** (0.016)	0.149*** (0.017)	0.077*** (0.006)	0.144** (0.017)
Adjusted R-Squared	0.173	0.269	0.307	0.310	0.277	0.316

*p < 0.010, **p < 0.05, ***p < 0.01. *Notes* This table regresses moisture growth and various additional controls (listed in the leftmost column) on annualized urban population growth. Regressions 2–6 employ state fixed effects. Urban population is calculated as the sum of the population of Class I–VI cities located within a district

Table A.5 Regression results: rural population growth, non-winsorized

Independent Variable	(1)	(2)	(3)	(4)	(5)	(6)
Log(Baseline urban share)	−1.66e−09*** (6.01e−10)	−6.15e−10 (5.39e−10)	3.39e−10 (9.17e−10)	9.29e−10 (9.91e−10)	−6.61e−10 (5.25e−10)	8.51e−10 (8.86e−10)
Baseline moisture	−0.0006 (0.0008)	−0.002* (0.001)	−0.002* (0.001)	−0.002* (0.001)	−0.002 (0.001)	−0.002 (0.001)
Coastal	−0.008** (0.001)	−0.003*** (0.0009)	−0.003** (0.0009)	−0.003** (0.0009)	−0.003*** (0.0009)	−0.003*** (0.0009)
Annualized moisture growth	−0.116 (0.072)	−0.134* (0.072)	−0.133 (0.071)	0.133** (0.068)	−0.152** (0.063)	0.151*** (0.060)
Log(Baseline industry emissions)	–	–	−0.0008 (0.0006)	−0.001** (0.0007)	–	−0.001** (0.0006)
Annualized industry emissions growth	–	–	–	0.050*** (0.015)	–	0.047*** (0.014)
Baseline precipitation	–	–	–	–	−9.71e−06 (0.00001)	−9.00e−06 (0.00001)
Annualized precipitation growth	–	–	–	–	0.018 (0.027)	0.012 (0.027)
Baseline air temperature	–	–	–	–	−0.001 (0.002)	−0.001 (0.002)
Annualized air temperature growth	–	–	-	–	−0.058 (0.038)	−0.049 (0.036)
State fixed effects	NO	YES	YES	YES	YES	YES
Constant	0.019*** (0.001)	0.019*** (0.001)	0.005 (0.011)	−0.006 (0.012)	0.020*** (0.001)	−0.004 (0.011)
Adjusted R-Squared	0.179	0.540	0.545	0.558	0.547	0.562

*p < 0.010, **p < 0.05, ***p < 0.01. *Notes* This table regresses moisture growth and various additional controls (listed in the leftmost column) on annualized rural population growth. Regressions 2–6 employ state fixed effects. Rural population is calculated as total district population minus urban population

Table A.6 Regression results: city population growth, non-winsorized

Independent Variable	(1)	(2)	(3)	(4)	(5)	(6)
Baseline urban share	−0.0001 (0.0007)	−0.0002 (0.0007)	−0.0002 (0.0007)	0.0003 (0.0008)	0.0002 (0.0007)	0.0003 (0.0008)
Baseline moisture	0.002 (0.001)	0.0006 (0.002)	0.0005 (0.002)	0.0004 (0.002)	0.0003 (0.002)	0.0004 (0.002)
Coastal	−0.0008 (0.002)	0.002 (0.002)	0.003 (0.002)	0.002 (0.002)	0.002 (0.002)	0.002 (0.002)
Annualized moisture growth	0.097 (0.101)	0.400*** (0.142)	0.398*** (0.143)	0.396*** (0.145)	0.354*** (0.143)	0.349** (0.146)
Log(Baseline industry emissions)	–	–	−0.0007 (0.0006)	−0.001** (0.0005)	–	−0.001** (0.0005)
Annualized industry emissions growth	–	–	–	0.059** (0.024)	–	0.060** (0.025)
Baseline precipitation	–	–	–	–	−7.23e−06 (0.00001)	−3.08e−06 (0.00001)
Annualized precipitation growth	–	–	–	–	0.031 (0.032)	0.027 (0.032)
Baseline air temperature	–	–	–	–	0.002** (0.0008)	0.002** (0.0008)
Annualized air temperature growth	–	–	–	–	0.081 (0.052)	0.092* (0.053)
State fixed effects	NO	YES	YES	YES	YES	YES
Constant	0.021*** (0.007)	0.021*** (0.008)	0.010 (0.025)	0.002 (0.013)	0.020* (0.008)	−0.003 (0.14)
Adjusted R-Squared	0.0037	0.086	0.087	0.093	0.088	0.095

*p < 0.010, **p < 0.05, ***p < 0.01. *Notes* This table regresses moisture growth and various additional controls (listed in the leftmost column) on annualized city population growth. Regressions 2–6 employ state fixed effects. All regressions cluster standard errors at the district level

References

Abeysiriwardana, H.D., Muttil, N., Rathnayake, U.: A comparative study of potential evapotranspiration estimation by three methods with FAO penman–monteith method across Sri Lanka. Hydrology **9**(11), 206 2022. https://doi.org/10.3390/hydrology9110206

Abhishek, N., Jenamani, M., Mahanty, B.: Urban growth in Indian cities: are the driving forces really changing? Habitat Int. **69**(September 9), 48–57 (2017). https://doi.org/10.1016/j.habitatint.2017.08.002

Abiy, A.Z., Melesse, A.M., Seyoum, W.M., Abtew, W.: Drought and climate teleconnection and drought monitoring. In: Melesse, A.M., Abtew, W., Senay, G. (eds.) Extreme Hydrology and Climate Variability: Monitoring, Modelling, Adaptation and Mitigation, p. 2019. Elsevier, Cambridge, MA (2019)

Adger, W.N., Crépin, A.-S., Folke, C., Ospina, D., Stuart Chapin III, F., Segerson, K., Seto, K.C., Anderies, J.M., Barrett, S., Bennett, E.M., Daily, G., Elmqvist, T., Fischer, J., Kautsky, N., Levin, S.A., Shogren, J.F., van den Bergh, J., Walker, B., Wilen, J.: Urbanization, migration, and adaptation to climate change. One Earth **3**(4), 396–399 (2020). https://doi.org/10.1016/j.oneear.2020.09.016

Alam, S.M., Markandey, K.: Consequences of unplanned growth: a case study of metropolitan Hyderabad. In: Thakur, R., Dutt, A., Thakur, S., Pomeroy, G. (eds.) Urban and Regional Planning and Development. Springer, Cham (2020). https://doi.org/10.1007/978-3-030-31776-8_13

Angel, S., Lamson-Hall, P., Blanco, Z.G.: Anatomy of density: measurable factors that constitute urban density. Build. Cities **2**(1), 264–282 (2021). https://doi.org/10.5334/bc.91

Asher, S., Campion, A., Gollin, D., Novosad, P.: The long-run development impacts of agricultural productivity gains: evidence from irrigation canals in India. In: Structural Transformation and Economic Growth, Working Paper 004 (2022). https://steg.cepr.org/publications/long-run-development-impacts-agricultural-productivity-gains-evidence-irrigation

Barrios, S., Bertinelli, L., Strobl, E.: Climatic change and rural–urban migration: the case of Sub-Saharan Africa. J. Urban Econ. **60**(3), 357–371 (2006). https://doi.org/10.1016/j.jue.2006.04.005

Bohra-Mishra, P., Oppenheimer, M., Hsiang, S.M.: Nonlinear permanent migration response to climatic variations but minimal response to disasters. Proc. Natl. Acad. Sci. U.S.A. (PNAS) **111**(27), 9780–9785 (2014). https://www.jstor.org/stable/23802664

Buis, A.: Steamy relationships: how atmospheric water vapor amplifies earth's greenhouse effect. In: National Aeronautic and Space Administration, Global Climate Change: Vital Signs of the Planet (2022). https://climate.nasa.gov/ask-nasa-climate/3143/steamy-relationships-how-atmospheric-water-vapor-amplifies-earths-greenhouse-effect

Bustos, P., Caprettini, B., Ponticelli, J.: Agricultural productivity and structural transformation: evidence from Brazil. Am. Econ. Rev. **106**(6), 1320–1365 (2016). https://www.jstor.org/stable/43861124

Bustos, P., Garber, G., Ponticelli, J.: Capital accumulation and structural transformation. In: Global Poverty Research Lab, Working Paper No. 17-107 (November) (2019). Available at SSRN: https://doi.org/10.2139/ssrn.3126453

Census of India.: Decadal variation in population 1901–2011. Office of the Registrar, General & Census Commissioner, India (2022). https://censusindia.gov.in/nada/index.php/catalog/43333

Census of India. [n.d.].: Towns and urban agglomerations classified by population size class in 2011 with variation between 1901 and 2011—Class I, II, III, IV, V, & VI. Office of the Registrar, General & Census Commissioner, India. https://censusindia.gov.in/nada/index.php/catalog/?page=1&sk=Towns%20and%20urban%20agglomerations%20classified%20by%20population%20size%20class%20in%202011&sort_by=popularity&sort_order=desc&ps=15

Christiaensen, L., De Weerdt, J., Todo, Y.: Urbanization and poverty reduction: the role of rural diversification and secondary towns. Agric. Econ. **44**(4–5), 435–447 (2013). https://doi.org/10.1111/agec.12028

de Sherbinin, A., Levy, M., Adamo, S., MacManus, K., Yetman, G., Mara, V., Razafindrazay, L., Goodrich, B., Srebotnjak, T., Aichele, C., Pistolesi, L.: Global Estimated Net Migration Grids by Decade: 1970–2000. Dataset. NASA Socioeconomic Data and Applications Center (SEDAC), Palisades, NY (2015). https://doi.org/10.7927/H4319SVC

Ellis, P., Roberts, M.: Leveraging Urbanization in South Asia: Managing Spatial Transformation for Prosperity and Livability. The World Bank Group, Washington, DC (2016). https://openknowledge.worldbank.org/entities/publication/f1ba8d07-e6d1-5b8e-8a16-899838f1c884

Glaeser, E.L.: Triumph of the City: How Our Greatest Invention Makes Us Richer, Smarter, Greener, Healthier and Happier. Penguin Books, New York (2012)

Gornall, J., Betts, R., Burke, E., Clark, R., Camp, J., Willett, K., Wiltshire, A.: Implications of climate change for agricultural productivity in the early twenty-first century. Philos. Trans. R. Soc. B Biol. Sci. **365**(1554), 2973–2989 (2010). https://royalsocietypublishing.org/doi/10.1098/rstb.2010.0158

Harari, M.: Cities in bad shape: urban geometry in India. Am. Econ. Rev. **110**(8), 2377–2421 (2020). https://doi.org/10.1257/aer.20171673

Hatab, A.A., Ravula, P., Nedumaran, S., Lagerkvist, C.-J.: Perceptions of the impacts of urban sprawl among urban and peri-urban dwellers of Hyderabad, India: a Latent class clustering analysis. Environ. Dev. Sustain. **24**, 12787–12812 (2021). https://doi.org/10.1007/s10668-021-01964-2

Henderson, J.V., Storeygard, A., Deichmann, U.: Has climate change driven urbanization in Africa? J. Dev. Econ. **124**(January), 60–82 (2017). https://doi.org/10.1016/j.jdeveco.2016.09.001

IPCC (Intergovernmental Panel on Climate Change).: Climate change 2022: impacts, adaptation and vulnerability. In: Pörtner, H.-O., Roberts, D.C., Tignor, M., Poloczanska, E.S., Mintenbeck, K., Alegría, A., Craig, M., Langsdorf, S., Löschke, S., Möller, V., Okem, A., Rama, B. (eds.) Contribution of Working Group II to the Sixth Assessment Report of the Intergovernmental Panel on Climate Change. Cambridge University Press, Cambridge, UK (2022). https://report.ipcc.ch/ar6/wg2/IPCC_AR6_WGII_FullReport.pdf

Liu, M., Shamdasani, Y., Taraz, V.: Climate Change, Structural Transformation, and Infrastructure: Evidence from India (2020). https://www.dropbox.com/s/8hyyk9x0sg2tfwb/LST_updated.pdf?dl=0

Kumar, R., Gautam, H.R.: Climate change and its impact on agricultural productivity in India. J. Climatol. Weather Forecast. **2**(1), 1000109 (2014). https://doi.org/10.4172/2332-2594.1000109

Menashe-Oren, A., Bocquier, P.: Urbanization is no longer driven by migration in low- and middle-income countries (1985–2015). Popul. Dev. Rev. **47**(3), 639–663 (2021). https://doi.org/10.1111/padr.12407

Mitra, A., Tripathi, S.: Rural non-farm sector: revisiting the census towns. Environ. Urban. ASIA **12**(1), 148–155 (2021). https://doi.org/10.1177/0975425321990324

Monforti-Ferrario, F., Oreggioni, G., Shaaf, E.: Fossil CO_2 and GHG Emissions of All World Countries—2019 Report. European Commission, Joint Research Center. Publications Office of the European Union, Luxembourg (2019). https://doi.org/10.2760/687800

Monforti-Ferrario, F., Crippa, M., Guizzardi, D., Muntean, M., Shaaf, E., Lo Vullo, E., Solazzo, E., Olivier, J., Vignati, E.: EDGAR v6.0 Greenhouse Gas Emissions. Dataset. European Commission, Joint Research Centre (2021). http://data.europa.eu/89h/97a67d67-c62e-4826-b873-9d972c4f670b

Neumann, B., Vafeidis, A.T., Zimmermann, J., Nicholls, R.J.: Future coastal population growth and exposure to sea-level rise and coastal flooding—a global assessment. PloS ONE **10**(3), e0118571. https://doi.org/10.1371/journal.pone.0118571

Olivier, J., Guizzardi, D., Schaff, E.: Fossil CO_2 and GHG Emissions of All World Countries—2021 Report. Publications Office of the European Union, Luxembourg (2021). https://op.europa.eu/en/publication-detail/-/publication/45c88a84-2d65-11ec-bd8e-01aa75ed71a1

Ozturk, U., Bozzolan, E., Holcombe, E.A., Shukla, R., Pianosi, F., Wagener, T.: How climate change and unplanned urban sprawl bring more landslides. Nature **608**(7922), 262–265 (2022). https://www.nature.com/articles/d41586-022-02141-9

Peri, G., Sasahara, A.: The Impact of Global Warming on Rural-Urban Migrations: Evidence from Global Big Data. National Bureau of Economic Research, Working Paper 25728 (April) (2019). http://www.nber.org/papers/w25728

Rigaud, K.K., de Sherbinin, A., Jones, B., Bergmann, J., Clement, V., Ober, K., Schewe, J., Adamo, S., McCusker, B., Heuser, S., Midgley, A.: Groundswell: Preparing for Internal Climate Migration. The World Bank Group, Washington, DC (2018). https://openknowledge.worldbank.org/bitstream/handle/10986/29461/WBG_ClimateChange_Final.pdf

Shepard, D.: A two-dimensional interpolation function for irregularly-spaced data. In: Proceedings of the 1968 23rd ACM National Conference (January), pp. 517–523 (1968). https://dl.acm.org/doi/10.1145/800186.810616

Shuttleworth, W.J., Wallace, J.S.: Evaporation from sparse crops—an energy combination theory. Q. J. R. Meteorol. Soc. **111**(469), 839–855 (1985). https://doi.org/10.1002/qj.49711146910

Small, C., Nicholls, R.J.: A global analysis of human settlement in coastal zones. J. Coast. Res. **19**(3), 584–599 (2003). https://www.jstor.org/stable/4299200

Stannard, D.I.: Comparison of Penman-Monteith, Shuttleworth-Wallace, and Modified Priestley-Taylor evapotranspiration models for wildland vegetation in semiarid rangeland. Water Resour. Res. **29**(5), 1379–1392 (1993). https://doi.org/10.1029/93WR00333

Tacoli, C., McGranahan, G., Satterthwaite, D.: Urbanisation, rural-urban migration and urban poverty. In: International Institute for Environment and Development, IIED Working Paper (March) (2015). https://www.jstor.org/stable/pdf/resrep01308?acceptTC=true&coverpage=false&addFooter=false

Thornthwaite, C.W.: An approach toward a rational classification of climate. Geogr. Rev. **38**(1), 55–94 (1948). https://doi.org/10.2307/210739

Tripathi, S.: Towards sustainable urban system through the development of small towns in India. Reg. Sci. Policy Pract. **13**(April), 777–797 (2020). https://doi.org/10.1111/rsp3.12424

Tuholske, C., Caylor, K., Evans, T., Avery, R.: Variability in urban population distributions across Africa. Environ. Res. Lett. **14**(8), 085009 (2019). https://iopscience.iop.org/article/10.1088/1748-9326/ab2432

Tuholske, C., Caylor, K., Funk, C., Verdin, A., Sweeney, S., Grace, K., Peterson, P., Evans, T.: Global urban population exposure to extreme heat. Proc. Natl. Acad. Sci. U.S.A. (PNAS) **118**(41), e2024792118 (2021). https://doi.org/10.1073/pnas.2024792118

van Duijne, R.J., Nijman, J.: India's emergent urban formations. Ann. Am. Assoc. Geogr. **109**(6), 1978–1998 (2019). https://doi.org/10.1080/24694452.2019.1587285

Willmott, C.J., Rowe, C.M., Philpot, W.D.: Small-scale climate maps: a sensitivity analysis of some common assumptions associated with grid-point interpolation and contouring. Am. Cartogr. **12**(1), 5–16 (1985). https://doi.org/10.1559/152304085783914686

Willmott, C.J., Feddema, J.J.: A more rational climatic moisture index. Prof. Geogr. **44**(1), 84–88 (1992). http://climate.geog.udel.edu/~climate/publication_html/Pdf/WF_ProfGeog_92.pdf

Willmott, C.J., Matsuura, K.: Willmott and Feddema's Moisture Index Archive: Gridded Monthly Climatologies. Version 1.01 (1999). http://climate.geog.udel.edu/~climate/html_pages/README.im.html

Willmott, C.J., Matsuura, K.: Terrestrial Air Temperature: 1900—2017 Gridded Monthly Time Series. Dataset, Version 5.01 (2018a). http://climate.geog.udel.edu/~climate/html_pages/Global2017/README.GlobalTsT2017.html

Willmott, C.J., Matsuura, K.: Terrestrial Precipitation: 1900—2017 Gridded Monthly Time Series. Dataset, Version 5.01 (2018b). http://climate.geog.udel.edu/~climate/html_pages/Global2017/README.GlobalTsP2017.html

Willmott, C.J., Matsuura, K.: Willmott and Feddema's Moisture Index Archive: Monthly Time Series (1900–2017). Dataset, Version 4.01 (2018c). http://climate.geog.udel.edu/~climate/html_pages/Global2017/README.GlobalImTs2017.html

World Bank.: Advancing Climate Action through an Urban Lens. World Bank Group (2021). https://www.worldbank.org/en/topic/urbandevelopment/brief/climate-action-through-an-urban-lens

UN Habitat (United Nations Human Settlements Programme).: Climate Change. United Nations (n.d.). Accessed April 12, 2022. https://unhabitat.org/topic/climate-change

UNFCC (United Nations Framework Convention on Climate Change).: Initiatives in the Area of Human Settlements and Adaptation: Summary Report by the Secretariat. Subsidiary Body for Scientific and Technological Advice, Forty-Sixth Session (2017). https://unfccc.int/documents/9664

Zhang, T., Mahmood, R., Lin, X., Pielke Sr, R.A.: Irrigation impacts on minimum and maximum surface moist enthalpy in the central great plains of the USA. Weather Clim. Extremes **23**(March), 100197 (2019). https://doi.org/10.1016/j.wace.2019.100197

The Interface Between Growing Informal Settlements and the Environment: How to Plan for Taboo Spaces?

Danielle Hill

Abstract

Global South cities have been characterised by the proliferation of informal settlements. Due to the speed and development of informal settlements being more 'prone' to settle on environmentally sensitive areas, such as wetlands and setbacks of agricultural areas which ultimately have knock-on effects on water quality, estuaries, management of informal settlements, in dialogue with sustainability becomes critical when thinking about city futures. Informal settlements fundamentally challenge the core of urban planning. What happens when planners' earmark environmentally sensitive (taboo) areas on their spatial plans, only to find that informal settlement communities are illegally occupying these areas? How is contemporary urban planning equipped to harness the necessary skill when dealing with the protection of the environment on one hand and assisting informal settlement residents on the other? Through exploring the management of Thembalethu Phase 1 of Upgrading of Informal Settlements Programme (UISP) in George, South Africa, the paper seeks to determine how built environment professionals involved in upgrading of the settlement challenged the role of professional planners and their ability to steer 'informal' occupation and development.

1 Introduction

For almost three decades, planning has been criticised by urban planning scholars for its inability to be flexible (Kamete 2009, 2012; Watson 2009a, b) in ways planners choose to incorporate social processes into their rigid plans. For the most part, this idea of flexibility within planning is addressed as topic specific, which in turn makes the addressing of interfaces cross topic difficult. There is a need for a holistic approach to urban development that incorporates a multi-layered approach to how we understand flexibility within planning discourses. Planning is experiencing increasing pressure from both bottom-up and top-down directions to think more innovatively when it comes to planning for the consequences of rapid urbanisation. Some scholars (Pinson, 2004: 503) argue that the multidisciplinary nature of planning may weaken urban planning as a discipline and for the need to rethink and re-establish planning's identity in order to implement a fertile transdisciplinarity contribution. I argue that the dynamic nature of the current proliferation of informal settlements provides a window of opportunity to reconsider the contemporary role of urban planning as a potentially flexible governance tool in global South cities.

The relationship between the environment and informal settlement is not new. This relationship probes questions of planning's value system, i.e., whose needs are more important and require prioritising. I would argue that the rapid growth of informal settlements in recent decades mirrors peculiar insights into the relationship between humans and non-human species/the environment. This is because many informal settlements

> … occupy land that is unsuitable for development, which has resulted in the destruction of environmentally sensitive areas. In many cities, the informal occupation of areas near water reservoirs, areas prone to landslides and flooding, or protected forests is another looming problem (Kekana et al., 2023: 3711).

D. Hill (✉)
School of Architecture, Planning & Geomatics, University of Cape Town, Cape Town, South Africa
e-mail: danielle.hill@up.ac.za

Department of Town and Regional Planning, University of Pretoria, Pretoria, South Africa

P. Mohammad et al. (eds.), *Geographic Approaches to Climate Change and Mitigation: Urban and Rural Perspectives (Volume 1)*, Advances in Science, Technology & Innovation, https://doi.org/10.1007/978-3-031-92119-3_2

As a result of this, Kekana et al. (2023: 3711) argues that 'understanding the dynamics of human settlements is a prerequisite for sustainable development and environmental management'.

How we understand the dynamics of human settlements as a core prerequisite for sustainable development is crucial when thinking about sustainable city futures. Simply put, on an un-nuanced level, the bulk of the scholarly focus when it comes to discussing informal settlements has largely to do with either uncovering the reasons for their formation, in other words, asking who is to blame, or how the urban poor is being pro-active by managing their poverty (Abagissa 2019; Nassar and Elsayed 2018), or critiquing upgrading processes and policy (Kamete 2009, 2013; Huchzermeyer 2006; Smit 2017). However, what is less spoken about is how professionals are equipped, or not, when confronted with the knowledge that informal communities will, and often do, settle in environmentally sensitive areas, how do/will planners 'prepare' for this, and how do/will these impact their 'formal' plans? (Hill 2023).

Since inception, South Africa's democratic housing agenda has, to a significant extent, been a politically sensitive matter. Smit (2017) documents the process of how, since the 1980s, South Africa's has attempted to address informal settlements. The description of this forty-year journey traces and identifies familiar themes of eradication changing to a now more incremental approach to upgrading of informal settlements. The incremental four-phased structured approach called the Upgrading of Informal Settlements Programme (UISP) is an incremental aim towards formalisation. The upgrading process happens at a municipal level and is a team effort, which includes different built environmental professionals who work collectively on the upgrading project. Even though all four phases are expected by municipalities to be implemented, many of them only reach stage 2.[1] Due to the dynamic and fluid nature of informal settlements, the UISP becomes at times challenging to implement as it is caught between the impatience from the communities involved and the tedious bureaucratic timelines of municipal approvals and funding.

I explore a sample of planners' role by looking at a pilot (Phase 1), UISP project in Thembalethu, municipality of George, Western Cape Province. Even though the UISP is a housing policy rather than a planning tool, the UISP is actively designed to address and upgrade informal settlements by following this four-phased approach to address broader socio-economic challenges. Planning is institutionalised and is a legislative tool in South Africa, with the majority of decision-making occurring within planning processes still being done by elite technocrats. I consider the UISP as a planning instrument and Thembalethu UISP Phase 1 project as a planning exercise. This is because, in order for the implementation of a UISP project to be approved by provincial government on a municipal level, the project needs to strategically align with the various planning frameworks of the municipality (Integrated Development Plan (IDP), the Housing Development Plan, and Spatial Development Framework (SDF)). Furthermore, the UISP project needs to adhere to community-based participatory planning facilitation structures and processes. In the process of exploring the Thembalethu UISP, I examine the degree to which planners (their perceptions and how other professionals on the team perceive their role) are able to intervene and manage the complexities and contradictions inherent in informal settlement upgrading processes, and the management of informal growth. I attempt to derive what this could mean for the relationship between the environment and informal settlement development. Thus, the paper seeks to determine how the built environment professionals involved in upgrading of the settlement challenged the role of professional planners and their ability to steer 'informal' occupation and development. I also look closely at the ways in which their perceptions and realities of the UISP mirror (1) planners ability to 'handle' informal settlements growth and (2) their ability to manage the 'infringement' of informal settlements on environmentally sensitive areas and what this means for planning as governance practice. I discuss the gap within planning literature concerning full engagement with the complexities that informal settlements embody spatially and the long-term consequences of these for the environment.

1.1 The Contested Role of Planning

Despite urban planning's shift in the last two decades, from the conventional blueprint, technocratic, modernist-inspired models, to more relational flexible models (Albrechts 2015; Kamete 2013), the recent 'resurgence' (Varley 2013) of informality has been found to impede the 'make-up' and functionality of urban planning. Recent studies are increasingly focusing on, and actively engaging with, the relationship between urban informality and urban planning (Amin and Cirolia 2018; Dastidar 2007; Recio 2015; Roy 2005; Wade 2009). This focus is premised on an increasing understanding of the heterogeneous nature of informality, and how it manifests and translates within planning processes, in so doing, exposing the inability of planners to 'handle' the multiple forms of urban life that present themselves.

Todes (2011) envisages the development of an international quest to revitalise urban and regional planning. This stems from a shift in the past from planning which focused narrowly on spatial control and order, to urban planning as a profession that manages the growth of cities and promotes

[1] Informal Settlement Upgrading: An Institutional Map For NGOs. Find online: https://isandla.org.za/en/resources/item/download/121_5228c2c0de1099ad6e2bd91e1185b908 (10/12/2023).

integrated development together with sustainability. The critical issue associated with this new approach is that planning becomes a mechanism to surface and negotiate conflicting rationalities and concomitant priorities. This kind of urban planning, therefore, moves away from a control impulse to one focused on mediation and facilitation, while retaining a strong normative footing. According to Wade (2009: 2), this shift is towards what appears to be a more sensitive, multi-layered and integrated planning approach. Wade (2009) argues that 'informality' runs the risk of becoming another catchphrase like 'participation', which, if used loosely, can distort the transformational essence necessary to disrupt dominant epistemological frames and thinking around informality. From this, and from other research, the possibility appears to exist of a global south epistemological orientation towards a better and more comprehensive understanding of urban planning challenges, one that can be characterised and associated with everyday practices (Amin 2014; Cirolia and Scheba 2019; Pieterse 2011). However, if urban planning systems are primary institutional entry points to advancing an alternative agenda (Pieterse 2008: 151), there exists a gap in the literature of studies which have the potential to provide passages of interconnection and interaction between everyday practices, their diverse cultural characteristics, and the institutional make-up of planning systems.

Kamete (2013: 645) describes how urban planning in sub-Saharan Africa has sought to 'cleanse urban spaces [from informal settlements] from spatial pathologies'. This approach attributes and shapes how informality is perceived and constructed by the state and urban planners. The continuous and pervasive nature and 'pathology' of informality, and its manifestation within urban space, has become the new norm. Kamete's work uses the word 'normal' in his characterising of a lens through which to examine urban planning and informality. To him, urban planning is one of many instruments the state uses to reinforce its authority, and spatial/'normalising' technologies, standards, and discourse(s) are utilised to correct/normalise what critics of these practices, and their particular discourse(s) would describe as informal spatial abnormalities. He further argues that addressing informality has reduced itself to technical interventions, as opposed to interventions premised on social factors, economic governance, and political dynamics. More importantly, he argues that it is important for planners to abandon their continuous efforts to find a technical solution to informality and to recognise the dynamic and multi-layered nature of evolving urban spaces. In this context, planners need to embrace the idea that informality is 'normal', and to 'take informality back to the broader political and social sphere and insert it into the broader debates about social justice and economic and political governance' (Kamete 2013: 648). Adding to Kamete's argument, Gonzales (2008) reminds us that:

> Informality is not an exotic transplant from Latin America or a manifestation of legal 'underdevelopment.' Rather, informality is a rational response to poverty and inequality, and will likely flourish in both the North and the South as long as the underlying economic causes persist (Gonzales, 2008: 252).

What becomes apparent within the literature, which focuses on navigating the relationship between the role of planning/er in managing informal settlements, is the existence of the many ways there should be/is a rightful preoccupation with understanding the social processes, political and social spheres embodied in and contributing to informal settlement development. However, what is less considered are the broader environmental implications of increasing informal settlement development and how this further adds to the multi-layered nature of evolving urban spaces. Thus, this gives rise to questioning the 'how' in the process of planners, planning instruments and processes should manage or regulate informal settlements to conserve environmentally sensitive areas. Even with recent research, Connolly and Wiggle (2017: 183) caution, through their examination of 'the introduction of land-use planning requirements into the regularisation process of informal settlements in areas designated as "conservation land" in Mexico City', that the 'tenuous distinction between formal and informal urban development shows how the state uses this trope to reproduce social class divisions in managing access to land and housing' (195). They emphasise the role of the framing of informality in planning discourses. Framings—and the discourse used—include informal settlements as being 'chaotic' or the ways in which informal settlements 'disrupt' ecological spaces and the environmental services provided by conservation land. The result of this is the justification of politicised

> procedural machinations which will decide the fate [of informal communities]: whether they will be allowed to follow the conventional path of regularization, land-titling and assimilation into the urban fabric, or whether they will continue to be classified as "irregular" and thus deprived of access to basic amenities and urban services" (195).

Thus navigating planning pathways for planners to find ways for both the informal settlement communities and the environment to co-exist remains a challenge.

1.2 The 'Progressive' UISP

Given the many social inequalities created and exacerbated by apartheid, with the dawn of democracy it was pertinent for the redress approach, as specified in the Constitution's Bill of Rights, to housing delivery to provide dignified and sufficient housing, together with dignified, healthy living conditions. This intent was harnessed in the late 1990s within the Reconstruction and Development Programme (RDP), a programme which materialised through the provision of formal housing on an extensive scale. This programme was critiqued by urban scholars for its design and implementation: it was seen to reinforce apartheid spatial disparities, while at the same time urban informality continued to exist. Even though there had been a global shift in policy reform away from informal eradication to in situ upgrading in the early 1970s (Huchzermeyer 2011), in 2004 South Africa introduced the Breaking New Ground Policy (DoH 2004). This policy represented the introduction to informal settlement upgrading with the funding support of Upgrading of Informal Settlements Programme (UISP). This shift was characterised by the prioritisation of infrastructure services (Parikh et al. 2020). The UISP has four stages[2] and activities for each stage include, Stage 1—community participation, land acquisition, preplanning interim engineering services; Stage 2—surveying/participation/registration, land acquisition, finalising social compacts, suitability/pre-planning studies, interim engineering services; Stage 3—project management capacity, tenure/occupation/other land issues, installation of permanent services, installation of social services/amenities, housing support, relocation assistance; Stage 4—project management capacity, finalising township establishment housing support services, tenure/occupation other land issues, construction of top-structures, construction of outstanding social services.

Maina (2013) argues that, despite the progressive policy discourse, translation of the UISP has in fact replicated the RDP model, a model that provided a rationale for the removal—eradication—of settlements and the redevelopment of cleared sites. Even though the UISP was originally envisaged in terms of supporting upgrading and maintaining community networks, as well as minimising disruption, with the noble intention to 'secure tenure and access to health and security and the empowerment of settlement residents through participative processes', implementation of the UISP remains a challenge throughout the country. Moreover, in the initial phases (Phase 1, 2), settlement-level assessments and categorisation are done, and initial scoping of the geotechnical suitability of the land, and a desktop scoping of the environmental suitability of the area for upgrading, is done. Like planning, the environmental role is marginal. According to Huchzermeyer (2021: 45), in her assessment of UISP deliberations involving all three spheres of government in South Africa, she gives emphasis to the 'displaced' role of planning within the UISP. She refers to planning's role within the UISP as an 'additional obstacle, despite the supportive clauses in SPLUMA [Spatial Planning and Land Use Management Act 16 of 2013]'. Moreover, Huchzermeyer (2021), Misselhorn (2017), and other scholars focusing on UISPs in South Africa, attribute UISP's modest implementation progress to the lack of effective budget allocation, political will, and institutional support, as well as to the setting of unrealistic targets.

1.3 Thembalethu Phase 1 UISP Project and Method

Thembalethu is a historic 'formal' black township that has expanded informally (reasons for informal expansion detailed in Hill 2023). Thembalethu is situated in George Municipality, Western Cape Province, South Africa, with a total population reported in 2011 by STATSA of approximately 43,000.[3] Selection of Thembalethu as a case study emanates from George's Municipality's history of having Thembalethu, which at its inception was a municipal response to informal settlements in George pre-1994 (Hill 2023) on their housing agenda since before the overturn to democracy.

In 2010, the George Municipality Informal Settlement Master Plan (ISMP) prioritised Thembalethu as a vital start off location for upgrading. Thembalethu has always been significant to the municipal housing agenda and continues to be, as it continues to grow informally. In 2021, the George municipality, with the help of Aurecon (since 2020 known as Zutari), rolled out Phase four of the Upgrading Informal Settlements Programme (UISP) structured approach in Thembalethu. According to interviewees, there are more phases in the pipeline as informal settlements are continually increasing in Thembalethu. My focus is on Phase 1, which took place from 2010 to 2014.

My study adopted a qualitative case study research design approach, as I was interested in the perceptions of built environment professionals. The reason for choosing UISP Phase 1 is because Phase 1 was approached as an in situ upgrading project, and due to the phases being well-documented, it allowed me to follow through and assess all three of the stages of the UISP. Data were collected through semi-structured interviews with 11 professionals (planners and built environment professionals) who administered and were responsible for, the upgrading project, together with field observation. Due to the small sample size and anonymity, I am unable to detail professionals' roles. Data were analysed using a deductive thematic analysis technique. My approach to analysing the data was intended to help unpack the diverse experiences of those individuals who were involved with the Thembalethu Phase 1 UISP project, together with their assertions. The themes which emerged from the data, with specific emphasis on planners role in informal growth management as it pertains to aspects of informal settlements 'infringing' on environmentally sensitive areas. Key to data collection and analysis was understanding how a sample of planners intervene and manage the complexities and contradictions inherent in

[2] UISP stages explained: https://mbuisp.org/homepage/project-overview/uisp-explained/ (14/6/2022).

[3] http://www.statssa.gov.za/?page_id=4286&id=240 (27 August 2021).

informal settlement upgrading processes, and management of informal settlements.

In the case of the Thembalethu UISP project, Aurecon was appointed (through a tendering process) as a multidisciplinary implementing agent for the multi-year/phased UISP implementation and was specifically responsible for the application UISP process. In July 2010, Aurecon drew up a project feasibility report and implementation plan whose objective was to 'upgrade' all informal settlements in Thembalethu. Twenty-eight informal settlements, and approximately 4350 geotagged (surveyed) households, were identified and prioritised to kick start Phase 1 (Areas 4A, 4C, and 4B) of the Thembalethu UISP project. In Phase 1, 28 pockets of informal structures were identified. These are referred to as 28 informal settlements. These 28 pockets were prioritised because informal structures were characterised by being erected on steep gradients/slopes, and within environmentally hazardous conditions, some of these pockets were characterised (west of these areas is a watercourse) for communal farming practices. However, due to rapid informal growth, water resources and agricultural land were polluted. Of the cumulative 4350 households that were counted when the site analysis was done, only 1746 serviced sites were constructed in Phase 1: 4A = 823, 4B = 287, 4C = 636. As a result, only 1746 households of the 4350 could be accommodated in Phase 1. Community participation was conducted through a company Aurecon outsourced.

In order for the Phase 1 UISP project to commence, certain entities approvals were necessary. These included the National Home Builders Registration Council (NHBRC), environmental authorisation, Land Use and Planning Ordinance, 1985 (LUPO) authorisation, and funding approval. Both planning and environmental approval took approximately two years from the time of application, and as a result, UISP implementation could only take place after these approvals. Consequently, it was evident from professionals that, when they had to implement UISP, more than the 4350 households had settled on that area, mirroring the gap between UISP housing programme and informal settlement growth, which exacerbates residents displacement as a result of the UISP implementation process and potential further informal development. Further details of the project can be found in Hill's (2023) PhD thesis.

2 Findings/Discussion

Key findings revolved around themes relating to the perceptions of the planners by other built environment professionals, and the planners' perceptions of their 'own' role in the UISP process and the unique role informal settlements play in the broader urban development landscape. Examining these perception is pertinent because the UISP becomes a mirror for how planners 'manage' informal settlements through upgrading. It was clear from the reported perceptions that the UISP is a 'quick' fix but does not directly address the challenge of informal growth. The UISP was also criticised for not being a sustainable model due to its single-story outcomes in conjunction to the issue of land scarcity in George. Planners were also displayed as not 'influencing' the development process (echoed in Adams 2012). This questions the 'power' planning has in spatially managing informal settlement growth onto environmentally sensitive areas. The UISP is the only governance tool to address the challenge of informal settlements in South Africa, due to the extensive period it takes to approve layouts, to conduct environmental assessments, and collect all the necessary approvals. There are no other mechanisms in place to 'stop' or 'prevent' consequences or impact of informal settlement development on environmentally sensitive areas. Thus, 'prevention' happens only through the UISP, which leaves the essential gap concerning how to plan the taboo spaces in urban environments amidst fast-growing informal settlements, when there is only one tool that addresses informal settlements, and this after residents have already settled as opposed to a tool that is pro-active in its approach.

3 Perceptions of the Role/Influence of Planners

The study sought to gain insight into what various professionals responsible for the upgrading project understood by the role of the planner in the context of upgrading and informal settlement development. While there was a consensus among the professionals themselves that the planners play a measurable role in the upgrading process, there were mixed responses in terms of the degree of their significance and value to the process, and their role in managing informal settlement development. I mention and detail a few below.

One of the policy managers questioned the role of the planner: he communicated the need for planners to be more assertive with their plans and role, alluding to the planners' forecasting role to drive development. For him, within this context, informal development was/is setting the development agenda as opposed to planning doing so. In many ways he hinted at the victimhood of planning. One of the project managers (also a town planner) was of the opinion that planners had played, and tend to play, a marginal role in the upgrading process. This is because many of their roles and functions are outsourced. One of the engineers expressed his categorical view that town planning is the success or downfall of a housing project. Many of these professionals, communicated a high regard for the role of planners, and for the role they can play in providing spatial direction.

The role of planning within Thembalethu Phase 1 UISP project was decentralised. The various actors included the implementing agent, planners that were outsourced (responsible for layouts) by the implementing agent, and municipal planners. Planners who were outsourced recognised their limitations in terms of technical layouts, their role being budget bound. Municipal planners had stronger opinions about their role, detailing how they have the 'power' to say no to housing projects that do not comply with urban design guidelines. Less was spoken about their role in managing the growth of informal settlements.

As mentioned earlier, when the policy manager questioned planners assertiveness in 'getting ahead' of the informal development, he used an example to demonstrate his point. He explained how, in one of the neighbouring towns, informal settlement growth reached the point of 'encroaching' on the town's reservoir. He further stated that, because the informal settlements were located there, the municipality had no choice but to provide sites and services there, on land that was not necessarily suitable for development, thus risking the water quality of the town. The municipal planner further detailed her position on managing informal settlements:

> I think informality can work if it doesn't, if it doesn't induce health and security risks. Uhm if it doesn't compromise biodiversity. So, within those boundaries, I think there are certain boundaries we need to draw. And you can still structure informality in a way.

When taking a closer look at the informal sites earmarked for upgrading for this project, the municipal planner mentioned a more serious concern: planners were not just there to plan for informal settlements; there are other pressing urban needs needing to be attended to. One of these issues was the impact of informal growth on the environment. The municipal planner emphasised the need to consider sanitation, the cleanliness, and the dignity of residents residing in informal settlements, while at the same time not compromising the environment.

The municipal planner further mentioned that, within George municipality, there are a few mountains close to key corridors and which have a natural network of rivers, catchments and wetlands that connect them. Consequently, if damage is done to the wetland (up there), or the water source (up there), the water source and the riparian area around the water source would be compromised (down there). Management of these riparian areas would be critical as it determines what happens in the water itself and the quality of the water. She further alluded to how, when measuring the quality of the water in the areas closer to the mountain and comparing this with the quality of the water down in Thembalethu, Scarkop River, the Meul River, a significant amount of pathogens are found in that water than in the town's water. The main reason for this is because the informal communities have been established too close to the watercourse. She admits that there had not been any mitigation that would allow for some form of filtering, especially because a sewer runs off into that space, and as a result, the water is contaminated.

In addition, there were challenges with using this water as irrigation for the farmlands. There are agricultural sites immediately below Thembalethu, between the ocean and Thembalethu. Some of these farmlands still belong to the Western Cape provincial government. With government's initiative to establish small farmers on these farmlands, one of the challenges they faced was the pathogens in the water. Unfortunately, the water was too contaminated to launch this project intended to help communities in Thembalethu. The municipal planner stressed the comprehensive aspects to planning, and the importance of the municipality to consider all of the above factors involved because they want every aspect of that community and the quality of its members' lives to function well.

Although broad planning themes for consideration were highlighted, the spatial direction, or the 'how' of getting ahead of informal settlement development remained absent from the process. Planners in this project were open to the idea of the permanence of informal settlements, while at the same time showing great concern for the environment. Although informal settlements were being established near water sources and farmlands, spatial development frameworks (SDF) could not prevent this, but the municipal SDF did help motivate reasons for upgrade in the area. In part, due to South Africa's politicised housing history, this exacerbates the unintended political nature of upgrading projects. What becomes clear is how planners do in fact consider comprehensive elements of urban development, yet are still faced with the institutional limitations of formal planning processes (partly echoed in Kamete's work and literature above). Moreover, because planning is a formal instrument that thrives as a formal process, I question how we could open pathways for informal and, in many ways, disruptive realities to interface with municipal planning processes, without pathologising these areas. In addition, in what ways do we hold space for both the environment and informal settlements realities?

While this paper looked at the UISP and acknowledges its well-intentioned contribution, one which was project-focused and funding-dependent, it brought insights into state interventions and how the fixed instruments provided are still there to 'solve' the issue of informal settlements, which is exactly what the RDP sought to achieve, but failed. The UISP remains a blanket approach to addressing a single layer of informal settlements, which still seeks to include both participation of residents and environmental assessments as tick box exercises, when there remains that gap prior to the implementation of the UISP, during UISP approvals and after UISP intervention, and the aftermath of displacements, how to steer informal growth from these taboo areas. This could represent the potential opportune turn to rejuvenate (Todes 2011) urban planning spatial imaginaries in the twenty-first century.

More importantly, many of these formal planning tools (SDF's or IDP's) are not accessible to communities living in these settlements, to help inform them where to settle, on a site that is environmentally 'safe'. Even though there has been a growing body of literature (Amoako 2018; Campos et al. 2019; Dobson et al. 2015), that focuses on bottom-up initiatives, bringing more attention to planning 'with' communities, there is a need to open up the custodianship of the planning processes, if we desire to improve the relationship between the environment and informal settlement residents. Maybe the contemporary role of planning in balancing (Beall 1996) both the needs of the community and the environment lies in moving away from managing urban consequences to exploring the discomforting 'how' to help steer a holistic approach to urban development.

4 Conclusion

Even though planners in the Thembalethu UISP phase 1 project were faced with multiple challenges, and thought about the implications of these challenges, there exists a need for planners to start envisioning city futures again. However, this envisioning requires a solidifying and affirming of their identity, which in turn requires building a toolkit that can harness a strong contribution to multi-layered urban realities. The case of Thembalethu UISP phase 1 project, foregrounds the need to rethink ways in which we 'address' informal settlement growth and recognise the 'influencing power' planners should develop (through existing or new planning instruments and community partnerships) in the pro-active space (before UISP intervenes) when planning for taboo areas.

References

Abagissa, J.: Informal settlements in addis ababa: extent, challenges and measures taken. J. Public Adm. Financ. Law **15**, 7–30 (2019)

Adams, D.: Urban Planning and the Development Process. Routledge (2012)

Albrechts, L.: Ingredients for a more radical strategic spatial planning. Environ. Plann. b. Plann. des. **42**(3), 510–525 (2015)

Amin, A.: Lively infrastructure. Theory Cult. Soc. **31**(7–8), 137–161 (2014)

Amin, A., Cirolia, L.R.: Politics/matter: governing Cape Town's informal settlements. Urban Stud. **55**(2), 274–295 (2018)

Amoako, C.: Emerging grassroots resilience and flood responses in informal settlements in Accra, Ghana. GeoJournal **83**, 949–965 (2018)

Beall, J.: Building a city for all: notes after Habitat II. City **1**(5–6), 133–141 (1996)

Campos, M.J.Z., Barinaga, E., Kain, J.H., Oloko, M., Zapata, P.: Organising Grassroots Initiatives for a More Inclusive Governance: Constructing the City from Below (2019)

Cirolia, L.R., Scheba, S.: Towards a multi-scalar reading of informality in Delft, South Africa: weaving the 'everyday' with wider structural tracings. Urban Stud. **56**(3), 594–611 (2019)

Connolly, P., Wigle, J.: (Re) constructing informality and "doing regularization" in the conservation zone of Mexico City. Plan. Theory Pract. **18**(2), 183–201 (2017)

Dastidar, S.: Planning Spaces for Informality. Essay Submitted to the Development Planning Unit (2007). https://www.academia.edu/2045744/Planning_Spaces_for_Informality

Department of Housing (DoH).: Breaking New Ground: A Comprehensive Plan for the Development of Sustainable Human Settlements. Pretoria: Department of Housing (2004). http://www.dhs.gov.za/sites/default/files/documents/26082014_BNG2004.pdf

Dobson, S., Nyamweru, H., Dodman, D.: Local and participatory approaches to building resilience in informal settlements in Uganda. Environ. Urban. **27**(2), 605–620 (2015)

Gonzales, C.G.: Squatters, pirates, and entrepreneurs: Is informality the solution to the urban housing crisis? Univ. Miami Inter-Am. Law Rev. **40**(2), 239–259 (2008)

Hill, D.G.: The role of professional urban planners in understanding and managing the dynamism of informal settlements. Ph.D. thesis, University of Cape Town, Cape Town (2023)

Huchzermeyer, M.: The new instrument for upgrading informal settlements in South Africa: contributions and constraints. Informal Settl. Perpetual Challenge **41**, 61 (2006)

Huchzermeyer, M.: Cities with 'Slums': From informal settlement eradication to a right to the city in Africa. Juta and Company Ltd. (2011)

Huchzermeyer, M.: A critical Lefebvrian perspective on planning in relation to informal settlements in South Africa. Town Reg. Plann. **79**, 44–54 (2021)

Kamete, A.Y.: In the service of tyranny: debating the role of planning in Zimbabwe's UrbanClean-up'operation. Urban Stud. **46**(4), 897–922 (2009)

Kamete, A.Y.: Interrogating planning's power in an African city: time for reorientation? Plan. Theory **11**(1), 66–88 (2012)

Kamete, A.Y.: Missing the point? Urban planning and the normalisation of 'pathological' spaces in southern Africa. Trans. Inst. Br. Geogr. **38**(4), 639–651 (2013)

Kekana, H.N., Ruhiiga, T.M., Ndou, N.N., Palamuleni, L.G.: Environmental justice in South Africa: the dilemma of informal settlement residents. GeoJournal 1–17 (2023)

Maina, M.M.: Challenges in policy transition: *In situ* upgrading of informal settlements in Johannesburg and Nairobi, Ph.D. thesis, Wits University, Johannesburg (2013). https://wiredspace.wits.ac.za/bitstream/handle/10539/12813/Miriam%20M.%20Maina_Dissertation_May%202013.pdf?sequence=2

Misselhorn, M.: A Programme Management Toolkit for Metros: Preparing to scale up informal settlement upgrading in South Africa-a city-wide approach, vol. 20, pp. 4–5 (2017). https://csp.treasury.gov.za/Resource%20_Centre/Conferences/Documents/Informal%20Settlements%20Upgrading%20Workshop

Nassar, D.M., Elsayed, H.G.: From informal settlements to sustainable communities. Alex. Eng. J. **57**(4), 2367–2376 (2018)

Parikh, P., Bisaga, I., Loggia, C., Georgiadou, M.C., Ojo-Aromokudu, J.: Barriers and opportunities for participatory environmental upgrading: case study of Havelock informal settlement, Durban. City Environ. Interact. **5**, 100041 (2020)

Pieterse, D.E.: City Futures: Confronting the Crisis of Urban Development. Bloomsbury Publishing, London (2008)

Pieterse, E.: Rethinking African urbanism from the slum. Power **26**(14.1), 40–48 (2011)

Pinson, D.: Urban planning: an 'undisciplined' discipline? Futures **36**(4), 503–513 (2004)
Recio, R.B.: Engaging the 'ungovernable': urban informality issues and insights for planning. J. Urban Reg. Plann. **2**(1), 18–37 (2015)
Roy, A.: Urban informality: toward an epistemology of planning. J. Am. Plann. Assoc. **71**(2), 147–158 (2005)
Smit, W.: Informal settlement upgrading: international lessons and local challenges. In: Ciriola, L.R., Görgens, T., Donk, M., Smit, W., Drimie, S. (eds.) Upgrading Informal in South Africa: A Partnership-Based Approach, pp. 26–48. University of Cape Town Press, Cape Town (2017)
South Africa (SA).: Spatial Planning and Land Use Management Act 16 of 2013 (2013)
Todes, A.: Reinventing planning: critical reflections. Urban Forum **22**(2), 115–133 (2011)
Varley, A.: Postcolonialising informality? Environ. Plann. D Soc. Space **31**(1), 4–22 (2013). https://doi.org/10.1068/d14410
Wade, A.: Urbanism & The Production of Knowledge: The Case of Informality. Development Planning Unit, University College London, London (2009)
Watson, V.: Seeing from the South: refocusing urban planning on the globe's central urban issues. Urban Stud. **46**(11), 2259–2275 (2009a)
Watson, V.: 'The planned city sweeps the poor away…': urban planning and 21st century urbanisation. Prog. Plan. **72**(3), 151–193 (2009b)

Adaptive Ventilation Strategy for Climate Control in an Unheated Cathedral in Northern Italy

Harold Enrique Huerto-Cardenas, Niccolò Aste, Claudio Del Pero, Stefano Della Torre, Fabrizio Leonforte, Camille Luna Stella Blavier, and Felice De Francesco

Abstract

Unheated historic churches often contain objects of significant historical and cultural value, such as organs and paintings. Such buildings, with their features (massive envelope, dimensions, etc.) can often mitigate external climate fluctuation, providing a more stable indoor environment. However, although artefacts and visitors can often acclimatize to indoor conditions, the alterations caused by climate change might induce stressful events for the preserved materials and worsen the comfort levels of visitors and worshippers. In this context, the main goal of this study is to assess the implementation of an adaptive ventilation system, in a relevant religious building in Milan, through the re-adaptation of existing ventilation ducts distributed in the vaults to mitigate the indoor climate conditions, without jeopardizing valuable artworks. In particular, the proposed solution includes the replacement of existing Knapen atmospheric siphons, installed in the ventilation ducts, with fans that allow the increase of the air exchange with outside during specific occasions. Simulation analyses have shown that the proposed system might provide an average reduction in internal temperatures up to 1.7–2 °C in summer, while for relative humidity, it has been observed that the system can effectively reduce the excessive moisture content, without leading risk of deterioration for the artworks.

Keywords

Historic building · Cathedral · Adaptive ventilation strategy · Indoor climate · Climate control · Artworks conservation

Nomenclature

HVAC	Heating, Ventilation and Air Conditioning;
T	Air temperature [°C];
RH	Relative air humidity [%];
MR	Mixing ratio or humidity ratio [g/kg];
CFD	Computational Fluid Dynamics;
CPNV	Climatic Potential for Natural Ventilation [%];
ACH	Air Changes per Hour;
SHGC	Solar Heat Gain Coefficient;
τv	Visible trasmittance;
TMY	Typical Meteorological Year.

1 Introduction

The integrity of historical buildings and the preservation of their contained artefacts are intrinsically linked to the management of their indoor climate. This is especially relevant in historical buildings which normally are not provided by a heating, ventilation, and air conditioning

H. E. Huerto-Cardenas (✉) · N. Aste · C. Del Pero · S. Della Torre · F. Leonforte · C. L. S. Blavier
Architecture, Built Environment and Construction Engineering Department, Politecnico di Milano, 20133 Milan, Italy
e-mail: haroldenrique.huerto@polimi.it

N. Aste
e-mail: niccolo.aste@polimi.it

C. Del Pero
e-mail: claudio.delpero@polimi.it

S. Della Torre
e-mail: stefano.dellatorre@polimi.it

F. Leonforte
e-mail: fabrizio.leonforte@polimi.it

C. L. S. Blavier
e-mail: camilleluna.blavier@polimi.it

F. De Francesco
ZH Srl, Via Ariberto 20, 20123 Milan, Italy
e-mail: felice.defrancesco@zhspinoff.com

P. Mohammad et al. (eds.), *Geographic Approaches to Climate Change and Mitigation: Urban and Rural Perspectives (Volume 1)*, Advances in Science, Technology & Innovation, https://doi.org/10.1007/978-3-031-92119-3_3

(HVAC) system, where the control of the indoor environmental conditions depends mainly on the building's inherent architectural features (Boeri et al. 2022; Fabbri and Pretelli 2014; Napp and Kalamees 2015).

Studies have shown that massive structural envelopes that normally characterize historical buildings can help to maintain more stable internal conditions (Boeri et al. 2022; Napp and Kalamees 2015), preserving their materials and artworks (Camuffo 2013). Furthermore, the thermal mass of these buildings plays a very important role in improving the thermal comfort of occupants (Vella et al. 2020). Nonetheless, climate change has led to an increase in extreme weather events, such as heatwaves (REACHout Project s.d.), resulting in higher indoor temperatures and potential risks to both the building materials and visitor comfort, and prior research underscores the importance of adapting to the evolving conditions (Huerto-Cardenas et al. 2021). In such regard, the search for suitable climate strategies that align with conservation principles becomes pivotal. One approach gaining traction is the employment of natural ventilation techniques (Bay et al. 2022), which include passive solutions, such as the strategic use of existing architectural features to increase the airflow, as well as forced adaptive ventilation systems, where technology such as fans can be integrated to enhance air exchange without compromising the building's historical features (Napp and Kalamees 2015).

In this context, different studies suggested that the combined effect of high thermal mass and natural ventilation implemented through passive or hybrid strategies can be pivotal to improving thermal comfort and energy saving (Gagliano et al. 2016; Hao et al. 2020), particularly in hot and humid climates (Iskandar et al. 2024). Similarly, in a massive building located in Rome, the authors suggest that alternatively to the ventilation through simply opening of windows, it can be adopted active ventilation system, employing ducts and aspirators that ensure an exchange of air between indoor and outdoor environments (Asdrubali et al. 2022). However, as stressed by Iskandar et al. (2024) these strategies should not be invasive or compromise the integrity of the building. In a traditional massive building located in the south of Italy, the adoption of night ventilation is fundamental for the improvement of the thermal performance of massive buildings during summer and to reduce the energy demand for cooling to about 30% (Gagliano et al. 2016). Furthermore, in a review paper focused on the implications of climate change for retrofitted historic buildings, the authors highlight that the combined effects of thermal mass and night ventilation may be an effective strategy to remove the exceeded heat to maintain a comfortable indoor temperature during summer (Hao et al. 2020).

In the past, some studies have investigated the exploitation of adaptive ventilation in historical buildings, not only to reduce the indoor temperature. For example, in 2007, in a Castle in the Netherlands, the authors (Neuhaus and Schellen 2007) analysed the implementation of a combined conservation heating and ventilation system in order to maintain within a suitable range the relative humidity (RH) of the indoor spaces, by monitoring the mixing ratio (MR) inside and outside. Such strategies have shown that conservation conditions can be improved without the use of humidification or dehumidification. Similarly, in 2011 (Broström et al. 2011) evaluated the implementation of an adaptive ventilation system in an old stone church in Sweden, in order to reduce moisture-related problems. In this case, it was seen how the ventilation system if properly activated can keep the indoor environment within the acceptable risk thresholds suggested by standards. In 2022 (Bay et al. 2022) evaluated the study of different natural ventilation strategies in a historic building in Texas to mitigate temperature and relative humidity conditions through the use of energy simulations and computational fluid dynamics (CFD). In detail, the results of the analysis have shown that night ventilation can be one of the most effective ventilation scenarios to improve occupant comfort and reduce excessive loads during the summer period, leading to a reduction in indoor temperature peaks by about 2 °C. A similar analysis was done by Fiorito et al. (2022) in a building located in southern Italy. In this case, through energy simulation of different current and future scenarios, the implementation of natural ventilation has been evaluated.

This work has shown that for all the scenarios analysed, ventilation through window opening during the summer period can decrease the discomfort conditions from 30 to 50%. In another study, Vázquez-Torres et al. (2022) implemented the ventilation system in a religious building by modifying some windows with a shutter system that can be programmed to remain open during summer and closed in winter. The results obtained by the authors have shown that in a subhumid temperate climate, the people's comfort in historical religious buildings can be improved and the energy demand for cooling can be reduced.

Despite the growing body of research, there is a lack of practical solutions for climate control in historical buildings, mainly due to conservation constraints that limit the use of standard climate control technologies. This lacuna in the literature and practical application underscores the need for novel, alternative solutions that align with conservation principles and effectively address climate adaptation (Blavier et al. 2023).

This study aims to address this gap by investigating the impacts of implementing an adaptive ventilation system, employed to decrease internal temperatures and control relative humidity within Milan Cathedral, a prime example of Italy's architectural heritage. This research is being conducted in Milan, a city that has been experiencing increasingly frequent heatwaves in recent summers. In light

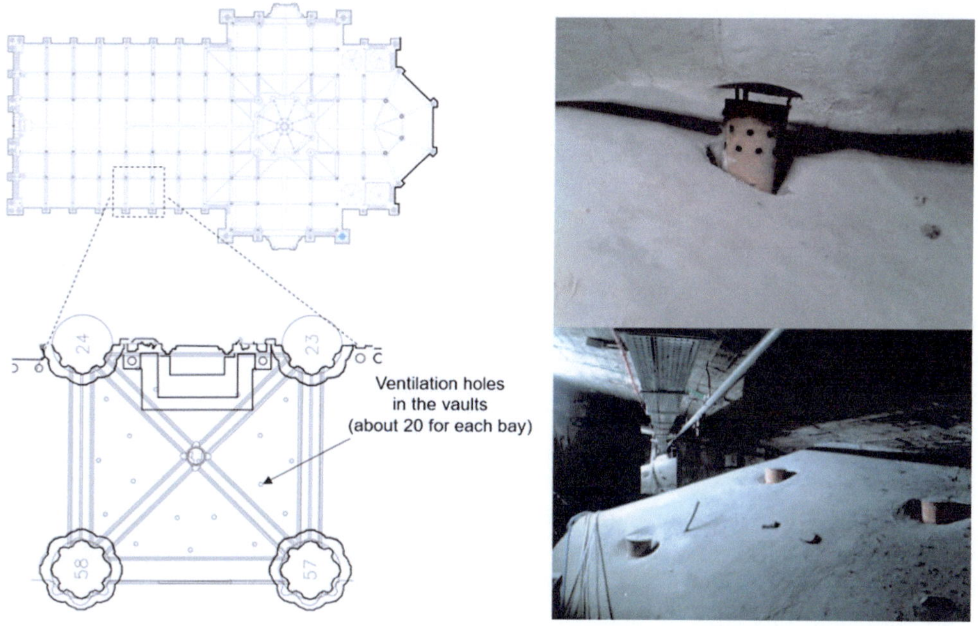

Fig. 1 Left: Schematic representation of the ventilation duct system located within the cathedral vaults. Right: Close-up photographs of the ventilation ducts as seen from the attic space (By authors, 2024)

of historic building preservation, the ventilation system will be even analysed from a material conservation perspective, assessing the risks that potential alterations in internal conditions may lead to artworks and finishes. Moreover, the feasibility and effectiveness of integrating a modern ventilation system, by readapting disused ducts distributed on the vaults, will provide an example of an adaptation and non-invasive solution in the context of climate change.

2 Case Study

This study focuses on the Milan Cathedral, a key element of Italy's architectural heritage. The Cathedral encompasses an indoor area of approximately 8500 m^2 and a volume of about 300,000 m^3. Its architectural grandeur is marked by the varying heights of its naves: 20 m for the lateral naves and 45 m for the central nave, while the cupola reaches 65 m. Constructed with masonry walls of variable thicknesses and a double-layered brick vault roof system, it features over 50 stained glass windows with grisaille paintings.

The Cathedral's interior hosts an array of artistic and historical artefacts spanning different periods, including historic organs, marble sculptures, wooden benches and confessionals, and extensive painted panels. Additionally, its aisles' altars house numerous artefacts, contributing significantly to its cultural and historical significance.

More details are provided in previous publications (Aste et al. 2019; Huerto-Cardenas et al. 2021, 2023).

About natural ventilation, it is managed during the year through the doors and some windows located in the ambulatory and the transept. Moreover, air exchanges outdoors are even provided by the numerous ventilation ducts distributed on the vaults and connected to the attic spaces under the roof, as shown in Fig. 1.

In detail, as illustrated in Fig. 1, each span of the structure contains approximately 20 ventilation ducts, each with a diameter of about 20 cm and equipped with Knapen atmospheric siphons, distributed uniformly. These objects were created around the first half of the twentieth century to reduce the humidity in the vaults and enhance ventilation within the Cathedral. However, in the current condition, not all these vents remain open or connected to the Cathedral's interior environment.

In this regard, as will be described in Sect. 3.4, the current work proposes the re-utilization of the existing ventilation ducts by the integration of fans to control the air exchanges with outdoors, aiming to mitigate the climatic conditions inside the Cathedral during certain times of the year or under specific operational circumstances.

3 Method

This investigation is focused on an adaptive ventilation system conceptualized to exploit the natural ventilation capabilities of the Milan Cathedral. The system's design required the implementation of the existing ventilation ducts, by installing fans to control the air exchange with outside.

In detail, the methodology incorporates an extensive monitoring campaign to record the cathedral's indoor conditions, essential for the calibration and validation of a simulation model developed for the ventilation strategy analysis. Such model has been developed through the geometrical and material information collected from documents and drawings provided by the Veneranda Fabbrica del Duomo, while the calibration and validation process of the model has been done by comparing the simulated results with the measured data collected from dataloggers.

Further, the analysis of the potentially suitable conditions for the exploitation of natural ventilation in Milan has been done through a method proposed by (Causone 2016), through the analysis of a Typical Meteorological Year from 2007 to 2021, and some recent years (2019, 2020, 2021, and 2022), where temperatures during the summer period in Milan have become excessive due to an increase in heat waves.

Finally, the proposed intervention aims to modulate the cathedral's internal climate selectively throughout specific periods or in response to particular environmental conditions. In detail, this adaptive approach intends to counter summer peaks through night ventilation and to maintain relative humidity levels by augmenting air exchange whenever the outdoor mixing ratio (MR) falls below the indoor levels. The control mechanisms of this ventilation system adhere to the EN 15759-2 standard (EN 15759-2 2018), ensuring compatibility with heritage conservation guidelines.

3.1 Measurements

The research encompassed the systematic collection of environmental data spanning four consecutive years, from 2019 through 2022. This was achieved utilizing a non-invasive approach, employing datalogger placed strategically at specific locations within the Cathedral, that safeguard the integrity of the finishes and artefacts while not impeding the routine activities carried out within the sacred space.

The deployed dataloggers (HOBO MX1101 model) guarantee a precision of ±0.21 °C in the temperature range of 0 to 50 °C and ±2% for relative humidity measurements between 20 and 80%.

Conversely, the assessment of external weather conditions relied on data sourced from the proximally located Brera meteorological station operated by (ARPA s.d.), providing a comprehensive overview of the external climatic variables influencing the Cathedral's interior environment.

3.2 Simulation Model Characterization

In this study, EnergyPlus software was adopted to carry out the environmental simulations. The model is made of 20 thermal zones, divided into the main indoor space of the Cathedral, the basement, the attic spaces beneath the roof, and the two sacristies. Since the building is not provided by an HVAC system, the simulations were performed in free-floating conditions.

Model calibration, i.e., the process of reducing the uncertainty of a model by comparing the simulated output and the measured data under the same set of conditions (Huerto-Cardenas et al. 2020), represents a pivotal step in obtaining reliable results. Such process has been done by comparing microclimate data (T and RH). The process to reduce the discrepancy between measured and simulated data has been done by adjusting several input parameters, for example:

- providing a more detailed thickness of the building envelope instead of an initial average thickness;
- optimizing internal gains due to people and artificial lighting based on real conditions;
- optimizing ACH due to infiltration and ventilation based on the real conditions;
- adding the internal mass of the numerous and massive columns present in the internal space;
- creating a specific weather file for the year 2019 adopted in the calibration process, by downloading data collected from the ARPA meteorological station. Such step is crucial for comparing measured and simulated data under the same boundary conditions.

After several simulations, the fine-tuning of the input parameters involved in the calibration process provided satisfactory results. In such regard, the optimized parameters adopted in the simulations are reported below.

About the opaque envelope characteristics assumed in the model, the Cathedral is made of masonry walls, with U-values estimated to range from 0.29 to 0.92 W/m^2K, depending on the thickness. The roof system is characterized by a double system of vaults made of solid bricks with a U-value estimated to be about 1.17 W/m^2K.

Regarding the transparent envelope, the stained glass windows were simplified and modelled adopting a U-value, solar heat gain coefficient (SHGC), and visible transmittance (τv) equal to 5.8 W/m^2K, 0.15 and 0.1, respectively.

The air change per hour (ACH) due to ventilation and infiltration of the current conditions of the building has been estimated to be about 0.33–0.37 h^{-1} in the summer months and about 0.22–0.26 h^{-1} for the rest of the year.

For the internal heat contributions due to people presence, an occupancy density of 0.11 $people/m^2$ has been assumed based on the tickets sold during the weekdays, while for the internal contributions due to artificial lighting, an average value of 2 W/m^2 has been assumed for the lighting system of the Cathedral.

Regarding the reliability of the simulation model, its accuracy has been evaluated by comparing the simulated and measured T and RH through the calculation of the root mean squared error (RMSE) and the squared Pearson correlation coefficient (r^2). In particular, the validation process provides an RMSE equal to 0.52 °C and 4.21% and an r^2 equal to 0.993 and 0.904, respectively, for T and RH. Further details about the building model are provided in (Huerto-Cardenas et al. 2021).

Then, the validated model was implemented with the airflow rate provided by the activation of the fans, considering an additional ACH activated with the conditions provided in Sect. 4.2 through the integration of an hourly schedule.

3.3 Calculation of Climatic Potential for Natural Ventilation

To assess when it may be possible to exploit natural ventilation in Milan, have been calculated the "Climatic Potential for Natural Ventilation" (CPNV) index proposed by Causone (2016).

In the context of material conservation, this index is particularly useful as it allows for determining the periods during the year when natural ventilation can be utilized without compromising the preservation of materials typically found in historic buildings. This is possible through the adoption of thresholds for external air temperature and relative humidity established from standard EN 15759-2, which is focused on ventilation management for the protection of cultural heritage buildings and collection.

This index quantifies the percentage of hours in a year conducive to natural ventilation, based on outdoor climatic conditions. The CPNV is calculated using Eq. 1 reported hereafter, where $h_{nv,i}$ denotes the number of hours suitable for natural ventilation, and h_{tot} represents the total annual hours.

$$\text{CPNV} = \sum_{i=1}^{n} \frac{h_{nv,i}}{h_{\text{tot}}} \times 100 \quad (1)$$

As previously stated for the estimation of the CPNV index, temperature and relative humidity thresholds have been defined according to the EN 15759-2 standard. In detail, these limits fall within 18–26 °C for temperature and 35–75% for relative humidity. Ventilation is deemed beneficial when outdoor conditions align with these parameters. Furthermore, to comprehensively represent the spectrum of possible conditions, as proposed by Causone (2016), can be identified nine different climate areas based on various temperature and humidity combinations relative to these thresholds. These conditions are graphically depicted in Fig. 2 with the optimal scenario marked with the green colour and labelled as "optimum".

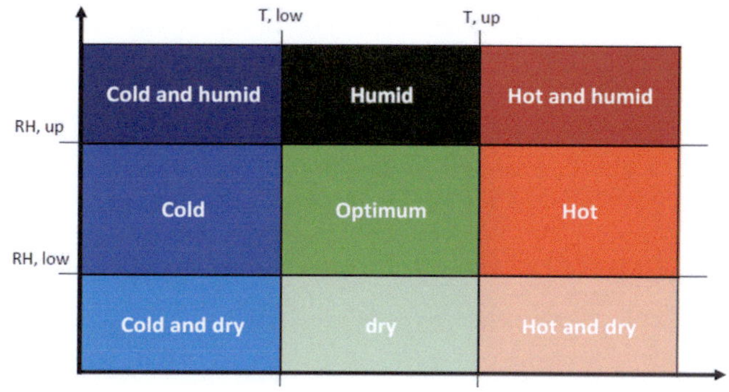

Fig. 2 Range of potential external climatic conditions for natural ventilation exploitation based on CPNV index. The optimal conditions are indicated by the green area labelled "optimum" in the centre, which falls within the specified thresholds: 18–26 °C for temperature and 35–75% for relative humidity (By authors, 2024)

3.4 Description of the Proposed Adaptive Ventilation System and Strategies for the Activation

In this first analysis, four fans per span have been considered, which corresponds to a total of about 360 fans (28% of the total ducts) uniformly distributed in the vaults of the Cathedral.

As already described, the fans are characterized by the same diameter of the ventilation duct, and each of them can extract a maximum airflow of 1460 m^3/h. Considering the entire system, the proposed configuration yields a maximum flow rate of 525,600 m^3/h, which is equal to an additional air change rate of approximately 1.75 h^{-1}. It should be noted that the number of fans can be further reduced since the air changes that can be exploited with the proposed configuration are more than sufficient. Therefore in the next insights, the number of fans will be optimized according to the noise that can be triggered as a result of the combined activation of the fans, the effect on thermal stratifications caused by the large airflow from outside in a short time and the indoor air velocities generated. More specifications of the fans are provided in Table 1.

Since the ventilation ducts are directly connected to the attic, to prevent re-circulation of air extracted from indoors, the fans are designed to be provided with tubes that connect them directly with the outdoor environment. Moreover, when

Table 1 Main specifications of the fans adopted for the analysis

Nominal diameter [cm]	20
Maximum power consumption [W]	119
Weight [kg]	12.6
Maximum airflow [m^3/h]	1460
Sound power [dB(A)]	58

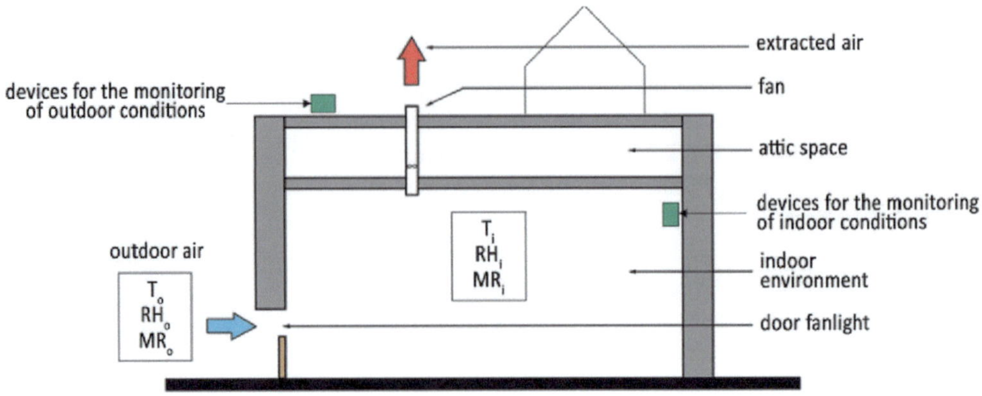

Fig. 3 Layout of the ventilation fan system in the Cathedral (By authors, 2024)

the fans are activated, the air from outside comes from the fanlight of around 8 m^2, located above the four main facade doors.

The proposed system requires real-time monitoring instruments that allow the activation of the fans according to indoor and outdoor conditions. In such regard, the existing indoor and outdoor monitoring instrumentation must be integrated with sensors able to collect the main environmental parameters such as air temperature, relative humidity, mixing ratio, and CO_2 levels. Additionally, surface temperature devices and anemometers can be strategically placed inside to assess the impact on material preservation. A layout of the proposed ventilation system is shown in Fig. 3.

3.5 Conservation Risk Assessment

Within the context of historic buildings, which typically host important artworks, it is crucial to assess the potential risks caused by the implementation of the ventilation system proposed. One of the most significant deterioration risks which can arise from the rapid fluctuations of the indoor condition is the physical damage to organic hygroscopic materials (such as wood and painting on canvas). To determine whether the implemented system may threaten the conservation of such kind of materials, the possible risks have been evaluated using the methodology outlined in the standard EN 15757: 2010—"Conservation of Cultural Property—Specifications for temperature and relative humidity to limit climate-induced mechanical damage in organic hygroscopic materials" (EN 15757 2010). Such method relies on the concept of acclimatized materials to indoor microclimate conditions over a prolonged period. This means that the materials have adapted themselves to the seasonal and daily fluctuations of temperature and relative humidity. However, it should be noted that the prolonged adaptation process may not be beneficial for the materials, as it can induce repeated internal stress. In this sense, further deformations or stress, especially caused by high RH fluctuations, could lead existing internal cracks to spread and become visible (Camuffo et al. 2022). In such regard, a safe target range for relative humidity has been determined based on EN 15757: 2010 to verify the possible risks associated with the implementation of the ventilation system. Moreover, such kind of analysis might allow for optimisation of the activation set points of outdoor T and RH to activate the system, as will be discussed in Sect. 5.

4 Results

The following sections present the findings from the potential for exploiting natural ventilation in Milan and the effectiveness of implementing a night ventilation strategy within Milan Cathedral over the summer periods.

4.1 Identification of Periods for the Exploitation of Natural Ventilation in Milan

The calculated CPNV index for the years 2007–2021, derived from Milan's TMY data, reveals key periods favourable for natural ventilation, aligning with optimal temperature and humidity ranges for both human comfort and material preservation.

Figure 4 illustrates that the best conditions for natural ventilation predominantly occur during the afternoon hours of March, April, May, September, and partially in October. Conversely, during the summer, afternoon and evening temperatures are often too high for effective natural ventilation. Ideal conditions in summer months shift to night-time, roughly from 10:00 p.m. to 9:00 a.m., except in August when high humidity levels can reduce the exploitation of night ventilation.

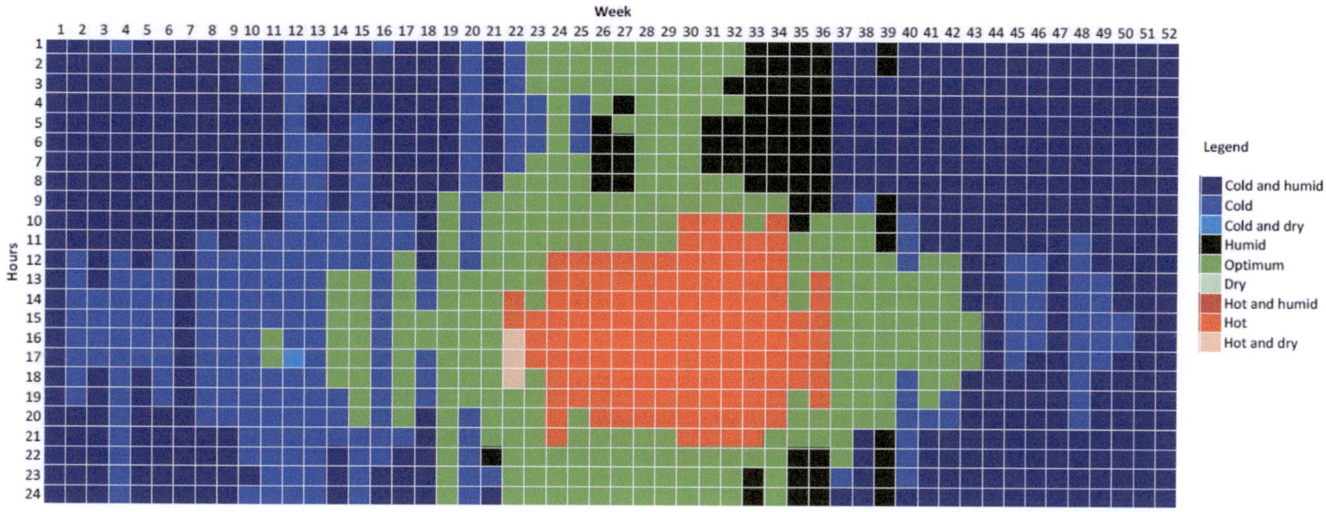

Fig. 4 Thermal climate map of Milan showing periods suitable for natural ventilation, reference climatic file: TMY 2007–2021 (By authors, 2024)

Table 2 Exceedance metrics (%) in the five selected reference years

	TMY 2007–2021	2019	2020	2021	2022
Cold and humid	42	24	27	25	25
Cold	20	35	33	35	28
Cold and dry	0	2	1	2	2
Humid	5	1	1	2	1
Optimum	22	22	26	25	24
Dry	0	1	1	0	1
Hot and humid	0	0	0	0	0
Hot	10	13	12	10	14
Hot and dry	0	1	0	1	4

In recent years, the heatwaves in the city of Milan have been increasing, as observed in the project (REACHout Project s.d.). A demonstration of the increase in peak temperatures especially in the summer period is given by the temperatures inside the Cathedral, which exceeded 30 °C during the summer period in the last years (Huerto-Cardenas et al. 2023). In such respect, the evaluation of the potential exploitation of natural ventilation was also done for the years 2019, 2020, 2021, and 2022, for which the results of the CPNV index are shown in Table 2. As can be seen from the results shown in the table, the optimum conditions for the exploitation of natural ventilation have been estimated to be about 22–26% in recent years, in line with the analysis made for the TMY 2007–2021. Moreover, as can be seen from the percentage of hours with outdoor weather conditions falling between in the "hot" range, in recent years the percentage is slightly higher, especially for 2022.

4.2 Exploitation of Night Ventilation During the Summer Period

The analysis of outdoor climatic conditions for natural ventilation potential led to evaluating the impact of a night ventilation strategy during summer. As detailed in Sect. 3.4, the model incorporates the airflow rate from the ventilation fan activation, adding an ACH of 1.75 h^{-1}. Under these parameters, the fan system operates for an average of about 500–600 h for the whole summer period. The conditions for activating the ventilation system defined in the simulations were:

- fan system activation from 11:00 p.m. to 07:00 a.m., from mid-May to mid-September;
- in such periods during the daytime, the air exchanges with the outside have been limited;

- outdoor T must be at least 2 °C lower than indoor;
- outdoor T must be between 18 and 26 °C;
- outdoor RH must be between 35 and 75%;
- outdoor MR must be less than indoor.

In such regard, the simulation carried out using the climatic file TMY 2007–2021 shows a maximum indoor temperature reduction of about 2 °C during the visiting hours for tourists and worshippers (from 9:00 a.m. to 7:00 p.m.), as shown in Fig. 5. This finding is significant, considering that recent years have detected peaks in indoor temperatures that exceed 30 °C during July and August. In detail, the estimated average reduction of temperature during the whole period analysed is about 0.92 °C.

Similarly, the simulation was done for 2019, 2020, 2021, and 2022, which represents some recent intensely hot years in Milan. The results, shown in Fig. 6, provide a reduction of the average temperature in the indoor climate for the whole period considered of about 0.76 °C, 0.65, 0.72 °C, and 0.85, respectively, for 2019, 2020, 2021, and 2022, with a maximum peak for all the years analysed above 1.7 °C.

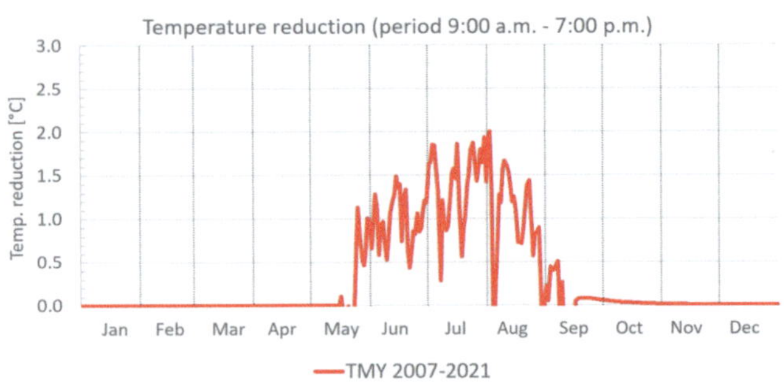

Fig. 5 Reduction of indoor temperatures from 9:00 a.m. to 7:00 p.m. achieved through night ventilation exploitation for the TMY 2007–2021 (By authors, 2024)

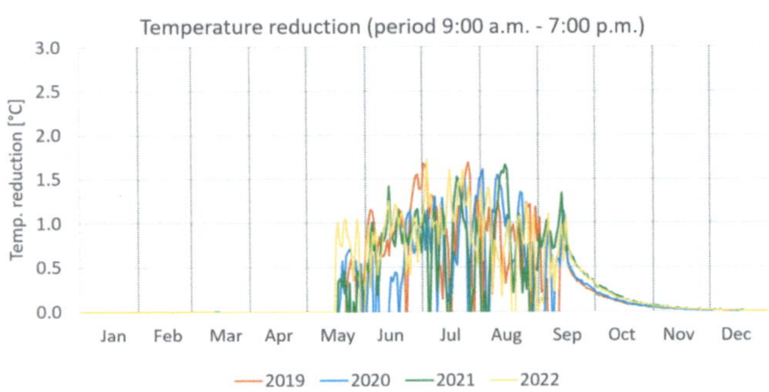

Fig. 6 Reduction of indoor temperatures from 9:00 a.m. to 7:00 p.m. achieved through night ventilation exploitation for 2019, 2020, 2021, and 2021 from Brera weather station a few hundred meters from the Cathedral (By authors, 2024)

These outcomes can be considered favourable to improve the comfort conditions for the visitors and limit the potential risks for the artworks caused by the intensification of the high temperatures.

5 Effects on the Conservation Condition of the Materials Due to the Activation of the Ventilation System

Assessing the potential for reducing indoor temperatures during summer, the study also examined the impact of climatic changes caused by fan system utilization on the preservation of materials and artworks within the Cathedral.

Analyses covering various phenomena, such as salt efflorescence, chemical risk, surface condensation, and biodeterioration (from mould and insects), indicated that the internal climate alterations due to fan operation were negligible. Nevertheless, the possibility of mechanical damage to hygroscopic materials, potentially triggered by intense relative humidity fluctuations following fan activation, was identified as a significant concern.

To address this risk, an evaluation was conducted adopting the European standard EN 15757: 2010. This regulation helped to establish safe relative humidity ranges to prevent mechanical damage.

As illustrated in Fig. 7 (the first image), the use of fans occasionally during July, August, and September causes fluctuation that leads to exceeding the relative humidity safe ranges defined according the EN 15757. Therefore, incorporating a control system into the fans or employing a more stringent external relative humidity threshold can be mitigated the risks of exceeding safety thresholds (as shown in the second image of Fig. 7).

6 Conclusion and Further Work

This work focused on the assessment and implementation of an adaptive ventilation system in Milan Cathedral, aiming to mitigate indoor climate conditions without jeopardizing valuable artworks. Here is a summary of the main findings:

- the CPNV index was calculated for Milan, using data from 2007 to 2022. This analysis identified optimal periods for natural ventilation, primarily in the afternoon hours of March, April, May, September, and partially October. During summer, the night-time hours were more suitable due to lower temperatures and relative humidity levels;
- the proposed ventilation system involved approximately 360 fans in the vaults. These fans, each with a 20 cm diameter, collectively offered a maximum airflow of

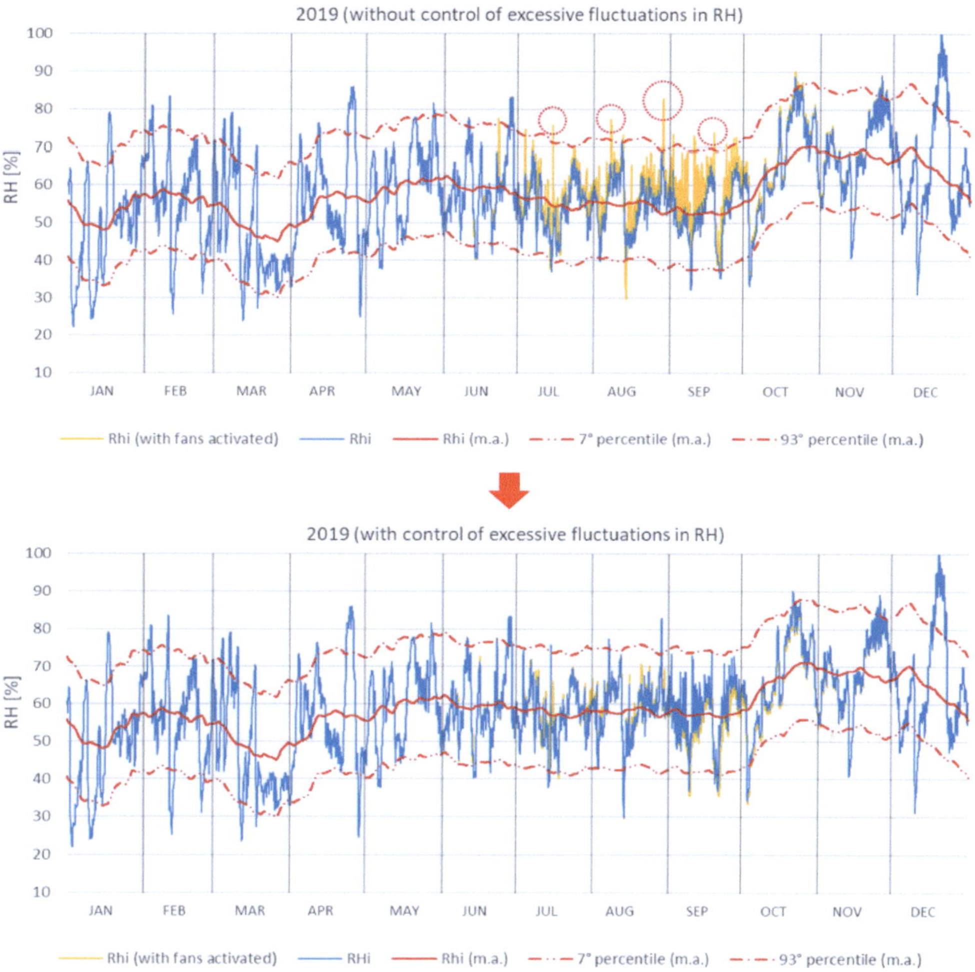

Fig. 7 Risk assessment of mechanical damage in hygroscopic materials through EN 15757. This graph shows occurrences where fan utilization leads to exceeding the relative humidity thresholds. Integrating these safety ranges into fan control logic or using stricter external humidity thresholds can mitigate such events (By authors, 2024)

525,600 m^3/h, contributing to an air change rate of about 1.75 h^{-1}. It is important to highlight that such value represents the maximum achieved through the combined utilization of fans;

- the fan system's operational strategy was set for the night-time hours in summer (from 11:00 p.m. to 07:00 a.m.), with specific conditions regarding temperature and humidity levels. This strategy was expected to operate for an average of 500–600 h in summer, reducing the average indoor temperatures up to 1.7–2 °C during visiting hours (9:00 a.m. to 7:00 p.m.);
- potential risks of mechanical damage to hygroscopic materials due to excessive humidity fluctuations were evaluated, adopting the EN 15757: 2010 standard. A preliminary analysis shows that the impact of fans on the preservation of artworks if properly set up can avoid risks to these materials.

In summary, the study also identifies areas for future in-depth analysis, such as:

- assessing potential thermal stratifications caused by the ventilation system, using CFD modelling;
- examining the risks to material preservation due to thermal stratifications, particularly concerning the related relative humidity changes;
- evaluating the perceived sound pressure at floor level with all fans operational;
- optimized the air change rate with the outdoors to minimize the noise generated by active fans while also preventing excessive indoor air velocities;
- assessing user comfort in the context of fan utilization, through adaptive comfort models.

References

ARPA.: Richiesta dati misurati (2021, January 17) (s.d.). https://www.arpalombardia.it/Pages/Meteorologia/Richiesta-dati-misurati.aspx

Asdrubali, F., Evangelisti, L., Guattari, C., Roncone, M., Fontana, L., Salerno, G., Tonelli, C., Vitale, V.: Modeling and measurements in natural ventilation of massive buildings: a case study. Build. Simul. Appl. 411–416 (2022)

Aste, N., Adhikari, R.S., Buzzetti, M., Della Torre, S., Del Pero, C., Huerto C, H.E., Leonforte, F.: Microclimatic monitoring of the Duomo (Milan Cathedral): risks-based analysis for the conservation of its cultural heritage. Build. Environ. **148**(October 2018), 240–257 (2019). https://doi.org/10.1016/j.buildenv.2018.11.015

Bay, E., Martinez-Molina, A., Dupont, W.A.: Assessment of natural ventilation strategies in historical buildings in a hot and humid climate using energy and CFD simulations. J. Build. Eng. **51**(November 2021), 104287 (2022). https://doi.org/10.1016/j.jobe.2022.104287

Blavier, C.L.S., Huerto-Cardenas, H.E., Aste, N., Del Pero, C., Leonforte, F., Della Torre, S.: Adaptive measures for preserving heritage buildings in the face of climate change: a review. Build. Environ. **245**(September) (2023). https://doi.org/10.1016/j.buildenv.2023.110832

Boeri, A., Longo, D., Fabbri, K., Pretelli, M., Bonora, A., Boulanger, S.: Library Indoor microclimate monitoring with and without heating system. A bologna university library case study. J. Cult. Herit.herit. **53**, 143–153 (2022). https://doi.org/10.1016/j.culher.2021.11.012

Broström, T., Hagentoft, C.-E., Wessberg, M.: Humidity control in historic buildings through adaptive ventilation: a case study. In: NSB 2011: 9th Nordic Symposium on Building Physics, January, 1–8 (2011)

Camuffo, D.: Microclimate for cultural heritage: conservation and restoration of indoor and outdoor monuments (2013)

Camuffo, D., Della Valle, A., Becherini, F.: The European Standard EN 15757 concerning specifications for relative humidity: suggested improvements for its revision. Atmosphere **13**(9) (2022). https://doi.org/10.3390/atmos13091344

Causone, F.: Climatic potential for natural ventilation. Archit. Sci. Rev. **59**(3), 212–228 (2016). https://doi.org/10.1080/00038628.2015.1043722

EN 15757: 2010—Conservation of cultural property. Specifications for temperature and relative humidity to limit climate-induced mechanical damage in organic hygroscopic materials, 18 (2010)

EN 15759-2: 2018 Conservation of cultural heritage—Indoor climate. Part 2: Ventilation management for the protection of cultural heritage buildings and collections (2018)

Fabbri, K., Pretelli, M.: Heritage buildings and historic microclimate without HVAC technology: Malatestiana Library in Cesena, Italy, UNESCO Memory of the World. Energy Build. **76**, 15–31 (2014). https://doi.org/10.1016/j.enbuild.2014.02.051

Fiorito, F., Vurro, G., Carlucci, F., Campagna, L.M., De Fino, M., Carlucci, S., Fatiguso, F.: Adaptation of users to future climate conditions in naturally ventilated historic buildings: effects on indoor comfort. Energies **15**(14) (2022). https://doi.org/10.3390/en15144984

Gagliano, A., Nocera, F., Patania, F., Moschella, A., Detommaso, M., Evola, G.: Synergic effects of thermal mass and natural ventilation on the thermal behaviour of traditional massive buildings. Int. J. Sustain. Ener. **35**(5), 411–428 (2016)

Hao, L., Herrera-Avellanosa, D., Del Pero, C., Troi, A.: What are the implications of climate change for retrofitted historic buildings? A literature review? Sustainability (Switzerland) **12**(18), 1–17 (2020). https://doi.org/10.3390/su12187557

Huerto-Cardenas, H.E., Aste, N., Del Pero, C., Della Torre, S., Leonforte, F.: Effects of climate change on the future of heritage buildings: case study and applied methodology. Climate **9**(8) (2021). https://doi.org/10.3390/cli9080132

Huerto-Cardenas, H.E., Aste, N., Pero, C.D., Leonforte, F., Stella-Blavier, C.L., Torre, S.D.: Effects of visitor influx on the indoor climate of the Milan Cathedral. Atmosphere **14**(4)(743), 21 (2023). https://doi.org/10.3390/atmos14040743

Huerto-Cardenas, H.E., Leonforte, F., Aste, N., Del Pero, C., Evola, G., Costanzo, V., Lucchi, E.: Validation of dynamic hygrothermal simulation models for historical buildings: state of the art, research challenges and recommendations. Build. Environ. **180**(June), 107081 (2020). https://doi.org/10.1016/j.buildenv.2020.107081

Iskandar, L., Bay-Sahin, E., Martinez-Molina, A., Toker Beeson, S.: Evaluation of passive cooling through natural ventilation strategies in historic residential buildings using CFD simulations. Energy Build. **308**(August 2023), 114005 (2024). https://doi.org/10.1016/j.enbuild.2024.114005

Napp, M., Kalamees, T.: Energy use and indoor climate of conservation heating, dehumidification and adaptive ventilation for the climate control of a mediaeval church in a cold climate. Energy Build. **108**, 61–71 (2015). https://doi.org/10.1016/j.enbuild.2015.08.013

Neuhaus, E., Schellen, H.: Humidistat-controlled heating and ventilation systems to create preservation conditions in historic buildings in the Dutch climate. In: Proceedings of the 9th Clima 2007 WellBeing Indoors, pp. 2718–2728 (2007)

REACHout Project.: Thermal Assessment Tool (TAT). Recuperato 27 dicembre 2023, da (s.d.). https://thermal-assessment.urban.tecnalia.dev/

Vázquez-Torres, C.E., Beizaee, A., Bienvenido-Huertas, D.: The impact of human occupancy in thermal performance of a historic religious building in sub-humid temperate climate. Energy Build. **259** (2022). https://doi.org/10.1016/j.enbuild.2022.111912

Vella, R.C., Martinez, F.J.R., Yousif, C., Gatt, D.: A study of thermal comfort in naturally ventilated churches in a Mediterranean climate. Energy Build. **213**, 109843 (2020). https://doi.org/10.1016/j.enbuild.2020.109843

Earth Observation Analysis and Microclimate Simulations of Mitigation Scenarios to Support Urban Planning: The MIRIFICUS Project

Giulia Guerri, Gennaro Albini, Alessandra Casali, Angela Cimini, Luca Congedo, Alfonso Crisci, Arnaldo A. De Benedetti, Stefano De Corso, Pasquale Dichicco, Michele Munafò, and Marco Morabito

Abstract

The existing urban settlement dynamics, as well as climate change effects and extreme events, are considerable challenges for the sustainable development of cities. There is an urgent need to consider the effects of climate change on urban areas, rethinking settlements with a view to urban microclimatic regeneration, with particular attention to mitigation strategies involving nature-based solutions (NBS) to mitigate the typical Urban Heat Island phenomenon. The present contribution shows the research activities from the MIRIFICUS project (Satellite Remote Sensing for Measuring Reforestation Interventions for Urban Heat Island), funded by the Italian Space Agency (ASI), with the scientific direction of the National Research Council of Italy (CNR), in collaboration with the Institute for Environmental Protection and Research of Italy (ISPRA). MIRIFICUS aims at promoting the operational development of innovative services for sharing data and services obtained by integrating Earth Observation (EO) satellite systems, such as those from Copernicus (Sentinel), NASA (Landsat), and ASI (Prisma) missions. Using the medium–high resolution of satellite data, EO data analysis allows the processing of urban, vegetation, surface thermal indicators, and the detection of surface thermal anomalies of the Italian territory at the municipal level. The monitoring and assessment of potential mitigation interventions will be carried out at urban scale within the cities of Rome and Florence, by analyzing the urban parameter variations considering the ex-ante and post-potential mitigation interventions. The mitigation effects provided by NBS will be quantified by performing specific microclimate simulations of urban planning scenarios through environmental modelling based on weather-climate characterization. The ultimate project goal is the development of a WebGIS tool helpful for Public Administrations, integrated by useful information for planning urban interventions to mitigate thermal anomalies and evaluate the effectiveness of reforestation actions carried out in recent years in the studied cities.

G. Guerri (✉) · G. Albini · A. Crisci · M. Morabito
Institute of Bioeconomy, National Research Council of Italy (CNR-IBE), Rome, Italy
e-mail: giulia.guerri@ibe.cnr.it

G. Albini
e-mail: gennaro.albini@ibe.cnr.it

A. Crisci
e-mail: alfonso.crisci@cnr.it

M. Morabito
e-mail: marco.morabito@cnr.it

A. Casali · A. Cimini · L. Congedo · A. A. De Benedetti · S. De Corso · P. Dichicco · M. Munafò
Institute for Environmental Protection and Research of Italy (ISPRA), Rome, Italy
e-mail: alessandra.casali@isprambiente.it

A. Cimini
e-mail: angela.cimini@uniroma1.it

L. Congedo
e-mail: luca.congedo@isprambiente.it

A. A. De Benedetti
e-mail: arnaldo.debenedetti@isprambiente.it

S. De Corso
e-mail: stefano.decorso@isprambiente.it

P. Dichicco
e-mail: pasquale.dichicco@isprambiente.it

M. Munafò
e-mail: michele.munafo@isprambiente.it

P. Mohammad et al. (eds.), *Geographic Approaches to Climate Change and Mitigation: Urban and Rural Perspectives (Volume 1)*, Advances in Science, Technology & Innovation, https://doi.org/10.1007/978-3-031-92119-3_4

Keywords

Mitigation strategies · Forestry interventions · Urban heat island · Microclimate simulations · Nature-based solutions

Abbreviations

ST	Surface Temperature.
NDVI	Normalized Difference Vegetation Index.
LCC	Land Consumption Coverage.
TC	Tree Coverage.
GC	Grass Coverage.

1 Introduction

The current settlement dynamics, as well as the effects of climate change and the increase in the frequency of extreme thermal and rainfall events, represent a significant challenge to ensure the resilience of cities. More than half of the world's population lives in urban areas, and the global rate of urbanization is expected to reach 70% by 2030, a level already reached in many European and Italian cities. Urbanization adversely affected the environment mainly through the production of pollution, changing the physical and chemical properties of the atmosphere, and altering the surface coverage of the soil. In particular, in Italy, as stated by the 2023 Report on "Land consumption, territorial dynamics, and ecosystem services" (Munafò 2023) provided by the Italian Institute for Environmental Protection and Research (ISPRA), cities are affected by a significant increase in land consumption, defined as the change from non-artificial land cover (unconsumed land) to artificial land cover (consumed land) mainly characterized by artificial waterproof materials.

Therefore, soil consumption greatly affects the urban microclimate, contributing to the typical phenomena of cities such as Urban Heat Island (UHI) and Surface Urban Heat Island (SUHI), and, in particular, by creating a tissue of intra-urban surface thermal anomalies. It was widely demonstrated that the surface temperature greatly influences the air temperature and contributes to the UHI formation (Voogt and Oke 2003; Zhou et al. 2017).

The SUHI phenomenon was defined as the core (city center) surface temperature higher than the surrounding areas (e.g., rural areas) and has its maximum during summer daytime (Oke 1995; Sismanidis et al. 2022; WMO 2023). This phenomenon is favored by the presence of predominantly impervious surfaces, especially in central urban areas (mainly affected by asphalt and concrete), which has adverse thermal properties for heat release, which is instead absorbed and stored during the daytime, generating an increase in surface temperature, affecting also nighttime temperatures (WMO 2023).

On the contrary, the permeable surfaces covered by vegetation (grass, shrub, and trees) showed favorable properties for humans and ecosystems in general because, through the evapotranspiration process and shading, they contributed significantly to air and surface temperature decrease, as well as promoted better air quality.

In particular, the land surface temperature (LST) is an important geophysical parameter used to understand the thermal pattern in numerous studies for the evaluation of the effects of SUHI (Guha et al. 2018; Morabito et al. 2018, 2021) and assess the impacts on population health (Morabito et al. 2015). Nowadays, the availability of a large amount of Earth Observation (EO) data, recorded by satellites in orbit in space, allows the estimation and analysis of the LST at a global level and with a maximum spatial resolution between 30 and 100 m, up to 1 km, as in the case of the satellites Landsat-8 and 9, Aster and Modis of NASA missions.

Leveraging open EO data, numerous research studies based on thermal remote sensing data were developed to investigate the surface thermal pattern (Guerri et al. 2021; Mavrakou et al. 2018; You et al. 2021; Zargari et al. 2024), detecting intra-urban surface thermal anomalies which spread beyond the urban center. The developed approaches showed important effects on societal and policy benefits, allowing the spatial identification of thermal hot-spot areas and facilitating the SUHI mitigation intervention planning.

Among the most used geospatial statistical approaches, the hot-spot method by using the Getis-Ord Gi* statistic is widely applied to investigate the thermal pattern by identifying statistically significant thermal clusters of high (hot-spots) and low (cold-spots) LST values at local scale.

A recent study applying the hot-spot method to the city of Teheran (Zargari et al. 2024) observed that the SUHI is not limited to urban areas but has spread beyond the urban core with the highest daily LST values observed in the suburbs, highlighting a close relationship between the latter and the spatial–temporal changes in grey and green areas. As a matter of fact, it was observed the contribution of grey, green and blue urban infrastructures to mitigate the urban heat (Kumar et al. 2024). In addition, it is worth highlighting that, as confirmed by several studies (Xu et al. 2024; Yao et al. 2022; Zargari et al. 2024), the urban thermal pattern is extremely complex, resulting in dynamic interaction of numerous urban landscape elements and surface properties (e.g., land cover, urban morphology, surface albedo, and vegetation indices).

This contribution aims to show the research framework and results from the MIRIFICUS project, financed by the Italian Space Agency (ASI) and coordinated by the Institute of Bioeconomy of the National Research Council of Italy (CNR-IBE), in partnership with the Italian Institute for

Environmental Protection and Research (ISPRA). The project involves public institutions at national and municipal levels, the National Association of Italian Municipalities (ANCI), and the municipal institutions of Rome and Florence, to provide useful geospatial information to support mitigation strategies for climate change. In particular, the combination of different spatial layers used in this project provides a comprehensive description of the urban characteristics affecting the thermal model in multiple different urban contexts, including the twenty Italian municipalities investigated.

2 Materials and Methods

2.1 Study Area

The study was conducted on the Italian territory with an extension of about 302.000 km^2 and about 59 million inhabitants. In particular, twenty regional capital municipalities were selected to carry out geospatial analysis on the surface thermal pattern and urban cover characterization with municipal detail. According to the Köppen-Geiger climate classification (Rubel and Kottek 2010), all municipalities revealed two climate subgroups: the northern area mainly showed humid subtropical (Cfa) and temperate subtropical highland (Cfb) climates, whereas the center and the southern regions were characterized by hot-summer Mediterranean climate (Csa). Among the Italian regional capitals, 70% of municipalities were characterized by a plain territory (considering an elevation of less than 300 m a.s.l.), equally distributed among inland and coastland cities.

The location of the study area, including the twenty Italian regional capital municipalities, is shown in Fig. 1.

2.2 Research Framework and Methodology

The MIRIFICUS project is focused on the SUHI monitoring and evaluation of the urban reforestation interventions to mitigate the SUHI phenomenon, by analyzing EO data and performing environmental modelling and microclimate simulations. The aim is to provide an innovative monitoring service of urban reforestation interventions by developing a series of useful geospatial indicators, available at national and municipal levels, to quantify the effects of mitigation interventions on the SUHI phenomenon.

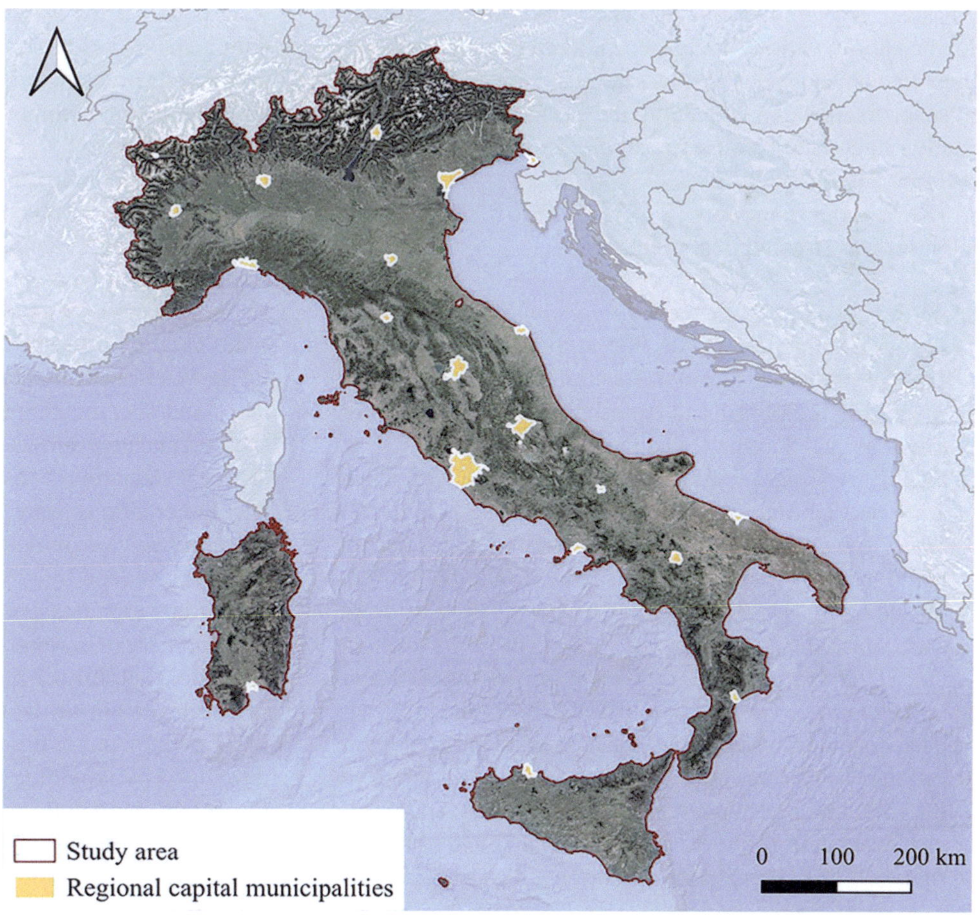

Fig. 1 The study area (*Source* By author)

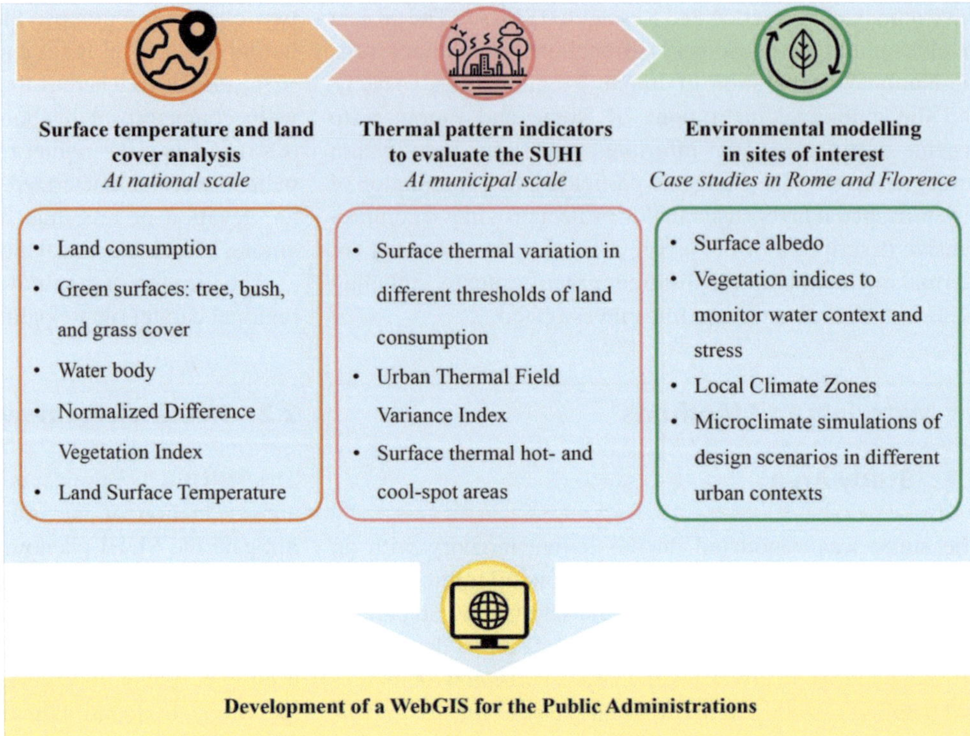

Fig. 2 The study framework. (*Source* By author)

MIRIFICUS allowed the characterization at the municipal level, throughout the national territory, of the distribution/density of various urban elements influencing the SUHI phenomenon, considering the seasonal periods of the last 10 years, to allow the evaluation of the effectiveness of reforestation interventions on some studies in the Municipalities of Rome and Florence.

The research framework (Fig. 2) of the first-year research activities was structured in 4 sections:

1. *Analysis of the surface temperature and land cover in the Italian municipalities.* The first section consisted of the analysis of thermal and multispectral remote sensing data, specifically Landsat-8 and Landsat-9 data from the NASA missions, and Sentinel-1 and Sentinel-2 data of the Copernicus Programme. Several layers related to surface temperature and land cover (including land consumption, tree cover, grassland, and water body) were obtained for the entire national territory, with a spatial resolution between 10 and 60 m.
2. *Development of thermal pattern indicators to evaluate the SUHI phenomenon in the Italian regional capital municipalities.* The summer surface temperature layer was used to quantify the SUHI phenomenon in areas with different soil consumption densities corresponding to different urbanization levels (urban, suburban, and rural), also calculating the Urban Thermal Field Variance Index (UTFVI) and carrying out an analysis of thermal anomalies (Hot-spot analysis). The latter geospatial layers were obtained for the 20 regional capital municipalities, with a spatial resolution of 60 m. Further analyses were performed to characterize the surface thermal pattern in terms of average summer surface temperature and land cover distribution.
3. *Environmental modelling in sites of interest to evaluate interventions of urban reforestation in Florence and Rome case-studies.* This section is at a preliminary phase and its completion is planned by the end of March 2025. High-resolution remote sensing data (PRISMA and Cosmo Sky-Med) are planned to be used to characterize Florence and Rome municipal areas to provide Local Climate Zone maps and specific characterization regarding green (tree species) and grey surfaces (surface albedo). In this section, different scenarios of microclimate simulations of mitigation interventions were selected in an industrial area in Florence as "a sample" to show the simulated thermal comfort provided by green interventions in a critical area affected by significant surface thermal anomalies.
4. *Development of a geographic informative tool (WebGis).* This tool, helpful for public administrations involved in the management of the territory and the safeguarding of the environmental, cultural, and landscape heritage, will be developed by providing useful information to plan SUHI mitigation interventions and evaluating the effectiveness of reforestation interventions carried out in recent years on the cities under study.

However, the present contribution was focused on showing the methodology and results related to the second and third sections, previously mentioned.

2.3 Urban Cover Analyses

The following layers characterizing the urban surfaces were considered for the entire national territory:

1. Land consumption. This binary layer was extracted from the national dataset provided by ISPRA and the National System for Environmental Protection (SNPA), resulting from the processing of large amounts of optical satellite data and SAR data (such as Sentinel-1 and Sentinel-2) by fusion, classification, and photointerpretation techniques.
2. *Land cover: tree, grassland, and water body coverages.* The classification system was derived from documents produced at the European level by the EAGLE group (EIONET Action Group on Land Monitoring in Europe, https://land.copernicus.eu/eagle). Land cover layers were obtained from Sentinel-1 and Sentinel-2 data, by adapting a recent methodology (De Fioravante et al. 2021), developed under the agreement between ISPRA and ASI for the Habitat Mapping project.
3. *The Normalized Difference Vegetation Index (NDVI).* The NDVI is an important indicator to investigate the effects of vegetation on the urban microclimate and the SUHI. It is a dimensionless index that describes the difference between the visible and the near-infrared reflectance of plant cover: the calculations for each pixel return values between -1 and 1, where 0 means no vegetation and close to +1 indicates the maximum possible density of green leaves. Urban and natural surfaces were detected by four NDVI thresholds: water for NDVI $\leqslant -0.20$; impervious for NDVI between -0.20 and 0.20; non-irrigated pervious for NDVI between 0.20 and 0.60; and irrigated pervious surfaces NDVI $\geqslant 0.60$. In this study, the seasonal NDVI layers were obtained from multispectral images (level 2-A) acquired by Sentinel-2 of the Copernicus Program (level-2A), considering the 2015–2023 period. A multitemporal stacking processing was performed to select the seasonal dataset through a cloud cover of less than 10%, and the subsequent application of the cloud mask on the band of cloud cover information (QA60), as indicated by the literature (European Space Agency 2015).

Considering the availability of land consumption and land cover for the 2012–2022 period, in this study, the 2022 datasets were selected to focus on the most recent urban characteristics. However, multitemporal analyses were performed (but not shown in this contribution) to analyze possible variations in each municipal area.

All layers were collected at 10 m spatial resolution that meets the needs of the MIRIFICUS project for mapping urban cover and monitoring SUHI at the municipal level.

2.4 Surface Thermal Pattern Characterization

The surface thermal pattern was analyzed from the Landsat-8 and Landsat-9 data of the 2013–2023 summer period (June–August). Through the Google Earth Engine platform, surface temperature (ST) layers of Landsat-8 and Landsat-9 Collection 2 imageries were selected with a cloud cover of less than 60%, in addition to MODIS dataset which was used to fill unavailable data in the Landsat collection in some areas. ST calculation was carried out using the methodology developed by the U.S. Geological Survey Agency, used in numerous international research, which involves the application of a specific scale factor.

The final layer is the average summer surface temperature with a spatial resolution of 60 m, previously resampled by 30 m and 1 km size (respectively the original spatial resolutions of Landsat and MODIS). The ST values were masked by the water body layer to focus on urban surfaces and avoid bias related to water effects.

Three different indicators were calculated on the summer ST dataset to evaluate the summer surface thermal pattern:

1. *The Surface Urban Heat Island (SUHI).* Urbanization level data of 2021, provided by ISPRA at 10 m spatial resolution for the entire Italian territory, was used. This data provided three levels of urbanization, based on different thresholds of land consumption coverage (hereafter named LCC): urban (LCC > 50%), suburban (LCC between 10 and 50%), and rural (LCC < 10%). SUHI was calculated by extracting average, minimum, and maximum ST values for each urban, suburban, and rural level.
2. *The Urban Thermal Field Variance Index (UTFVI).* This index was calculated to quantify the impact of surface temperature on the quality of life and to proceed with the ecological assessment of summer urban thermal conditions (Liu and Zhang 2011). The UTFVI index was determined by the following formula: (LST − LSTmean)/LSTmean, where LST is the recorded surface temperature in each pixel and LSTmean is the mean value of the selected area (each municipality domain) of the period 2013–2023. This index distinguishes 6 classes of ecological assessment and intensity of the SUHI phenomenon (Vimayak et al., 2022): (1) UTFVI < 0.000: absent SUHI; (2) UTFVI between 0.000 and 0.005: weak SUHI; (3) UTFVI between 0.005 and 0.010: moderate SUHI; (4) UTFVI between 0.010 and 0.015: strong SUHI; (5) UTFVI between 0.015 and 0.020: stronger SUHI; (6) UTFVI >

0.020: strongest SUHI. In this study, adverse ecological conditions were defined for UTFVI $\geqslant$ 0.010, namely from strong to strongest SUHI intensity.
3. *The Thermal Surface Hot-spots.* The Getis-Ord Gi* statistic, a local spatial autocorrelation statistical method, was applied to detect summer thermal anomalies. Thermal clusters of statistically significant summer hot- and cool-spots, considered as points characterized by abnormal values (high or low) and surrounded by points with similar anomalous values, within a search radius defined by the spatial resolution of the input data (Mavrakou et al. 2018). The Getis-Ord Gi* application provided 6 classes of hot- and cool-spots by confidence levels of 99, 95, and 90%, in which different thresholds of summer average LST were recorded.

Three geospatial thermal pattern layers related to the previously mentioned approaches were obtained for each regional capital municipality with a 60 m spatial resolution. The SUHI, UTFVI, and hot-spot areas were further characterized by zonal statistics of ST, NDVI, and percentage coverage of land consumption, tree, and grassland.

2.5 Environmental Modeling and Microclimate Simulations of Design Scenarios

An industrial area in the city of Florence, affected by surface thermal anomalies, was selected as a case study to perform environmental modeling to simulate design scenarios to mitigate the thermal hot-spots.

Analyses of the urban morphology, shading, and urban spatial characterization were carried out in the Florence industrial area, involving the detection of geometry, surface materials, and height of buildings and trees, as well as the collection of microclimatic monitoring data of the 2021 summer period.

The microclimate situation of the ex-ante scenario and tree design scenarios, by using 5- and 10-m high trees, were simulated to compare microclimate conditions among the current situation and the designed ones.

Particular attention was paid to the design scenarios in considering the appropriate tree distances, tree height, and crown size in order to allow the regular performance of the work activities and movement of the working vehicles.

In addition, the thermal pattern was analyzed by simulating the following parameters during the morning and afternoon periods (at 9:00 a.m. and 3:00 p.m.): air temperature, mean radiant temperature, surface temperature, and Universal Thermal Climate Index. QGIS and the Envi-met software were used to perform environmental and microclimate analyses and collect geospatial data on the thermal pattern. Further details about this case study are available in a recent study (Guerri et al. 2023).

3 Results

The present contribution showed the project results with municipal detail, with particular regard to Milan, Rome, and Florence municipalities, placed in the northern and central Italian regions. The main reason for this selection is to include the two major Italian cities for resident distribution (Milan and Rome, with about 1.3 and 2.8 million inhabitants), in addition to providing a focus on the case studies (Rome and Florence) selected by the research project to perform the monitoring and evaluation analyses of reforestation interventions.

3.1 Surface Urban Heat Island (SUHI) Characterization

Results of the SUHI analysis and urban characterization are shown in Table 1 and Fig. 3.

The urban area of the selected cities, characterized by impervious surfaces higher than 50%, represented, respectively, 40–60% of Florence and Milan municipal areas, and slightly less than about 20% of Rome. The highest ST values (43–44 °C), a vegetation coverage of less than 22% (with about 5% of tree coverage), and an NDVI of about 0.3 were observed in all urban areas.

The average ST differences between urban and rural areas were about 7 °C, 6 °C and 2 °C in Milan, Florence and Rome respectively. The average ST differences between urban and suburban areas were about 4 °C in Florence and Milan and about 1 °C in Rome. In addition, about a 70% increase in vegetation coverage (grass and tree) was observed between urban and rural areas in all cities.

3.2 Urban Thermal Field Variance Index (UTFVI) Application: The SUHI Evaluation

The results of the UTFVI application on the ST summer dataset of 2013–2023 and the urban characterization are shown in Table 2 and Fig. 4. In this study, adverse thermal conditions were considered for the UTFVI classes from "strong" to "strongest", while favorable conditions were from "absent" to "moderate" levels.

Table 1 Summer Surface Temperature (ST), Normalized Difference Vegetation Index (NDVI) values, and urban cover characterization in urbanization levels

City	Zonal statistics	Urbanization levels [Land Consumption Coverage %]		
		Urban [LCC > 50%]	Suburban [LCC 10%-50%]	Rural [LCC < 10%]
Milan	Area (%)	64.5	26.5	9.0
	Summer ST (°C) Mean ± sd [min–max]	43.1 ± 1.7 [30.1–52.8]	39.3 ± 2.4 [30.4–48.7]	36.2 ± 2.4 [30.6–42.7]
	Summer NDVI Mean ± sd	0.3 ± 0.2	0.5 ± 0.2	0.6 ± 0.2
	TC (%)	6.5	29.3	27.6
	GC (%)	11.2	40.6	66.1
Florence	Area (%)	39.2	48.9	11.9
	Summer ST (°C) Mean ± sd [min–max]	44.1 ± 2.0 [35.5–55.2]	40.2 ± 2.3 [32.5–49.3]	37.9 ± 2.6 [31.8–44.8]
	Summer NDVI Mean ± sd	0.3 ± 0.2	0.5 ± 0.2	0.6 ± 0.2
	TC (%)	5.2	31.0	50.0
	GC (%)	13.5	43.2	43.0
Rome	Area (%)	18.7	34.8	46.5
	Summer ST (°C) Mean ± sd [min–max]	43.7 ± 2.3 [30.9–57.2]	42.7 ± 2.8 [29.7–52.3]	41.9 ± 4.4 [29.5–51.6]
	Summer NDVI Mean ± sd	0.3 ± 0.2	0.4 ± 0.2	0.4 ± 0.2
	TC (%)	4.6	17.2	27.5
	GC (%)	17.8	53.8	69.0

Notes sd is the standard deviation value; ST is the Surface Temperature of the summer 2013–2023 period; NDVI is the Normalized Difference Vegetation Index (NDVI) of the summer 2015–2023 period; TC is the tree cover; GC is the grassland cover

Adverse conditions affected about 20% of the municipality areas in Florence and Rome, and 10% in Milan, showing an average ST value of about 47 °C (ranging between 46.5 and 49.9 °C). Concerning the average thermal pattern difference within areas characterized by adverse and favorable conditions, an ST variation of about 6 °C was typically observed in all cities, and a higher average ST variation of about 8 °C was recorded in the minimum ST dataset. However, the highest ST variation of about 10 °C was uniformly recorded between the "absent" and "strongest" levels in all cities.

Moreover, areas affected by adverse conditions were, on average, characterized by less than 10% of vegetation cover (specifically, less than 1% of TC), and about 90% of urban levels (i.e., over 50% of LCC) in Milan and Florence. A different situation was observed in Rome, which mostly showed adverse conditions in rural areas characterized by rural (LCC less than 10%) and a grass cover higher than 60% compared to the other cities.

3.3 Hot-Spot Method Application: The Surface Thermal Anomalies

The surface thermal anomalies, obtained from the hot-spot analysis by the Getis-Ord method application, and the urban characterization are shown in Table 3 and Fig. 5.

About 12% of the municipal areas were affected by surface thermal anomalies: hot-spot areas characterized between 3 and 5% of the studied areas, while cool-spots were between 6 and 10%.

Considering the average surface thermal anomaly pattern, about 14 °C of ST variation was recorded between hot- and cool-spots in all selected cities: 49 °C and 35 °C were, on average, observed in the hot-spot and cool-spot areas, respectively. In particular, the highest average ST value (52 °C, recording 57 °C as the maximum value) was observed in the hot-spot (level 3) in Rome, mainly characterized by urban level, and also showing the highest ST variation of 19 °C

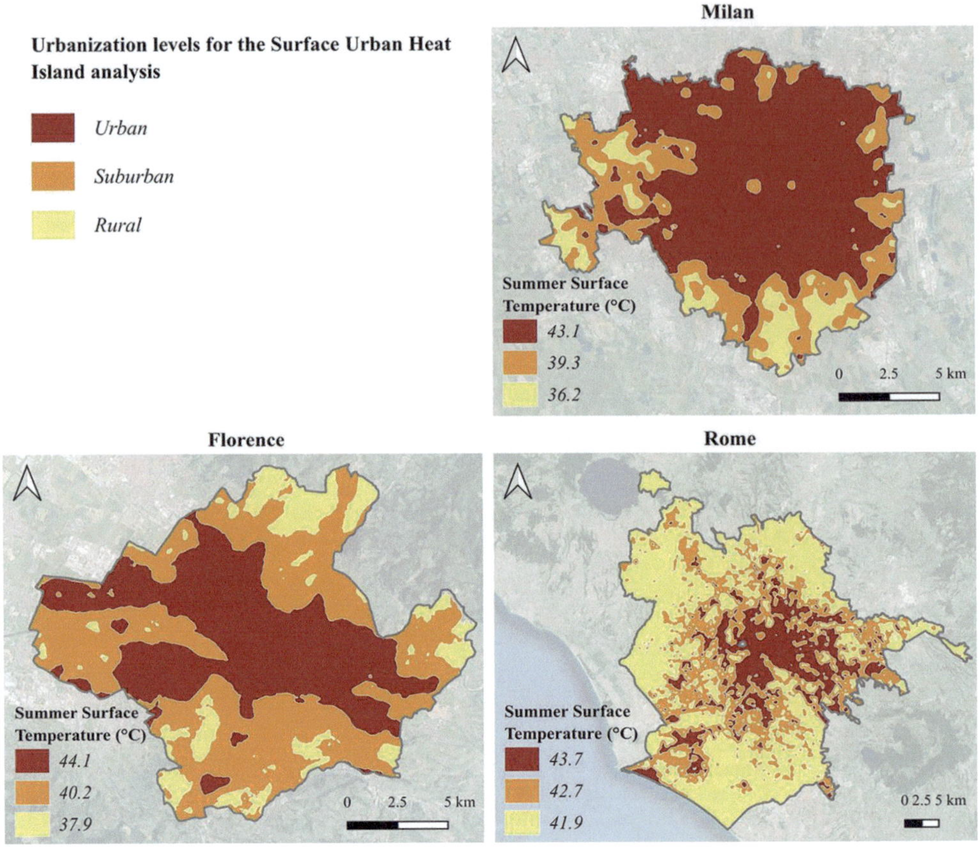

Fig. 3 Surface Urban Heat Island: surface temperature variations in urbanization levels, retrieved from the 2013–2023 summer dataset (*Source* By author)

compared to the cool-spot (level 3). Furthermore, the lowest ST (32.4 °C) was recorded in the cool-spot (level 3) of Rome, covered by about 98% of vegetation surfaces (mainly TC).

In addition, a homogeneous pattern was observed in the minimum and maximum datasets where about 15 °C of ST difference was recorded between hot- and cool-spots. The hot-spot areas in Florence and Milan were mainly characterized by 90% urban level, no rural levels and TC, and a GC of less than 10%. Different urban characteristics were observed in the hot-spots of Rome: respectively, about 46% and 36% of the urban and rural levels, and 50% of GC were recorded. However, the cool-spot areas showed no urban level, and the prevalence of rural level: about 90% of vegetation coverage, mainly 80% TC, was observed, except for Milan characterized by about 40% TC. In addition, the prevalence of imperviousness in hot-spot and of pervious surfaces in cool-spot was also demonstrated by the average NDVI values of 0.1 and 0.7, respectively.

3.4 Microclimate Simulations to Mitigate Thermal Hot-Spots: The Florence Case Study

The selected case study was 250,000 m^2 in size, mainly covered by asphalt road pavements (55%) and low vegetation (5%). Some sample illustrations of the ex-ante and ex-post scenarios are included in Fig. 6.

The results of the tree scenario simulations indicated that trees are important elements for urban mitigation strategies, especially for mitigating surface thermal hot-spot areas by shading and improving the thermal comfort of outdoor workers in the selected industrial areas. The best-planned tree scenario in terms of reduction in thermal parameters involved about 130 trees with a height and a canopy diameter of 10 m and providing a 10% increase in the road surface cover. In particular, the designed trees with a height of 10 m showed the best cooling effect on these surfaces by favoring significant decreases of 4 °C in ST compared with the ex-ante situation. This tree scenario also improved the urban morphology and solar access, underlining the importance of tree height and canopy cover dimension to provide significant variations of urban microclimate in a complex urban area such as industrial and commercial zones.

4 Discussion

This study allowed a detailed description of the surface thermal pattern variation within contiguous urban areas and with a precise geographical location, concerning the distribution of the urban landscape features.

Table 2 Summer Surface Temperature (ST), Normalized Difference Vegetation Index (NDVI) values, and urban cover characterization in 6 SUHI intensity level classes, defined by the Urban Thermal Field Variance Index (UTFVI)

City	Urban characterization	UTFVI class (SUHI intensity level)					
		Absent	Weak	Moderate	Strong	Stronger	Strongest
Milan	Area (%)	41.0	24.8	23.7	8.2	1.7	0.6
	Summer ST (°C) Mean ± sd [min–max]	38.6 ± 2.4 [30.1–41.5]	42.3 ± 0.5 [41.5–43.1]	43.8 ± 0.4 [43.1–44.6]	45.2 ± 0.4 [44.6–46.2]	46.9 ± 0.4 [46.2–47.8]	48.7 ± 1.0 [47.8–52.8]
	Summer NDVI Mean ± sd	0.5 ± 0.2	0.3 ± 0.2	0.2 ± 0.2	0.2 ± 0.1	0.1 ± 0.1	0.1 ± 0.1
	Urban level (%)	25.8	85.6	95.9	97.5	97.5	99.3
	Suburban level (%)	53	14.1	4.1	2.5	2.5	0.7
	Rural level (%)	21.2	0.3	0	0	0	0
	TC (%)	29.3	7.4	2.2	1.0	0.5	0.3
	GC (%)	42.1	17.3	8.4	5.1	4.2	3.1
Florence	Area (%)	49.3	16.5	17.1	12.1	3.8	1.2
	Summer ST (°C) Mean ± sd [min–max]	38.8 ± 1.9 [31.8–41.4]	42.2 ± 0.5 [41.4–43]	43.8 ± 0.5 [43–44.6]	45.3 ± 0.4 [44.6–46.1]	46.7 ± 0.4 [46.1–47.7]	49.2 ± 1.5 [47.7–55.2]
	Summer NDVI Mean ± sd	0.6 ± 0.2	0.4 ± 0.2	0.3 ± 0.2	0.2 ± 0.2	0.2 ± 0.1	0.1 ± 0.1
	Urban level (%)	6.5	44.9	75.8	91.5	91.8	85.8
	Suburban level (%)	70.8	51.3	23.3	8.5	8.2	14.2
	Rural level (%)	22.7	3.8	0.9	0	0	0
	TC (%)	42.3	10.1	3.4	1.3	0.4	0.1
	GC (%)	37.7	41.5	25.3	12	8.1	8.5
Rome	Area (%)	43.0	21.0	18.5	12.0	4.6	0.9
	Summer ST (°C) Mean ± sd [min–max]	39.4 ± 3 [29.5–42.5]	43.3 ± 0.5 [42.5–44.1]	44.9 ± 0.5 [44.1–45.7]	46.4 ± 0.4 [45.7–47.3]	47.9 ± 0.4 [47.3–48.8]	49.6 ± 0.8 [48.8–57.2]
	Summer NDVI Mean ± sd	0.5 ± 0.2	0.3 ± 0.2	0.3 ± 0.1	0.3 ± 0.1	0.2 ± 0.1	0.2 ± 0.1
	Urban level (%)	13.5	25.8	23.2	17.1	19.1	39.0
	Suburban level (%)	35.3	40.0	34.7	29.4	23.3	15.9
	Rural level (%)	51.2	34.2	42.1	53.5	57.6	45.1
	TC (%)	39.7	8.4	3.8	1.3	0.3	0.1
	GC (%)	40.3	56.6	65.1	75.5	76.1	56.1

Notes sd is the standard deviation value; ST is the Surface Temperature of the summer 2013–2023 period; NDVI is the Normalized Difference Vegetation Index (NDVI) of the summer 2015–2023 period; TC is the tree cover; GC is the grassland cover

The three selected approaches for the thermal pattern analysis represented the basis for the SUHI monitoring by using EO data with an adequate spatial resolution at municipal detail.

The summer SUHI analysis showed the highest variability in cities where the urban core is the mainly represented area related to the municipality area, such as in Milan and Florence. In fact, the average ST differences between urban and rural areas were about 7 °C, 6 °C, and 2 °C in Milan, Florence, and Rome, respectively. This study highlighted that the SUHI intensity is directly proportional to the urban core extension of the studied cities.

This aspect is confirmed by recent studies (Xu et al. 2024; Yao et al. 2022) which demonstrated that the SUHI footprint

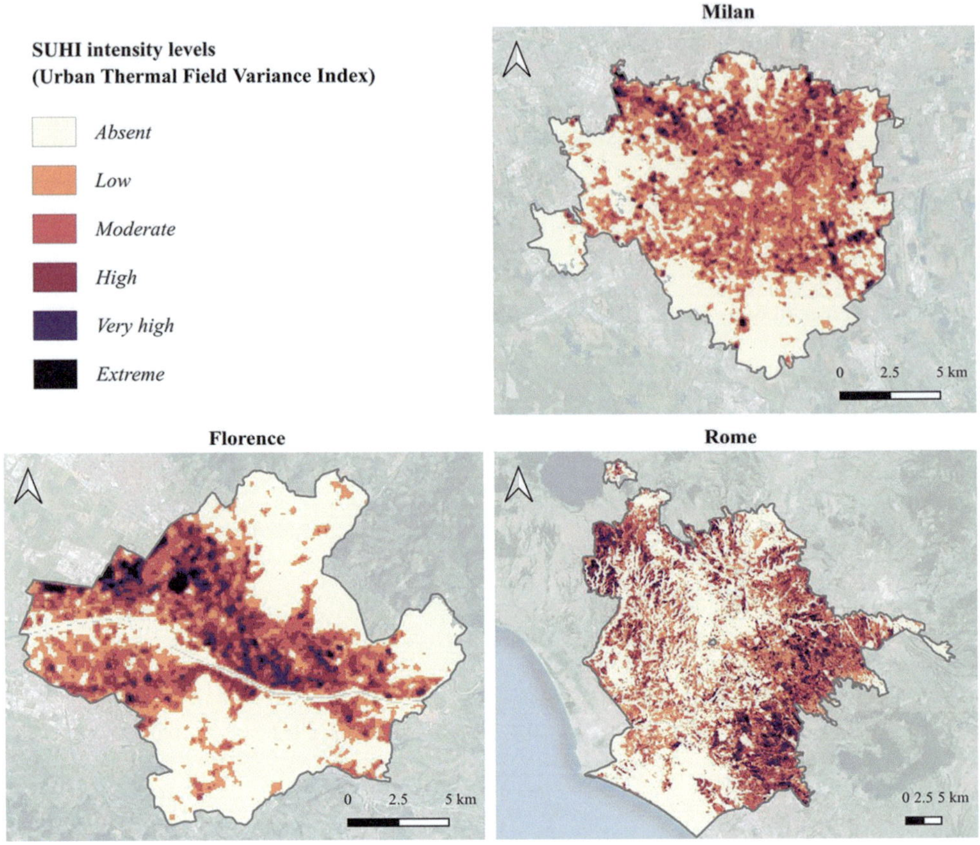

Fig. 4 SUHI intensity classes, detected by the Urban Thermal Field Variance Index (UTFVI) applied to the summer surface temperature dataset of 2013–2023 (*Source* By author)

area (corresponding to the extent of the SUHI effect) tended to increase as the city size increased. In particular, it was observed a positive correlation during daytime between this footprint and built-up area, and population, showing that the development of the SUHI extent was closely related to the increase in built-up areas and population growth.

In addition, the correct identification of thermal anomalies was extremely important in urban thermal studies because the ST values of a given location are not independent, but they are self-related spatially with those of neighboring areas due to heat flows of the soil surface (Song et al. 2014). The originality of these approaches lies in the processing of ST data, obtained from available satellite images (Landsat-8 and Landsat-9), through the hot-spot method already applied in previous works (Guerri et al. 2021; Zargari et al. 2024).

A further novel element of this study is the combination of different data sources, from satellite remote sensing data, in situ measurements, and environmental modeling. The use of a combined approach is necessary to monitor and analyze the thermal pattern fluctuations across various scales, ranging from national to local levels. Although remote sensing-based measurements provide spatially detailed information for an extensive coverage, their accuracy need to be validated with ground measurements in order to decrease uncertainty due to spatial resolution. In addition, microclimate modeling also needs to be validated with in situ monitoring to offer an efficient methodology for evaluating the NBS performance.

As widely stated by recent studies (Kumar et al. 2024; Zargari et al. 2024), this combined approach provides a comprehensive and accurate framework to evaluate the efficiency of NBS to mitigate urban heat, offering spatial and temporal coverage, multidimensional scales, and cost-effective data collection.

These relevant elements are necessary for the MIRIFICUS project, characterized by numerous challenges involving the Italian cities, especially considering the current urban structure complexity due to the multiple presence of historical tissue, economic activities, and densely built-up areas.

Considering the complexity of the Italian city structure, the next project phase related to the mitigation intervention simulations will require to include multiple scenarios linked to NBS and cool materials, depending on both the urban functions and the microclimate characteristics. Furthermore, it is important to conduct assessments to determine the feasibility and viability of implementing mitigation measures in hot-spot areas, particularly in densely populated urban areas lacking green spaces.

Among the green–blue-grey infrastructures (GBGI), street trees were reported to have the highest number of co-benefits in addition to their potential effectiveness in heat mitigation and adaptation, followed by green roofs and

Table 3 Summer Surface Temperature (ST), Normalized Difference Vegetation Index (NDVI) values, and urban cover characterization in the surface thermal anomalies (Hot- and Cool-spots)

City	Urban characterization	Surface thermal anomalies					
		Cool-spot (Level 1)	Cool-spot (Level 2)	Cool-spot (Level 3)	Hot-spot (Level 1)	Hot-spot (Level 2)	Hot-spot (Level 3)
Milan	Area (%)	2.6	3.9	3.6	1.4	1.1	0.2
	Summer ST (°C) Mean ± sd [min–max]	36.5 ± 0.3 [35.8–37.1]	35.2 ± 0.5 [33.9–36.2]	33.6 ± 0.5 [30.1–34.4]	46.4 ± 0.3 [46–47.1]	47.6 ± 0.5 [46.9–49]	49.7 ± 1 [48.6–52.8]
	Summer NDVI Mean ± sd	0.6 ± 0.2	0.7 ± 0.1	0.7 ± 0.1	0.1 ± 0.1	0.1 ± 0.1	0.1 ± 0.1
	Urban level (%)	0.9	0.3	0.2	98.1	98.2	99.9
	Suburban level (%)	65.6	50.2	25.7	1.9	1.8	0.1
	Rural level (%)	33.5	49.5	74.1	0	0	0
	TC (%)	48.7	41.6	26.5	0.6	0.3	0.2
	GC (%)	41.0	51.4	70.1	3.9	4.7	1.6
Florence	Area (%)	2.6	2.4	1.3	3.1	1.6	0.6
	Summer ST (°C) Mean ± sd [min–max]	36.3 ± 0.3 [35.6–36.9]	35.1 ± 0.5 [33.8–35.9]	33.3 ± 0.6 [31.8–34.1]	46.5 ± 0.3 [46–47]	47.6 ± 0.5 [46.9–48.8]	50.3 ± 1.5 [48.7–55.2]
	Summer NDVI Mean ± sd	0.7 ± 0.2	0.7 ± 0.1	0.8 ± 0.1	0.2 ± 0.1	0.2 ± 0.1	0.1 ± 0.1
	Urban level (%)	0.1	0.1	0	93.5	84.0	94.3
	Suburban level (%)	60.3	35.1	8.6	6.5	16.0	5.7
	Rural level (%)	39.6	64.8	91.4	0	0	0
	TC (%)	73.9	83.7	92.2	0.4	0.2	0
	GC (%)	17.8	10.4	5.1	6.8	12.8	3.1
Rome	Area (%)	1.8	2.4	4.1	2.1	0.7	0.1
	Summer ST (°C) Mean ± sd [min–max]	36.7 ± 0.3 [35.7–37.3]	35.1 ± 0.6 [33.5–36.2]	32.4 ± 1.2 [29.5–34.2]	48.3 ± 0.3 [47.8–49.2]	49.6 ± 0.5 [48.9–51.2]	51.7 ± 0.8 [50.9–57.2]
	Summer NDVI Mean ± sd	0.7 ± 0.2	0.7 ± 0.2	0.8 ± 0.1	0.2 ± 0.1	0.2 ± 0.1	0.1 ± 0.1
	Urban level (%)	1.9	0.5	0	20.3	37.4	80.7
	Suburban level (%)	32.7	18.5	3.2	20.9	15.6	15.9
	Rural level (%)	65.3	80.9	96.8	58.8	47	3.4
	TC (%)	69.6	81.9	93.7	0.2	0.1	0
	GC (%)	23.2	13.5	5.1	75	57.7	13.9

Notes sd is the standard deviation value; ST is the Surface Temperature of the summer 2013–2023 period; NDVI is the Normalized Difference Vegetation Index (NDVI) of the summer 2015–2023 period; TC is the tree cover; GC is the grassland cover

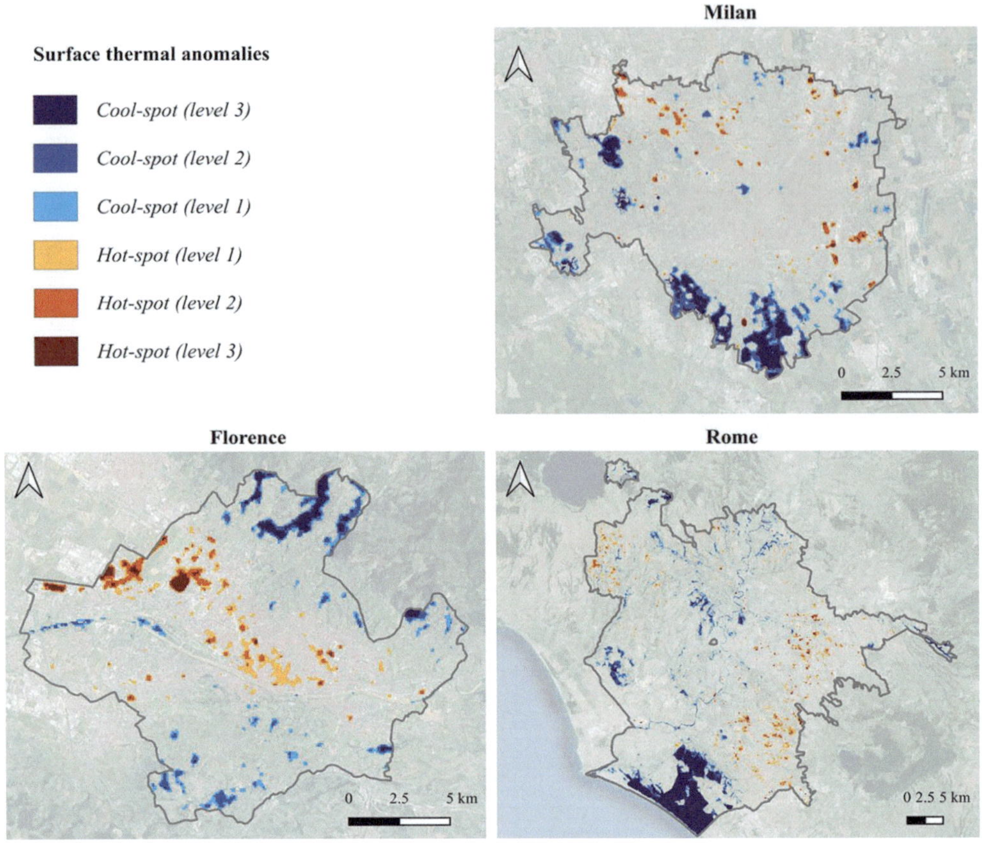

Fig. 5 Thermal surface anomalies, detected by the Getis-Ord Gi* statistic method applied to the summer surface temperature dataset of 2013–2023 (*Source* By author)

botanical gardens (Kumar et al. 2024). Tree planting along streets was the most effective GBGI category, and it was observed that the modeling results showed a temperature decrease with very relatively low uncertainty. In this study, preliminary design scenarios involving street trees were applied to a thermal hot-spot area in Florence, a relevant commercial area covered by low vegetation (less than 5% of grass and no tree cover), with particular regard to consider both microclimate conditions and the current urban functions.

5 Limitations and Future Research Developments

This study focused on the 2013–2023 average surface thermal pattern and urban landscape distribution in order to provide information at both national and city scales. There are some limitations in this study that need to be mentioned. Nighttime pattern and multitemporal thermal variation were not investigated.

About the nighttime data, it is worth mentioning that, according to several studies, it was widely demonstrated that the SUHI is a phenomenon typically occurring during summer daytime, showing its maximum intensity during this summertime (Oke 1995; Sismanidis et al. 2022; WMO 2023). Considering the aim of this research, the multitemporal thermal dimension was not considered because the study focus was on the average surface thermal distribution, with particular regard to the daytime livability in urban areas of the last ten years.

In addition, a further limitation is linked to microclimate modeling carried out by the Envi-met software which offers detailed spatio-temporal information, but with uncertainties due to assumptions.

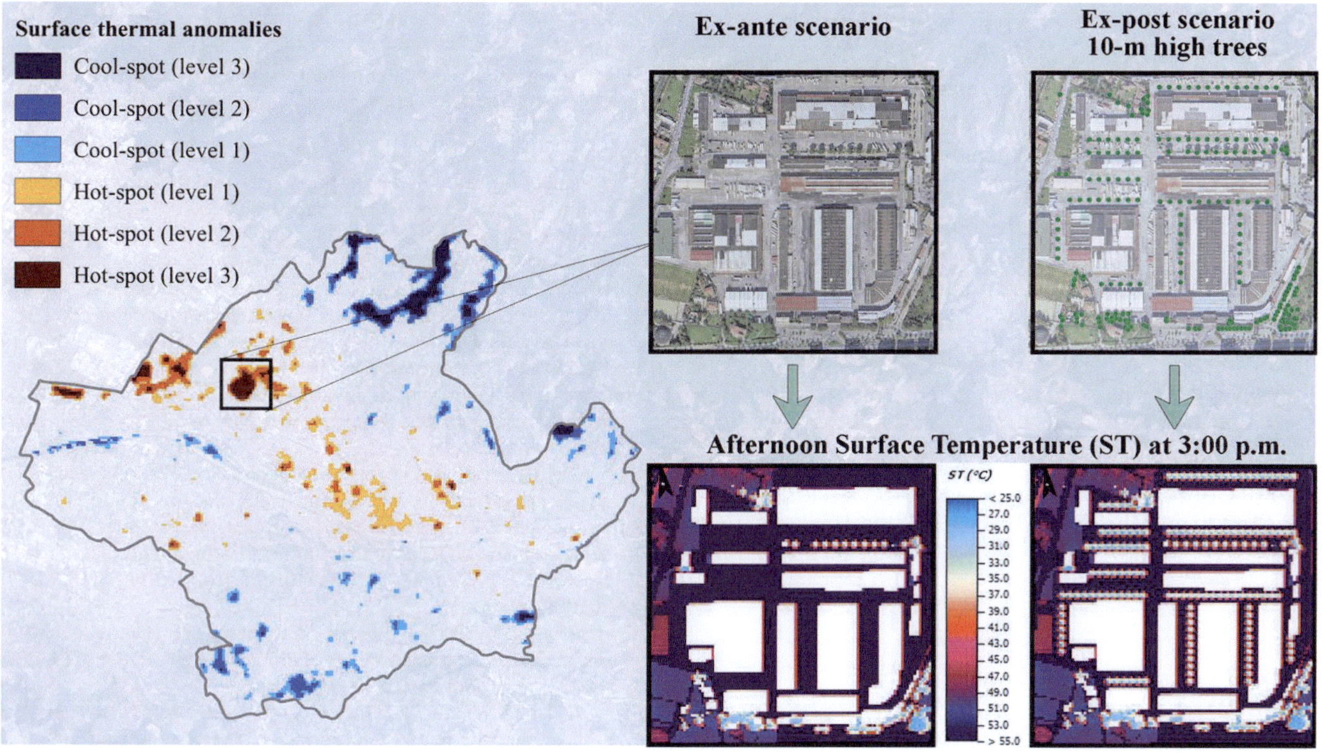

Fig. 6 Ex-ante and tree design scenario in the hot-spot area of Florence (*Source* By author)

Moreover, relevant elements affecting the urban model, the surface albedo and urban morphology, previously investigated in the Italian cities (Guerri et al. 2021, 2023), will be included in the next MIRIFICUS project stages to quantify the contribution on the thermal pattern and to develop microclimate simulations of design scenarios in the cities of Florence and Rome.

However, future studies will focus on the multitemporal variation of the thermal pattern, as well as involving population and built-up area data in order to provide a comprehensive study about the relation between SUHI intensity, population distribution and city size over the years.

6 Conclusion

Our research highlighted that urban thermal pattern is a complex phenomenon resulting in coupling interaction between the human–earth complex systems: the combination of several urban elements contributing to the production of urban heat is required, including grey, green, and blue infrastructures, the urban morphology and further information on vegetation stress and vigor.

It is well known that for effective climate-resilient urban intervention detection, it is crucial to analyze the diverse urban landscape to spatially detect urban critical characteristics (Kumar et al. 2024). The integration of thermal characteristics and land cover data was a key operation for urban projects, not only to provide detailed information on urban characteristics that contributed to influencing the microclimate, but also to spatially identify the priority areas that needed mitigation interventions.

Furthermore, the environmental modeling and the microclimate simulation of mitigation scenarios, under the current climate conditions, are helpful in programming mitigation interventions in complex urban contexts.

The scientific significance of this project is its multidimensional scale approach. Moreover, the study was applied to geospatial data from the national to the municipality scales, allowing the detection of specific case study to develop microclimate simulations of design scenarios to mitigate urban heat at local scale, based on the previous detection of critical urban characteristics. This particular project showed important effects on societal and policy benefits as it aims at providing information (data, maps, urban microclimate scenarios) and tools (such as the environmental microclimate modeling of different urban scenarios) shared through the WebGis developed guiding policy-makers and urban planners to identify the best urban mitigation strategies based on green–blue infrastructures and cool materials, improving thermal comfort and the whole livability in urbanized areas.

An additional relevant element of the MIRIFICUS project is linked to the stakeholder engagement, highlighting the importance of including diverse perspectives, fostering a sense of shared ownership and collective responsibility into a collaborative process to combat the effect of urban heat hotspots. The project will involve all relevant stakeholders and key players including a wider range of researchers, government bodies, and local communities in the co-creation of urban design initiatives in the cities of Florence and Rome.

Funding This work was funded by the Italian Space Agency (ASI), Call for Ideas "Effects of climate change and extreme events"—"I4DP_PA (INNOVATION FOR DOWNSTREAM PREPARATION FOR PUBLIC ADMINISTRATIONS)"—MIRIFICUS Project (Monitoraggio degli Interventi di RIForestazione per l'Isola di Calore Urbana tramite i Satelliti).

References

De Fioravante, P., Luti, T., Cavalli, A., Giuliani, C., Dichicco, P., Marchetti, M., Chirici, G., Congedo, L., Munafò, M.: Multispectral Sentinel-2 and SAR Sentinel-1 integration for automatic land cover classification. Land **10**(6), 611 (2021). https://doi.org/10.3390/land10060611

European Space Agency.: Sentinel-2 User Handbook; Revision 2; ESA Standard Document; ESA: Paris, France (2015). https://sentinels.copernicus.eu/documents/247904/685211/Sentinel-2_User_Handbook

Guerri, G., Crisci, A., Messeri, A., Congedo, L., Munafò, M., Morabito, M.: Thermal summer diurnal hot-spot analysis: the role of local urban features layers. Remote Sens. **13**(3), 538 (2021). https://doi.org/10.3390/rs13030538

Guerri, G., Crisci, A., Morabito, M.: Urban microclimate simulations based on GIS data to mitigate thermal hot-spots: tree design scenarios in an industrial area of Florence. Build. Environ. **245**, 110854 (2023). https://doi.org/10.1016/j.buildenv.2023.110854

Guha, S., Govil, H., Dey, A., Gill, N.: Analytical study of land surface temperature with NDVI and NDBI using Landsat 8 OLI and TIRS data in Florence and Naples City, Italy. Eur. J. Remote Sens. **51**(1), 667–678 (2018). https://doi.org/10.1080/22797254.2018.1474494

Kumar, P., Debele, S.E., Khalili, S., Halios, C.H., Sahani, J., Aghamohammadi, N., Andrade, M., Athanassiadou, M., Bhui, K., Calvillo, N., Cao, S., Coulon, F., Edmondson, J.L., Fletcher, D., Dias de Freitas, E., Guo, H., Hort, M., Katti, M., Kjeldsen, T., Lehmann, S., Locosselli, G., Malham, S.K., Morawska, L., Parajuli, R., Rogers, C.D.F., Yao, R., Wang, F., Wenk, J., Jones, L.: Urban heat mitigation by green and blue infrastructure: drivers, effectiveness, and future needs. The Innovation **5**, 100588 (2024). https://doi.org/10.1016/j.xinn.2024.100588

Liu, L., Zhang, Y.: Urban heat island analysis using the landsat TM data and ASTER data: a case study in Hong Kong. Remote Sens. **3**(7), 1535–1552 (2011). https://doi.org/10.3390/rs3071535

Mavrakou, T., Polydoros, A., Cartalis, C., Santamouris, M.: Recognition of thermal hot and cold spots in urban areas in support of mitigation plans to counteract overheating: application for Athens. Climate **6**(1), 16 (2018). https://doi.org/10.3390/cli6010016

Morabito, M., Crisci, A., Gioli, B., Gualtieri, G., Toscano, P., Di Stefano, V., Orlandini, S., Gensini, G.F.: Urban-hazard risk analysis: mapping of heat-related risks in the elderly in major Italian cities. PLoS ONE **10**(5), e0127277 (2015). https://doi.org/10.1371/journal.pone.0127277

Morabito, M., Crisci, A., Georgiadis, T., Orlandini, S., Munafò, M., Congedo, L., Rota, P., Zazzi, M.: Urban Imperviousness effects on summer surface temperatures nearby residential buildings in different urban zones of Parma. Remote Sensing **10**(1), 26 (2018). https://doi.org/10.3390/rs10010026

Morabito, M., Crisci, A., Guerri, G., Messeri, A., Congedo, L., Munafò, M.: Surface urban heat islands in Italian metropolitan cities: tree cover and impervious surface influences. Sci. Total. Environ. **751**, 142334 (2021). https://doi.org/10.1016/j.scitotenv.2020.142334

Munafò (Ed.): Consumo di suolo, dinamiche territoriali e servizi ecosistemici. Edizione 2023. Report SNPA 37/23 (2023). ISBN 978-88-44-1178-5

Oke, T.R.: The heat island of the urban boundary layer: characteristics, causes and effects. In: Wind Climate in Cities, pp. 81–107 (1995)

Rubel, F., Kottek, M.: Observed and projected climate shifts 1901–2100 depicted by world maps of the Köppen-Geiger climate classification. Meteorol. z. **19**, 135–141 (2010). https://doi.org/10.1127/0941-2948/2010/0430

Sismanidis, P., Bechtel, B., Perry, M., Ghent, D.: The seasonality of surface urban heat islands across climates. Remote Sens. **14**, 2318 (2022). https://doi.org/10.3390/rs14102318

Song, J., Du, S., Feng, X., Guo, L.: The relationships between landscape compositions and land surface temperature: quantifying their resolution sensitivity with spatial regression models. Landsc. Urban Plan. **123**, 145–157 (2014). https://doi.org/10.1016/j.landurbplan.2013.11.014

Vimayak, B., Lee, H., Gedam, S., Latha, R.: Impacts of future urbanization on urban microclimate and thermal comfort over the Mumbai metropolitan region, India. Sustain. Cities Soc. **79**, Article 103703 (2022). https://doi.org/10.1016/j.scs.2022.103703

Voogt, J.A., Oke, T.R.: Thermal remote sensing of urban climates. Remote Sens. Environ. **86**, 370–384 (2003). https://doi.org/10.1016/S0034-4257(03)00079-8

World Meteorological Organization.: Guidance on Measuring, Modelling and Monitoring the Canopy Layer Urban Heat Island (CL-UHI). WMO-No. 1292 (2023). ISBN 978-92-63-11292-2

Xu, H., Li, C., Hu, Y., Wang, H., Wen, D., Li, Z., Ping, X., Wang, Q., Li, Q.: Spatiotemporal evolution and influencing factors of surface urban heat island footprint across different-sized cities. Urban Clim. **54**, 101852 (2024). https://doi.org/10.1016/j.uclim.2024.101852

Yao, L., Sun, S., Song, C., Wang, Y., Xu, Y.: Recognizing surface urban heat 'island' effect and its urbanization association in terms of intensity, footprint, and capacity: a case study with multi-dimensional analysis in northern China. J. Clean. Prod. **372**, 133720 (2022). https://doi.org/10.1016/j.jclepro.2022.133720

You, M., Lai, R., Lin, J., Zhu, Z.: Quantitative analysis of a spatial distribution and driving factors of the urban heat island effect: a case study of Fuzhou Central Area, China. Int. J. Environ. Res. Public Health **18**, 13088 (2021). https://doi.org/10.3390/ijerph182413088

Zargari, M., Mofidi, A., Entezari, A., Baaghideh, M.: Climatic comparison of surface urban heat island using satellite remote sensing in Tehran and suburbs. Sci. Rep. **14**, 643 (2024). https://doi.org/10.1038/s41598-023-50757-2

Zhou, B., Rybski, D., Kropp, J.P.: The role of city size and urban form in the surface urban heat island. Sci. Rep.-UK **7**, 4791 (2017). https://doi.org/10.1038/s41598-017-04242-2

Climate Maps and Urban Planning in Search of a Reliable Framework for Adapting Cities and Communities to the Impacts of Climate Change

Rosalba D'Onofrio, Giorgio Caprari, Roberta Cocci Grifoni, Francesco Conti, Graziano E. Marchesani, Simone Malavolta, Ilenia Pierantoni, and Massimo Sargolini

Abstract

The urban environment is particularly vulnerable to the catastrophic effects of extreme weather events, making it crucial to have a reliable framework in place. Numerous studies have shown how urban geometry, building materials, and vegetation can affect indoor and outdoor comfort. Integrating climate assessments into urban planning can be challenging. This involves linking climate issues identified in an urban context with broader urban planning considerations and tools. The PNRR-Ecosystem of Innovation—Vitality Project aims to explore the connections between urban climatology, urban form, and social relations. The project seeks to use climate maps to inform urban planning and design. This contribution documents the development of knowledge that has led to the investigation of climate maps using two documented methodologies: UCMap (Urban Climate Map) and LCZ (Local Climate Zones). The goal is to identify strengths and weaknesses to create a potential path for innovation in classifying elementary climate units (from LCZ to CAU.us—Climate Adaptive Urban Units) and develop guidelines and design proposals at the urban and detailed climate-proof scale.

Keywords

Local climate zones · UCMap · Urban planning · Climate adaptation · Urban Design

R. D'Onofrio (✉) · G. Caprari · R. C. Grifoni · F. Conti · G. E. Marchesani · S. Malavolta · I. Pierantoni · M. Sargolini
School of Architecture and Design, University of Camerino, Camerino, Italy
e-mail: rosalba.donofrio@unicam.it

G. Caprari
e-mail: giorgio.caprari@unicam.it

R. C. Grifoni
e-mail: roberta.coccigrifoni@unicam.it

F. Conti
e-mail: francesco.conti@unicam.it

G. E. Marchesani
e-mail: graziano.marchesani@unicam.it

S. Malavolta
e-mail: simone02.malavolta@unicam.it

I. Pierantoni
e-mail: ilenia.pierantoni@unicam.it

M. Sargolini
e-mail: massimo.sargolini@unicam.it

1 Introduction

Urban planning has a significant impact on climate change-related actions (Abougendia et al. 2020; Carter et al. 2015; Hurlimann et al. 2021). However, the application of urban climatology in urban planning is still limited (Aslam and Rana 2022). The existence of this gap in the literature can be attributed to (1) the different language and knowledge between urban climatologists and urban planning academics and professionals (Eliasson 2000), (2) the difficulty in finding morphological and climatic data at the urban scale, and (3) the limited application tools for translating climate knowledge to the operational scale (Eliasson 2000; Yin et al. 2022). These obstacles are compounded by (4) the specificity of research findings and (5) the paucity of experience attempting to standardize processes to quantitatively describe urban climate (Chien et al. 2016).

Currently, scientific research includes two conceptual frameworks for assessing urban climate that may be relevant to urban planning: the Urban Climatic Map (UCMap) system (Ren et al. 2011) and local climate zones (LCZs). (Stewart and Oke 2009, 2012).

P. Mohammad et al. (eds.), *Geographic Approaches to Climate Change and Mitigation: Urban and Rural Perspectives (Volume 1)*, Advances in Science, Technology & Innovation, https://doi.org/10.1007/978-3-031-92119-3_5

The UCMap (Urban Climate Map) is a cartographic representation of climate information and the translation of climate assessment results into a planning language. The methodology behind it is the work of the Association of German Engineers (VDI), which in the 1990s published a national standard, VDI 3787-Part 1 (VDI 1997), to guide the application of the UCMap in local practice. The UCMap system usually consists of two components: the Urban Climate Analysis Map (UC-AnMap) and the Urban Planning Recommendations Map (UC-ReMap). The UC-AnMap collects meteorological, planning, land use, topographic and vegetation information. Their interrelationships and impacts are spatially analysed and evaluated (Ren et al. 2013) through the identification of climatic areas called "climatopes". The UC-ReMap includes climate planning recommendations for wind, vegetation, water and urban morphology (Ren et al. 2013). The UCMap uses mesoscale and microscale weather simulation models or empirical data from field measurements (Chen et al. 2017; Zhongli and Hanqiu 2016). There are two main problems with the generation of UCMaps. First, simulation efficiency is limited by computational accuracy and complicated operational techniques. Second, measured empirical data usually cannot fully represent the different surface conditions associated with multiple urban morphologies (Yin et al. 2022). To address these limitations, it seems essential to rely on a better technical system using a land surface classification scheme suitable for urban planners and a more standardized urban climate assessment model (Jin et al. 2020).

The Local Climate Zones (LCZs) proposed by Stewart and Oke (2012) are aligned with this objective. They represent a relatively novel conceptual framework for urban climatology, as they have succeeded in representing both the characteristics of climate and urban morphology at the city level worldwide. The LCZ system is based on the application of generalized knowledge of building forms and land cover and comprises 17 distinct classes. The thermal climate of each class is determined by its surface characteristics, including thermal, radiative, geometric, metabolic and surface cover properties.

LCZs 1-10 are designated for "built" typologies, which may be defined as the expression of urban form (Eldesoky et al. 2022). LCZs A-G are designated for "natural" typologies, which represent permeable land cover, water bodies, and vegetation. The standardization of LCZ definitions and classifications allows for comparisons to be made within and between cities. The LCZ system operates on a scale between cities and neighbourhoods and strongly correlates with land use and building functionality. This allows for the representation of different local climates (Stewart and Oke 2012). The advent of big data and GIS spatial processing techniques has led to a continuous evolution in research methods, resulting in a multitude of experiments conducted at varying scales (Quan and Bansal 2021; Zheng et al. 2018). To date, almost 300 cities worldwide have conducted urban climate studies in the context of LCZs (Zhou et al. 2022). Notable projects such as the World Urban Database and Access Portal Tools - WUDAPT (Demuzere et al. 2021; Ren et al. 2016), MapUCE (Masson et al. 2015), and URBAN-FLUXES (http://urbanfluxes.eu/) have consistently demonstrated advancements in data optimization. Moreover, the development of sophisticated techniques, including deep learning algorithms, pattern simulation, and neural networks (Lau et al. 2023), has led to the integration of these methods in LCZ mapping.

Nevertheless, there are still some shortcomings that require attention. Despite the LCZ's efficacy in educating urban planners and designers on the impact of urban form and function on urban temperatures and heat wave risks, there remains a paucity of substantial experiences to inform urban planning (Lelovics et al. 2014; Ren et al. 2011). One of the most critical aspects is the size of the LCZ methodology's local units, which serve as the minimum units of investigation and evaluation (Perera and Emmanuel 2018). Traditional urban planning is conducted at multiple levels, from the neighbourhood to the urban and regional scales. At present, the most prevalent approach to LCZs entails the categorization of urban blocks as homogeneous units (Quan and Bansal 2021). Accordingly, to investigate the impact on extensive urban areas, it may be necessary to identify a compromise between block-level urban areas and larger planning zones (Perera and Emmanuel 2018). In comparison to the Urban Climatic Map system, LCZs offer a number of notable advantages. These include the requirement for less data, the presence of universal classification criteria, the capacity for comparisons between cities, and enhanced accessibility for non-professionals (Chen et al. 2019; Zhongli and Hanqiu 2016). To date, no studies have been conducted to compare and contrast the two methodologies, to ascertain their respective differences, to explore their potential for combination and to identify any potential mutual benefits (Yin et al. 2022).

This article proposes an analysis of the literature on the experiments related to climate maps, with the goal of:

- developing a reflection on the performance of climate maps created with UCMap and LCZ methodologies, with a view to verifying their usefulness for urban planning and the opportunity for their hybridization;
- proposing a methodology for the construction of a climate map that will be tested in the PNRR Vitality project, Spoke 6 "Innovation and safeness in living environments in the digital and green transition", Work Package 1: "CONTEXT: INDOOR AND OUTDOOR AT THE CENTER OF THE DIGITAL AND GREEN TRANSITION OF LIVING ENVIRONMENTS". This will support the Project in providing urban planning and urban design

with a useful and complete tool for tackling climate adaptation issues by combining the LCZ methodology with some of the UCMap's capabilities.

2 Materials and Methodology

A systematic review was conducted using the Scopus database with the following keywords: "UC-AnMap/UC-Map and urban planning", "UC-ReMap and urban planning", "Climatopes and urban planning" and "LCZ Map and urban planning". This resulted in the selection of 188 scientific articles. The initial search, conducted with the keywords "UC-AnMap; UC-ReMap and urban planning", yielded a total of 15 articles. Subsequently, an additional 13 articles were identified through the extension of the search to include the keywords "climatopes and urban planning". A search for "LCZmap and urban planning" yielded 160 articles.

Following a review of the abstracts and the exclusion of articles not written in English, a total of 56 articles were selected based on the content of the introduction. A comprehensive reading of the full papers enabled the selection of an even smaller number of significant articles, totalling 21. From these, four were selected for the UCMap methodology and only three for the LCZ methodology. The restricted selection process involved the exclusion of articles that solely focused on the expression of interest in the relationship between urban planning instruments and climate surveys, without establishing a clear and effective connection between the two.

The UCMap experiments that we have gathered include those conducted in the cities of Arnhem (Ren et al. 2012), Lisbon (Alcoforado et al. 2009), Kaohsiung (Ren et al. 2012) and Beijing (He et al. 2015). In the case of LCZ methodology experiments, the experiences that we considered most significant were those of some Chilean cities involved in the Research Projects "Conicyt/Fondecyt 1,130,305 (2013–2016)" and "ANID/Fondecyt initiation 11,180,990", the city of Bogotá (Jiménez-Ariza et al. 2023) and the city of Colombo in Sri Lanka (Perera and Emmanuel 2018). To develop this process, we used the following keys:

- objectives of climate maps;
- climatic and urban planning analyses fielded;
- classification and sizing of climate areas;
- relationship with urban planning.

At the conclusion of the study, the findings were employed to set a potential methodology for the delineation of the contents of a morpho-climatic map. It was proposed that this map could exert influence and effect upon urban planning and urban design within the context of the Vitality Project. Indeed, one of the objectives of this PNRR (National Plan for Recovery and Resilience) initiative is to delineate pivotal design parameters that emphasize the interrelation between the building scale and the "climate-adaptive urban unit scale" (CAU.u). This will be achieved through the application of quantifiable factors to facilitate the development of mitigation and adaptation strategies that enhance the conditions of comfort, safety, energy efficiency and environmental sustainability of indoor and outdoor spaces within the settlement systems of Central Italy.

3 Results

In examining the cities that have employed the UCMap methodology, it became evident that a common objective exists, namely, to gain insight into the existing urban climatic and environmental conditions through the utilization of a range of climatic parameters. This information is subsequently employed and interpreted from an urban planning perspective, with the objective of creating the city's climate map and generating urban planning recommendations. In terms of climate objectives, the aim is to reduce the heat-island effect and improve urban ventilation. For instance, the cities of Lisbon, Beijing and Arnhem prioritize these objectives, whereas Kaohsiung is focused on reducing the effects of the heat island and rising temperatures in general. A comprehensive cognitive framework is initially established in order to construct the UC-An Map. This includes the collection of meteorological data (such as average temperatures, precipitation, wind speed and solar radiation), as well as data on land use, topography and vegetated areas. Subsequently, the second phase combines the aforementioned information levels to understand their effects on the energy and heat balance at the meso- and/or micro-climatic level, depending on the scale of interest. Additionally, other information concern urban morphology parameters, such as building density, which affects the roughness of the urban soil, which in turn conditions ventilation and the built environment's capacity to absorb heat; the percentage of land cover; proximity to open spaces, etc. In the case of Arnhem, the analyses covered a number of key areas, including land use, the construction of the digital terrain model, basic meteorological data with particular reference to wind conditions, and the role of urban morphology in relation to thermal aspects. In the case of Lisbon, the cognitive analyses were oriented towards the construction of a map of the different built densities (high, medium, low, and very low) and a ventilation map. Subsequently, a co-occurrence matrix was constructed between the various land cover/urban density/ventilation classes with the objective of identifying climate-response units (HCR). In the case of Beijing, the cognitive apparatus is based on the recognition of the "thermal load map," which expresses the intensity of heat stored or emitted by urban areas. This parameter is contingent upon the topographic height, the sky factor, the presence of green space and water. The presence of these components is then entered into a geographic information system (GIS) and subjected to assessment. A comparable methodology is employed in the construction of the ventilation potential map, whereby a multitude of fundamental layers are integrated, encompassing building height, roughness length, building coverage, and woodland terrain. A comparable climate assessment was conducted in Kaohsiung. In this instance, the cognitive analyses encompassed the collation of meteorological data, the examination of land use and planning, the analysis of population data, the assessment of UHI intensity, the investigation of natural landscapes, and the evaluation of water bodies.

A climate assessment matrix was developed based on the impacts of these topics on thermal load (TL) and dynamic potential (DP). This matrix reports the effects on thermal load, dynamic potential, wind, etc. In the case of Arnehm, the areas covered by the climate assessments included: transportation areas (railway, road, airport); built-up areas (residential, commercial and catering, public space, socio-cultural facilities, building site); semi-built-up areas (landfill, cemetery, building site and others); leisure (park, sports ground, holiday resort); agriculture; forest and natural areas; water bodies (Rhine-Meuse river, recreational water areas, alluvial areas, wet field water channel). In the case of Lisbon, the areas subject to climate assessment are defined by their physical morphologies, including hilltops, the northern plateau, the southern plateau, the main valley beds, the Tagus bank, green areas, and so forth. Regarding Beijing, the urban areas were classified into seven categories, with the prevailing environmental and settlement characteristics corresponding to the climatic features. In the case of Kaohsiung, three areas of the city were identified, corresponding to districts for which the principal environmental characteristics and climatic risks are summarized.

With regard to the construction of the UC-AnMap, the city of Arnhem has identified four distinct climate zones, which correspond to built-up areas, green spaces, water, and transportation infrastructure. The potential for thermal load (TL) and dynamic potential (DP) have been identified for these areas. In the case of Lisbon, the identification of HCR units is based on delineating seven distinct areas, which are characterized by specific topographical features, population density, vegetation cover, and climatic characteristics. In the case of Beijing, a platform of climatic information for planning was created with the objective of assisting decision-making. The five "Planning Zones" are distinguished as follows: Zones of High Climate Value (with respect to fresh air ventilation paths); Zones of Climate Value (green wedge effects); Moderately Climate Climate-Sensitive Zones (densely built-up areas with a high heat load and low ventilation); and Highly Climate-Sensitive Zones (densely built-up areas with a very high heat load and extremely low ventilation). In conclusion, the recommendations for urban planning in the case of Arnhem present a summary of the main climatic conditions and the major problems and actions to be promoted for each area type. To illustrate, the area type designated as "High thermal load and very low dynamic potential" is indicative of commercial and residential zones. In this instance, the primary issue identified is the urban heat island effect. To address this, the proposed measures include the creation of an air path to facilitate the passage of cool breezes through the city centre, as well as the control of further urban expansion.

In the case of Lisbon, the recommendations for urban planning with reference to the different HCr Units are particularly detailed. To illustrate, the HCr Unit "High-Density Urban Areas of the Southern Plateau," which encompasses the concentrated urban fabric of the historic centre and a significant portion of the city constructed before the 1950s, is guided by the following primary urban planning principles: (1) it is recommended that an increase in built density in the valley bottoms be avoided without a detailed study of the consequences on the ventilation of the valley; (2) it is recommended that the H/W ratio (i.e., the ratio between the height (H) of buildings and the width (W) of streets) be kept below 1 in cases where urban development is currently taking place; (3) it is recommended that vegetated areas, including roof gardens, be maximized; (4) it is recommended that light colours and materials with low thermal admittance be selected in the case of building renovation. A similar level of detail can be observed in the case of Beijing, with reference to the different planning areas. In Kaohsiung, three levels of planning actions were developed: city, urban, and neighbourhood scales. For each urban district, potential control measures were devised and summarized according to criteria such as greenery, shading, albedo, anthropogenic heat release, air exchange, and air pollution. The recommended course of action comprises three distinct categories: improvement actions, mitigation actions, and protection actions.

In the case of the LCZ methodology, the cases examined concern the case study of 21 Chilean cities involved in the Research Project "Conicyt/Fondecyt 1,130,305 (2013–2016) and ANID/Fondecyt initiation 11,180,990 project", as well as the city of Bogotá and the city of Colombo (Sri Lanka). In the case of Chilean cities, the primary objective was to mitigate the intensification of extreme events, including daytime heat associated with increased heat waves, nighttime heat, and tropical nights. Local climate zones (LCZs), delineated by urban design variables such as building density and height, were correlated with other urban attributes influencing climate, including surface temperature and vegetation. The recognition of LCZs was conducted through the photointerpretation of images accessible in Google Earth Pro software (2021), in accordance with the recognition criteria and utilization of colours proposed by Wudapt. Subsequently, an investigation was conducted into the correlation between surface temperature and vegetation within the designated LCZs, utilizing the Normalized Vegetation Difference Index (NDVI). LCZ3 is the most prevalent local climate zone in urban areas worldwide. It is typified by residential urban areas comprising detached houses with one to three storeys, constituting between 20 and 45% of the urban area in different cities. This LCZ was subsequently selected to investigate the correlation between temperature and vegetation activity in six cities situated at varying latitudes. This is due to the fact that it is one of the typologies that is least affected by the presence of shadows cast by buildings. To enhance the climatic conditions of urban areas, a series of urban planning measures are currently being developed. These measures aim to modify the climatic behaviour of cities by modifying the LCZ categories to which they belong. In this regard, a number of factors have been identified which urban planning tools should address. These include the height of buildings, which is associated with the proportion of visible sky and, therefore, with the amount of sunlight and shadows that affect a given area over a year; the building coefficient, which influences the density of buildings; the distance between adjacent buildings, which determines the amount of available space and is therefore associated with sunlight and shade; and the role that vegetation plays in reducing temperatures. Instead of this, the recommendations for urban design at the detailed scale are organized in a table based on three principal parameters: colours, materials and density. The potential impacts of these three parameters are identified, and recommendations are made to avoid the most harmful effects.

In the case of Bogotá, the necessity to address the issue of heavy rainfall and urban water drainage has prompted the utilization of LCZs and LCZSs (cell subdivisions of 20 × 20 m). This approach enables the assessment of the requirements and potential for the provision of ecosystem services across the city's diverse locations, with the aim of identifying the most suitable NBSs for water runoff management. The territory is classified into six LCZs and 18 LCSZs based on eight variables with a significant relationship with temperature. These variables include altitude, residential use, tree canopy, NDVI, building density, water table, distance to main roads and distance to transport stations. For each subdivision of the LCZs (which were divided into three cells based on the amount of annual rainfall), recommendations are defined in accordance with the potential benefits and feasibility of implementation. Firstly, priority and opportunity scores are defined for each NBS type in each cell, considering the NBS type's potential to contribute to ecosystem services. The analysis provides a qualitative score on a scale of zero to three, based on an assessment of the feasibility of the results. Subsequently, the NbS are filtered by identifying potential impediments to their implementation in the same cell and selecting the types with a higher score if area requirements or similar characteristics and functions might limit simultaneous use. This process yields a list of NbS for each cell, which assists in identifying stormwater management processes (i.e., detention, retention, transport and infiltration) by considering the characteristics and information provided in the literature (Woods-Ballard et al. 2016). In the case of Colombo, Sri Lanka, the approach used involved matching LCZs to urban blocks as the area of survey and climate assessment. Each block was mapped and merged with neighbouring blocks exhibiting similar values for sky view factor (SVF), aspect ratio (H/W), building surface fraction (BSF), impervious surface fraction (ISF), pervious surface fraction (PSF), height of roughness elements (HRE), terrain roughness class, surface admittance, albedo and anthropogenic heat flux. Subsequently, each isolate was assigned the corresponding LCZ and, in some cases, a subclassification was also required. A preliminary subclassification of 46 LCZ combinations, including primary LCZs, was conducted. Subsequently, to avoid excessive fragmentation, 31 subclasses were employed, comprising 16 primary and 15 subclasses. The simplified mapping of the land use-based urban system, with some consideration of urban activities, allows for a more straightforward interpretation of the climate. It is proposed that LCZs should be designated as a fundamental unit for urban planning. This is aimed at promoting sustainable urban development and preventing the occurrence of intensive urban development under urban planning instruments that encourage land use (Table 1).

4 Discussion

The principal objective of the intervention strategies and urban planning recommendations outlined in a climate map is to facilitate a transformation in urban settlement characteristics or the preservation of specific territorial conditions that can safeguard them from the catastrophic consequences of climate change. Both described methodologies aim to create climate maps, but they have different practical outcomes for urban planning. UCMaps provides climate information at the regional or urban scale for planners and policymakers. This is achieved by grouping the data according to the scale of different climate zones, which are called

Table 1 Summary comparison between UCMap and LCZ methodologies

	Urban climate map (UCMap)	Local climate zones (LCZ)
Definition	A cartographic representation of climate information, integrating climate assessments into urban planning	A classification system representing urban morphology and climate characteristics using 17 standardized classes
Components	Consists of Urban Climate Analysis Map (UC-AnMap) and Urban Planning Recommendations Map (UC-ReMap)	Composed of 17 classes, including built and natural typologies, based on building forms and land cover
Scale	Operates on regional or urban scales, focusing on broader climate zones called "climatopes"	Functions at city and neighbourhood levels, allowing detailed local climate representation
Data requirements	Requires detailed meteorological, planning, land use, topographical, and vegetation data	Uses generalized data on building forms and land cover
Methodology	Utilizes mesoscale and microscale weather simulation models or empirical data from field measurements	Employs standardized classification criteria, often using satellite data and GIS techniques
Current use and potential for innovation	• Used for creating climate maps and generating urban planning recommendations to reduce heat islands and improve ventilation • Provides comprehensive climate information for planners and policymakers • Established tradition in urban planning	• Useful for predicting urban fabric transformations and assessing climate adaptation needs • Standardized morphological classification allows comparisons across cities • Easier for non-professionals to understand • Could be enhanced by integrating more parameters for a holistic urban planning approach
Limitations	• Simulation efficiency is limited by calculation accuracy and operational complexity • Empirical data may not fully represent diverse urban morphologies • Primarily serves as an informational base with limited action prioritization	• Limited significant experiences in guiding urban planning • Determination of local unit dimension for traditional urban planning scales • May require a compromise between block-level and larger planning zones for broader impact
Studies and applications	Arnhem (Ren et al. 2012), Lisbon, Kaohsiung (Ren et al. 2012) and Beijing	Chilean cities involved in the Research Projects "Conicyt/Fondecyt 1,130,305 (2013–2016)" and "ANID/Fondecyt initiation 11,180,990", Bogotá (Jiménez-Ariza et al. 2023), Colombo (Perera and Emmanuel 2018)

climatopes. This approach allows for a reduction in thermal load and an increase in the dynamic potential of a particular climate zone. UCMaps have a well-established tradition in urban planning, yet their use provides only an informational basis. The generated maps are regarded as a significant visual instrument for planners to convey information to political leaders and the general public (Ren et al. 2011). The recommendations provided are largely confined to a single set of potential courses of action. A further limitation is that they rarely contemplate a classification of actions based on priority in terms of effectiveness and feasibility relative to the urban context. Conversely, LCZs have yet to be widely employed in urban planning, and their potential for enhancing urban climate resilience is significant. If they could be used to describe the climate performance of homogeneous territorial areas, they could provide additional knowledge and assessment for urban planning, particularly in the context of zoning. The utilization of LCZs offers a distinctive advantage, as exemplified by the city of Colombo. This lies in the capacity to anticipate potential transformations of urban fabrics by elucidating the inherent risks and opportunities associated with diverse urban policies, both extant and prospective (Perera and Emmanuel 2018). Furthermore, as evidenced by the case of Chilean cities, a change could be planned in the more problematic LCZ categories, which would have significant implications for urban planning. In the case of Bogotá, the designation of "planning" units can facilitate the decision-making process for implementing climate adaptation and mitigation actions utilizing NBS (nature-based solutions). These are selected based on the priorities identified in the various LCZs. Furthermore, they can assist in identifying strategic intervention areas and formulating potentially replicable environmental improvement interventions in other contexts (Jiménez-Ariza et al. 2023). The experiences of the LCZs discussed in this article illustrate the importance of selecting appropriate boundaries for the division of areas (Perera and Emmanuel 2018). It can, therefore, be concluded that

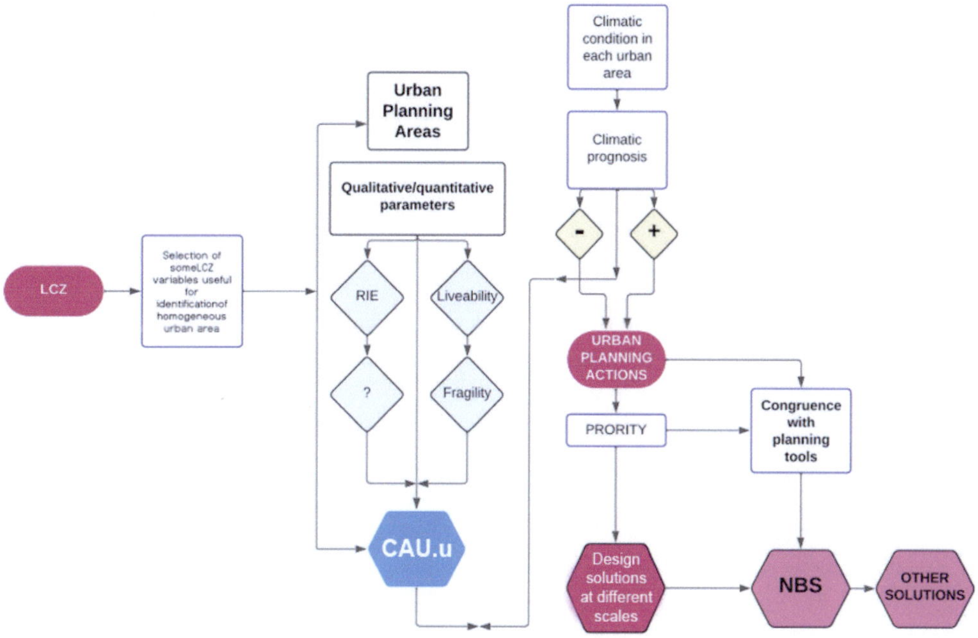

Fig. 1 Proposed methodological flowchart for integration of LCZs as urban planning. (By author)

the LCZ methodology is not particularly effective when attempting to assess the impact of climate change on extensive urban areas. A compromise between the urban and territorial scales and the block scale is likely to be necessary, as evidenced by the cases of Bogotá and Colombo. Furthermore, it would be beneficial for the LCZ methodology to be more closely aligned with the specific political and social contexts at the local level. It is recommended that studies on urban temperature focus on human activities rather than solely on climate classification based on physical properties (Yu et al. 2021).

In light of the findings of this comparison, which is by no means exhaustive, we have defined and proposed a potential further contribution to the debate regarding urban planning and design in the context of the Research Project Vitality. The proposal is to create a morpho-climatic map that integrates the standardization system of the LCZ (Local Climate Zone) map with various social, economic, environmental, and energy factors. This integrated map will help in gaining a comprehensive understanding of urban environments as complex ecosystems. It will enable the development of customized urban planning solutions to address the specific challenges and requirements of different urban contexts as shown in Fig. 1 representing the integration of cAU.u in the urban planning and design process. The LCZs will be clustered and reconfigured as CAU.us (Climate Adaptive Urban Units), and a combination of intricate parameters and indicators will be employed to assess the urban environment, thereby enabling the formulation of bespoke urban planning solutions that can address the specific challenges and meet the distinctive needs of different urban areas. The parameters mentioned above encompass the quality of building structures; the proportion of residential, commercial, and industrial areas, green spaces, urban mobility and road infrastructure parameters, environmental parameters, and energy and building indicators. Conventional data-gathering methods will be integrated with sophisticated technologies such as Geographic Information Systems (GIS), parameterised processes and remote sensing to identify and quantify these parameters within the CAU.us framework. The CAU.us will be grouped, as shown in Fig. 2, according to their analogous behavioural patterns with respect to morpho-climatic characteristics, social and health vulnerability, primary urban functions, vegetation quality, energy efficiency, and other potential parameters. The resulting clusters will constitute discrete urban areas, and priority levels will be assigned on the basis of the identified vulnerabilities, feasibility, and compatibility with the local social and economic conditions. Subsequently, the requisite adaptation measures will be implemented.

Fig. 2 Clusterization process from LCZs to CAU.us in Ascoli Piceno (Italy) historical centre. (By author)

5 Conclusions

Climate maps have the potential to play a pivotal role in urban planning at various stages of the design process. From the experiences examined, UCMaps have been primarily used for recognition and information. However, it is possible that the LCZs, which urban planners currently underuse, could play a central role in designing adaptation actions if they incorporate a series of parameters and indicators that value other components of the urban system, such as functions, social and demographic components, in addition to physical ones. The PNRR Vitality Project can utilise LCZs categorized as CAU.us to direct urban regeneration or restructuring initiatives, thereby enhancing their efficacy. CAU.us can be employed to identify the primary critical issues and, consequently, select a series of beneficial actions to mitigate extreme climate events. It can also be used to ascertain the suitability of specific locations based on their climate, social, and environmental vulnerability. Furthermore, it can serve as a reliable instrument for evaluating the sustainability of land governance strategies. In conclusion, CAU.us has the potential to offer a more comprehensive and effective approach to urban planning. This could help in managing the environmental impact of urban areas, improving energy efficiency, and enhancing the quality of life and well-being of urban inhabitants.

Acknowledge Financial Support From PNRR MUR project ECS_00000041-VITALITY

Author Contributions Conceptualization, R.D. and R.C.G.; Formal analysis, R.D., G.C., R.C.G., F.C., G.E.M., S.M., I.P., M.S.; Methodology, R.D., R.C.G.; Supervision: R.D. and R.C.G., Writing—review and editing, R.D., R.C.G and I.P. All authors have read and agreed to the published version of the manuscript.

References

Abougendia, S.M., Ayad, H.M., El-Sayad, Z.T.: Classification framework of local climate zones using world urban database and access portal tools: case study of Alexandria city, Egypt, pp. 309–322 (2020). https://doi.org/10.2495/sdp200251

Alcoforado, M.-J., Andrade, H., Lopes, A., Vasconcelos, J.: Application of climatic guidelines to urban planning. Landsc. Urban Plan. **90**(1–2), 56–65 (2009). https://doi.org/10.1016/j.landurbplan.2008.10.006

Aslam, A., Rana, I.A.: The use of local climate zones in the urban environment: a systematic review of data sources, methods, and themes. Urban Clim. **42**, 101120 (2022). https://doi.org/10.1016/j.uclim.2022.101120

Carter, J.G., Cavan, G., Connelly, A., Guy, S., Handley, J., Kazmierczak, A.: Climate change and the city: building capacity for urban adaptation. Prog. Plan. **95**, 1–66 (2015). https://doi.org/10.1016/j.progress.2013.08.001

Chen, Y.-C., Lin, T.-P., Lin, C.-T.: A simple approach for the development of urban climatic maps based on the urban characteristics in Tainan, Taiwan. Int. J. Biometeorol. **61**(6), 1029–1041 (2017). https://doi.org/10.1007/s00484-016-1282-0

Chen, Y.-C., Lo, T.-W., Shih, W.-Y., Lin, T.-P.: Interpreting air temperature generated from urban climatic map by urban morphology in Taipei. Theoret. Appl. Climatol. **137**(3–4), 2657–2662 (2019). https://doi.org/10.1007/s00704-018-02764-x

Chien, L.-C., Guo, Y., Zhang, K.: Spatiotemporal analysis of heat and heat wave effects on elderly mortality in Texas, 2006–2011. Sci. Total. Environ. **562**, 845–851 (2016). https://doi.org/10.1016/j.scitotenv.2016.04.042

Demuzere, M., Kittner, J., Bechtel, B.: LCZ generator: a web application to create local climate zone maps. Front. Environ. Sci. **9**, 637455 (2021). https://doi.org/10.3389/fenvs.2021.637455

Eldesoky, A.H., Gil, J., Pont, M.B.: Combining environmental and social dimensions in the typomorphological study of urban resilience to heat stress. Sustain. Cities Soc. **83**, 103971 (2022). https://doi.org/10.1016/j.scs.2022.103971

Eliasson, I.: The use of climate knowledge in urban planning. Landsc. Urban Plan. **48**(1–2), 31–44 (2000). https://doi.org/10.1016/S0169-2046(00)00034-7

He, X., Shen, S., Miao, S., Dou, J., Zhang, Y.: Quantitative detection of urban climate resources and the establishment of an urban climate map (UCMap) system in Beijing. Build. Environ. **92**, 668–678 (2015). https://doi.org/10.1016/j.buildenv.2015.05.044

Hurlimann, A., Moosavi, S., Browne, G.R.: Urban planning policy must do more to integrate climate change adaptation and mitigation actions. Land Use Policy **101**, 105188 (2021). https://doi.org/10.1016/j.landusepol.2020.105188

Jiménez-Ariza, S.L., Rey, C.V., Rodríguez, J.P., Guzmán-Ramírez, M.: Multi-criteria decision analysis inputs for planning the implementation of nature-based solutions in urban contexts. ACE Architect. City Environ. **18**(52) (2023). https://doi.org/10.5821/ace.18.52.11871

Jin, L., Pan, X., Liu, L., Liu, L., Liu, J., Gao, Y.: Block-based local climate zone approach to urban climate maps using the UDC model. Build. Environ. **186**, 107334 (2020). https://doi.org/10.1016/j.buildenv.2020.107334

Lau, T.-K., Chen, Y.-C., Lin, T.-P.: Application of local climate zones combined with machine learning to predict the impact of urban structure patterns on thermal environment. Urban Clim. **52**, 101731 (2023). https://doi.org/10.1016/j.uclim.2023.101731

Lelovics, E., Unger, J., Gál, T., Gál, C.: Design of an urban monitoring network based on Local Climate Zone mapping and temperature pattern modelling. Clim. Res. **60**(1), 51–62 (2014). https://doi.org/10.3354/cr01220

Masson, V., Hidalgo, J., Amossé, A., Bélaïd, F., Bocher, E., Bonhomme, M., Bourgeois, A., Bretagne, G., Caillerez, S., Cordeau, E., Demazeux, C., Faraut, S., Gallato, C., Haouès-Jouve, S., Lambert, M.-L., Lemonsu, A., Lestringant, R., Lévy, J.-P., Long, N., Vye, D.: Urban climate, human behavior & energy consumption: From LCZ mapping to simulation and urban planning (the MapUCE project) (2015)

Perera, N.G.R., Emmanuel, R.: A "Local Climate Zone" based approach to urban planning in Colombo, Sri Lanka. Urban Clim. **23**, 188–203 (2018). https://doi.org/10.1016/j.uclim.2016.11.006

Quan, S.J., Bansal, P.: A systematic review of GIS-based local climate zone mapping studies. Build. Environ. **196**, 107791 (2021). https://doi.org/10.1016/j.buildenv.2021.107791

Ren, C., Ng, E.Y., Katzschner, L.: Urban climatic map studies: a review. Int. J. Climatol. **31**(15), 2213–2233 (2011). https://doi.org/10.1002/joc.2237

Ren, C., Spit, T., Lenzholzer, S., Yim, H.L.S., Heusinkveld, B., Van Hove, B., Chen, L., Kupski, S., Burghardt, R., Katzschner, L.: Urban climate map system for Dutch spatial planning. Int. J. Appl. Earth Obs. Geoinf. **18**, 207–221 (2012). https://doi.org/10.1016/j.jag.2012.01.026

Ren, C., Lau, K.L., Yiu, K.P., Ng, E.: The application of urban climatic mapping to the urban planning of high-density cities: the case of Kaohsiung, Taiwan. Cities **31**, 1–16 (2013). https://doi.org/10.1016/j.cities.2012.12.005

Ren, C., Cai, M., Wang, R., Xu, Y., Ng, E.: Local Climate Zone (LCZ) Classification Using the World Urban Database and Access Portal Tools (WUDAPT) Method: A Case Study in Wuhan and Hangzhou (2016)

Stewart, I.D., Oke, T.: A new classification system for urban climate sites. Bull. Am. Meteor. Soc. **90**, 922–923 (2009)

Stewart, I.D., Oke, T.R.: Local climate zones for urban temperature studies. Bull. Am. Meteor. Soc. **93**(12), 1879–1900 (2012). https://doi.org/10.1175/BAMS-D-11-00019.1

Woods-Ballard, B., Kellagher, R., Martin, P., Jeffries, C., Bray, B., Shaffer, P.: The SuDs Manual. CIRIA (2016)

Yin, S., Ren, C., Zhang, X., Hidalgo, J., Schoetter, R., Kwok, Y.T., Lau, K.K.-L.: Potential of synthetizing climatopes and local climate zones for urban climatic planning recommendations: a case study in Toulouse, France. Cybergeo (2022). https://doi.org/10.4000/cybergeo.39417

Yu, Z., Jing, Y., Yang, G., Sun, R.: A new urban functional zone-based climate zoning system for urban temperature study. Remote Sens. **13**(2), 251 (2021). https://doi.org/10.3390/rs13020251

Zheng, Y., Ren, C., Xu, Y., Wang, R., Ho, J., Lau, K., Ng, E.: GIS-based mapping of Local Climate Zone in the high-density city of Hong Kong. Urban Clim. **24**, 419–448 (2018). https://doi.org/10.1016/j.uclim.2017.05.008

Zhongli, L., Hanqiu, X.: A study of urban heat island intensity based on "local climate zones": a case study in Fuzhou, China. In: 2016 4th International Workshop on Earth Observation and Remote Sensing Applications (EORSA), pp. 250–254 (2016). https://doi.org/10.1109/EORSA.2016.7552807

Zhou, X., Yamamoto, M., Yan, S., Ishida, Y., Cai, M., Ji, Q., Makvandi, M., Li, C.: Exploring the impacts of heat release of vehicles on urban heat mitigation in Sendai, Japan using WRF model integrated with urban LCZ. Sustain. Cities Soc. **82**, 103922 (2022). https://doi.org/10.1016/j.scs.2022.103922

Climate-Related Changes of Energy Flows in the Major Cities of Europe

Marina A. Arshinova, Tatiana V. Komarova, and Svetlana N. Mikheeva

Abstract

Cities that accumulate more than 55% of the world's population always were and still are the centers of the most powerful transformation of the natural environment. They change natural and produce artificial matter and energy flows including those of water, food, wastes, etc. Our study is based on the data for the largest cities of the European region with the population numbers over 1 mln inhabitants. First of all we evaluated human-related flow of energy against the natural flow represented by the radiation balance. The indicator of anthropogenic energy flow was the consumption of energy by urban dwellers. Correlation of radiation balance and energy consumption allows identifying the groups of major European cities according to the degree of the natural energy flow transformation. The anthropogenic energy load within the urban territories of Europe averages 30–40% of the natural radiation balance being even higher in the largest cities, such as London or Moscow. We also analyzed climate changes within urban areas against the rural background. The rise of annual temperatures in European cities is extremely diversified. The maximum changes are in the cities of Denmark, southern Finland and Romania. If compared with temperature changes over the rural areas the most pronounced correlation is also for Denmark and Finland. Despite certain fluctuations the general trend of urban air temperature increase is obvious since the 1950s.

Keywords

Energy flows · Natural radiation budget · Anthropogenic energy load · Temperature rise · Urban/rural correlation

M. A. Arshinova (✉) · T. V. Komarova · S. N. Mikheeva
Lomonosov Moscow State University, Moscow, Russia
e-mail: amari_geo@mail.ru

1 Introduction

Cities accumulate more than 55% of the world's population. They occupy less than 2.5% of the Earth's land area, but despite rather small territorial dimension urban geosystems are particularly complicated in terms of their structure, functions and socio-economic features. Moreover, particular manifestations of their common features vary according to historical, social, cultural, and other city-forming factors (Conke and Ferreira 2015).

Cities always were and still are the centers of the most powerful transformation of natural environment. They are areas of specific anthropogenic and natural-anthropogenic processes (Kurbatova et al 2004; Romanova and Arshinova. 2021; Bai 2016; Yanitsky 2013). They change natural and produce artificial matter and energy flows including those of water, food, wastes, etc. (Pincetl et al. 2012; Shahrokni et al. 2015).

Based on their origin and the degree of technogenic impact, natural-anthropogenic processes in cities could be classified into two groups. The first one includes variously transformed natural-anthropogenic processes that already existed in indigenous landscapes, for example, climatogenic processes. The second group includes autonomous processes, i.e., material and energy flows that have never existed in the natural environment, such as concentration of food or solid wastes within a limited area of a city or emission of greenhouse gases. The parameters of such processes depend on the number of residents, the area of a city and the main functions of an urban geosystem.

Increasing urbanization of human societies combined with intense energy demands of modern economies requires the integrated study of energy and urban systems to quantify energy consumption and use patterns in urban environments. The aim of our study is to analyze energy flows in the largest cities of the European region in order to reveal the scale of their

P. Mohammad et al. (eds.), *Geographic Approaches to Climate Change and Mitigation: Urban and Rural Perspectives (Volume 1)*,
Advances in Science, Technology & Innovation, https://doi.org/10.1007/978-3-031-92119-3_6

transformation as compared to energy parameters of the background natural landscapes, and the intensity of their current transformation against the rural background. The study is based on the data for the European cities with population numbers over 1 mln inhabitants.

There are 33 such cities in Europe, 11 of them situated in the European part of Russia. Apart from Moscow (13.1 mln), London (9.6 mln) and St. Petersburg (5.6 mln) the population numbers are from 1 to 3 mln (World Urbanization Prospects 2019). The population of an average "millionaire" city in the European Russia doesn't exceed 1.3 mln inhabitants. An average land area of a "millionaire" city in Europe is about 667 km^2 varying from 101 km^2 (Barcelona) to 2562 km^2 (Moscow). The largest cities in terms of population numbers do not necessarily have vast land areas, therefore urban population densities and population pressure on the urban territories could be particularly high.

The majority of cities under study are national capitals (15) or regional capitals (9, mainly in Russia). Others are large industrial and/or transportation centers. Usually a European country has only one city with the population exceeding 1 mln inhabitants. However, there are several largest cities in Germany (4), Italy (2), Spain (2), Ukraine (3) and United Kingdom (2). As for the geographical location the cities are predominately within the temperate climatic zone, only 4 of them have subtropical (Mediterranean) type of climate. Millionaire cities of the European Russia are dispersed from the boreal forest to the steppe natural zones.

2 Materials and Methods

The energy flows in urban geosystems were analyzed in two aspects. First of all we evaluated human-related flow of energy against the natural energy flow. The latter was accounted by the values of the annual radiation balance (kcal per 1 cm^2) interpolated from the corresponding map (Resources and Environment 1998). The anthropogenic energy flow was characterized by the consumption of energy by urban dwellers (Gosudarstvennyj doklad 2018; Energy use). The initial data on energy consumption (kg of oil equivalent, kg o.e.) was recalculated into the comparable units (kcal) and the total amount of energy consumption in a city was calculated using the corresponding population numbers (see below the calculation algorithm for Moscow as an example). Then we estimated the anthropogenic energy load over the land area of a city (kcal per 1 cm^2) and compared these values with the natural radiation balance.

Calculation Algorithm

Moscow:

Population—13,104,177 inhabitants.

City area—2562 km$^{2.}$

Energy consumption—3460 kg o.e. per capita per year (1 kg o.e. = 10002.4 kcal).

- 3460 kg o.e. = 34.608 mln kcal (per capita per year)
- Total energy consumption = 34.608 mln kcal • 13.104177 mln inh. = 453404524.2 mln kcal per year
  (*total anthropogenic energy load*)
- *Specific anthropogenic energy load* = 453404524.2 mln kcal per year: 2562 km^2 = 176972.9 mln kcal per 1 km^2 per year = 17.7 kcal per 1 cm^2 per year
- *Natural radiation budget*—28 kcal per 1 cm^2 per year
- *Ratio* 17.7: 28 = 0.63 = 63%.

Secondly, we analyzed climate changes within urban areas against the rural background. There are a lot of graphs and cartographic interpretations of data on the temperature changes in Europe. For our analysis we used a schematic map showing the trend of urban air temperature rise during the last decades (Europe is getting warmer… 2018). It was produced at the European Center for Medium-Range Weather Forecasts (ECMWF) based on the reanalysis of over 100 million data points. We compared the temperature rise in the cities with that of their rural surroundings using maps from the EEA database (Trends in annual temperature 2020). Another source of data about temperature changes in the cities of Europe was the climatic reference databases for several time periods, namely 1930–1960 (Klimaty Zapadnoj Evropy 1979), 1960–1990 (Klimadiagramme weltweit), and the recent decade (Climate analyses (year)).

3 Results and Discussion

The natural energy flow within the territory of Europe varies significantly in line with the solar radiation supply. The radiation budget gradually increases from nearly 20 kcal/cm^2 per year in the northern parts of Scandinavia to 70 kcal/cm^2 per year in the southernmost areas of the Iberian Peninsula. As for the largest cities of Europe the radiation budget of their so-called "embracing" landscapes is nearly 60 kcal/cm^2 per year in Madrid, Sofia and Barcelona, and lies within the range of 30–40 kcal/cm^2 per year in most cities of Western and Central Europe. The highest radiation budget values within the European part of Russia are for the cities of Krasnodar (50 kcal/cm^2 per year), Volgograd and Rostov-on-Don (42 kcal/cm^2 per year). The majority of the Russian "millionaire" cities under study are characterized by the radiation budget exceeding 25 kcal/cm^2 per year.

The per capita energy consumption in European cities is practically the same as in the rural areas, due to almost

Table 1 Regional features of energy flows in the major European cities

Region	Average energy consumption, kg o.e. per capita per year	Average anthropogenic energy load, kcal/cm^2 per year	Average percentage of the natural radiation budget, %
Western Europe (8 cities)	3450	23.5	42
Central and Eastern Europe (10 cities)	2464	10.2	23
Southern Europe (4 cities)	2526	24.0	43
European Russia (11 cities)	4417	10.6	33
Europe (33 cities)	3361	15.2	33

similar lifestyles of urban and rural dwellers. This is also true for most regions of the European part of Russia. However the actual values of per capita energy consumption show considerable differences resulting from both natural conditions and the efficient power management. The average energy consumption is about 2500 kg o.e. in the cities of Central and Southern Europe and nearly 3500 kg o.e. for the Western Europe cities (Table 1). The average energy consumption in the largest cities of the European Russia is nearly 4500 kg o.e. Russian cities are also the absolute leaders: the northernmost city of Perm (7675 kcal/cm^2 per year) and the Volga region cities of Samara, Ufa, Kazan, etc. (4500–6000 kcal/cm^2 per year).

The anthropogenic energy load over the urban territories depends on several factors, such as the total energy consumption within the urban area, the spatial size of particular city, as well as the overall level of energy supply. The results of our calculations reveal several trends in the spatial pattern of the indicator. The largest cities, i.e., Moscow, London, Madrid and Berlin, are among the leaders because of their high population numbers, while smaller cities, i.e., Barcelona, Birmingham or Milan, with rather small urban areas are also characterized by higher values of specific anthropogenic energy load. Average values of the indicator are practically the same for the cities of Central Europe (10.2 kcal/cm^2 per year) and the European part of Russia (10.6 kcal/cm^2 per year) (see Table 1). Average values for the cities of Western (23.5 kcal/cm^2 per year) and Southern (24 kcal/cm^2 per year) Europe are more than two times higher than those in the eastern part of the European region.

Correlation of the radiation budget and the energy consumption allows identifying the groups of major European cities according to the degree of the natural energy flow transformation (Fig. 1). Minimum anthropogenic energy load as related to the natural radiation budget is characteristic of several cities, from Rome to Volgograd and Voronezh, mainly in the south-east of the region where the radiation budget values are relatively high. The majority of cities under study (16) produce anthropogenic energy flows which equal 20–40% of the natural radiation budget. And there are two clusters of cities with high anthropogenic energy load in relation to the radiation budget, i.e., in the western part of the region (including London, Birmingham, Paris, Madrid and Barcelona) and in the north-eastern belt (Minsk, Moscow, N.Novgorod and Samara). The high percentage (above 40 and even 50%) is due to generally lowering values of the radiation balance, as well as rather small areas of such cities as Barcelona and Birmingham. On the average the anthropogenic energy load is 23% of the natural radiation budget in 10 cities of Central and Eastern Europe (see Table 1). It averages 42–43% in the cities of Western and Southern Europe, resulting from the higher energy consumption related to a city area. The anthropogenic energy load in 11 cities of the European part of Russia averages 33% of their natural radiation budgets.

Under climate change of recent decades the urban air temperatures are gradually rising in almost all cities of Europe. According to the reanalysis data provided by the European Center for Medium-Range Weather Forecasts (ECMWF) (Europe is getting warmer… 2018) the maximum changes are in the cities of Denmark (about 1.3 °C), southern Finland and Romania (up to 1.5 °C). The urban air temperature rise of 1.2–1.5 °C is also reported for Central and South-Eastern Europe and the central and southern regions of the Iberian Peninsula. The minimum air temperature increase is in the cities of the British Isles and the Southern Italy (less than 0.5 °C).

In order to trace the trend of the urban air temperature changes we analyzed climate data for several time periods, i.e., 1930–1960, 1960–1990, and 2010–2020, using a number of climatic reference databases. The rise of annual air temperatures in the largest European cities is significantly diversified (Fig. 2). The minimum rise since the middle of the 20th century is characteristic of Birmingham (0.6 °C), Munich (0.8 °C), and Cologne (0.9 °C). During the same time period the air temperature increased by 2.1 °C in Belgrade and even by 2.9 °C in Vienna. And it is only in Prague and Rome that the annual air temperatures remain practically the same over the recent decades.

The increasing tendency in urban air temperatures is even more pronounced for several cities if we compare data on the annual air temperatures for 1960–1990 and 2010–2020 time periods. The difference is 1.5 °C in Sofia and Prague (i.e., about 0.5 °C/decade), 2.2 °C for London (0.7 °C/decade),

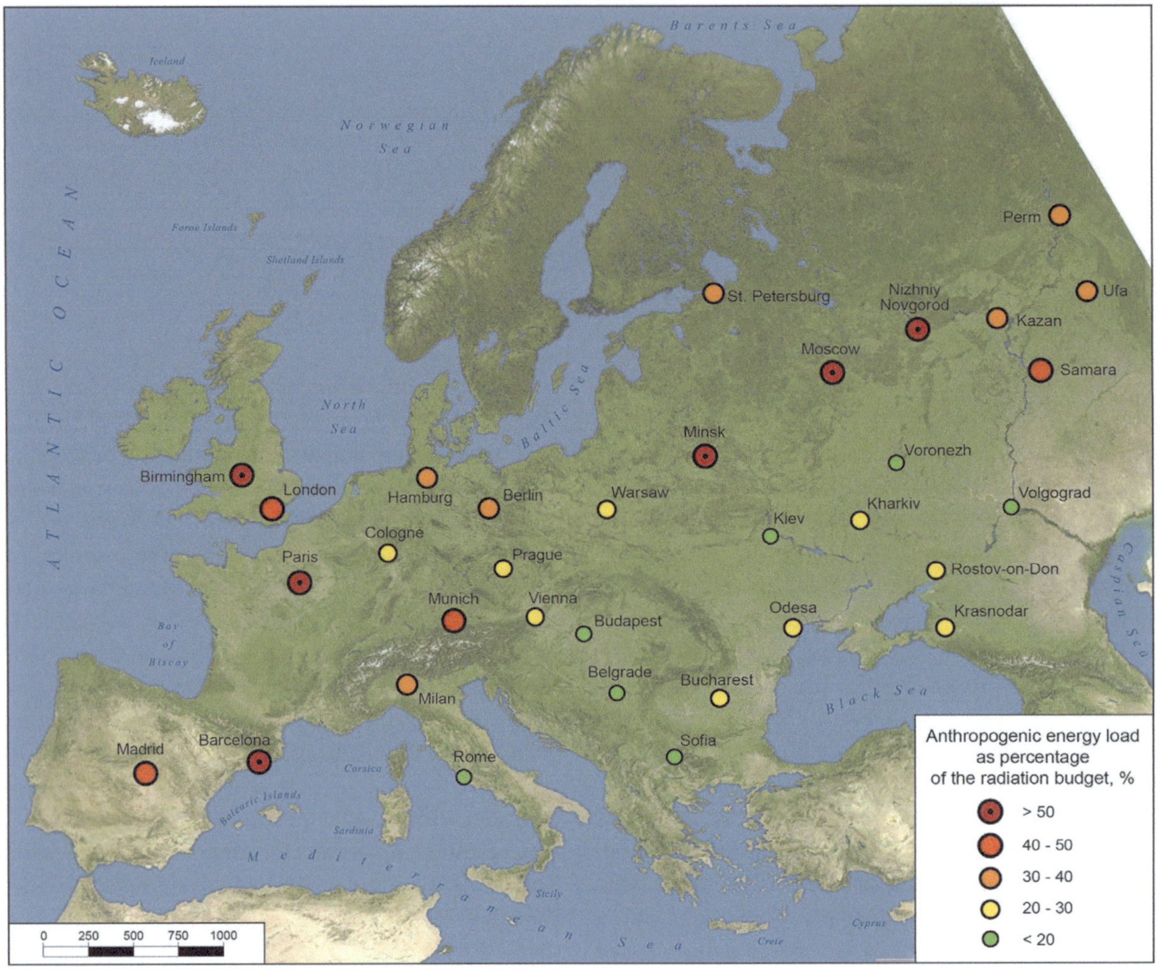

Fig. 1 Anthropogenic energy load versus natural radiation budget in the largest cities of Europe (compiled by the authors)

and even 3.3 °C for Milan (1.1 °C/decade). This is obviously the result of combining the general climate warming and the growing UHI effect under the increasing anthropogenic energy load over the urban territories.

The results of the ECMWF reanalysis are somehow lower than those of direct observations at meteorological stations. The center attributed this to the effect of the Urban Heat Island stating that "absolute values might differ from data collected at weather stations directly, but the overall trends are the same" (Europe is getting warmer… 2018). It means that despite certain fluctuations the general trend of air temperature increase in the major cities of Europe since the 1950s is obvious both from the reanalysis data and the comparison of direct observation results.

Information on the overall changes of annual air temperature over Europe between 1960 and 2017 is provided by the EEA (Trends in annual temperature 2020). The spatial pattern of changes to a certain extent differs from the urban air temperature rise. The areas of the most pronounced temperature rise are Southern France (up to 0.4 °C/decade), Southern Scandinavia and Central and Southern Finland (0.35 °C/decade). The lowest temperature rise is characteristic of Central Balkans and the Southern Italy.

Compared to the rural environs urban areas are always characterized with locally increased air temperatures. The effect is known as Urban Heat Island (UHI) (Susca 2020). The most recent UHI values for several major cities of Europe (Manoli et al. 2019) are in between 1 °C in Prague and 2.0–2.2 °C in Paris, Rome and Sofia. Average annual air temperatures in Bucharest, Budapest, Berlin and Munich are 1.66–1.76 °C higher than over the surrounding countryside. The UHI values in Birmingham and Hamburg are about 1.4 °C. The summertime urban–rural difference in air temperatures is even higher, i.e., up to 3.0 °C.

Among the factors influencing the UHI spatial distribution there are urban climate, urban distribution characteristics, urban spatial distribution, vegetation, and urban thermal environment (Jiang et al. 2022). Another model links population,

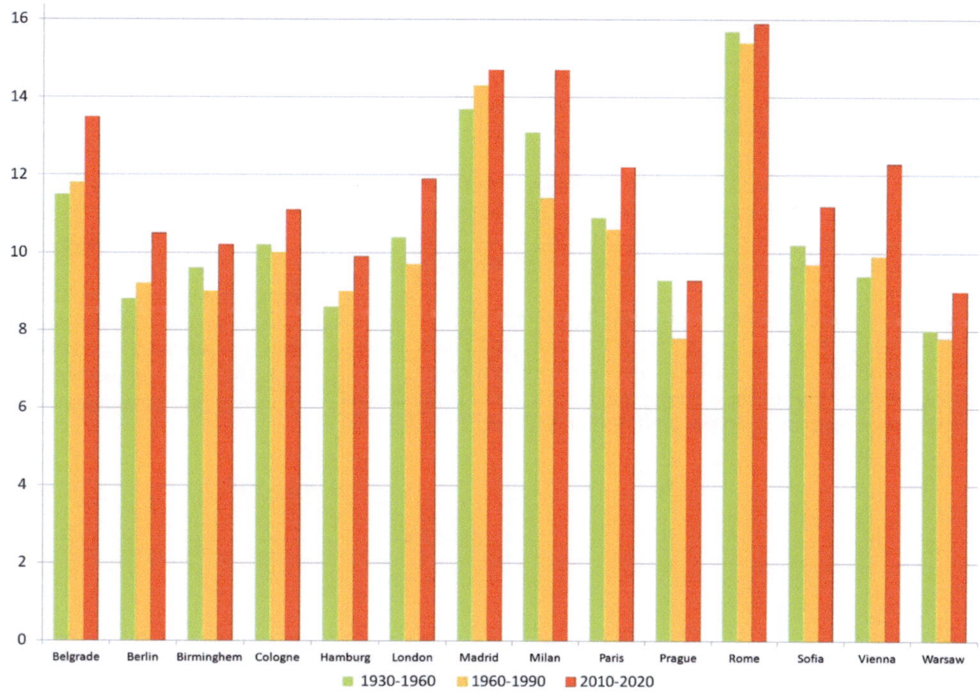

Fig. 2 Mean annual air temperatures in several cities of the European region (°C) (compiled by the authors)

background climate, and UHI intensity concluding that mitigation strategies aimed at increasing green cover and albedo are more efficient in dry regions (Manoli et al. 2019).

If we compare urban air temperature changes with those over the rural areas of Europe the most pronounced correlation is for Denmark and Finland, as well as for Southern Italy. Our calculations suggest that the temperature rise for urban territories, especially for the largest European cities, is often higher than over the adjacent rural areas.

4 Conclusions

The major cities of the European region provide local but powerful impact on their "embracing" landscapes by changing natural material and energy flows. It was found that specific anthropogenic energy load over urban territories is the highest in the cities with the largest population numbers, as well as in the cities with rather small urban areas. The highest average anthropogenic energy load is typical for the cities of Western and Southern Europe. Correlation of specific anthropogenic energy load against the natural radiation budget made it possible to evaluate the degree of energy flow transformation within urban territories. Minimum transformation is characteristic of the cities located in the south-eastern part of the European region, while the maximum one is typical for the westernmost cities, as well as for the cities located in the central part of the European Russia. The general trend of increasing urban air temperatures since the middle of the twentieth century was revealed by comparison of climatic data for particular time periods. Further research is supposed to concentrate on studying the diversity of anthropogenic energy load spatial distribution and the UHI influencing factors in relation to natural (zonal) and socio-economic features of urban territories.

References

Bai, X.: Eight energy and material flow characteristics of urban ecosystems. Ambio **45**, 819–830 (2016)

Conke, L.S., Ferreira, T.L.: Urban metabolism: measuring the city's contribution to sustainable development. Environ. Pollut. **202**, 146–152 (2015)

Gosudarstvennyj doklad o sostoyanii energosberezheniya i povyshenii energeticheskoj effektivnosti v 2017 godu (2018) [State report on power saving and increasing of energy efficiency in 2017] Minenergo RF, Moscow (in Russian)

Jiang, J., Zhou, Y., Guo, X., Qu, T.: Calculation and expression of the urban heat island indices based on GeoSOT grid. Sustainability **14**(5), 2588 (2022). https://doi.org/10.3390/su14052588

Klimaty Zapadnoj Evropy [Climates of Western Europe] (1979) Leningrad, Gidrometeoizdat Publ. (In Russian)

Kurbatova, A.S., Bashkin, V.N., Myagkov, M.S., Savin, D.S.: Ekologicheskie resheniya v Moskovskom megapolise [Environmental solutions in the Moscow metropolitan area] Smolensk, Madzhenta Publ. (in Russian) (2004)

Manoli, G., Fatichi, S., Schläpfer, M., et al.: Magnitude of urban heat islands largely explained by climate and population. Nature **573**, 55–60 (2019). https://doi.org/10.1038/s41586-019-1512-9

Pincetl, S., Bunje, P., Holmes, T.: An expanded urban metabolism method: toward a systems approach for assessing urban energy processes and causes. Landsc. Urban Plan. **107**, 193–202 (2012)

Resources and Environment.: World Atlas. I. Plates 47–49. Ed. Holzel, Vienna (1998)

Romanova, E.P., Arshinova, M.A.: Geoekologicheskaya sistematika i klassifikatsiya gorodskih geosistem na osnove analiza materialnyh i energeticheskih potokov [Geoecological systematization and classification of urban geosystems based on the analysis of material and energy flows] *Vestnik Mosk. Un-ta Ser.5. Geografiya*, 6: 3–12 (In Russian) (2021)

Shahrokni, H., Lazarevic, D., Brandt, N.: Smart urban metabolism: towards a real-time understanding of the energy and material flows of a city and its citizens. J. Urban Technol. **22**, 65–86 (2015)

Susca, T.: Climate change and European cities. Eur. J. Clim. Ch. **02**(01), 01–02 (2020)

World Urbanization Prospects: The 2018 Revision (ST/ESA/SER.A/420).: United Nations, Department of Economic and Social Affairs, Population Division. United Nations, New York (2019)

Yanitsky, O.N.: Metabolicheskaya kontseptsiya sovremennogo goroda [Metabolic concept of modern city]. Sociol. Sci. Soc. Pract. **3**, 16–32 (in Russian) (2013)

Web-Sources

Climate analyses (year). https://www.weatheronline.co.uk/. Accessed on 22 Nov 2023

Energy use (kg of oil equivalent per capita). https://data.worldbank.org/indicator/EG.USE.PCAP.KG.OE. Accessed on 22 Nov 2022

Europe is getting warmer… EDJNet 2018. https://www.europeandatajournalism.eu/cp_data_news/europe-is-getting-warmer-and-it-s-not-looking-like-it-s-going-to-cool-down-anytime-soon/. Accessed on 22 Nov 2023

Klimadiagramme weltweit. https://www.klimadiagramme.de/. Accessed on 22 Jan 2024

Trends in annual temperature across Europe between 1960 and 2018. EEA, 2020. https://www.eea.europa.eu/data-and-maps/figures/decadal-average-trends-in-mean-9. Accessed on 22 Nov 2023

Fusion of Airborne and Spaceborne Thermal Imagery for Temperature Monitoring in Urban Areas

Raniero Beber, Thomas Maffei, Salim Malek, Daniela Poli, and Fabio Remondino

Abstract

Land use and surface properties directly influence and affect events and phenomena, such as flash flooding due to heavy rains, Urban Heat Island (UHI) intensification during heat waves, biodiversity reduction, etc. To deeply study such phenomena within urban areas, remotely sensed data are essential to describe, investigate, and model them and drive actionable insights. Aerial sensors mounted on airplanes allow for the acquisition of high-resolution data in terms of geometry (up to 5 cm with multispectral cameras) and spectral content (hundreds of narrow bands with wavelengths from visible to short-wave and thermal infrared). In urban applications, images are generally acquired during aerial surveys planned on demand by the municipalities. Satellite images, on the other hand, feature a consistent revisit time thanks to their elliptic sun-synchronous orbits. Satellite images with low spatial granularity, often acquired through international public missions (i.e., Copernicus), are publicly accessible. Within the USAGE—Urban Data Space for Green Deal—EU project [https://www.usage-project.eu/], we investigate the integration of multi-modal and multi-resolution data from aerial and satellite platforms in urban areas for environmental analyses. In this paper, the activities specifically realized in thermal analysis are discussed. Two pilot cities are considered, Graz (Austria) and Ferrara (Italy), where aerial multispectral, thermal, hyperspectral, and LiDAR data, as well as thermal satellite images, are available (Beber et al., Int. Arch. Photogramm. Remote Sens. Spatial Inf. Sci., XLVIII-1/W3-2023, 9–16, 2023). Firstly, the land surface temperatures (LST) are calculated from thermal aerial images (0.5–1 m spatial resolution), also with the support of aerial hyperspectral images (VNIR and SWIR ranges, 1 m spatial resolution) to retrieve information on the type of surface material, thus on the surface emissivity. The LST values are then compared to those retrieved using the Landsat TIRS sensor, in order to characterize the representativeness of the Landsat pixel over the urban landscape. Moreover, a deep learning algorithm to downscale a Landsat product to airborne resolution is presented. Finally, LST maps are coupled with population density to highlight areas with higher risks. The proposed methodology could be replicated also in other similar cities.

Keywords

Thermal · Hyperspectral · Airborne · Satellite · Data-fusion

1 Introduction

The European Union (EU) is actively promoting the development of a European single market for data aimed at facilitating horizontal cross-sector data sharing, which is crucial for driving innovative solutions and supporting the decarbonization of the energy sector. To realize this vision, it is vital to implement the Findability, Accessibility, Interoperability, and

R. Beber (✉) · S. Malek · F. Remondino
3D Optical Metrology (3DOM) Unit, Bruno Kessler Foundation (FBK), Trento, Italy
e-mail: rbeber@fbk.eu

S. Malek
e-mail: smalek@fbk.eu

F. Remondino
e-mail: remondino@fbk.eu

T. Maffei · D. Poli
AVT Airborne Sensing Italia Srl, Trento, Italy
e-mail: t.maffei@avt.at

D. Poli
e-mail: d.poli@avt.at

P. Mohammad et al. (eds.), *Geographic Approaches to Climate Change and Mitigation: Urban and Rural Perspectives (Volume 1)*,
Advances in Science, Technology & Innovation, https://doi.org/10.1007/978-3-031-92119-3_7

Reuse (FAIR) principles (GO-FAIR 2024). Several initiatives have been launched to foster the establishment of a Green Deal data space (Gutierrez David et al. 2023) and to enhance the green transition in urban environments (Amado and Poggi 2022). Among these, the EU project "Urban Data Space for Green Deal" (USAGE) is particularly focused on cities, given that they are the primary consumers of energy resources (Gago et al. 2013) and face greater vulnerability to climate change impacts compared to other areas (Wouters et al. 2019). As part of USAGE, a comprehensive collection of geospatial, thematic, and additional datasets has been compiled from existing sources or newly generated to evaluate solutions aimed at (i) deepening our understanding of the changing climate and environmental trends (Rüegg 2019), and (ii) investigating the human factors contributing to these changes, including adaptive behaviors and mitigation efforts (Foshag et al. 2020). The solutions proposed in USAGE are shaped by the diverse needs of stakeholders and rely on inter- and transdisciplinary collaboration, leveraging geospatial data alongside local knowledge (Adler et al. 2018). It is essential for public authorities, city planners, and urban stakeholders engaged in green transitions to have access to straightforward yet effective ICT tools, geospatial solutions, strategies, and methodologies for efficient energy monitoring, management, renewable energy integration, and climate change adaptation (Nowacka and Remondino 2018). The use cases addressed in the USAGE project aim to implement and comply with European Green Deal policies within pilots' urban planning and improvement actions. Among these, actions like energy efficiency, management of climate change events (such as heat waves or flooding from heavy rainfall), and urban biodiversity monitoring, were proposed to be implemented to bring benefits to urban areas.

This paper focuses on urban temperature analysis from thermal aerial and satellite images on two pilot cities where suitable data are available, i.e., Ferrara (Italy) and Graz (Austria). Remotely sensed data are essential to describe, investigate, and model climate change phenomena, such as Urban Heat Island (UHI) intensification during heat waves, and drive actionable insights (Guo et al. 2024; Jumari et al. 2023; Milesi and Churkina 2020; Mandanici et al. 2016).

The main contributions of the article include: (i) Land Surface Temperature (LST) retrieval from airborne thermal and hyperspectral imaging; (ii) Comparison between airborne and satellite thermal imagery; (iii) Downscaling of satellite thermal images to airborne resolution; (iv) Multi-modal data analysis to produce decision-ready information.

2 Airborne Land Surface Temperature (LST) Retrieval

High-resolution aerial thermal images (Yang et al. 2021) are the most suitable source of data to recover land surface temperature (LST) values on an entire city for the analysis of heat islands. Within USAGE activities, aerial thermal images were acquired with a Dual-DigiTherm camera[1] which consists of two uncooled microbolometer FPA-detectors delivering single band images (1024 × 768 pixels) in the thermal infrared range (7.5–14 μm). In each of the two pilot cities (Ferrara ad Graz), two summer flights were performed by AVT Airborne Sensing, one in the afternoon at the peak of heat and one around nighttime. The raw thermal images were geometrically processed, orthorectified on a digital surface model (DSM) and mosaicked following standard photogrammetric workflows. The final orthomosaics had a resolution of 0.5 m in Ferrara and 1 m in Graz (Beber et al. 2023). The information in the orthomosaic refers to the temperature measured at the sensor and, in order to retrieve the temperature at the surface level (LST—Fig. 1), further processing is needed. The recent work of Ru et al. (2023) on LST retrieval from the Spaceborne Thermal Radiometer Experiment on Space Station (ECOSTRESS) thermal infrared (TIR) data shows the role of Land Surface Emissivity (LSE) and urban geometry and volumes in the temperature estimation. Indeed, following the radiative balance schema from Byrnes and Schott (1986), the total spectral radiance reaching a sensor installed on an aircraft (l_S) depends not only on the radiance emitted by the surface l_B, which is a function of the surface temperature (our unknown) according to Planck law, but also by other components depending on the material emissivity, atmospheric properties, and the surface geometry. Using the nomenclature and notation in Bitelli et al. (2015), the relation is expressed by:

$$l_s = \tau \varepsilon l_B + l_u + \rho \tau F l_d + \rho \tau (1 - F) l_{sr} \quad (1)$$

where

- ε is the surface emissivity, being the emissivity of a material and its effectiveness in emitting energy as thermal radiation;
- τ is the atmosphere transmittance;
- l_u is upwelling radiance (radiance emitted upward by the atmosphere layer between surface e sensor);
- l_d is the downwelling radiance (radiance from the entire atmosphere column);
- l_{sr} is the radiance emitted by the surrounding objects;
- ρ is the surface reflectance, i.e., its effectiveness in reflecting radiant energy;

[1] https://www.igi-systems.com/dual-digitherm.html.

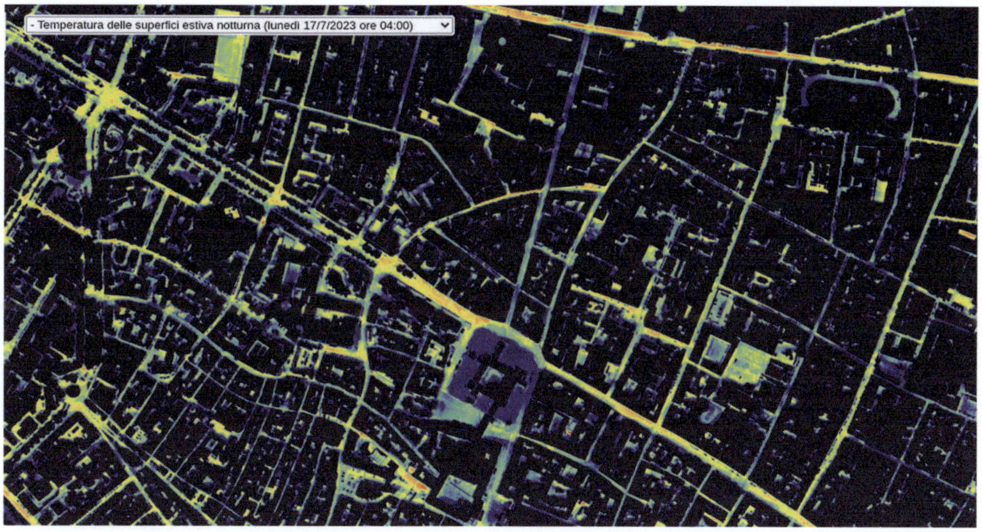

Fig. 1 Example of airborne night LST for Ferrara (Italy)[2] (by author)

- F is the Sky View Factor (SVF), defined as the ratio of radiation received by a planar surface to that received from the entire hemispheric radiating environments, and expressed as the radiation coming from the visible part of the sky divided by the total radiation from the entire hemisphere (Zhu et al. 2013).

The atmospheric transmittance as well as the upwelling and downwelling radiances can be estimated on a pixel basis using the MODTRAN® (MODerate resolution atmospheric TRANsmission) model (Berk et al. 2014). The Sky View Factor (Kakšek et al. 2011) is derived from the digital surface models (Lindberg and Grimmond 2010) and it ranges from zero (obstructed spaces) up to one (free spaces).

On the other hand, in order to derive the correct emissivity values at pixel level, it is necessary to know in advance the type of surface's material, both on the ground and on the building roofs. In our pilot locations, the information on the material is retrieved using hyperspectral aerial images (Jia et al. 2020) available in both cities (Beber et al. 2023). The images are acquired with the AISAFenix384 sensor (SPECIM 2024) and contain 364 bands in the VNIR (visible near-infrared) and SWIR (shortwave infrared) spectral ranges. The hyperspectral flight over Ferrara was performed on two cloudless days on July 19th and 20th 2022 at a flying height of 667 m above ground in North–South direction with a Ground Sample Distance of 1.0 m. On the other hand, the flight over Graz was performed on October 20th 2021 with a Ground Sampling Distance (GSD) of 1.8 m. The raw hyperspectral images are firstly geometrically and radiometrically corrected and then an atmospheric correction is applied. For the image classification and surface material identification, a supervised pixel-wise hierarchical approach based on Random Forest is applied. At first, five macro classes are considered: *Built-up, Shadow, Soil, Vegetation e Water*. Then the *Built-up* class is further analyzed to retrieve the single classes of man-made cover materials, using annotations provided by the local municipalities. As input to the model, the first 50 bands after Minimum Noise Fraction reduction (Luo et al. 2016) are used together with a normalized DSM (nDSM). A similar approach is used to refine the *Soil* class into *sand, mixed soil,* and *bare soil*. The macro class *vegetation* is refined based on thresholds on the nDSM: *grass* < 0.2 m, 0.2 m > *low vegetation* < 3 m, and *high vegetation* for > 3 m.

For the Ferrara pilot, 24 classes are overall identified while 23 classes are used for Graz (Fig. 2). Emissivity value for each material class is then determined and fitted over the spectral range of the Dual-DigiTherm thermal camera. To smooth the emissivity map, a moving window of 3 × 3 pixels is applied with a mean function. Once all the components of the radiative balance are calculated, the correct radiance at surface level (L_B in Eq. 1) is retrieved and the LST back calculated. Figure 3 shows a zoom on the Graz's orthophoto, the original thermal image (temperature at sensor) for the same area, the material classification map, the emissivity map, the corrected thermal image (surface temperature) and the difference between the sensor and surface temperature values. It can be observed that before and after the correction, the temperature values change up to 20 °C in correspondence with surfaces with low emissivity, like metal roofs.

The same workflow was followed for the two thermal flights executed on Ferrara to estimate the LST on July 16th 2023 between 14:35 and 16:00 local time and on July 17th

[2]https://sit.comune.fe.it/allegati/mappe/usage/RasterCOG.html.

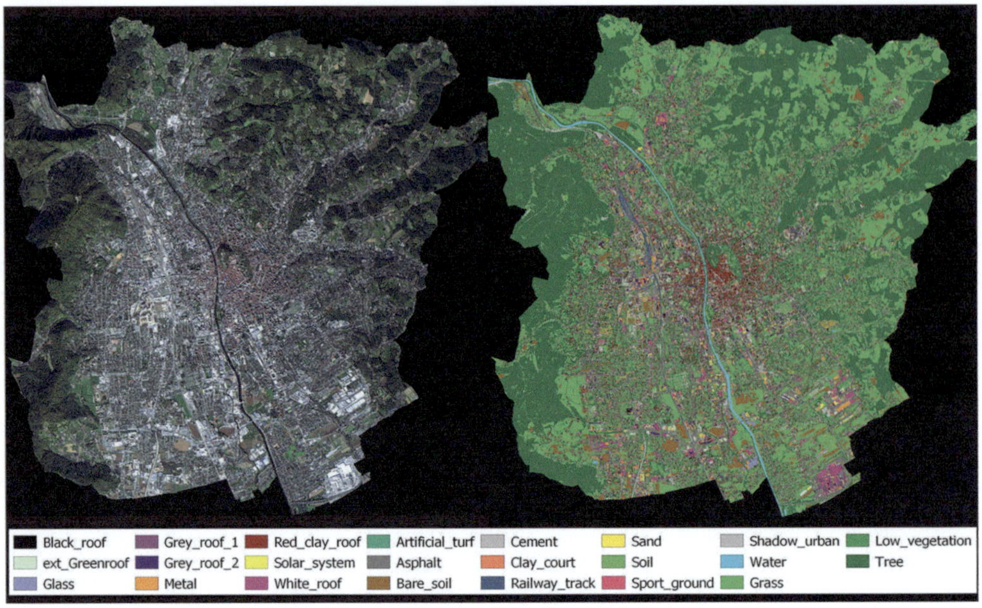

Fig. 2 Classification of surface materials in Graz (Austria): hyperspectral mosaic visualized with RGB visible bands (right) and classification map with 23 classes (left) (by author)

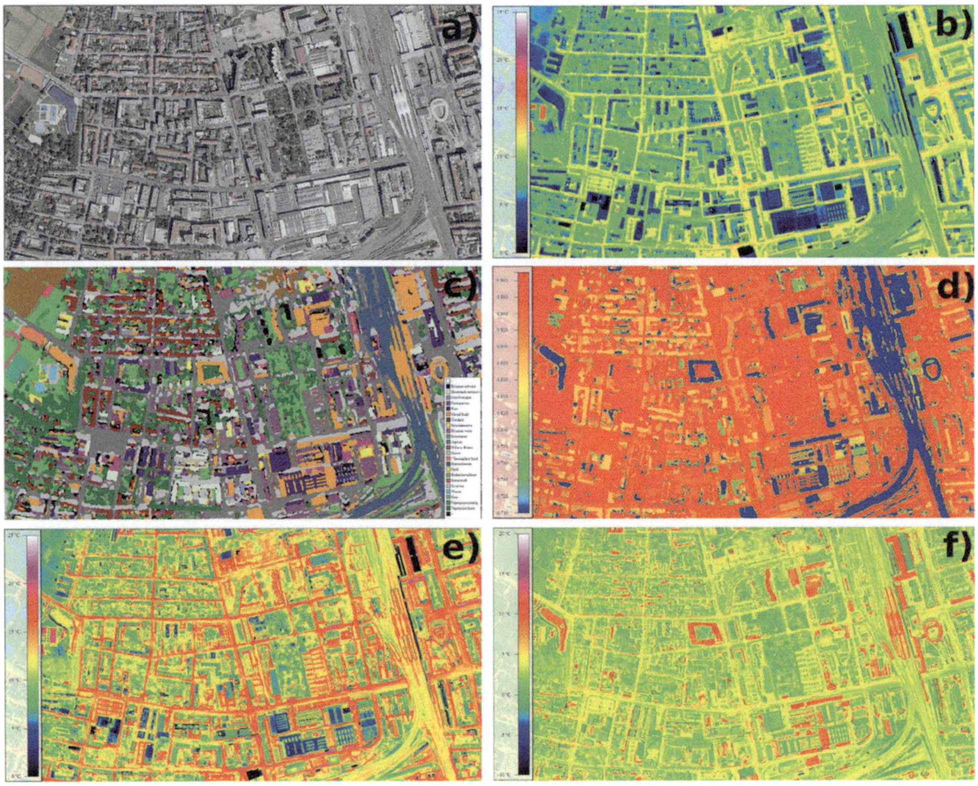

Fig. 3 Close-up views of Graz orthoimage (**a**), uncorrected day thermal image, i.e., temperature at sensor (**b**), surface material classification map (**c**), emissivity map (**d**), derived LST (**e**) and difference between sensor and land surface temperature (**f**) (by author)

2023 between 03:40 and 05:08 local time. The distribution of the temperature differences between day and night acquisitions shown in Fig. 4 reports values up to 56 °C and the presence of urban spots with high values. Thanks to the availability of the surface material map from the classification of hyperspectral aerial images, cross-analysis are ongoing in USAGE to model the relationship between temperature differences and materials. It can be noticed, for example, the clear mitigation effect of water on heating.

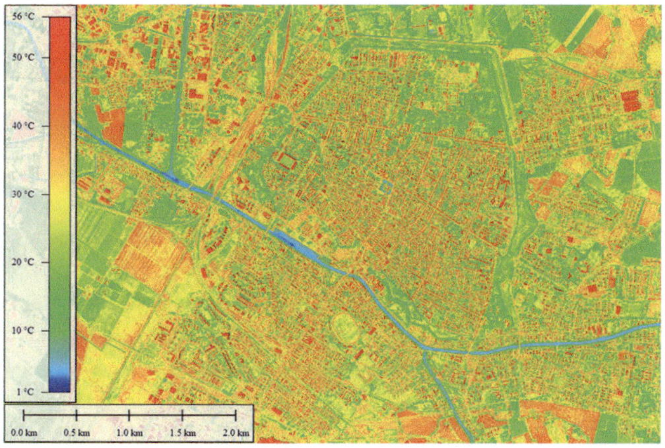

Fig. 4 Temperature differences in Ferrara between LST in the day (17.07.2023 afternoon) and night (16.07.2023 night) (by author)

3 Satellite LST

Copernicus Sentinel 3-SLSTR sensor and USGS Landsat 8 and Landsat 9-TIRS sensors acquire thermal images with a resolution of 1000 m and 100 m, respectively. Landsat data are then resampled to 30 m resolution before distribution. Similarly to thermal aerial images, the temperature values provided by satellite sensors need to be corrected for the atmosphere effects on the satellite platform (Ermida et al. 2020). The LST is computed using the Statistical Mono-Window (SMW) algorithm (Duguay-Tetzlaff et al. 2015) and, exploiting archival images, hotspots based on LST can be performed, e.g., for the last decade (Fig. 5).

4 Airborne and Satellite LST Comparison

Airborne and satellite LST products have different but complementary characteristics. The LST map from aerial thermal images has the great advantage of providing information with high granularity (in our case up to 0.5 m) and temperature accuracy. This is essential for detailed analysis, at single feature level. However aerial data are available on demand, thus, despite their high accuracy and resolution, are not suitable for high-frequency temporal monitoring. On the other hand, satellite thermal missions provide images that have low geometric resolution but with a regular revisit time (from one to 8 days). Additionally, archived long-lasting time series are available to monitor the historical development, when cloud cover allows. Figure 6 shows the granularity of the aerial and satellite LST products in Graz.

As a basis for airborne and satellite thermal products comparison and combination, the surface material classification map of Graz (Fig. 2) represents a useful spatial clustering of the city. A class-wise comparison of the LST retrieved

Fig. 5 LST monthly median of Landsat 8 acquisitions from 2013 to 2023 over Ferrara. The considered months are those where UHI is more likely (by author)

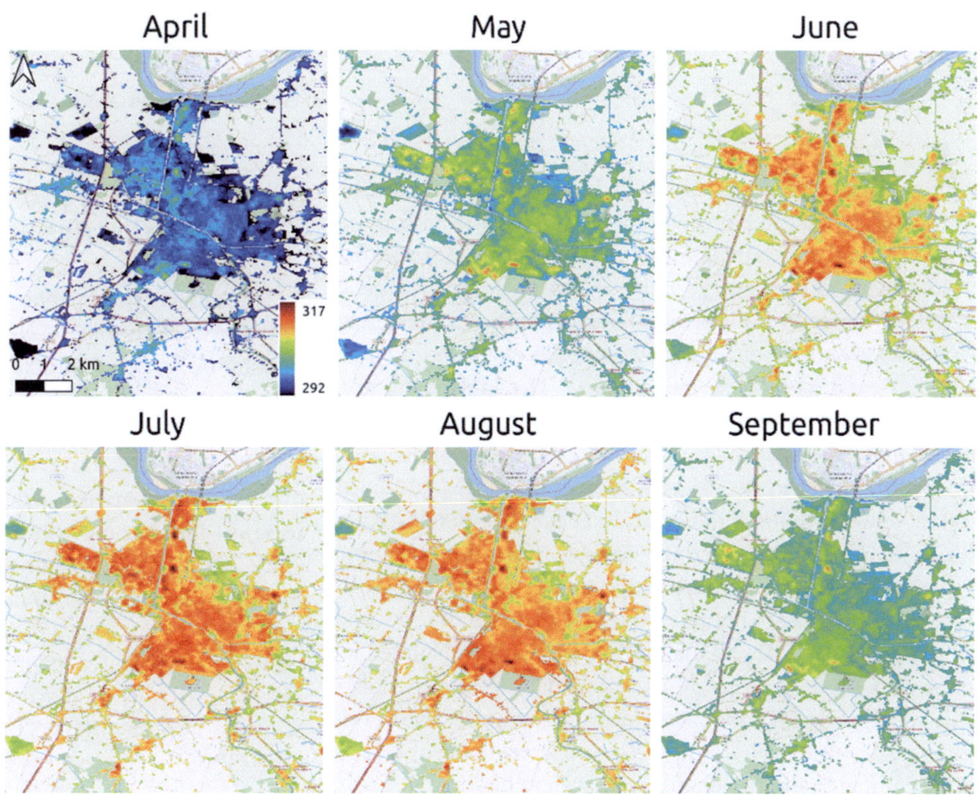

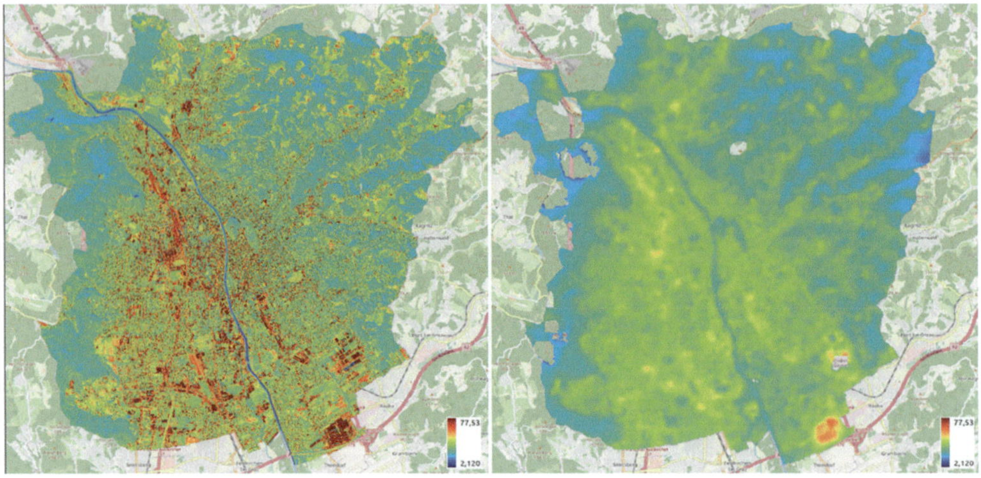

Fig. 6 Comparison of granularity between September 9th 2021 airborne (left) and September 12th 2021 satellite (right) derived LST maps in Graz (by author)

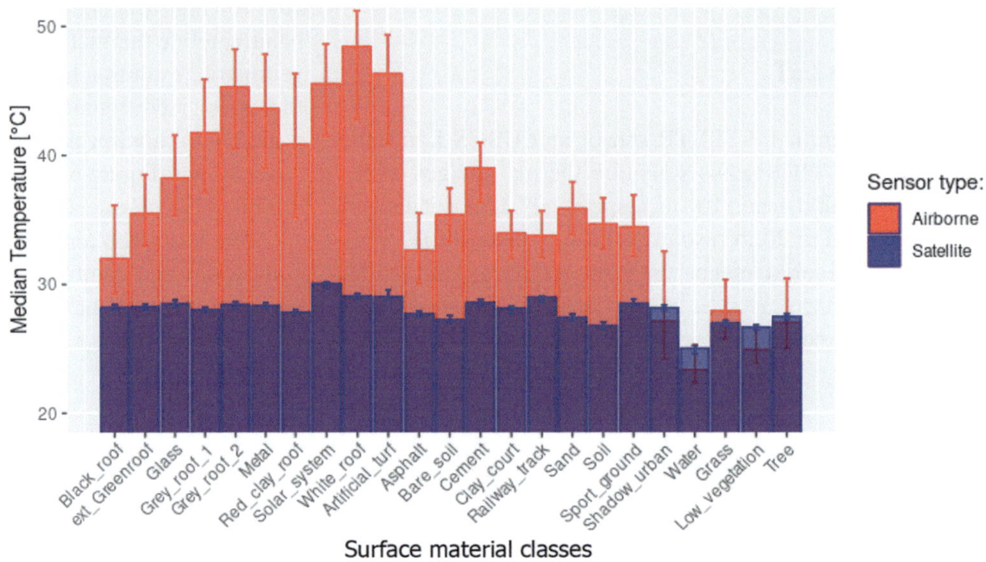

Fig. 7 LST median temperature comparison across surface material classes between airborne and satellite LST maps in Graz (by author)

from airborne on September 9th 2021 and an almost concurrent scene of Landsat 8 on September 12th 2021 is performed (Fig. 6). Firstly, the median temperature from both sensors is computed within every enclosed uniform area for each of the 23 classes of surface materials. Figure 7 shows, for each sensor and for each surface material class, the median LST retrieved as well as its error bar. The high-resolution thermal airborne flight can measure subtle temperature differences among the man-made materials covering the urban area, meanwhile the satellite, with its coarser native resolution of 100 m, tends to smooth those differences in the pixel mixing during acquisition. At the same time, the error bars from the airborne acquisition are larger due to the variety of conditions that the high-resolution is able to spot, particularly when small patches of different materials are present or isolated its temperature is even more influenced by its neighboring materials. Satellite data distinguishes macro classes effectively. For both sensors appreciable temperature differences are observed between man-made, vegetation, and water classes.

5 Downscaling Satellite Thermal Imagery to Airborne Resolution

To take advantage of the high geometric and temperature resolution of the airborne-based products and the high revisit time of satellite images, a downscaling procedure to improve the geometric resolution of satellite thermal data is developed. The method would allow, accepting some error on the estimation of LST, to retrieve airborne-like thermal products for every cloudless revisit acquisition of a thermal satellite. We used a U-Net-like convolutional neural network (Ronneberger et al. 2015), tailored with a loss function based on a combination of the mean absolute error (MAE) and the Structural Similarity Index Measure (SSIM) (Brunet et al. 2011). The network is trained with the last available airborne thermal products as well as the orthophoto of the same resolution, with the aim to infer LST at 0.5 from the satellite-derived LST (30 m). The model is trained using couple of tiles taken

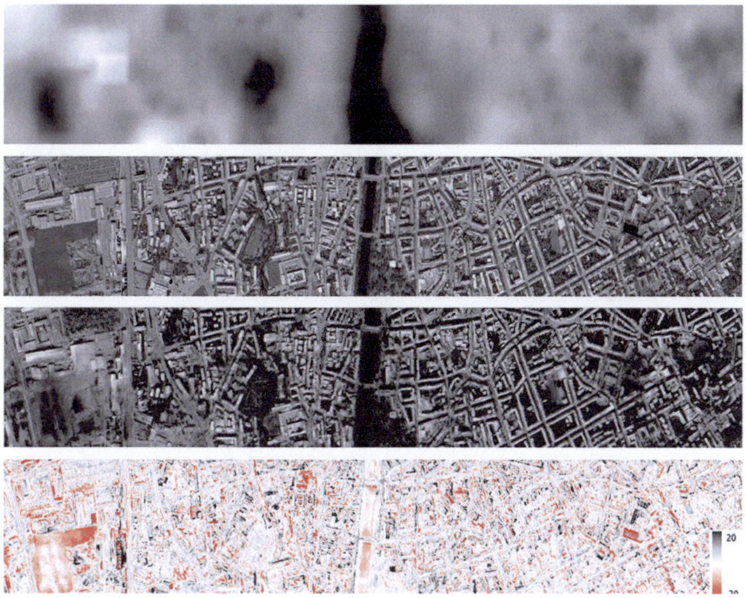

Fig. 8 From top to bottom: satellite and airborne thermal images over Graz, inference from U-Net CNN, difference error map (by author)

from the two LST products. Thanks to the introduction of the orthophotos in the training process, the model is also capable of learning urban morphology. As shown in Fig. 8, results are promising and the learning-based downscaling process offers consistent results. Nevertheless, the temperature difference between the ground truth and the inference, even having a Gaussian distribution centered on zero, has some difference up to 20°, therefore more training data are necessary. The proposed neural network can be used to infer relative spatial thermal patterns in the city rather than correct absolute values and to improve the use of low-resolution satellite thermal imagery.

6 Application of LST and Hotspots

The prioritization of hotspot mitigation can be performed by introducing aggregated census data available at the municipality level (e.g., https://dati.comune.fe.it/dataset/distribuzione-popolazione-2022-gdpr). The July (2013–2023) monthly median surface temperature in Ferrara displays some areas visibly hotter (hotspots in yellow in Fig. 9c). The population density (in hexagons—Fig. 9b) highlights areas where more (or less) inhabitants are located. Overlapping LST and population density (Fig. 9d), the risk by the resident population can be displayed. Dark hexagons represent lower risk (i.e., low residential density) whereas bright hexagons represent higher risk due to the combination of high population density and hotspot. The lack of hexagon reticulate means either that census data are not available or that there is no population resident. For example, the industrial area on the top-left represents a hotspot but is poorly inhabited, thus presenting almost no risk.

7 Conclusions

In this work the application of remotely sensed thermal data from airborne and satellite platforms for urban temperature analysis is discussed. Using two pilot cities of the USAGE project as demonstration, the characteristics of the thermal data and the processing for the retrieving land surface temperature (LST) is discussed. The LST data acquired by airborne and satellite platforms have clear complementary characteristics: if the airborne-based data have high spatial and temperature resolution but low availability, long-time and frequent series of spaceborne thermal products are openly available but with low or very low spatial resolution. The data were suitably combined to allow monitoring of urban environments at different scales in time and to derive decision-ready information for policy makers. Additionally, the initial results of a learning-based thermal downscaling method for satellite thermal images using airborne thermal data showed the potential for multi-temporal analysis.

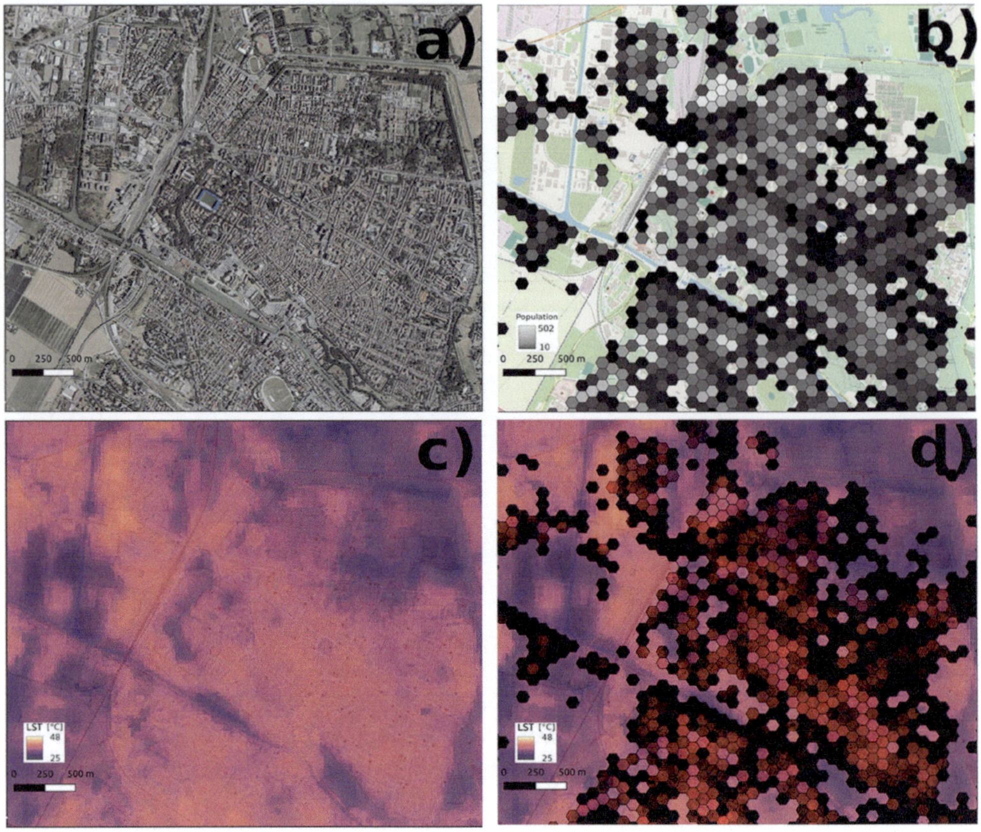

Fig. 9 Orthoimage of Ferrara (**a**); population density displayed with hexagons (**b**); monthly median LST for the period 2013–2023 (**c**); LST with superimposed population risk, from dark (low risk) to bright (high risk: densely populated and high median LST) (by author)

Acknowledgements The work is funded by the EU project USAGE—Urban Data Space for Green Deal (https://www.usage-project.eu/) which has received funding from the European Union's Horizon Europe Framework Programme for Research and Innovation under the Grant Agreement no 101059950—call HORIZONCL6-2021-GOVERNANCE-01-17 (IA).

References

Adler, C., Hirsch Hadorn, G., Breu, T., Wiesmann, U., Pohl, C.: Conceptualizing the transfer of knowledge across cases in transdisciplinary research. Sustain. Sci. **13**(1), 179–190 (2018)

Amado, M., Poggi, F.: Sustainable Energy Transition for Cities. Elsevier (2022). ISBN 978-0-12-824277-3

Beber, R., Perda, G., Takhtkeshha, N., Remondino, F., Maffei, T., Poli, D., Moe, K., Cipriano, P., Ciliberti, M.: Multi-modal geospatial and thematic data to foster green deal applications. Int. Arch. Photogramm. Remote Sens. Spatial Inf. Sci. XLVIII-1/W3-2023, 9–16 (2023)

Berk, A., Conforti, P., Kennett, R., Perkins, T., Hawes, F., van den Bosch, J.: MODTRAN6: a major upgrade of the MODTRAN radiative transfer code. In: Proceedings of SPIE 9088, Algorithms and Technologies for Multispectral, Hyperspectral, and Ultraspectral Imagery XX, 90880H (2014)

Bitelli, G., Conte, P., Csoknyai, T., Franci, F., Girelli, V.A., Mandanici, E.: Aerial thermography for energetic modelling of cities. Remote Sens. **7**(2), 2152–2170 (2015)

Brunet, D., Vrscay, E.R., Wang, Z.: On the mathematical properties of the structural similarity index. IEEE Trans. Image Process. **21**(4), 1488–1499 (2011)

Byrnes, A.E., Schott, J.R.: Correction of thermal imagery for atmospheric effects using aircraft measurement and atmospheric modeling techniques. Appl. Opt. **25**(15), 2563–2570 (1986)

Duguay-Tetzlaff, A., Bento, V.A., Göttsche, F.M., Stöckli, R., Martins, J.P., Trigo, I., Kunz, H.: Meteosat land surface temperature climate data record: achievable accuracy and potential uncertainties. Remote Sens. **7**(10), 13139–13156 (2015)

Ermida, S.L., Soares, P., Mantas, V., Göttsche, F.M., Trigo, I.F.: Google earth engine open-source code for land surface temperature estimation from the Landsat series. Remote Sens. **12**(9), 1471 (2020)

Foshag, K., Aeschbach, N., Höfle, B., Winkler, R., Siegmund, A., Aeschbach, W.: Viability of public spaces in cities under increasing heat: a transdisciplinary approach. Sustain. Cities Soc. **59**, 102215 (2020)

Gago, E.J., Roldan, J., Pacheco-Torres, R., Ordóñez, J.: The city and urban heat islands: a review of strategies to mitigate adverse effects. Renew. Sustain. Energy Rev. **25**, 749–758 (2013)

GO-FAIR.: https://www.go-fair.org/fair-principles/ (visited on 5/4/2024) (2024)

Guo, Y., Unger, J., Khabibolla, A., Tian, G., He, R., Li, H., Gál, T.: Modeling urban air temperature using satellite-derived surface temperature, meteorological data, and local climate zone pattern—a case study in Szeged, Hungary. Theor. Appl. Climatol. 1–19 (2024)

Gutierrez David, M., Dietrich, M., Raczko, N., Denvil, S., Santoro, M., Chatzikyriakou, C., Borejko, W.: Towards the European green deal data space. In: Proceedings of EGU General Assembly, EGU23-8788 (2023)

Jia, J., Wang, Y., Chen, J., Guo, R., Shu, R., Wang, J.: Status and application of advanced airborne hyperspectral imaging technology: a review. Infrared Phys. Technol. **104**, 103115 (2020)

Jumari, N.A.S.K., Ahmed, A.N., Huang, Y.F., Ng, J.L., Koo, C.H., Chong, K.L., Sherif, M., Elshafie, A.: Analysis of urban heat islands with landsat satellite images and GIS in Kuala Lumpur Metropolitan City. Heliyon **9**(8) (2023)

Kakšek, K., Oštir, K., Kokalj, V.: Sky-view factor as a relief visualization technique. Remote Sens. **3**, 398–415 (2011)

Lindberg, F., Grimmond, C.S.B.: Continuous sky view factor maps from high resolution urban digital elevation models. Clim. Res. **42**(177–183), 2010 (2010). https://doi.org/10.3354/cr00882

Luo, G., Chen, G., Tian, L., Qin, K., Qian, S.E.: Minimum noise fraction versus principal component analysis as a preprocessing step for hyperspectral imagery denoising. Can. J. Remote. Sens. **42**(2), 106–116 (2016)

Mandanici, E., Conte, P., Girelli, V.A.: Integration of aerial thermal imagery, LiDAR data and ground surveys for surface temperature mapping in urban environments. Remote Sens. **8**(10), 880 (2016)

Milesi, C., Churkina, G.: Measuring and monitoring urban impacts on climate change from space. Remote Sens. **12**(21), 3494 (2020)

Nowacka, A., Remondino, F.: Geospatial data for energy and low carbon cities—overview, experiences and new perspectives. ISPRS Int. Arch. Photogramm. Remote Sens. Spatial Inf. Sci. **XLII-4** (2018)

Ronneberger, O., Fischer, P., Brox, T.: U-net: convolutional networks for biomedical image segmentation. In: Proceedings of Medical Image Computing and Computer-Assisted Intervention–MICCAI 2015, part III 18, pp. 234–241. Springer (2015)

Ru, C., Duan, S.B., Jiang, X.G., Li, Z.L., Huang, C., Liu, M.: An extended SW-TES algorithm for land surface temperature and emissivity retrieval from ECOSTRESS thermal infrared data over urban areas. Remote Sens. Environ. **290**, 113544 (2023)

Rüegg.: Simultaneous heat waves caused by anthropogenic climate change. PhysORG, **3** (2019)

SPECIM.: Spectral Imaging Ltd. AISAFENIX 384 sensor. https://www.specim.com/wp-content/uploads/2020/03/AisaFENIX-ver1-2020.pdf. Accessed on 5 Apr 2024 (2024)

Wouters, B., Gardner, A.S. and Moholdt, G.: Global glacier mass loss during the GRACE satellite mission (2002–2016). Front. Earth Sci. **7**, 96 (2019)

Yang, J., Shi, Q., Menenti, M., Wong, M.S., Wu, Z., Zhao, Q., Xu, Y.: Observing the impact of urban morphology and building geometry on thermal environment by high spatial resolution thermal images. Urban Clim. **39**, 100937 (2021)

Zhu, S., et al.: Influence of sky temperature distribution on sky view factor and its applications in urban heat island. Int. J. Climatol. **33**, 1837–1843 (2013)

Land Use Compatibility and Renewable Energy Siting—The Case of Greece

Chatzopoulou Ioanna

Abstract

Renewable energy sources refer to those that derive energy from natural sources such as the wind, the sun, the water, and geotherm. During the last few years, renewable energy has been promoted in the European Union as well as in Greece, serving as a vital contributor to the fight against climate change. It is worth mentioning that the revised Renewable Energy Directive EU/2023/2413 provides an increase of the share of renewable energy sources to at least 42.5% by 2030. Exploiting these sources also provides a remedy to the problem of depletion of fossil fuels and non-renewable sources (i.e., coal, natural gas, oil, and nuclear energy). Nevertheless, the over-promotion of renewable resources comes at a cost as it may have negative implications not only on the ecosystems, such as forest, mountainous, and island ecosystems but also on the landscape conservation itself. Apart from the above-mentioned implications, renewable energy resources might impact positively or negatively the local economy and other activities like tourism. This article examines the existing policies and legal tools in the EU and in Greece which aim at the sustainable use of renewable energy resources. More specifically, it focuses on the legal tools in Greece associated with the special planning of these resources and examines their sufficiency to ensure both the sustainable development of the renewable energy sector and its compatibility with other land uses (i.e., alternative forms of low-impact tourism, like ecotourism and green tourism).

C. Ioanna (✉)
Department of Tourism Management, University of Patras, Patras, Greece
e-mail: ichatzop@upatras.gr

Keywords

Renewable energy · Sustainable development · Ecosystems · Spatial planning · Tourism · Climate change · Decarbonization

1 Introduction

Greece possesses a naturally high potential for renewable energy sources, due to its climate and geographical conditions (the weather is sunny for more than 250 days a year, while the wind capacity is strong not only in the Aegean Sea but also in the mainland, which is covered mainly by mountains) (ENTERPRISE GREECE. Invest & Trade n.d.; Spanos et al. 2015). The utilization of these domestic energy sources is rapidly increasing all over Greece. More specifically, the Greek government has announced the further expansion of the national energy system by the siting of an additional 1150 turbines on Greek islands or offshore over the next two decades, capable of generating 12 GW of wind power, and by speeding up clean energy technologies, whose contribution to the national energy production is expected to reach 80% by 2030 (Euronews 2024). Furthermore, the country's installed wind power capacity is planned to reach 6500 MW within the next three years (RESINVEST 2023)

Nevertheless, without ignoring the contribution of renewable energy source potential utilization towards energy security and climate neutrality, it is to say that the overuse of renewable energy resources, without rational spatial planning can cause environmental degradation (Barthelmie and Pryor 2014). For instance, wind energy can have a negative impact on birds, through collision mortality and habitat degradation (Anderson et al. 1999; Dimalexis et al. 2010; Balotari-Chiebao et al. 2023), interrupt the landscape with negative impacts on other uses, like ecotourism and consequently on

P. Mohammad et al. (eds.), *Geographic Approaches to Climate Change and Mitigation: Urban and Rural Perspectives (Volume 1)*, Advances in Science, Technology & Innovation, https://doi.org/10.1007/978-3-031-92119-3_8

Fig. 1 The forest of Dadia in the Evros region, before and after the massive fires of 2023. *Source:* https://www.lifo.gr/now/greece/fotia-ston-ebro-dasos-tis-dadias-prin-kai-meta-apokardiotikes-eikones

the local economy, or harm biodiversity and sensitive ecosystems, like forest ecosystems, which are threatened every year by tremendous fires.

It is worth mentioning here, that after the disastrous fires in the Evros region in northern Greece last summer, whose impacts are depicted in Fig. 1, and the fires in northern Evia of 2021, which have destroyed millions of acres of forest land the Greek administration has announced the postponement of wind parks environmental licenses for 6 years so that the forest ecosystems can recover (Maria 2023).

This study examines whether the existing legal spatial planning tools in Greece are adequate to ensure the sustainability of the "green energy transition", with respect to the biodiversity, the sensitive forest, mountainous and island ecosystems, the landscape itself, but also the local economy.

2 Methodology

According to the aim of the study, which is to give a comprehensive overview of the existing policies and legal tools in the EU and in Greece on the sustainable use of renewable energy resources, a mix of methods is applied, including a literature review, and a review of the renewables promotion legal instruments in the European Union and in Greece, with emphasis on the legal tools in Greece, associated with the spatial planning of these resources.

The literature and legal documents review is used for responding to the following research questions:

Are the existing legal tools in the EU and in Greece adequate to safeguard the sustainability of renewable energy sector and its compatibility with other land uses?

What are the implications of the implementation of the renewable energy legal tools and policies in Greece?

What points can be improved or changed?

What are the remaining challenges?

3 The EU Policy on Renewable Energy

Within the last years, the EU has been promoting renewable energy use through the adoption of legislative measures and other policies, acknowledging its key role in achieving climate neutrality, reducing greenhouse gas emissions and environmental pollution, and tackling energy poverty, by securing the supply of sustainable, affordable energy, promoting technological development and innovation and boosting the economy and the regional development through the creation of new jobs and the support of local industries.

In this context, worth mentioning are, among others, the Directive (EU) 2018/2001 on the promotion of the use of energy from renewable sources—also known as the Renewable Energy Directive (REDII) (EUR-Lex n.d.), the Regulation (EU) 2018/1999, the Directive 98/70/EC, providing EU-wide standards for petrol and diesel used in cars, trucks and other off-road vehicles, the Council Directive (EU) 2015/652), the Council Regulation (EU) 2022/2577 of 22 December 2022, providing means and targeted measures towards a quick deployment of renewables across the EU (EUR-Lex n.d.), the "European Green Deal", presented by the European Commission in December 2019, an action plan, whose main scope is that Europe, will become the first climate-neutral continent by 2050 (EUR-Lex n.d.), by decarbonizing, inter alia, its energy system through the wide use of renewable sources (European Commission n.d.), and the relevant Regulation (EU) 2021/1119 of the European Parliament and of the Council (Regulation EU 2021/1119), which sets the above-mentioned objective of climate neutrality by 2050 and a 2030 EU-level target of reducing greenhouse gas emissions by at least 55% compared to 1990 levels and the Commission's "Energy Roadmap 2050".

Taking into account, that renewable energy is a key factor in achieving the above-mentioned objectives of climate neutrality, emission reduction, environmental protection, and security of energy supply, but also the need to achieve energy independence from Russia, the revised Renewable Energy Directive EU/2023/2413, which amends the previous Directive (EU) 2018/2001, provides the further promotion of the renewable energy sources, raising the renewable energy penetration in the European gross energy consumption from the previous 32–42.5% by 2030 and setting targets for renewable energy use in different sub-sectors, such as transport, heating and cooling, buildings and industry (European sources online n.d.).

4 Greece's Renewable Energy Policy

In order to secure the achievement of the EU's Energy Union targets, the Greek Ministry of the Environment and Energy has launched in December 2019 the National Energy and Climate Plan (NECP), a roadmap for the national energy and climate policy for the period 2021–2030.

More specifically, the NECP provides specific energy and climate objectives by 2030, namely reducing greenhouse gas emissions, raising the share of renewable energy sources (RES) in gross final energy to 35% by 2030, improving energy efficiency, and terminating the use of lignite for power generation in Greece by 2028.

To achieve the above objectives, the NECP sets out specific policy measures in seven different axes (1. Climate change, greenhouse gas emissions, 2. Renewable energy sources, 3. Energy efficiency improvement, 4. Energy supply security, 5. Energy market, 6. Agriculture, shipping, tourism, and 7. Research, innovation, and competitiveness).

As already mentioned above, regarding renewable energy sources (RES), the NECP raises the renewable energy penetration in the gross energy consumption to 35% by 2030, setting further quantitative objectives for the increase of renewable energy shares in different sub-sectors by that year, e.g. the RES share in gross final electricity consumption to reach at least 60%, the RES share in covering heating and cooling to exceed 40% and the RES share in the transport sector to exceed 14%.

In this context specific governmental policies and measures are being provided and promoted, such us simplifying and accelerating permitting processes for renewable energy, reducing administrative obstacles projects, providing incentives to renewable energy growth, such us grants and subsidies, income tax deductions, low-interest and guaranteed loans, increasing the RES installed capacity for power generation, by scheduling new RES projects for most of the relevant technologies (mainly wind farms, including offshore wind farms, like the ones shown in Fig. 2 and photovoltaics), securing optimal integration of RES in electricity networks, promoting RES systems in buildings, through autoproduction and the coverage of thermal needs with RES systems, investing in new technologies and applications, for wave energy utilization, hydrogen production, combined desalination facilities, dispersed generation systems, like small wind turbines, promoting storage systems and the gradual electrification of the transport sector, by supplying the natural gas

Fig. 2 *Source* https://eletaen.gr/en/teliiki-ekthesi-eletaen-owf-2/

network with hydrogen or methane, produced from renewable energy resources (HELLENIC REPUBLIC Ministry of the Environment and Energy 2019).

Furthermore, the utilization of RES potential as a substantial contributor towards environmental protection and energy security is also promoted by the regulatory framework, related to environmental and forest protection, but also by the relevant jurisprudence of the Greek supreme administrative court.

More specifically, the Greek environmental law 1650/1986 (Art. 2 of Law 1650/1986 "On the protection of the environment", amended by Law 3851/2010) provides the RES promotion as a key contributor to climate change mitigation and to the protection of the atmosphere and in compliance with the EU' relevant regulatory framework.

Furthermore, the forest legislation allows RES projects in forest areas, even in reforestable forests and forest expansions, with the exception of the photovoltaics, introducing a deviation from the constitutional prohibition of the alteration of the use of forests (unless there is a use imposed for the public interest and the national economy is promoted) and of the constitutional imperative for the absolute protection of the forests and forests expansions, that have been proclaimed as reforestable (Art. 24 and 117 par. 3 of the Greek Constitution of 1975, Art. 45, 46 and 57 of the Greek Forest Law 998/1979, amended by Law 4280/2014).

More specifically according to Art. 24 par. 1 of the Greek Constitution the alteration of the use of forests and forest expanses is prohibited, except where agricultural development or other uses imposed for the public interest prevail for the benefit of the national economy. Furthermore, Art. 117 par. 3 of the Greek Constitution provides that "3. Public or private forests or forest expanses which have been destroyed or are being destroyed by fire or have otherwise been deforested or are being deforested, shall not thereby relinquish their previous designation and shall compulsorily be proclaimed reforestable, the possibility of their disposal for other uses being excluded" (HELLENIC PARLIAMENT n.d.).

5 Renewable Energy Spatial Planning Regulation

5.1 The Renewable Energy Land Use Plan of 2008

After the annulment of several wind farm licenses by the Supreme Administrative Court of Greece due to the absence of spatial planning in the renewable energy sector, the "Committee responsible for the coordination of the spatial planning and sustainable development governmental policy" has approved the Renewable Energy Land Use Plan of 2008, which sets the basic rules for the RES spatial planning across Greece.

More specifically, as far as the wind farms' spatial planning is concerned, the Plan differentiates the mainland into two different categories based on the existing exploitable wind potential, namely "the wind energy priority areas" and the "wind energy suitable areas". Furthermore, it defines certain categories of areas, from which RES projects are excluded, while it provides a maximum allowed percentage of land coverage by wind farms.

5.2 The New Renewable Energy Land Use Plan

The advancements in renewable energy technologies, the development of new renewable energy types (like biomass and offshore wind energy), the climate change effects on the renewable energy sector, and the fact that the 2008 renewable energy land use plan have caused reactions and disputes with the local communities, due to the accommodation of the majority of the renewable power plants in specific areas, like the southern part of Evia and the central region of the country (Sterea Hellas), which, as depicted in Fig. 3, host 40% of the country's wind installations, based on their suitability for commercial renewable energy activities, and without taking into account environmental and user conflicts issues, have led

Fig. 3 Spatial distribution of wind capacity. *Source* https://resinvest.gr/2023/02/01/2022-annual-statistics-hellenic-wind-energy-association-2/

to a new renewable energy land use plan, which is going to be approved by the end of this year.

The new land use plan adopts the distinction of the previous one among wind energy priority and wind energy suitable areas, increases the areas where the renewable energy projects are not permitted, sets rules relating to the distances of renewable energy projects from monuments, settlements or regions of touristic interest and provides specific criteria for the sting of photovoltaics.

6 Implications of the Promotion of the Renewable Energy Sector in Greece

The promotion of the utilization of renewable energy sources in Greece has had benefits related to the country's energy security and its independence from fossil fuels. However, the RES promotion has caused in many cases conflicts with other land uses, like alternative forms of low-impact tourism, property issues due to the lack a cadaster and of a forest register, environmental issues due to the siting of wind power plants in sensitive ecosystems, like small islands, landscapes of outstanding natural beauty, significant bird habitats, untouched, roadless forests, large areas of whom have been destroyed for the constructing of several kilometers of new roads, leading to the top of the mountain, where the wind turbines are being sited. It is worth mentioning here, that as wind turbines increase in size, a recent law amendment in Greece provides the deforestation and the construction of even bigger roads through the Greek forests, in order to serve the wind turbine transportation. The increase in the wind turbines's size is illustrated in Fig. 4 (Office of Energy Efficiency and Renewable Energy 2024).

In order to achieve the objective of climate change mitigation through RES promotion without harming biodiversity, Ms. Vassiliki Kati, Professor of Biodiversity Conservation and Head of Biodiversity Conservation Lab at the University of Ioannina and her team have suggested an exclusion zone of ecologically sensitive areas, comprising 58.6% of the country, from windfarm development, depicted in Fig. 5, leaving the rest, comprising 41.4% of the country as a potential investment zone (Kati et al. 2020; Reporters United 2021).

7 Conclusions

As already mentioned above, Greece promotes green energy transition as a vital contributor towards climate neutrality and decarbonization.

However, in order to ensure the environmental, social, and economic sustainability of this transition a strategic national, planning framework for the rational RES use is needed, aiming to ensure the cover of national energy needs, with respect to environment, promoting not only the windfarm development, but also other domestic energy sources.

Moreover, addressing the pressing issue of environmental protection in the context of energy transition requires synergies and effective coordination and cooperation between all relevant stakeholders, like regional and local government bodies, the academia, the civil and the private sector, as well as the affected local communities, in order to address promptly and validly concerns and questions that always arise.

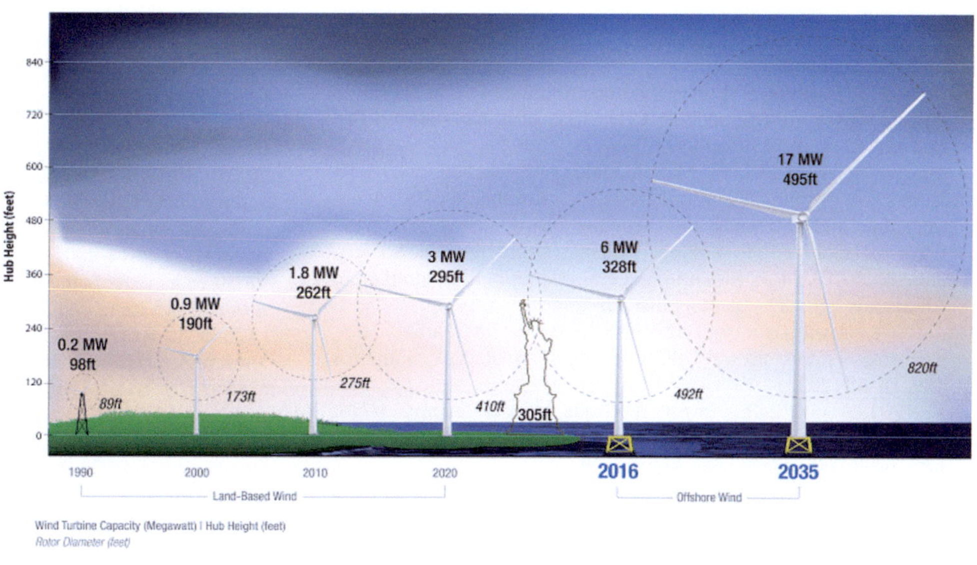

Fig. 4 Illustration of increasing turbine heights and blade lengths over time. *Source* https://www.energy.gov/eere/articles/wind-turbines-biggerbetter#:~:text=Larger%20rotor%20diameters%20allow%20wind,areas%20with%20relatively%20less%20wind

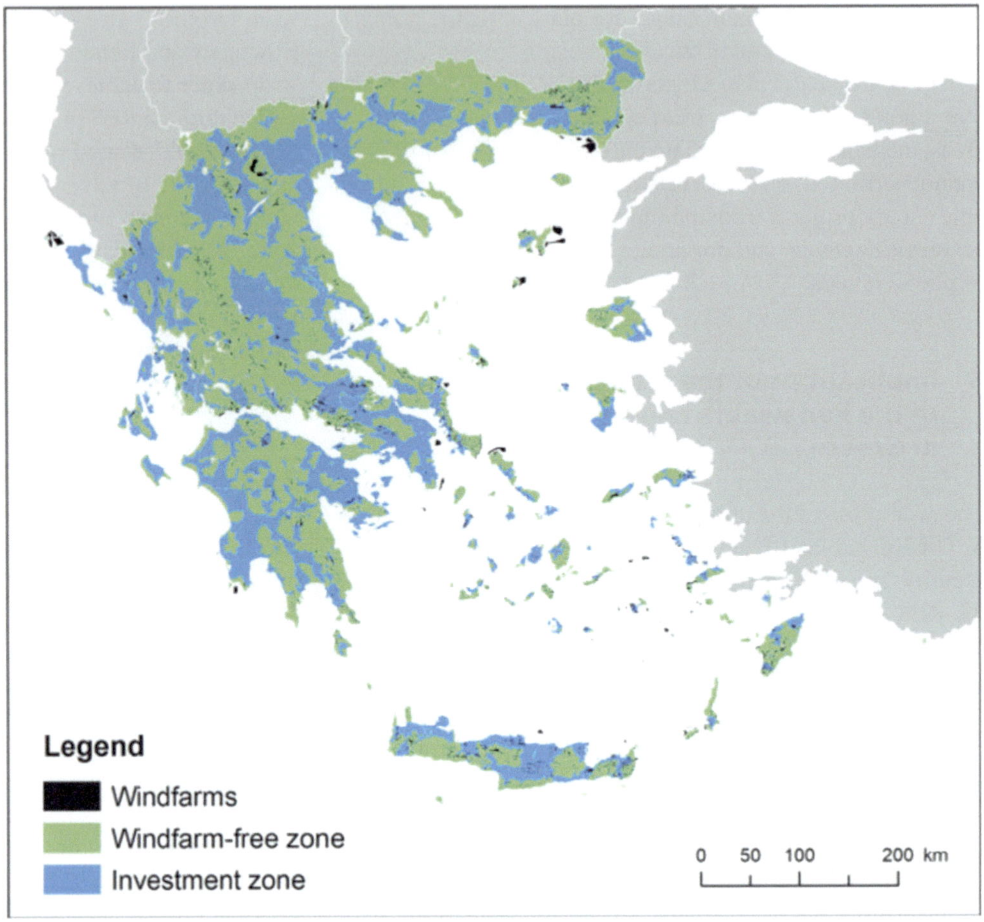

Fig. 5 Professor Kati's proposal would exclude ecologically sensitive areas (green) from windfarm development. The proposed investment zone (blue) comprises 41.4% of the country. 2020 *Source* Kati et al. ()

References

Anderson, R., Morrison, M., Sinclair, K., Strickland, D.: Studying Wind Energy/Bird Interactions: A Guidance Document, National Wind Coordinating Committee (1999)

Balotari-Chiebao, F., Santangeli, A., Sirke Piirainen, B.P.: Wind energy expansion and birds: identifying priority areas for impact avoidance at a national level. Biol. Conserv. **277** (2023). https://doi.org/10.1016/j.biocon.2022.109851

Barthelmie, R.J., Pryor, S.C.: Potential contribution of wind energy to climate change mitigation. Nat. Clim. Change **4**, 684–688 (2014). https://doi.org/10.1038/nclimate2269

Dimalexis, A., Kastritis, T., Manolopoulos, A., Korbeti, M., Fric, J., Saravia Mullin, V., Xirouchakis, S., Bousbouras, D.: Identification and Mapping of Sensitive Bird Areas to Wind Farm Development in Greece. Hellenic Ornithological Society, Athens (2010). http://files.ornithologiki.gr/docs/politiki/aiolika/prosdiorismos%20kai%20xartografisi.pdf

Maria, E.A.: Δασικές πυρκαγιές στην Ελλάδα και αναδασωτέες εκτάσεις = Forest fires in Greece and reforestated areas. Sinigoros **159**, 18–23 (2023)

ENTERPRISE GREECE. Invest & Trade.: Greek energy sector (n.d.). https://www.enterprisegreece.gov.gr/en/invest-in-greece/sectors-for-growth/energy. Accessed 13 Sept 2024

EUR-Lex.: Council Regulation (EU) 2022/2577 of 22 December 2022 laying down a framework to accelerate the deployment of renewable energy (n.d.). https://eur-lex.europa.eu/eli/reg/2022/2577/oj. Accessed 15 June 2024

EUR-Lex.: Directive (EU) 2018/2001 of the European Parliament and of the Council of 11 December 2018, preamble (n.d.). https://eur-lex.europa.eu/eli/dir/2018/2001/oj. Accessed 15 June 2024

EUR-Lex.: EU energy policy (n.d.). https://eur-lex.europa.eu/EN/legal-content/glossary/eu-energy-policy.html. Accessed 15 June 2024

Euronews.: Why are plans to build more wind farms in Greece so controversial? (2024). https://www.euronews.com/2024/04/12/why-are-plans-to-build-more-onshore-wind-farms-in-greece-so-controversial. Accessed 15 June 2024

European Commission.: Clean Energy. The European Green Deal (n.d.). Clean_energy_en.pdf

European sources online.: Directive (EU) 2023/2413 amending Directive (EU) 2018/2001, Regulation (EU) 2018/1999 and Directive 98/70/EC as regards the promotion of energy from renewable sources, and repealing Council Directive (EU) 2015/652 (n.d.). https://www.europeansources.info/record/proposal-for-a-directive-amending-directive-eu-2018-2001-regulation-eu-2018-1999-and-directive-98-70-ec-as-regards-the-promotion-of-energy-from-renewable-sources-and-repealing-council-directive. Accessed 15 June 2024

HELLENIC PARLIAMENT.: The Constitution of Greece (n.d.). https://www.hellenicparliament.gr/UserFiles/f3c70a23-7696-49db-9148-f24dce6a27c8/001-156%20aggliko.pdf. Accessed 17 June 2024

HELLENIC REPUBLIC Ministry of the Environment and Energy.: National Energy and Climate Plan (2019). https://energy.ec.europa.eu/system/files/2020-03/el_final_necp_main_en_0.pdf. Accessed 17 June 2024

Kati, V., Kassara, Ch., Vrontisi, Z., Moustakas, A.: The biodiversity-wind energy-land use nexus in a global biodiversity hotspot. Sci.

Total Environ. **768** (2020). https://doi.org/10.1016/j.scitotenv.2020.144471

Office of Energy Efficiency & Renewable Energy.: Wind Turbines: The Bigger, The Better (2024). https://www.energy.gov/eere/articles/wind-turbines-bigger-better#:~:text=Larger%20rotor%20diameters%20allow%20wind,areas%20with%20relatively%20less%20wind. Accessed 16 Sept 2024

Regulation (EU) 2021/1119 of the European Parliament and of the Council of 30 June 2021 establishing the framework for achieving climate neutrality and amending Regulations (EC) No 401/2009 and (EU) 2018/1999 ('European Climate Law') (OJ L 243, 9.7.2021, p. 1)

Reporters United.: Too much of a good thing? Wind power and the battle for Greece's wild heart (2021). https://www.reportersunited.gr/en/6557/too-much-of-a-good-thing-wind-power-and-the-battle-for-greeces-wild-heart/. Accessed 17 June 2024

RESINVEST.: 2022 Annual Statistics Hellenic Wind Energy Association (2023). https://resinvest.gr/2023/02/01/2022-annual-statistics-hellenic-wind-energy-association-2/. Accessed 15 June 2024

Spanos, I., Meliadis, I., Platis, P., Mantzanas, K., Samara, T., Meliadis, M.: Forest Land Ownership Change in Greece. COST Action FP1201 FACESMAP Country Report. European Forest Institute Central-East and South-East European Regional Office, Vienna (2015). http://facesmap.boku.ac.at/library/FP1201_Country%20Report_GREECE.pdf. Accessed 16 Sept 2024

The Role of Geography in Climate Change Mitigation

Territorial Development of Fragile Urban Contexts in the Face of Climate Change Challenges: A Comparative Analysis of Brindisi and Porto Marghera, Italy

Federica Epifani and Fabio Pollice

Abstract

The role played by fossil fuel companies in the territorialisation-deterritorialisation-reterritorialisation cycle provides a geographical interpretation of climate change. This helps in understanding the phenomenon, given the evolution of vertical (environment/human communities) and horizontal (industry, institutions, local communities, environmental movements, etc.) relationships across different scales. The establishment of fossil fuel companies in a region initiates irreversible transformation processes—both tangible (landscape, infrastructure, etc.) and intangible (production relations, programmatic trajectories, etc.). The impacts of such industries on the environment and climate play a dual role. On the one hand, they can be interpreted from a deterritorialisation perspective, especially when accompanied by harmful deindustrialisation forms. On the other hand, fossil fuel companies' acknowledged accountability for environmental degradation processes in their vicinity and contribution to climate change has sparked a diverse range of responses from multiple territorial actors, identifying various potential reterritorialisation perspectives. Based on these premises, this contribution focuses on two industrial urban contexts on the Italian Adriatic coast: Brindisi (Puglia, Southeast) and the area of Venice–Porto Marghera (Veneto, Northeast). Both areas host a petrochemical plant and other heavy industries, whose harmful effects on health, the environment, and the climate have led to recognising the two territories as sites of national concern. After a comparative analysis of the two areas' vulnerability profiles, carried out through the contextual vulnerability approach, the contribution delves into the development trajectories undertaken by the two territories to understand how environmental and climate issues have been addressed in local political agendas.

Keywords

Vulnerability · Fossil firms · Brindisi · Porto Marghera · Harmful deindustrialisation · Environmental justice

1 Introduction, Problem Statement, and Study Aim

Industrial development in Italy, particularly since the Second World War, has redefined not only the country's production trajectories but also its demography, socio-economic fabric, and trade relationships (Brugnoli and Fachin 2000; Caterina 2009; Muscarà 1967; Rota 2008). However, while these industrial strategies have bolstered Italian economic growth, enhanced competitiveness, and helped narrow the economic divide between the economically advanced North and the predominantly rural South, they have also resulted in immediate environmental and health consequences (Epifani 2024; Fatichenti 2018; Pellicano 2022).

Brindisi (Apulia, Southeast) and Venice–Porto Marghera (Veneto, Northeast) represent two emblematic examples of the Italian industrialist experience, currently facing a new cycle in which the environment, health, and climate change are not only strategic assets to invest in but unavoidable aspects on which to base future development paths.

The two areas have been for decades among the most relevant petrochemical poles in the country and Europe, attracting investments from the same companies (Montecatini, Enichem, Montedison, Eni Versalis, Dow Chemical, Syndial). The gas and power sectors are also paramount. In both cases, such industries' impacts on the environment and the health of people residing in the surrounding areas—especially those close to the petrochemical plants—caused the

F. Epifani (✉) · F. Pollice
Department of Human and Social Sciences, University of Salento, Lecce, Italy
e-mail: federica.epifani@unisalento.it

F. Pollice
e-mail: fabio.pollice@unisalento.it

P. Mohammad et al. (eds.), *Geographic Approaches to Climate Change and Mitigation: Urban and Rural Perspectives (Volume 1)*, Advances in Science, Technology & Innovation, https://doi.org/10.1007/978-3-031-92119-3_9

Italian government in 1998 to designate the two industrial areas as sites of national concern (SINs).

Geophysical and economic vulnerabilities are exacerbated by the persistence of hierarchic, multi-scalar centre–periphery relationships (Souza 2021), recalled by Shade (2015) as a "state-territorial strategy", according to which sacrificing a territory could be necessary and functional to preserve national interests. For this reason, we can refer to Brindisi and Venice–Porto Marghera as "sacrifice zones" (Bell 2016; Juskus 2023; Lerner 2012; Rosenbloom 2024; Souza 2021).

The peculiarity of the local scale and the territorial approach allows the observer to switch the focus from mitigation to adaptation, underscoring the relevance of environmental and climate emergencies to local agendas (Adger 2001; Adger et al. 2005; Argent 2019; Dannevig et al. 2012; Vincent 2023; Wilbanks and Kates 1999). Indeed, we argue that the local scale, and especially the urban dimension, represents a privileged point of view (Boyd and Juhola 2015; Dhar and Khirfan 2017; Kaplan and Holloway 2024; Kythreotis et al. 2023; Landauer et al. 2019). The big industrial companies, the main ones responsible for climate-changing emissions, choose their sites according to specific locational factors such as geomorphology, proximity to infrastructure, workforce availability, and political conditions. Once settled, an industrial plant acts as a territorialising driver, as it irreversibly shapes both tangible features (landscapes, infrastructures, etc.) and intangible characteristics (production relationships, planning, governance, etc.) Among the most evident transforming effects due to industrial development in both Brindisi and Venice–Porto Marghera, we can consider the boost of urbanisation, the migration flows from rural areas nearby, and changes in consumption habits (Ravenda 2018; Romeo 2024; Zazzara 2017). Nevertheless, degenerative processes caused by such activities' pressure on local ecosystems are also relevant (Bressan et al. 2023), and their effects can be detected both at the local scale—where, in the case of poorly resilient territorial systems, such processes end up in deterritorialisation—and at the global scale because of feedback mechanisms. As a consequence, heavily industrialised territories play a dual role, on the one hand, as hotspots of harmful effects on health, environment, and climate, and on the other hand, as areas that are among the first to suffer these direct or indirect effects, thus posing a problem of territorial vulnerability and environmental justice.

It is not rare that these vulnerable territories are also given a symbolic value, starting from which national and local institutions, environmental justice movements, but also the same industrial companies and other stakeholders become promoters of alternative development models, which can be interpreted from reterritorialising perspectives (Epifani 2024; Gonda et al. 2023; Knops and Vrydagh 2023). Concerning this, the two study areas show significant differences because the factors defining territorial resilience are different.

Hence, we argue that the concept of the sacrifice zone allows the comparative analysis to go beyond the mere description of the factors at play; instead, it supports a robust reflection on territorial vulnerability under the lenses of relatedness, both diachronic and synchronic, between the factors themselves.

Studies on sacrifice zones precede those on the territorial effects of climate change and rather address the impacts of industrial pollution. Nevertheless, as the UN stated (2022), "The chemical industry exacerbates the climate emergency by consuming more than 10% of fossil fuels produced globally and emitting an estimated 3.3 billion tons of greenhouse gas emissions annually. Global warming contributes to releasing and remobilising hazardous pollutants from melting glaciers and thawing permafrost". Indeed, links between pollution, territorial vulnerability, and climate change have been embedded for a long time by the environmentalist movements, most recently by the global political agendas; nowadays, they stand as one of the most urgent research questions.

After a short literature review and the methodological framework, we present the results of the comparative analysis between Brindisi and Venice–Porto Marghera. Following this, we critically discuss the evolutionary perspectives in the two territories, underscoring how, despite common factors shaping the two areas' industrial development paths, the combination of these factors with territorial peculiarities has resulted in different responses to the same type of impact and therefore different profiles of vulnerability and resilience strategies.

2 Literature Review

The interest of geography in climate change has been relevant in recent decades and has met an increasing demand for scientific support from global policy. In the European context, the launch of the so-called Green Deal has undoubtedly boosted the scientific debate. Bagliani et al. (2019) exposed the geographical nature of the topic and the relevance of geographical methodologies in studying climate change. They asserted that not only does climate change relate to spatial dynamics, but the impacts of climate change imbalance vary depending on the geographical scale. Hence, the geographical interpretation of climate change impacts allows us to read it under the lenses of the vertical and horizontal relationships developing in a territory, which, at the same time, embed the territory within a network of local–global ties.

On the other hand, as already mentioned, it is essential for this study to refer to the wide and consolidated scientific debate on sacrifice zones. There are at least two reasons for this. The first is the progressive inclusion of the climate crisis in the list of vulnerability factors to which already compromised territories and communities are subjected (Gonzalez 2021; Klein 2015); the second, which is more methodological, was clearly explained by Grasso and Delatin Rodrigues (2024) and consists of considering sacrifice zones as dynamic contexts where at least two groups of actors can be identified, namely the fossil machine and the transition actors, whose interactions determine spatial planning trajectories.

A huge scientific production has been conceived as a corollary of this interpretive framework, focusing mainly on critical positions. Because of the impossibility of encompassing the vast and lively scientific debate on the subject, it is possible to attempt to build a cognitive compass that starts with the main actors involved in these dialectical spatial interactions. In our study, this dialectical relationship involves, on the one hand, the major oil and gas companies and, on the other hand, the local communities, in dynamics that are always situated, thus dependent on the specific configurations they acquire within a particular territorial context. Researchers are investigating the roles of oil and gas companies as the primary responsible parties for climate-altering emissions (Grasso 2022; Grasso and Heede 2023), their interactions with other stakeholders, and the resulting territorial repercussions, particularly in terms of regulation and investment choices (Cojoianu et al. 2021).

Perception geography has significantly contributed to valorising the territorial dimension of observation, focusing attention on contextual conditions that determine how climate change is perceived (Weckroth and Ala-Mantila 2022). From this perspective, diverse and polarised geographies can be delineated, reflecting the varied attitudes and sensitivities developed by local communities toward climate change (Brieger 2019; Huddart-Kennedy et al. 2009; Marquart-Pyatt 2008; McCright et al. 2016).

Furthermore, the extensive debate on environmental justice is largely focused on the dialectic between local communities and companies. This debate has been analysed from various perspectives. Firstly, one perspective concerns risk distribution (Schlosberg 2013). Secondly, attention is given to recognising the damages suffered by local communities (Calderón-Argelich et al. 2021; Gemmiti et al. 2022). The political perspective on improving environmental conditions for communities holds equal relevance (Bryant 1995; Stephens and Willis 2017). Other aspects include transparency and inclusiveness in decision-making processes and the promotion of alternative territorial development strategies (Calderón-Argelich et al. 2021; Philippopoulos-Mihalopoulos 2014). Finally, a particularly challenging aspect is the dilemma-contradiction between local environmental justice and energy security (Huber 2015).

3 Methodology

The two study areas (Fig. 1. The study areas location map) have been analysed through the lenses of contextual vulnerability (Kelly and Adger 2000; O'Brien et al. 2007). According to this approach, territorial vulnerability concerning the effects of climate change depends on a complex combination of place-specific factors—environmental, political, economic, cultural, and ecological—on which the territory's response capacity and resilience rely. Moreover, the contextual vulnerability approach has the merit of promoting a dynamic reading of a territory's climate vulnerability, as it highlights the direct and indirect effects of multiple factors with a territorialising agent such as an industrial company in the different stages of the territorialisation–deterritorialisation–reterritorialisation (T–D–R) cycle.

The study is based on the joint analysis and interpretation of secondary data from multiple institutional sources (Italian Ministry of the Environment, ISPRA, Ministry of Economic Development, Ministry of Health, local statistical offices), official reports, and regulations. Also crucial was the review of the available historical–geographical scientific literature on Brindisi and Porto Marghera to contextualise the observed dynamics within a framework that corresponds to the T–D–R cycle. Reference was made to the most recent reports from the IPCC for the definition of adverse climate impacts. Furthermore, to reconstruct the main dynamics between the actors involved, archival research was conducted on the local print and digital press.

4 Results

4.1 Context

Porto Marghera covers over 2000 ha on the mainland encompassed within the municipality of Venice, the capital of the Veneto region in northeastern Italy (Fig. 2. Map of Venice–Porto Marghera). It is currently one of the largest industrial and commercial harbours in Europe. The area's infrastructure development started in 1917, aiming to modernise the island city and expand it to the mainland. It underwent several expansion projects, with the most significant one taking place after World War II to accommodate the petrochemical complex. The activation of the petrochemical complex in the early 1950s, which became one of the most important in Italy during the 1960s and 1970s, triggered the urban development of the surrounding area. Currently, there are over a thousand active businesses in the area, with approximately 45% belonging to the advanced tertiary sector and other service activities (Osservatorio Porto Marghera 2012–2023). The workforce amounts to about 11,000 employees, while the municipality

Fig. 9.1 The study areas location map

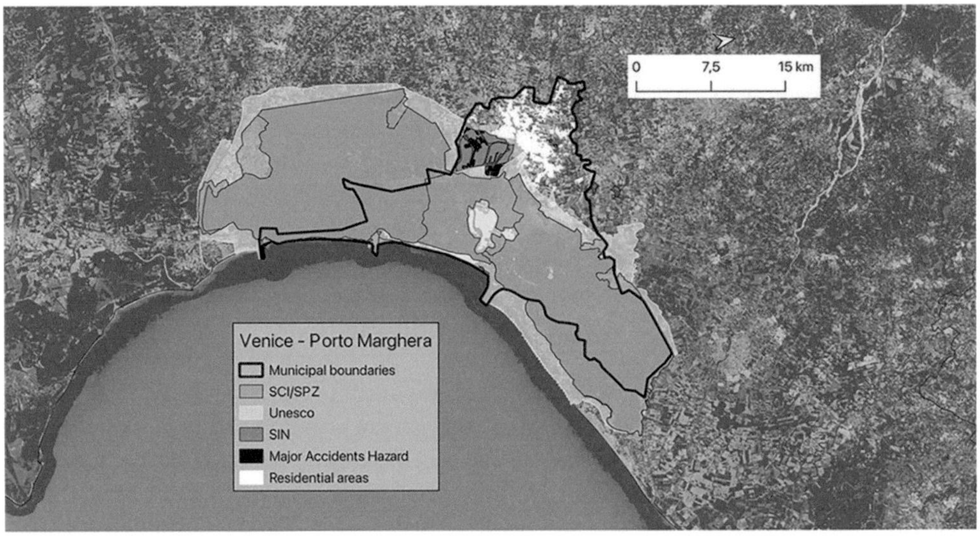

Fig. 9.2 Map of Venice–Porto Marghera

of Marghera is home to 28,000 residents, accounting for over 10% of the total population of Venice (Municipality of Venice 2021).

In 2000, the Venice–Porto Marghera SIN was established, covering an area of 5730 ha, including 513 ha of canals and 2200 ha of lagoon areas. In 2013, the perimeter was redefined, and current interventions are focused solely on land sites.

The industrial area of Brindisi (Fig. 3. Map of Brindisi), located in the southern part of the city's natural harbour, originated from the interventions promoted by the national government from 1957 onward to stimulate the industrialisation of Southern Italy according to the model of development poles. Within this framework, the same investor, Montecatini, established the Brindisi petrochemical complex a few years after the one in Venice. Currently, Brindisi is one of the most important industrial and energy hubs in the eastern Mediterranean; the industrial area covers 2255 ha and hosts 150 plants employing around 4000 workers (Consorzio Area Industriale di Brindisi 2023). The most significant sectors besides petrochemicals are aerospace and thermoelectric power generation.

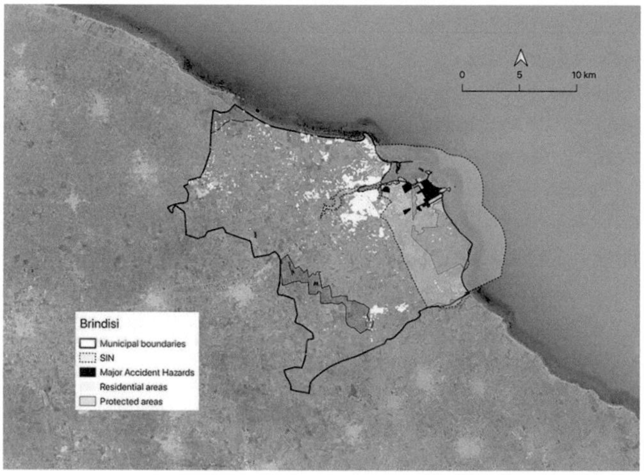

Fig. 9.3 Map of Brindisi

The Brindisi SIN is one of the largest in Italy, covering approximately 11,000 ha (5800 on land and 5600 at sea), adjacent to residential areas, and encompassing the city's entire industrial district. Within the SIN are also two protected areas. To the south stands the Enel "Federico II" coal-fired power plant, one of the largest and most polluting thermal power plants in Europe.

5 Vulnerability Profiles

Factors concurring to define the two areas' vulnerability profiles are classified into two categories: exposure to a climate-dependent extreme event—hence, the risk dimension—and susceptibility, namely the territorial system's capacity to cope with the harmful effects of such events—hence, the damage dimension (Bagliani et al. 2019; Cutter and Finch 2008).

Regarding exposure, the following aspects are considered:

- Geographical, locational, and contextual factors: Both Brindisi and Porto Marghera are situated in coastal areas. This is a primary locational factor for chemical, petrochemical, and thermoelectric plants, especially because both areas are characterised by port basins with sufficiently deep waters to allow the navigation and docking of tankers. However, coastal areas are inherently vulnerable (Doukakis 2005), and the high values of the coastal vulnerability index recorded in both Brindisi and Venice are due to intense industrialisation (Le Cozannet et al. 2020). In the case of Brindisi, it is important to note the industrial area's proximity to two protected areas, partly within the SIN perimeter. In the case of Porto Marghera, its proximity to the historic centre of Venice has been a primary concern since the inception of industrial activities. Besides the detrimental effects of emissions on the architectural fabric of the island, the opening of supplementary channels serving the industrial complex has been considered the cause of the island's increased exposure to the adverse effects of extreme climatic events, such as floods. It should be noted that the phenomenon of "*acqua granda*" (rising waters) characterises Venice and the entire Northern Adriatic. Still, its frequency increased in recent decades due to a general rise in sea levels (Lionello et al. 2009). In 1987, the historic centre of Venice and its lagoon were recognised as a UNESCO site. One of the criteria supporting such recognition states: "In the Mediterranean area, the Venice Lagoon represents an extraordinary example of a semi-lacustrine habitat made vulnerable by irreversible natural and climatic changes. In this interconnected ecosystem, where mudflats (marshy lands sometimes above, sometimes below sea level) are as important as islands, it is necessary to protect the houses built on stilts, fishing villages, and rice fields as much as palaces and churches" (UNESCO 2018).
- Social and economic factors: The socio-economic conditions of the analysed territories can be considered vulnerability factors since they played a fundamental role in the locational processes that attracted industrial firms. As mentioned earlier, the development of the Brindisi industrial complex needs to be contextualised within the framework of induced industrialisation aimed at stimulating the economic development of southern Italy. However, similar conditions of fragility affected Venice at the onset of its industrial development in the early twentieth century, when investments in the nascent Porto Marghera were promoted by companies external to the regional context, attracted by the opportunity to benefit from tax and regulatory incentives and the availability of labour. These were similar to the reasoning that 50 years later—and with the involvement of the same companies already operating in Marghera—would guide the induced industrialisation of Southern Italy (Zazzara 2017).
- Geo-economic and geopolitical aspects: Industrial plants like those operating in the two considered areas often have strategic significance. Consequently, locational choices are also determined by geo-economic and geopolitical factors. Similarly, the fate of these industries and, consequently, the territorial economies they influence are closely linked to international events. The oil crises of the 1970s contributed to the decline of the petrochemical industry and the consequent harmful deindustrialisation processes in both Brindisi and Porto Marghera.

Concerning susceptibility, some aspects to be considered are:

- Increase in neoplastic, pulmonary, and cardiovascular diseases: This trend emerges from a significant body of clinical studies conducted on the population of several contaminated areas such as SINs (Gianicolo et al. 2023; Pirastu et al. 2010; Stephens and Willis 2017), which have highlighted the correlation between the increase in such pathologies and proximity to industrial plants. This scientific evidence, moreover, has contributed to exacerbating the conflict, as it directly intervenes in delineating the responsibility profiles of industrial firms located in the area. According to one of the most recent reports on the environmental and health risks due to industrial pollution, such fragility is higher in the contaminated areas in southern Italy due to their worse socio-economic conditions. Such trends are confirmed by our case study, as discussed further below.
- Erosion of ecosystem services (Martinez-Alier et al. 2016), particularly biodiversity, agricultural, and, in the case of Brindisi, tourist services, due to soil consumption and contamination of land and water surrounding the industrial area: In the case of Venice, cultural aspects have been affected too, as Cavallo stated concerning the loss of the so-called "lagoon genre de vie" (Cavallo 2016; de la Blache 1911).
- Increase in average temperatures: Both Brindisi and Venice have registered an increase in average temperature. In particular, a recent study by OBC Transeuropa for EDJNet (2020) showed that over the last 60 years, the city of Brindisi has recorded an increase in average temperatures of 4°, compared to a regional average value of 2.9. At the European level, Brindisi is among the areas that have recorded the most relevant temperature increases. Concerning Venice, Spano et al. (2021) showed increasing trends in the average temperatures and the number of consecutive hot days, while the trend in the number of consecutive days without rainfall is decreasing.

6 Discussion

The petrochemical sector in the industrial experiences of Brindisi and Venice–Porto Marghera is a relevant aspect deserving of thorough analysis. The presence of these industrial complexes has played a crucial role in fostering environmental awareness both among workers and within the local community as a whole. This is especially true considering the recognition of the environmental impacts and harmfulness of industrial activities, which has guided the trajectories of post-industrial reterritorialisation in the two sites, leading to different outcomes.

One of the main reasons for the spread of environmental awareness is the discovery that certain substances used in the production cycle have highly harmful effects on health and the environment. This led to the genesis of environmental movements as early as the 1960s to ensure a safer and healthier working environment. A territorially rooted environmental movement has represented a territorial resilience device and has driven the development of place-based climate adaptation strategies.

In this regard, it is interesting to underline that the early movements in Brindisi and Venice–Porto Marghera share their origins in the claims against harmfulness made since the late 1960s by certain trade union currents supported by associations of physicians and scholars. It has already been observed how such a manifesto, by acting on both national and local scales, helped contacts and exchanges among activists from the two territories; activists, hence, acted as pivots with disruptive effects on the relationships with the fossil bloc and, consequently, on industrial planning (Epifani 2024). At the same time, in both Brindisi and Venice–Porto Marghera, the intermingling between the dilemmas of work/development and health/environment, coupled with the progressive spread of environmental awareness among local communities, has resulted in the polarisation of the environmental movement into two distinct, partially conflictual, factions (Feltrin 2022; Ravenda 2018): on the one hand, the workplace-centred camp, supporting investments to expand production while making it greener and safer, and on the other hand, the community-centred camp, anti-productivist, supporting a global decrease in plastic production (Feltrin 2022). The influence of these two different approaches has also affected the climate change–related claims: while the workplace-centred approach shows a certain favour toward a transitioning fossil bloc, the community-centred approach has embedded the requests of the ultimate climate change movements against the fossil bloc.

Another significant aspect is the progressive resizing and downsizing of the petrochemical sector since the late 1970s, resulting in harmful deindustrialisation effects on the affected territories. As already underlined, one of the most relevant traces of the harmful deindustrialisation process, from both the environmental and socio-economic points of view, has been the demarcation of SINs. Besides being a tangible sign of deterritorialisation, SINs have contributed to redefining the role of the same companies responsible for environmental damage, as such companies have been called upon to remediate and restore the contaminated areas, as well as sustainably reindustrialise them. Moreover, SINs have urged local administrations and stakeholders to re-orient the strategic agenda to address not only industrial downsizing but also the impairment of other production activities on the site. Nevertheless, the way the actors of the two territories have addressed the socio-economic and employment crisis resulting from the chemical sector's downturn has been different and representative of the importance of territorial factors in shaping territorial projects.

In Porto Marghera, excess labour has been gradually absorbed by small businesses and services, following a model of territorial reconversion oriented toward advanced tertiary sectors. Conversely, in Brindisi, the employment crisis has been addressed through the construction of a thermal power plant managed by ENEL. Even today, observing the data on operating companies in the two industrial zones considered (Osservatorio Porto Marghera, Comune di Venezia, ASI Brindisi), it is noted that, given a similar extension, the companies active in Brindisi amount to about 10% of those operating in Porto Marghera. Furthermore, since the late 1990s, Porto Marghera has undergone a process of reconversion of the productive fabric, focusing more on commercial activities and aiming at advanced tertiary sectors, following the same reconfiguration paths implemented in the Ruhr region. This reconversion process also involves large industrial plants. In the petrochemical complex, in 2022 the steam-cracking and aromatics plants of ENI Versalis were definitively closed, while since 2013, various companies have been investing in green chemistry.

Among new investments, it is worth indicating those undertaken by ENI itself, particularly the establishment of a bio-refinery—the first of its kind in Italy—through the revamping of the existing plants. Such a project is representative of the green branding strategy that the Italian company has recently embraced; the resulting narrative not only relocates ENI from the list of the main perpetrators of environmental damage to the list of actors most committed to energy transition processes and the fight against climate change but also tends to favour a reversal of the symbolic configuration of Marghera, which from a place of harmfulness becomes the emblem of a new production model oriented toward the energy transition, emissions reduction, and the circular economy.

Similarly, the coal-fired power plant, like the one in Brindisi managed by ENEL, is currently inactive and undergoing gas conversion.

It should be noted that the environmental issue in Venice is not resolved, and indeed it is the subject of territorial dispute on several fronts. Concerning Porto Marghera, the narrative endorsed by the fossil block, supported by a large part of trade union circles and workplace-centred environmentalism, is instead strongly contested by community-centred environmentalists who accuse ENI of not having adequately conducted reclamation activities, thus preventing the full reuse of the compromised areas (Tedesco 2023). More generally, the Venetian environmental movement perceives ENI's reindustrialisation initiatives as greenwashing actions, as evidenced by recent protests against the potential construction of a sludge incinerator, which ENI also promotes and the Italian National Institute for Health harshly criticises. Moreover, in addition to the future of the industrial complex of Porto Marghera, which remains a complex industrial crisis area according to the Government, other issues fuel environmental conflicts, such as the passage of large cruise ships through the lagoon or the construction of MOSE, a large mobile infrastructure designed to protect Venice from floods, but which would have serious impacts on the fragile lagoon ecosystem. However, the prospect of a green transition appears embedded in the programmatic actions of local governance and the relationships between local institutions and industrial players. Similarly, the climate issue is also present in the city's development strategies: the adhesion to the European Covenant of Mayors for Climate and Energy, signed in 2020, committed the local administration to draw up a Sustainable Energy and Climate Action Plan. The plan sets the ambitious goal of achieving carbon neutrality by 2050, with an intermediate target in 2030 of a reduction of almost 70% compared to the 2005 level of CO_2 emissions. The plan is currently being drafted; according to the desk research we have run on institutional channels, the drafting work is supposed to be transversal, as it involves all municipal offices in different capacities in the planning of mitigation and adaptation measures; however, this transversality does not seem to transcend the sphere of public administration—except for some preliminary consultancy activities attributed to a private party and a research consortium led by the local university—while at the time of writing this paper (October 2024), there is no knowledge of moments of participatory sharing with other territorial actors, such as associations, trade unions, and, above all, industries themselves.

Unlike Porto Marghera, Brindisi shows a higher level of dependence on exogenous investments. This dependence is due to the presence of a weak local entrepreneurial fabric, unable to absorb excess labour following the productive crises of sectors such as chemicals and energy, increasingly dependent on international geo-economic and geopolitical scenarios, and not able to support innovation processes—in both products and processes—that are significant enough to favour the reconversion of the productive fabric. It should also be noted that Brindisi's dependence on exogenous interventions has contributed to consolidating its condition as a sacrifice zone. The strategic interventions often entail significant environmental and health costs for the host territories: the thermal power plant, built in the 1980s and operational since the mid-1990s, is currently one of the most polluting industrial sites in Europe. Since the early 2000s, it has become a symbol of the first claims of environmental movements, including major transnational organisations such as Greenpeace, explicitly targeting the fossil fuel industry's responsibilities with respect to climate change.

It should be noted that, as in the case of Porto Marghera, the environmental movement in Brindisi has been active and cross-sectional since its inception, particularly within its community-centred component. On several occasions, this movement has managed to ignite particularly intense conflicts with local and national institutions, as well as with territorial actors supporting fossil fuel initiatives, especially trade unions. Notably, in the early 1980s, protests were held against the construction of the aforementioned coal-fired power plant by ENEL, and between 2000 and 2009, demonstrations were organised against a proposed LNG terminal. In this second case, from the mid-2000s onwards, there was an unprecedented alignment of views between environmental movements, the local population, and the newly renewed local, provincial, and regional administrations. This period coincided with the strengthening of Brindisi's environmental movement, which culminated, in 2018, in the election of a city council reflecting the movement's values. In 2022, the city adopted a Sustainable Energy and Climate Action Plan, the result of a complex stakeholder engagement process. However, despite explicitly mentioning the effects of industrial activities—particularly in the petrochemical sector—on the Brindisi area's increased environmental sensitivity, the Plan does not foresee any specific interventions in the industrial sector or the secondary sector as a whole. Instead, "in these sectors, coordinated action at the regional, national (or higher) level would certainly have a more significant impact" (Apulia Region 2022, p. 44).

In any case, the opposition of the environmental movement in Brindisi to a territorial development strategy based on large industry has had limited effects without being able, at the moment, to stimulate the launch of an effective action plan for the industrial reconversion of the city. On the other hand, the condition of endemic socio-economic distress has favoured the exacerbation of the conflict between labour and the environment/health in Brindisi. To further support this point, reference is made to the outgoing environmentalist administration's failure to be re-elected in the 2023 municipal elections, having been labelled "the administration of no to everything" (Quaranta 2023), considered too radical in managing the combined crisis resulting from the petrochemical sector's downsizing and coal's phase-out by 2025. On the other hand, the former green mayor emphasised that the administration had initiated, immediately after COVID-19, an intense negotiation process with ENEL, which led to the definition of a preliminary draft for the reconversion of the decommissioned plants. However, this process came to a halt in 2023 with the change in ENEL's leadership.

Therefore, while on the one hand, the need to create jobs has favoured the establishment and permanence of particularly impactful plants over the years, on the other hand, the condition of territorial conflict, which persists also within the local institutions, makes the hypothesis of a green transition process in Brindisi extremely complex: indeed, in addition to the cancellation of the project to reconvert ENEL's large coal-fired power plant, other projects proposed by players other than ENI or ENEL, which involved the installation of wind and photovoltaic parks, remain dormant. In other words, Brindisi is no longer considered a geographic optimum for exogenous investments with a driving effect on the rest of the industrial sector.

Finally, it is important to highlight the symbolic dimension of the T-D-R processes, focusing on how the industry has been incorporated into the two cities' symbolic identity systems and how this has influenced planning decisions.

Indeed, more than 60 years after the start of Venetian petrochemical activities (and more than a century since the founding of Porto Marghera), the industry has been embedded in Venice's place identity, alongside its historically rooted identity as a city of art. Compared to the latter, the industrial city identity is undoubtedly a traumatic and conflicted one (Pain 2021), but equally pervasive, as it forms an integral part of the inhabitants' lived space. Moreover, the city's industrial identity, confirmed and renewed in the territorial development plans proposed by actors from the fossil fuel side who have embraced the energy transition, also plays a central role in the initiatives of those promoting alternative development models. For these groups, the active memory of and about Marghera, reinforced by structured research activities, serves as the starting point for the implementation of new ways to inhabit and reuse dismissed industrial spaces, ranging from performing arts to industrial tourism.

On the contrary, Brindisi is still undergoing a phase of harmful deindustrialisation that seems to have no way out. Certainly, one of the causes lies in the extreme dependence of Brindisi's industrial fabric on exogenous investments from the major fossil fuel players. However, another aspect contributing to the weakening of territorial resilience is that Brindisi, unlike Venice, does not seem to have integrated industry into its identity. In other words, while Venice today can also be identified as an industrial city, Brindisi is rather a city *with* industry, as over the decades it has not developed a competitive, indigenous industrial network capable of absorbing the effects of cyclical crises. Additionally, there has been a lack of an integrated vision in which industrial development was functional to territorial development, supported by a network of actors capable of carrying forward long-term projects for the territory's ecological transition.

7 Conclusion

The analysis highlights the T–D–R cycle's usefulness in providing an interpretative framework to study the territorial dimension of environmental and climate issues, as well as intermingling among the actors at play.

According to this interpretation, industrial firms serve as drivers of territorialisation, while their downsizing results in deterritorialising effects that are marked by significant environmental costs, thereby affecting climate change resilience. The analysis of territorial factors has a dual significance. On the one hand, it allows for the identification of the conditions predisposing a territory to specific environmental and climatic risks directly or indirectly related to the presence of such industrial facilities. On the other hand, it allows territories' responses to be interpreted from a reterritorialisation perspective.

Territorial factors also determine the modalities through which individual territories operate within networks of local–global relations. The recent phase of downsizing and decommissioning of traditional industry has also been triggered by global policies aimed at mitigating climate change. Different territorial predispositions, particularly in terms of social and human capital, have determined specific industrial governance arrangements, which have guided the investment choices of major industrial players engaged in the innovative green transition. This is destined to redefine not only the industrial geography of the country but also that of territorial development and disparities, considering the challenges faced by territories where strong industrialisation coexists with endemic socio-economic distress: these territories in Italy are those that in the 1950s were affected by the establishment of development poles, predominantly in the southern part of the country.

Given the emerging evidence, we believe that, in heavily compromised contexts, a key factor in territorial structures' capacity to increase their resilience levels lies in the extent to which industry, not only as a mere productive activity but as a pervasive territorialising factor, has been integrated into the socio-symbolic and programmatic systems of the local community and consequently redefined within the territorial vision and mission. In the case of Brindisi and Venice, this means, in other words, innovatively rethinking a territorial legacy that, in the past, made these two cities emblematic of global environmental crises and today makes them emblematic of the global climate challenge, where the fragilities induced by industrial impacts are now intertwined with demands for climate justice. This goal is discernible in both cities' strategic planning trajectories, as well as the social capital represented by environmental movements, which have played a fundamental role in raising awareness within the communities, thereby challenging the dominant development model during the major phases of crisis. Compared to Brindisi, Venice seems to be able to count on greater long-term convergence among local actors. But most importantly, Venice has integrated into its territorial development trajectories the recognition of Marghera as a polysemic landscape, where the signs of environmental disaster are not erased but become part of collective memory and are incorporated into the city's strategic choices. From this perspective, the Venetian experience is significant and represents a holistic approach to managing vulnerable territories, one that could serve as a model for similar contexts.

Competing Interests The authors have no conflicts of interest to declare that are relevant to the content of this chapter.

References

Adger, W.N.: Scales of governance and environmental justice for adaptation and mitigation of climate change. J. Int. Dev. **13**(7), 921–931 (2001)

Adger, W.N., Arnell, N.W., Tompkins, E.L.: Successful adaptation to climate change across scales. Glob. Environ. Chang. **15**(2), 77–86 (2005)

Apulia Region.: Sustainable Energy and Climate Action Plan. Brindisi (2022). Retrieved Nov 6, 2024, from https://programming14-20.italy-croatia.eu/documents/275198/2777230/D5.3+SECAP+Brindisi_v2b.pdf/8aefb6dc-fbb7-a702-edb4-039649019e95?t=1661856301110

Argent, N.: Rural geography II: scalar and social constructionist perspectives on climate change adaptation and rural resilience. Prog. Hum. Geogr.. Hum. Geogr. **43**(1), 183–191 (2019). https://doi.org/10.1177/030913251774311

Bagliani, M.M., Antonella, P., Bonati, S.: Il cambiamento climatico in prospettiva geografica. Aspetti fisici, impatti, teorie. Il mulino (2019)

Bell, M.Z.: Spatialising procedural justice: fairness and local knowledge mobilisation in nuclear waste siting. Local Environ. **26**(1), 165–180 (2021). https://doi.org/10.1080/13549839.2020.1867841

Bell, S.E.: Fighting King Coal: The Challenges to Micromobilization in Central Appalachia. MIT Press (2016)

Boyd, E., Juhola, S.: Adaptive climate change governance for urban resilience. Urban Stud. **52**(7), 1234–1264 (2015). https://doi.org/10.1177/0042098014527483

Bressan, G., Roberta, G., Rosaria, P.M.: Ambiente e industria in Italia. Contaminazione e fragilità sociale nei Siti di interesse nazionale per la bonifica. In: Oltre la globalizzazione-Narrazioni/Narratives. Memorie Geografiche NS 23, pp. 513–518. Società di Studi Geografici (2023)

Brieger, S.A.: Social identity and environmental concern: the importance of contextual effects. Environ. Behav.behav. **51**(7), 828–855 (2019). https://doi.org/10.1177/0013916518756988

Brugnoli, A., Fachin, S.: Concentrazione e diffusione: la geografia industriale italiana 1981–1991. Economia e Politica Industriale (2000/107) (2000)

Bryant, B. (Ed.).: Environmental Justice: Issues, Policies, and Solutions. Island Press (1995)

Calderón-Argelich, A., Benetti, S., Anguelovski, I., Connolly, J.J., Langemeyer, J., Baró, F.: Tracing and building up environmental justice considerations in the urban ecosystem service literature: a systematic review. Landsc. Urban Plan.. Urban Plan. **214**, 104130 (2021). https://doi.org/10.1016/j.landurbplan.2021.104130

Caterina, G.: Teorie spaziali e sviluppo industriale del Mezzogiorno, vol. 1, pp. 1–94. Laveglia & Carlone (2009)

Cavallo, F.: La Laguna di Venezia, dispute territoriali e movimenti sociali. Rivista Geografica Italiana **123**(2), 125–139 (2016)

Cojoianu, T.F., Ascui, F., Clark, G.L., Hoepner, A.G., Wójcik, D.: Does the fossil fuel divestment movement impact new oil and gas fundraising? J. Econ. Geogr. **21**(1), 141–164 (2021). https://doi.org/10.1093/jeg/lbaa027

Consorzio Area Industriale di Brindisi. https://www.asi.br.it/ws/consorzio/area-industriale-di-brindisi/

Cutter, S.L., Finch, C.: Temporal and spatial changes in social vulnerability to natural hazards. Proc. Natl. Acad. Sci. **105**(7), 2301–2306 (2008). https://doi.org/10.1073/pnas.0710375105

Dannevig, H., Rauken, T., Hovelsrud, G.: Implementing adaptation to climate change at the local level. Local Environ. **17**(6–7), 597–611 (2012). https://doi.org/10.1080/13549839.2012.678317

de la Blache, P.V.: Les genres de vie dans la géographie humaine: Premier article. Annales De Géographie **20**(111), 193–212 (1911)

Dhar, T.K., Khirfan, L.: A multi-scale and multi-dimensional framework for enhancing the resilience of urban form to climate change. Urban Clim. **19**, 72–91 (2017). https://doi.org/10.1080/09640568.2016.1178107

Doukakis, E.: Coastal vulnerability and risk parameters. Eur. Water **11**(12), 3–7 (2005)

Epifani, F.: Environmental justice and transformative geographies in the discourses of the environmental movement in Brindisi. In: Albanese, V., Fanetti, S., Minazzi, R. (eds.) Social Mobilisation for Climate Change, pp. 32–50. Routledge (2024)

Fatichenti, F.: Per una valutazione dell'impatto socioambientale dell'industria. Proposte di indicatori dal caso paradigmatico di un centro cementifero in Umbria. Semestrale di studi e ricerche di geografia (1) (2018). https://doi.org/10.13133/1125-5218.14969

Feltrin, L.: Situating class in workplace and community environmentalism: working-class environmentalism and deindustrialisation in Porto Marghera, Venice. Sociol. Rev. **70**(6), 1141–1162 (2022). https://doi.org/10.1177/00380261221106895

Gemmiti, R., Sanna, V.S., Prisco, M.R.: La giustizia ambientale in Italia. Riscontri empirici e percorsi metodologici per l'analisi dei Siti di Interesse Nazionale per le bonifiche. Geotema 60–70 (2022)

Gianicolo, E.A.L., Russo, A., Portaluri, M.: SENTIERI reports and the national priority contaminated site of Brindisi (Apulia Region, Southern Italy). Epidemiologia e prevenzione **47**(1–2), 3 (2023). https://doi.org/10.19191/EP23.1-2.A618.021

Gjergji, O., OBC Transeuropa.: Glocal Climate Change [Interactive Map] (2020). https://climatechange.europeandatajournalism.eu/it/map

Gonda, N., Flores, S., Casolo, J.J., Nightingale, A.J.: Resilience and conflict: rethinking climate resilience through Indigenous territorial struggles. J. Peasant Stud. **50**(6), 2312–2338 (2023). https://doi.org/10.1080/03066150.2022.2161372

Gonzalez, C.G.: The sacrifice zones of carbon capitalism: race, expendability, and loss and damage. In: Doelle, M., Seck, S.L. (eds.) Research Handbook on Climate Change Law and Loss & Damage, pp. 43–59. Edward Elgar Publishing (2021)

Grasso, M., Delatin Rodrigues, D.: Rage against the fossil machine: the deactivation of fossil energy production in Italy. Energy Res. Soc. Sci. **115**, 103655 (2024). https://doi.org/10.1016/j.erss.2024.103655

Grasso, M., Heede, R.: Time to pay the piper: fossil fuel companies' reparations for climate damages. One Earth **6**(5), 459–463 (2023). https://doi.org/10.1016/j.oneear.2023.04.012

Grasso, M.: From Big Oil to Big Green: Holding the Oil Industry to Account for the Climate Crisis. MIT Press (2022)

Huber, M.: Energy and social power: from political ecology to the ecology of politics. In: The Routledge Handbook of Political Ecology, pp. 481–492. Routledge (2015)

Huddart-Kennedy, E., Beckley, T.M., McFarlane, B.L., Nadeau, S.: Rural-urban differences in environmental concern in Canada. Rural. Sociol.sociol. **74**(3), 309–329 (2009). https://doi.org/10.1526/003601109789037268

IPCC.: Climate Change 2022: Impacts, Adaptation and Vulnerability. Contribution of Working Group II to the Sixth Assessment Report of the Intergovernmental Panel on Climate Change. Cambridge University Press. Cambridge University Press (2022). https://doi.org/10.1017/9781009325844

Juskus, R.: Sacrifice zones: a genealogy and analysis of an environmental justice concept. Environ. Humanit. **15**(1), 3–24 (2023). https://doi.org/10.1215/22011919-10216129

Kaplan, D., Holloway, S.: Urban Geography. Wiley & Sons (2024)

Kelly, P.M., Adger, W.N.: Theory and practice in assessing vulnerability to climate change and Facilitating adaptation. Clim. Change. Change **47**(4), 325–352 (2000)

Klein, N.: This Changes Everything: Capitalism vs. The Climate. Simon and Schuster (2015)

Knops, L., Vrydagh, J.: Between hopes and systemic unsustainability: an analysis of citizens assemblies' potential on climate change. In: Reuchamps, M., Vrydagh, J., Welp, Y. (eds.) De Gruyter Handbook of Citizens' Assemblies, pp. 211–226. De Gruyter (2023)

Kythreotis, A.P., Jonas, A.E., Mercer, T.G., Marsden, T.K.: Rethinking urban adaptation as a scalar geopolitics of climate governance: climate policy in the devolved territories of the UK. Territ. Polit. Gov. **11**(1), 39–59 (2023). https://doi.org/10.1080/21622671.2020.1837220

Landauer, M., Juhola, S., Klein, J.: The role of scale in integrating climate change adaptation and mitigation in cities. J. Environ. Plann. Manage. **62**(5), 741–765 (2019). https://doi.org/10.1080/09640568.2018.1430022

Le Cozannet, G., Oliveros, C., Brivois, O., Giremus, A., Garcin, M., Lavigne, F.: Detecting changes in european shoreline evolution trends using Markov chains and the Eurosion database. Front. Mar. Sci. **7**, 326 (2020). https://doi.org/10.3389/fmars.2020.00326

Lerner, S.: Sacrifice Zones: The Front Lines of Toxic Chemical Exposure in the United States. Mit Press (2012)

Lionello, P., Baldi, M., Brunetti, M., Cacciamani, C., Maugeri, M., Nanni, T., Pavan, V., Tomozeiu, R.: Eventi climatici estremi: tendenze attuali e clima futuro sull'Italia. In: Castellari, S., Artale, V. (eds.) I cambiamenti climatici in Italia: evidenze, vulnerabilità e impatti, pp. 81–106. Bononia University Press (2009)

Marquart-Pyatt, S.T.: Are there similar sources of environmental concern? Comparing industrialized countries. Soc. Sci. Q. **89**(5), 1312–1335 (2008).https://doi.org/10.1111/j.1540-6237.2008.00567.x

Martinez-Alier, J., Temper, L., Del Bene, D., Scheidel, A.: Is there a global environmental justice movement? J. Peasant Stud. **43**(3), 731–755 (2016). https://doi.org/10.1080/03066150.2016.1141198

McCright, A.M., Marquart-Pyatt, S.T., Shwom, R.L., Brechin, S.R., Allen, S.: Ideology, capitalism, and climate: explaining public views about climate change in the United States. Energy Res. Soc. Sci. **21**, 180–189 (2016). https://doi.org/10.1016/j.erss.2016.08.003

Municipality of Venice (n.d.). https://www.comune.venezia.it/it

Muscarà, C.: La Geografia Dello Sviluppo. Sviluppo industriale e politica geografica nell'Italia del secondo dopoguerra. Edizioni di Comunità (1967)

O'brien, K., Eriksen, S., Nygaard, L.P., Schjolden, A.N.E.: Why different interpretations of vulnerability matter in climate change discourses. Clim. Policy **7**(1), 73–88 (2007). https://doi.org/10.1080/14693062.2007.9685639

Osservatorio Porto Marghera (n.d.). https://www.comune.venezia.it/it/osservatorioportomarghera

Pain, R.: Geotrauma: violence, place and repossession. Prog. Hum. Geogr.. Hum. Geogr. **45**(5), 972–989 (2021). https://doi.org/10.1177/0309132520943676

Pellicano, A.: Impatto ambientale e industria: il caso degli impianti cementiferi di Colleferro e Guidonia. Documenti geografici (2), 255–270 (2022). https://doi.org/10.19246/DOCUGEO2281-7549/202102_17

Philippopoulos-Mihalopoulos, A.: Spatial Justice: Body, Lawscape, Atmosphere. Routledge (2014)

Pirastu, R., Ancona, C., Iavarone, I., Mitis, F., Zona, A., Comba, P.: SENTIERI Project. Mortality study of residents in Italian polluted sites: evaluation of the epidemiological evidence. Epidemiol. Prev. **34**(5–6), 1–96 (2010)

Quaranta, N.: Brindisi, cinque mesi di Marchionna: dal contestato NO a tutto al NI pronunciato sempre. Quotidiano di Puglia (2023). https://www.quotidianodipuglia.it/brindisi/brindisi_sindaco_marchionna_primi_cinque_mesi_giunta_comunale_centrodestra-7729249.html

Ravenda, A.F.: Carbone. Inquinamento industriale, salute e politica a Brindisi. Meltemi (2018)

Romeo, S.: La Repubblica e il mare: La parabola delle aree portuali-industriali in Italia. Una prospettiva storica. Italia Contemporanea-Sezione Open Access, 305 (2024). https://doi.org/10.3280/IC305-oa2

Rosenbloom, J.: Sacrifice zones. Nevada Law J. **24**(3), 892–962 (2024)

Rota, M.: Sviluppo industriale e dualismo economico in Italia nella seconda metà del XX secolo. Rubbettino (2008)

Schlosberg, D.: Theorising environmental justice: the expanding sphere of a discourse. Environ. Polit. **22**(1), 37–55 (2013). https://doi.org/10.1080/09644016.2013.755387

Shade, L.: Sustainable development or sacrifice zone? Politics below the surface in post-neoliberal Ecuador. Extr. Ind. Soc. **2**(4), 775–784 (2015). https://doi.org/10.1016/j.exis.2015.07.004

Souza, M.L.: 'Sacrifice zone': the environment–territory–place of disposable lives. Commun. Dev. J. **56**(2), 220–243 (2021). https://doi.org/10.1093/cdj/bsaa042

Spano, D., Mereu, V., Bacciu, V., Barbato, G., Buonocore, M., Casartelli, V., Ellena, M., Lamesso, E., Ledda, A., Marras, S., Mercogliano, P., Monteleone, L., Mysiak, J., Padulano, R., Rafa, M., Ruiu, M.G.G., Serra, V., Villani, V.: Analisi del rischio. I cambiamenti climatici in sei città italiane (2021). Retrieved Nov 6, 2024, from https://www.cmcc.it/it/rischio-clima-citta-2021

Stephens, C., Willis, R.: Environmental justice and health. J. Med. **342**, 406–413 (2017)

Tedesco, L.: Nelle notti abitate da fuochi di chimica combusta: giustizia ambientale e pratiche di resistenza nella laguna veneta (1990–2022). L'altro Diritto. Rivista, 7 (2023)

Unesco.: Venezia e la sua laguna (2018). https://www.unesco.it/it/unesco-vicino-a-te/siti-patrimonio-mondiale/venezia-e-la-sua-laguna/

United Nations.: The right to a clean, healthy, and sustainable environment: Non-toxic environment (A/HRC/49/53). United Nations General Assembly (2022). Retrieved Nov 6, 2024, from https://undocs.org/en/A/HRC/49/53

Vincent, K.: Development geography II: community-based adaptation and locally-led adaptation. Prog. Hum. Geogr.. Hum. Geogr. **47**(4), 604–612 (2023). https://doi.org/10.1177/03091325231166076

Weckroth, M., Ala-Mantila, S.: Socioeconomic geography of climate change views in Europe. Glob. Environ. Chang. **72**, 102453 (2022). https://doi.org/10.1016/j.gloenvcha.2021.102453

Wilbanks, T.J., Kates, R.W.: Global change in local places: how scale matters. Clim. Change. Change **43**(3), 601–628 (1999)

Zazzara, G.: I cento anni di Porto Marghera (1917–2017). Italia Contemporanea-Sezione Open Access **284** (2017). https://doi.org/10.3280/ic284-oal

Analysis and Assessment of NDVI Variations in Response to Climate Change Among Wild Trees in the Zhongar Alatau Mountain Region

Marina Khusnitdinova, Aisha Taskuzhina, Nazym Kerimbek, Alexandr Pozharskiy, Gulnaz Nizamdinova, and Dilyara Gritsenko

Abstract

The fluctuations in climatic factors, such as precipitation and temperature, play a significant role in driving changes in the normalized difference vegetation index (NDVI). Our study aimed to assess and analyze the inter-annual variation in the NDVI of wild trees in the Zhongar Alatau mountain region, specifically within the Zhongar Alatau State National Natural Park (SNNP), spanning from 2006 to 2023. We focused on data collected in August to understand the seasonal dynamics of NDVI in this area. NDVI values and their classifications were derived from Landsat imagery (Landsat 4-5, 8-9), which were processed using QGIS software. We created maps depicting NDVI values for the Zhongar Alatau SNNP, as well as diagrams illustrating the inter-annual variation of NDVI, precipitation, and temperature. On average, NDVI values of wild trees in the study area ranged from 0.31 to 0.44, with maximum values reaching approximately 0.50–0.55. The findings of the study relationship between climatic factors and NDVI values that NDVI demonstrates a positive correlation with precipitation while showing minimal correlation with temperature by using simple linear regression, and correlation coefficient analysis. In specific years such as 2006, 2008, 2014, 2016, and 2021, notably low NDVI values were identified, as closely associated with precipitation levels. The short intervals between these unfavorable years may contribute to genetic erosion within wild tree populations. These results highlight the critical importance of conserving biodiversity among wild trees, particularly by considering the trends in vegetation responses to climate change.

Keywords

NDVI variations · Climate change · Wild trees · Kazakhstan · M. sieversii

1 Introduction

Natural forests harboring wild trees are esteemed as the most valuable plant communities in terms of their uniqueness, genetic potential, ecological, scientific, and practical significance (Aleksanyan et al. 2011). Wild trees and their hybrids hold significant value as a reservoir of genetic resistance against both biotic and abiotic stressors, playing a pivotal role in breeding programs worldwide (Kolchenko et al. 2023).

One of the most important trees growing in the mountains of Kazakhstan, including Zhongar Alatau region, is *Malus Sieversii. M. sieversii*, commonly known as the wild apple, is native to the mountainous regions of Central Asia, including parts of Kazakhstan, Kyrgyzstan, Tajikistan, and China (Wang et al. 2018). It is considered the primary ancestor of the domesticated apple (*Malus domestica*) and is renowned for its genetic diversity and resilience. *M. sieversii* plays a pivotal role in the biodiversity of Kazakhstan by serving as a keystone species, providing habitat and food for various wildlife species, including birds and mammals. Its ability to thrive in diverse environmental conditions, including high altitudes and extreme temperatures, underscores its adaptability and resilience (Tian et al. 2022). Additionally, the primary forest-forming species of Zhongar Alatau in deciduous forests

M. Khusnitdinova · A. Taskuzhina · N. Kerimbek · A. Pozharskiy · G. Nizamdinova · D. Gritsenko (✉)
Laboratory of Molecular Biology, Institute of Plant Biology and Biotechnology, Almaty, Kazakhstan
e-mail: d.kopytina@gmail.com

D. Gritsenko
Research Center AgriBioTech, Almaty 050040, Kazakhstan

P. Mohammad et al. (eds.), *Geographic Approaches to Climate Change and Mitigation: Urban and Rural Perspectives (Volume 1)*, Advances in Science, Technology & Innovation, https://doi.org/10.1007/978-3-031-92119-3_10

include birch and aspen, while coniferous forests are formed by Asian spruce and Siberian fir (Saparbaeva 2017). The conservation and rational use of wild plants will always be crucial. The timely response to challenges created by climate change depends on monitoring the dynamics of wild plant population development and predicting the level of danger they face (Vitkovsky 2003).

It is crucial to analyze and assess all changes in the development of wild tree populations. One such change can be observed through variations in NDVI. The variability of NDVI is directly linked to climate change, particularly changes in rainfall patterns. Remote sensing data plays a crucial role in analyzing this relationship, and the accuracy of NDVI change assessments depends on advancements in remote sensing technology (Yang et al. 2014; Zhang et al. 2024). Ordinary Least Squares (OLS) regression is applicable for analyzing the relationship between NDVI and climate change at both global and local scales (Ding et al. 2007).

In this study, NDVI and climatic factors are respectively utilized as the dependent and independent variables, with the value of NDVI depending on rainfall (Hou et al. 2015). Our study aimed to assess and analyze the inter-annual variation in NDVI among wild trees in the Zhongar Alatau mountain region, particularly within the SNNP, over the past 20 years. We mapped NDVI values from 2006 to 2023, with a focus on data collected in August.

2 Study Area

The Zhongar Alatau State National Natural Park is situated within the Aksu, Sarkand, and Alakol districts of the Almaty region in the Republic of Kazakhstan (Fig. 1). It was established by Decree No. 370 of the Government of the Republic of Kazakhstan on April 30, 2010. The Zhongar Alatau SNNP covers an area of 356,022 ha (Master plan for the development of the SNNP 2021; Taskuzhina et al. 2024).

In the country, ongoing research works are dedicated to the biomonitoring of wild trees in the Zhongar Alatau, encompassing silvicultural and environmental assessments of *M. sieversii* populations, alongside the advancement of reproduction methodologies. Furthermore, botanical-geographical, geobotanical, population genetic, and various other studies are being conducted in the wild forests of the *M. sieversii* trees. Additionally, efforts are underway to establish a living collection comprising elite *M. sieversii* genotypes, including those derived from the Zhongar populations (Ghebrezgabher et al. 2020; Master plan for the development of the SNNP 2021).

3 Data Collection and Methodology

3.1 Data Collection

Landsat images (4-5, 8-9) used in this study have a 16-day repeat cycle with an equatorial crossing time of 10:00 a.m. plus or minus 15 min. Moderate-resolution imagery, ranging from 15 to 100 m, covers Earth's land surface and Polar

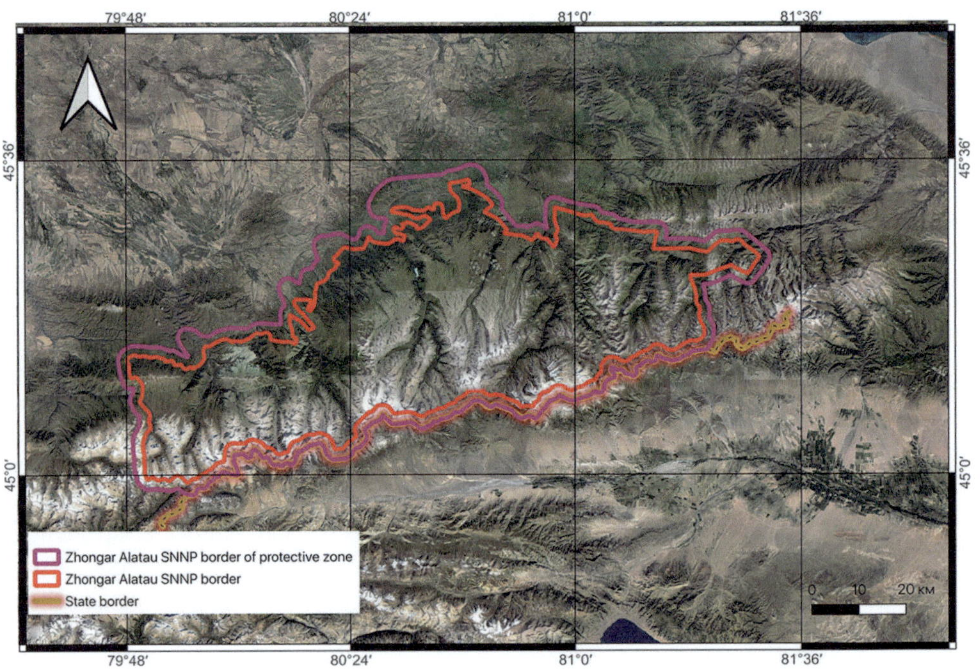

Fig. 1 Geographical location of the study area (by author)

Regions Landsat operates across the visible, near-infrared, short-wave infrared, and thermal infrared spectrums. Landsat images were downloaded from The U.S. Geological Survey (USGS) at https://earthexplorer.usgs.gov/.

Temperature and precipitation data for this study (2006–2023) were acquired from the gridded monthly time series available in the Kazakhstan Meteorological Database (http://ecodata.kz:3838/dm_climat_ru/). Specifically, data were extracted from the weather station situated in Sarkand (45.24 N, 79.54 E).

3.2 Methodology

In this study, the first step involved acquiring data, specifically satellite images. Next, to apply suitable corrections and calculate NDVI for Landsat images, we utilized QGIS 3.28 (Fig. 2). NDVI values for Landsat images were computed using bands 4 and 5, which correspond to Red and Near-Infrared (NIR) reflectance, respectively (Glenn and Tabb 2019; Master plan for the development of the SNNP 2021; Jacob et al. 2023). Equation (Index DataBase 2024) was calculated by Eq. 1:

$$NDVI = (\text{NIR} - \text{Red})/(\text{NIR} + \text{Red}) \quad (1)$$

where NIR = Near-Infrared region.

Red = Red/Visible region. The NDVI for a satellite/sensor is calculated using the respective bands corresponding to NIR and Red regions (Jacob et al. 2023).

After this, the NDVI values for the study area were extracted and categorized into five different classes. The resulting maps were then analyzed to assess the trend changes of NDVI among wild trees in the Zhongar Alatau mountain region in response to climate change.

In this study, an Ordinary Least Squares model was utilized to assess the impact of climatic variables on vegetation/NDVI. If the trend of any climate variable is found to be significant at the 0.05 or 0.01 level, it indicates a strong effect on NDVI. For example, a significant increase in annual mean precipitation would lead to a corresponding increase in NDVI values, and vice versa (Hou et al. 2015).

The OLS method is suitable for evaluating the spatial impact of climatic variability and altitude on vegetation growth (Yang et al. 2016). The OLS was calculated using Eq. 2:

$$NDVI = \beta_0 + \beta_1 P_1 + \beta_2 T_2 + \varepsilon \quad (2)$$

where 'NDVI' is the dependent variable, while P1 and T2 represent the variables corresponding to precipitation and temperature, respectively. β_0 is the intercept, β_1 and β_2 are coefficients and 'e' is residual and computed from the regression tool.

The inter-annual changes in NDVI with respect to climatic variables were analyzed using the linear regression method. In this analysis, time was considered as the independent variable, while NDVI values and climatic data served as dependent variables (Hao et al. 2012; Yang et al. 2016). Correlation and regression methods, such as the linear Pearson correlation coefficient (r), coefficient of determination (R-squared), and P-value, were utilized to analyze the relationship between NDVI and climatic variables (Yin et al. 2012). In this study, r greater than 0.98 indicates a very significant positive correlation, while a value below -0.26 indicates a very significant negative correlation. R-squared value above 0.95 indicates

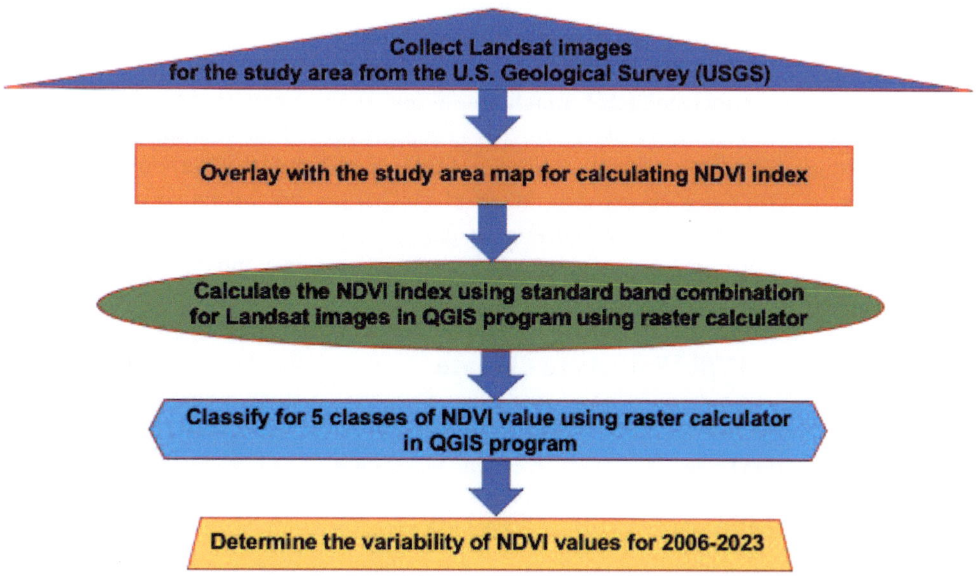

Fig. 2 Methodology flowchart (by author)

high approximation accuracy, suggesting that the model effectively describes the phenomenon being studied. In this study, with an R-squared value of 0.98, the model indicates that approximately 98% of the variability in the dependent variable can be explained by the independent variables (Table 1).

The P-value is a probability used to determine the significance of individual regression coefficients, indicating whether they are statistically significant in predicting the dependent variable. The P-value greater than 0.05 typically suggests that the corresponding independent variable has little effect on the dependent variable. In this study the extremely low P-value for precipitation (367E−14) has a significant effect on the dependent variable (NDVI), while the P-value for temperature (0.,024*) suggests a less significant effect.

4 Results and Discussion

The natural progression of vegetation development is affected by meteorological factors (Savinova et al. 2023). The onset of active vegetation growth typically occurs when the air temperature exceeds +10 °C. Additionally, precipitation plays a crucial role in influencing vegetation development, with increased precipitation leading to enhanced vegetation cover and phytomass production, while reduced precipitation may result in decreased vegetation cover (Savinova et al. 2023).

4.1 Inter-Annual Variation of NDVI and Climate Factors

The annual trend of NDVI and climate factors for the study area from 2006 to 2023 is depicted in Fig. 3. The results reveal that the trend of NDVI was relatively stable, ranging between 0.34 and 0.39 (Fig. 3a). The precipitation exhibited highly variable changes, fluctuating between 10.1 and 37.3 mm, with a maximum of 65.4 mm observed in 2009 and a minimum of 0 mm in 2006 (Fig. 3b). The trend of the annual average temperature remained stable, ranging between 21.2 and 22.9 °C (Fig. 3c). In the years 2006, 2014, and 2016, the NDVI values decreased due to low rainfall, whereas an improvement in NDVI was observed in 2009, 2011, 2013, and 2018, coinciding with periods of high rainfall, particularly notable in 2009 and 2011 (Figs. 3a, b and 4).

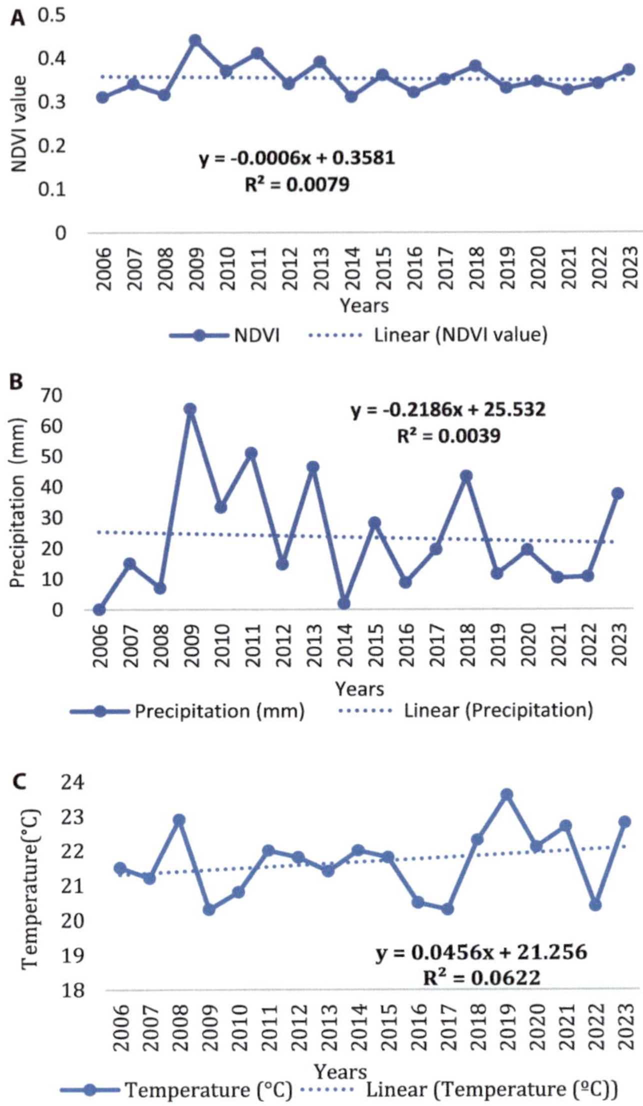

Fig. 3 The inter-annual variation of **a** NDVI, **b** precipitation, **c** temperature (2006–2023) (by author)

4.2 Responses of Wild Forests to Climate Change

Climate always plays an important role in vegetation dynamics with precipitation and temperature as the key factors (Wei et al. 2022). In this study, we conducted a simple linear regression analysis between NDVI and climatic factors (temperature and precipitation) according to the distributions of 18-year average (Fig. 5). The results indicated that NDVI was not correlated with temperature ($R^2 = 0.0679$) (Fig. 5b) but exhibited a positive correlation with precipitation ($R^2 = 0.9736$) (Fig. 5a). Figure 5a showed how NDVI strongly responds to changes in rainfall. In 2006, 2008, 2014, 2016, 2019, and 2021 years the effect of below-average precipitation resulted in gradually decreasing values of NDVI according to the data described in Table 1. The temperature trend doesn't show a significant impact on NDVI variability in the same time period (Fig. 5b).

This result agrees with the results obtained from other regions. (Kulesza and Hościło 2023) highlighted precipitation as the most important factor affecting forests in Poland, showing a strong correlation with NDVI. In the study by

Table 1 The inter-annual changes (August) of NDVI, temperature and precipitation data

Years	2006	2007	2008	2009	2010	2011	2012	2013	2014	2015	2016	2017	2018	2019	2020	2021	2022	2023
NDVI	0.31	0.34	0.315	0.44	0.37	0.41	0.34	0.39	0.31	0.36	0.32	0.35	0.38	0.33	0.345	0.325	0.34	0.37
T (°C)	21.5	21.2	22.9	20.3	20.8	22	21.8	21.4	22	21.8	20.5	20.3	22.3	23.6	22.1	22.7	20.4	22.8
P mm	0	15	6.9	65.4	33.2	50.9	14.7	46.3	1.8	28.1	8.6	19.5	43.2	11.4	19.3	10.1	10.5	37.3

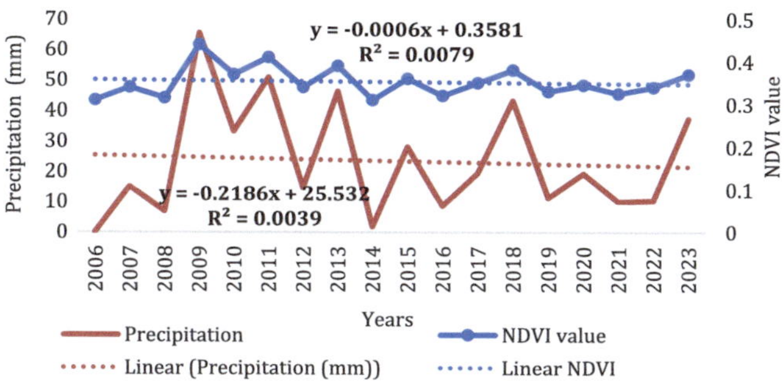

Fig. 4 The variations of NDVI and precipitation in Zhongar Alatau mountain region (2006–2023) (by author)

(Wang et al. 2023), while the time lag effect of rainfall on NDVI was found to be insignificant, the cumulative effect over time was more significant than that of temperature.

4.3 Mapping of NDVI Values

Using the described approaches and methodology, NDVI maps were generated and classified into 5 categories (<0, 0–0.1, 0.1–0.26, 0.26–0.51, and > 0.51) for the month of August from 2006 to 2023 (Fig. 6). The trends observed in NDVI time series serve as crucial indicators of the risk of declining growth in wild tree forests. Accordingly, we conducted an analysis of the spatial distribution of NDVI variations among wild trees and identified a relatively stable trend. In the years 2006, 2008, 2014, 2016 and 2021, low NDVI values were observed (Fig. 6). The short intervals between unfavorable years may contribute to genetic erosion within wild tree populations. These findings highlight the significance of preserving biodiversity among wild trees, especially by monitoring trends in vegetation responses to climate change and the importance of monitoring NDVI values.

Climate and human activities serve as fundamental driving forces that govern and influence the spatial distribution and change of vegetation. Examining the spatial and temporal variation patterns of vegetation cover using NDVI and assessing its relationship with climate change play pivotal roles in ecosystem regulation and enhancing the ecological environment (Wang et al. 2023). For example, the study by Pompa García et al. (2024) conducted at forest stands in northern Mexico highlights the importance of integrating NDVI analysis as a tool for forest management, considering the temporal variation of NDVI measurements.

The potential applications and relevance of NDVI maps lie in their use for monitoring wild trees and facilitating the rational management of environmental resources in the Zhongar Alatau mountain region. NDVI will serve as a reliable indicator of vegetation health in wild forests for the Zhongar Alatau State National Natural Park protected by the government. Monitoring NDVI over time will allow ecologists to assess changes in vegetation health, identify areas of stress or decline, and prioritize conservation interventions to restore degraded ecosystems. This capability is particularly crucial in remote and inaccessible forest areas where on-the-ground assessments may be challenging (Kulesza and Hościło 2023; Pompa García et al. 2024). Wild forests are also susceptible to various disturbances, including wildfires, insect outbreaks, and human activities. NDVI monitoring enables the rapid detection and assessment of such disturbances. Sudden decreases in NDVI values indicate areas

Fig. 5 The variability and trend of NDVI in response to **a** precipitation, **b** temperature (2006–2023) (by author)

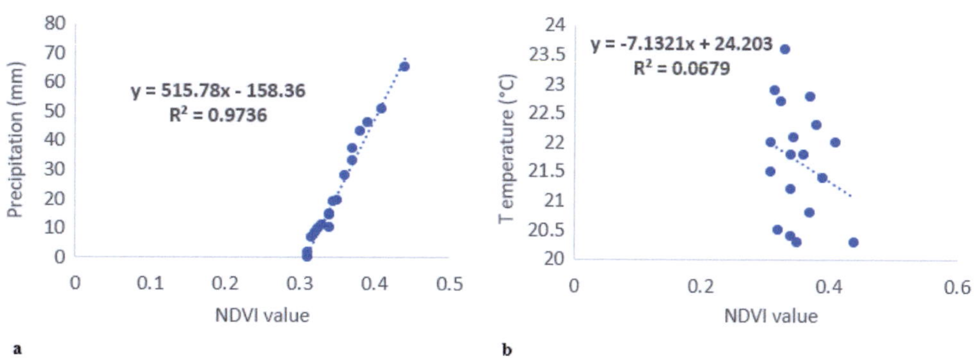

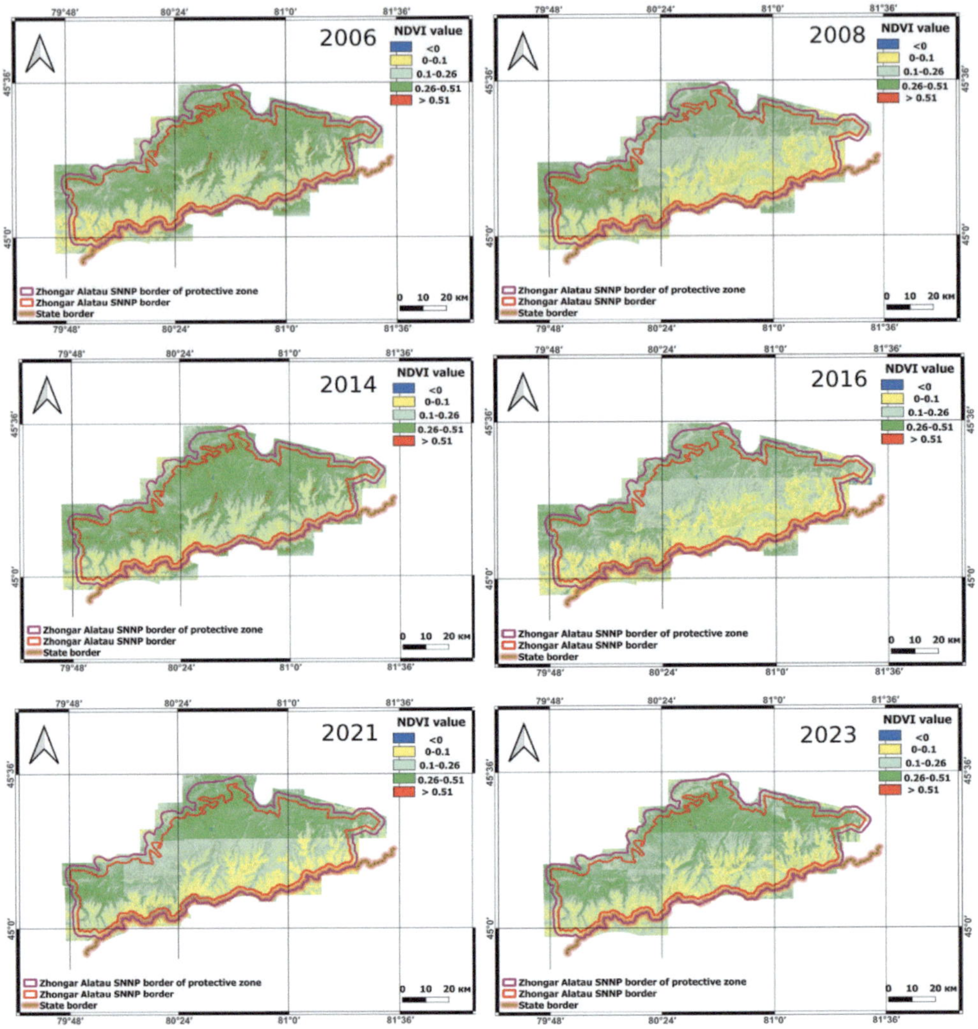

Fig. 6 NDVI trends among wild trees in the Zhongar Alatau mountain region (by author)

affected by disturbances, allowing for timely response and management actions. By identifying disturbance hotspots, NDVI helps minimize the ecological impact of disturbances, facilitate habitat recovery, and safeguard biodiversity in wild forests (Wang et al. 2023). Additionally, ecologists can indirectly assess changes in biodiversity levels, habitat fragmentation, and species distributions (Master plan for the development of the SNNP 2021; Kulesza and Hościło 2023; Pompa García et al. 2024). This information is critical for prioritizing conservation efforts, identifying key habitats for protection, including *M.sieversii* trees, and maintaining ecological connectivity in fragmented forest landscapes.

Limitation of the research: satellite images of appropriate quality are not consistently available for the required period of time (2003–2005) for study area, and the available images do not allow for accurate calculation of NDVI because of quality.

5 Conclusion

In this study, a close relationship was observed between NDVI and climate change, particularly regarding the NDVI of wild trees in the Zhongar Alatau mountain region, which significantly correlates with seasonal precipitation. Zhongar Alatau mountain region including Zhongar Alatau SNNP so results obtained can be also used for the rational management of environmental management. In the years 2006, 2008, 2014, 2016, and 2021, low NDVI values were observed, closely correlated with low precipitation levels. In this study, r for precipitation was greater than 0.98 indicates a very significant positive correlation, while r for temperature was −0.26 indicates a very significant negative correlation. R-squared value was 0.98, the model indicated that approximately 98% of the variability in the NDVI variable can be explained by the independent variables—precipitation. Obviously, the effect of precipitation on NDVI and NDVI decreased with rainfall. On the other side, the effect of temperature change on NDVI is minimal. The short intervals between these unfavorable years may lead to genetic erosion within wild tree populations. These findings emphasize the importance of preserving biodiversity among wild trees, especially by considering the trends in vegetation responses to climate change. Mapping the NDVI values over time (from 2006 to 2023) allows us to indicate the vital years for the development of wild forests, including the populations of *M.sieversii.*

Conflicts of Interest The authors have no relevant financial or non-financial interests to disclose.

Funding The work was funded by the Ministry of Science and Higher Education of the Republic of Kazakhstan, AP19679755 "Whole-genome molecular genetic study of Erwnina amylovora and fungal pathogens of Venturia genus in wild and cultivated fruit tree populations" and AP14871004 "Investigation of phytopathological and botanical aspects of wild apple populations growing in the Northern Tien Shan".

References

Aleksanyan, P.V., Burmistrov, L., Smekalova, T., Sorokin, A.: Modern Methods and International Experience of Preserving the Gene Pool of Wild Plants (on the Example of Wild Fruits). United Nations Development Program in Kazakhstan, Almaty (2011)

Ding, M., Zhang, Y., Liu, L., Zhang, W., Wang, Z., Bai, W.: The relationship between NDVI and precipitation on the Tibetan Plateau. J. Geogr. Sci. **17**(3), 259–268 (2007). https://doi.org/10.1007/s11442-007-0259-7

Ghebrezgabher, M.G., Yang, T., Yang, X., Eyassu Sereke, T.: Assessment of NDVI variations in responses to climate change in the Horn of Africa. Egypt J. Remote Sens. Space Sci. **23**(3), 249–261 (2020). https://doi.org/10.1016/j.ejrs.2020.08.003

Glenn, D.M., Tabb, A.: Evaluation of five methods to measure normalized difference vegetation index (NDVI) in apple and citrus. Int. J. Fruit Sci. **19**(2), 191–210 (2019). https://doi.org/10.1080/15538362.2018.1502720

Hao, F., Zhang, X., Ouyang, W., Skidmore, A.K., Toxopeus, A.G.: Vegetation NDVI linked to temperature and precipitation in the upper catchments of yellow river. Environ. Model. Assess. **17**(4), 389–398 (2012). https://doi.org/10.1007/s10666-011-9297-8

Hou, W., Gao, J., Wu, S., Dai, E.: Interannual variations in growing-season NDVI and its correlation with climate variables in the southwestern karst region of China. Remote Sens. **7**(9), 11105–11124 (2015). https://doi.org/10.3390/rs70911105

Index DataBase.: A database for remote sensing indices for Landsat (2024). https://www.indexdatabase.de/db/s-single.php?id=168. Accessed 30 Mar 2024

Jacob, P., Kommuru, P., Ruchitha, R., Kuracha, H., Taruni, T.: Comparative study of MODIS, LANDSAT-8, SENTINEL-2B, and LISS-4 images for precision farming using NDVI approach. E3S Web Conf. **405**, 01004 (2023). https://doi.org/10.1051/e3sconf/202340501004

Kolchenko, M., Nurtaza, A., Pozharskiy, A., Dyussembekova, D., Kapytina, A., Nizamdinova, G., Khusnitdinova, M., Taskuzhina, A., Kakimzhanova, A., Gritsenko, D.: Wild Malus niedzwetzkyana Dieck ex Koehne as a genetic resource for fire blight resistance. Horticulturae **9**(10), 1066 (2023). https://doi.org/10.3390/horticulturae9101066

Kulesza, K., Hościło, A.: Influence of climatic conditions on Normalized Difference Vegetation Index variability in forest in Poland (2002–2021). Meteorol. Appl. **30**(5), e2156 (2023). https://doi.org/10.1002/met.2156

Master plan for the development of the SNNP.: Adjustment of the feasibility study of the Zhongar-Alatau State National Nature Park, project planning in terms of the master plan for infrastructure development and functional zoning. (Almaty) **149** (2021)

Pompa García, M., Sigala Rodríguez, J.Á., Vivar Vivar, E.D., Rodríguez Flores, F.D.J., Rascón Solano, J.: Diurnal change of NDVI from UAV in trees of a temperate unavenged forest stand. Rev Mex Cienc For **15**(82), 50–68 (2024). https://doi.org/10.29298/rmcf.v15i82.1431

Saparbaeva: Distribution and species diversity of endemic plant species of the Dzhungar Alatau ridge. Bull. Karaganda Univ. Ser. Biol. Med. Geogr. **4**, 43–50 (2017)

Savinova, S.V., Bratkov, V.V., Murashova, I.D., Klyushin, P.V., Guseynova, N.O.: Assessment of changes in the normalized difference vegetation index (NDVI) of farmlands of Stavropol Territory, Russia, during the period of active vegetation in 2020 based on medium spatial resolution images. South Russ. Ecol. Dev. **18**(2), 82–91 (2023). https://doi.org/10.18470/1992-1098-2023-2-82-91

Taskuzhina, A., Khusnitdinova, M., Kerimbek, N., Kapytina, A., Pozharskiy, A., Moisseyev, R., Gritsenko, D.: Evaluation of normalized difference vegetation index in wild apple populations growing in southeast mountain regions of Kazakhstan. Eur. J. Appl. Biotechnol. (1), 27–33 (2024). https://doi.org/10.11134/btp.1.2024.2

Tian, Z., Song, H., Wang, Y., Li, J., Maimaiti, M., Liu, Z., Zhang, H., Zhang, J.: Wild apples are not that wild: conservation status and potential threats of malus sieversii in the mountains of central Asia biodiversity hotspot. Diversity **14**(6), 489 (2022). https://doi.org/10.3390/d14060489

Vitkovsky.: Fruit plants of the world (2003)

Wang, N., Jiang, S., Zhang, Z., Fang, H., Xu, H., Wang, Y., Chen, X.: Malus sieversii: the origin, flavonoid synthesis mechanism, and breeding of red-skinned and red-fleshed apples. Hortic. Res. **5**(1), 70 (2018). https://doi.org/10.1038/s41438-018-0084-4

Wang, S., Liu, X., Wu, Y.: Considering climatic factors, time lag, and cumulative effects of climate change and human activities on vegetation NDVI in Yinshanbeilu, China. Plants **12**(18), 3312 (2023). https://doi.org/10.3390/plants12183312

Wei, Y., Sun, S., Liang, D., Jia, Z.: Spatial–temporal variations of NDVI and its response to climate in China from 2001 to 2020. Int. J. Digit. Earth **15**(1), 1463–1484 (2022). https://doi.org/10.1080/17538947.2022.2116118

Yang, X., Liu, S., Yang, T., Xu, X., Kang, C., Tang, J., Wei, H., Ghebrezgabher, M.G., Li, Z.: Spatial-temporal dynamics of desert vegetation and its responses to climatic variations over the last three decades: a case study of Hexi region in Northwest China. J. Arid. Land **8**(4), 556–568 (2016). https://doi.org/10.1007/s40333-016-0046-3

Yang, X., Yang, T., Ji, Q., He, Y., Ghebrezgabher, M.G.: Regional-scale grassland classification using moderate-resolution imaging spectrometer datasets based on multistep unsupervised classification and indices suitability analysis. J. Appl. Remote. Sens. **8**(1), 083548 (2014). https://doi.org/10.1117/1.JRS.8.083548

Yin, H., Udelhoven, T., Fensholt, R., Pflugmacher, D., Hostert, P.: How normalized difference vegetation index (NDVI) trendsfrom advanced very high resolution radiometer (AVHRR) and Système Probatoire d'Observation de la Terre Vegetation (SPOT VGT) time series differ in agricultural areas: an inner mongolian case study. Remote Sens. **4**(11), 3364–3389 (2012). https://doi.org/10.3390/rs4113364

Zhang, H., Li, L., Zhao, X., Chen, F., Wei, J., Feng, Z., Hou, T., Chen, Y., Yue, W., Shang, H., Wang, S., Hu, M.: Changes in Vegetation NDVI and its response to climate change and human activities in the Ferghana basin from 1982 to 2015. Remote Sens. **16**(7), 1296 (2024). https://doi.org/10.3390/rs16071296

Examining the Effects of Coastal Erosion on Livelihoods: The Case of a Fishing Community in Kerala, India

Sheikh Omar Fye, Charles Aaron Adams Ekuban, David Brown, and Sudha Arlikati

Abstract

The research aims to examine the socio-economic impacts of coastal erosion on the livelihoods of the Alappad community and suggest mitigation strategies. Employing a qualitative Participatory Rural Appraisal (PRA) method, including resource mapping, transect walks, focus group discussions, and interviews with local stakeholders, six group discussions and twelve interviews were conducted alongside a literature review. Coastal erosion in Alappad, exacerbated by floods, sand mining, rising sea levels, and the 2004 Indian Ocean Tsunami, has degraded groundwater quality and heightened vulnerability to natural disasters. The primary livelihood source, fishing, has suffered decreased productivity, and the once vibrant beach area has significantly diminished. While providing valuable insights, the study's small qualitative design necessitates future quantitative panel studies to evaluate new mitigation strategies' long-term impacts. Participatory Rapid Appraisal (PRA) techniques facilitated community engagement, a crucial aspect often overlooked in policymaking, advocating for its inclusion in future policy endeavors. Nonetheless, further research is needed for a comprehensive understanding of the community's perspective, emphasizing community involvement. Community receptivity towards using tetrapods in seawall construction and nature-based solutions, despite limited exposure, underscores the need for a collaborative approach involving all stakeholders, particularly the community, to address Alappad's coastal erosion challenges.

Keywords

Coastal erosion · Livelihoods · Fishing community · Alappad · Kerala

1 Introduction

The sustainability of populations worldwide is seriously threatened by coastal erosion, especially those whose livelihoods depend on coastal ecosystems (Varea et al. 2022). According to recent studies, a mix of natural and man-made forces has caused coastal erosion to worsen in many places of the world. Coastlines are eroding more quickly as a result of rising sea levels, stronger storm surges, and human activities including urbanization and climate change. In India, the over-reliance of local communities on marine resources for economic purposes increases the vulnerability of coastal areas.

In southern areas, such as Kerala, an Indian coastal state with a sizable fishing community, this problem is particularly prevalent (Economic Review 2023; Haneefa 2023). Coastal erosion poses a major threat to these communities since they rely extensively on coastal and marine resources, especially for traditional fishing. According to Islam et al., erosion reduces fish stocks and interferes with fishing operations, which lowers revenue and threatens food security. Moreover, the negative consequences of coastal erosion also affect non-fishing industries including tourism and transportation. Communities in Kerala must prioritize and adopt adaptation strategies due to the increased frequency and severity of erosion brought on by climatic variability (Power et al. 2021).

S. O. Fye · C. A. A. Ekuban (✉) · S. Arlikati
Amrita School for Sustainable Futures, Amritapuri, Kollam, India
e-mail: am.id.dids22026@am.students.amrita.edu

S. O. Fye
e-mail: am.id.dids22058@am.students.amrita.edu

S. Arlikati
e-mail: sudhaarlikatti@am.amrita.edu

D. Brown
Central Queensland University, Rockhampton, Queensland, Australia
e-mail: d.brown@cqumail.com

P. Mohammad et al. (eds.), *Geographic Approaches to Climate Change and Mitigation: Urban and Rural Perspectives (Volume 1)*,
Advances in Science, Technology & Innovation, https://doi.org/10.1007/978-3-031-92119-3_11

The fishing population of Alappad is heavily dependent on marine resources for their livelihoods, making coastal erosion a serious danger. Fishermen suffer financial losses as a result of its direct impact on fishing productivity. Displacement brought on by erosion exacerbates social structures, customs, and cultural legacy, making the community's problems even more severe, including many other communities throughout the world, the Alappad community has responded by using adaptive measures including changing their fishing methods and building new homes. To lessen these effects, government initiatives, regulations, and assistance programs are crucial (Subramoniam and Suresh 2015; Santha 2014).

According to recent statistics, a considerable fraction of Alappad's fishermen are actively involved in marine fishing, demonstrating their great reliance on the maritime industry. Nevertheless, the effects of coastal erosion on the livelihoods of these fishermen have been largely disregarded, which has led to their poverty. In particular, 78.83% of active fishermen engage in marine fishing activities, despite the fact that 77% of all fishermen work in the marine industry (Pradeep Kumar et al.2022). The intricate connection between coastal erosion and the fishing industry has not been fully investigated, despite this dependence.

The purpose of this study is to investigate how coastal erosion affects the Alappad community's means of subsistence in order to close this gap.

1.1 Objective

1. To identify the socio-demographic profile of the residents in the coastal area.
2. To examine the awareness of coastal erosion and its impact on the community.
3. To examine socio-economic effects of coastal erosion on the livelihoods of Alappad communities.

The following Section begins with a review of Literature that focuses on Historical facts, Sand mining, Coastal erosion, and its consequences on livelihoods, and Coastal Erosion Mitigation Solutions.

2 Review of Literature

2.1 Historical Facts

Nestled between the Arabian Sea and the TS Canal, Alappad is a coastal community located in Kerala, India. The tiny strip of land measures 16 km in length. More than 20,000 people call it home, and many of them are from families that depend on fishing. The majority of people living in Alappad are Hindus, and the area is filled with temples. Due to strong waves during the southwest monsoon, this heavily populated coastal zone, including Alappad, suffers from significant beach and coastline erosion. Due to this, there are now serious environmental issues due to ongoing erosion, the devastation of coastal landforms, saltwater intrusion, loss of green belts and wildlife, and coastal flooding (Sharon n.d.; Punniyarajah 2022). A tsunami from an Indonesian earthquake on December 26, 2004, severely damaged parts of the village and shoreline. Alappad was the most severely damaged village in the state of Kerala, having lost over 100 lives as a result of the calamity (Economic Review 2023).

2.2 Sand Mining

The black sand in the Alappad coastal region is rich in minerals and has been the focus of heavy mining activity since the 1960s. Through campaigns like "Save Alappad, Stop Mining," the local community has been actively opposing this mining (Matovu et al. 2024).

Coastal erosion has been shown to have long-term consequences on the area and has a substantial impact on the livelihoods of the fishing community in the vicinity (Angnuureng et al. 2023; Irshad n.d.). Data spanning almost five decades (1973–2021) shows that erosion is a dominant force in the area despite the state's efforts to address the problem (Swathi 2018).

Installing detached breakwaters has been suggested as a potential way to lessen the effects of coastal erosion, especially during the monsoon season. The livelihoods of rural coastal communities may be threatened by these interventions, especially in light of climate change, and they may also increase debt levels. Therefore, it is crucial to take into account the environmental and economic costs associated with addressing coastal erosion (Angnuureng et al. 2023; Satish and Geetha 2021).

2.3 Coastal Erosion

Numerous elements, such as ocean currents, winds, waves, human activity, and natural disasters, cause coastal areas to constantly change. The degree to which stakeholders use the shore and the severity of outside influences they experience determine how long these ecosystems can last (Prieto 2022). However, there is a chance that external factors and excessive use will greatly worsen the effects of global warming.

Coastal erosion is significantly impacted by both climate change and overuse of coastal areas (Barua et al. 2020a, b). Due to the effect of oceanic phenomena like currents, waves, tides, and winds, as well as physical characteristics like rocks, shorelines, swamps, estuaries, and coral reefs, the coastal zone is highly dynamic and prone to disasters. Any change to these elements could have unintended implications, and authorities could only be able to return things to their original state if the disturbance was minimal (Monalisha and Panda 2018).

India is home to a wide variety of maritime and coastal environments, including sand dunes, mangroves, and coral reefs. The Bay of Bengal, the Arabian Sea, and the Indian Ocean all front the over 8,000 km^2 coastal region of India. With 66 coastal districts and a variety of coastal features, such as sandy beaches (43%), rocky coasts (11%), muddy flats (36%), and marshy coasts (10%), it includes 9 states and 2 union territories. It also has 34 small and big lagoons in addition to 97 large estuaries (Bai et al. 2022). A large section of the population's access to freshwater is at risk of being contaminated by saltwater due to rising sea levels and coastal erosion, which could contaminate groundwater sources. In the coastal region of India, coastal erosion is still a problem. With 6,632 km^2 of the Indian coastal region buried in the water between 1990 and 2016, the West Bengal coast is the most susceptible, according to a report by the National Centre for Coastal Research. Over the past three decades, frequent cyclonic storm surges from the Bay of Bengal have caused increased erosion in the eastern half of the Indian coastline area, whereas the western coast has demonstrated relative stability (Paul and Chowdhury 2021). 23% of the Indian mainland's coastline is currently affected by coastal erosion, and this percentage has been rising dramatically in recent years (Miyab et al. 2017; Dalezios et al. 2017). The elements causing coastal erosion are depicted in Fig. 1, which eventually leads to reduced livelihoods and a poorer standard of living.

2.4 Coastal Erosion Mitigation Solutions

2.4.1 Hard Seawall Structures

Tsunami mitigation strategies vary, depending on the environment, political landscape, and material availability. The basic strategies employed to reduce tsunami risk, as documented by the National Oceanic and Atmospheric Administration (NOAA), are shown in Fig. 1. (Wang and Cao 2020) In Alappad a blocking wall has been implemented along

Fig. 1 Reformulated by Barua et al. (2020a, b). *Source* Authors

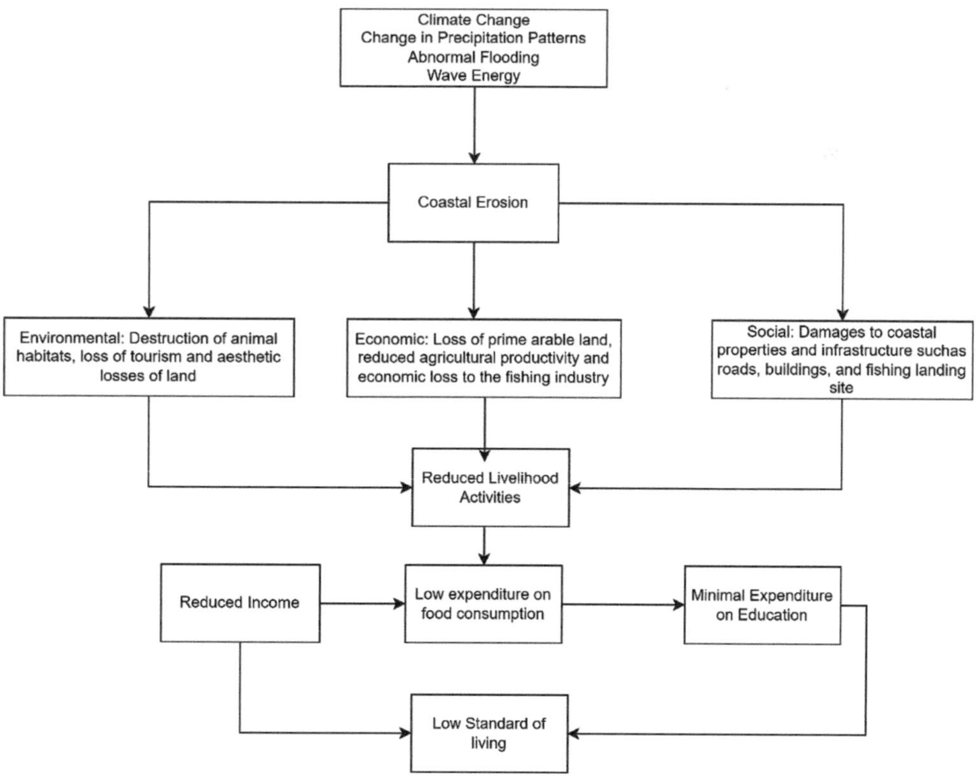

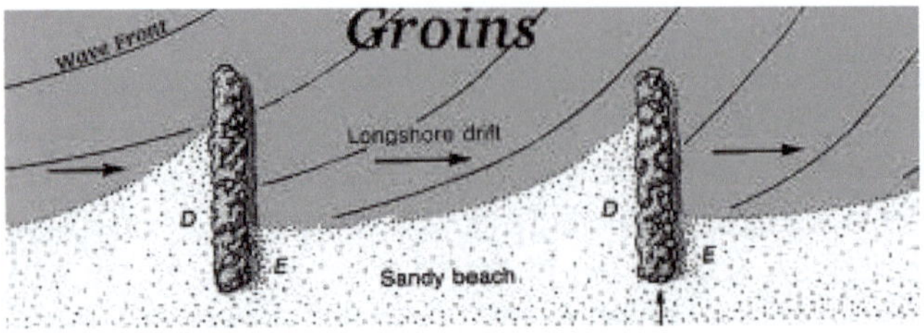

Fig. 2 Seawall groins. *Source* Authors

much of the coastline. Several seawall groins have also been constructed to act as wave-slowing elements.

In general, detached breakwater structures like the seawall groins as seen in Fig. 2 constructed in Alappad play a role in coastal defence and contribute to the restoration of eroded beach materials. However, careful consideration of their spacing is vital, as improper placement could potentially result in the formation of rip currents. (Vieira et al. 2021) (Fig. 3).

Most of the damage to seawall structures in Kerala is reportedly the result of overtopping. A review of this, along with a broader review, has demonstrated that a revision of seawall construction techniques needs to be completed to improve preparedness for future weather events in coastal communities in Asia. In particular, it was noted that there is a need to understand the interaction of tsunamis with countermeasures during the phases of wave impact, flooding, and possible backflows (Basheer Ahammed and Pandey 2019).

An integrated structural approach, as has been attempted in some parts of the coast of Alappad, using both breakwater structures and a seawall has the benefit of reducing wave energy and increasing sediment deposit, as well as preventing flooding and erosion.

Another factor that is highly regarded for structural mitigation of coastal erosion is the use of permeability in structural designs. It has been found that using staggered non-continuous configurations in mitigation structures can result in a reduction in hydrodynamic forces by up to 40% during wave impact, ongoing wave penetration, and backflow (Basheer Ahammed and Pandey 2022).

A tetrapod is a specific coastal engineering structure designed to counteract erosion caused by wind and waves. This type of structure is quickly becoming more popular given findings regarding its protective capabilities. Tetrapods provide remarkable stability in terms of hydraulic properties, material strength, and construction procedures. Constructed from concrete, tetrapods employ a tetrahedral shape to disperse the impact of incoming waves by permitting water to flow around them. Through the interlocking mechanism of their legs, tetrapods efficiently minimize displacement, allowing them to interlock and safeguard the surface of a structure. In terms of material strength, tetrapods are conveniently manageable even while the concrete is still curing. Breakage testing and on-site experience both demonstrated that steel reinforcement is unnecessary for tetrapods, presenting a significant advantage for marine construction (Pillai and Prasad 2020).

Tetrapods are typically placed in rows of at least two and placement pattern is a factor when it comes to their damage versus stability ratio. The two placement patterns, shown in Fig. 4, were tested and it was found that placement method 1 produced better results.

Fig. 3 Basic tsunami mitigation strategies (NOAA 2001). *Source* Authors

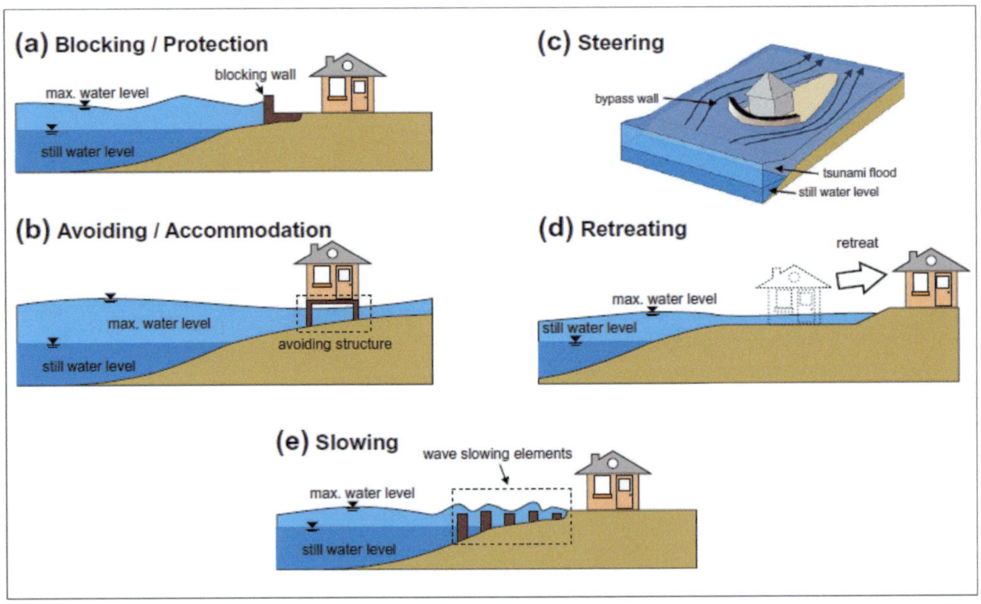

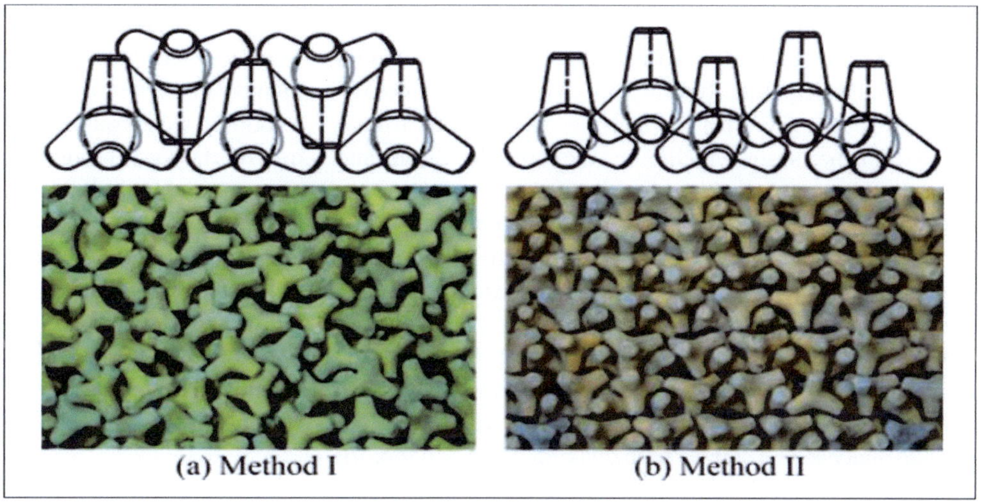

Fig. 4 Tetrapod placement methods

3 Methodology

3.1 Study Area

The study was conducted in the village of Alappad, situated in the Kollam district of Kerala, India. The geographical coordinates of Alappad panchayath are latitude 9.1160°N and longitude 76.4751°E, and its location is depicted in Fig. 5.

The village is positioned on a narrow strip of land that lies between the Arabian Sea and the TS Canal, measuring approximately 16 km in length, with its narrowest point being as thin as 33 m (Subramoniam and Suresh 2015). As per the Census of India, 2011, the estimated population of Alappad is 21655, comprising 10689 males and 10966 females. The average population in 2023 is projected to be 26,419 (Onda et al. 2019). Following the 2004 Tsunami, two additional bridges, Azheekal and Cheriyazhekkal, were constructed, in addition to the existing southern bridge. The village boasts abundant mineral resources such as ilmenite, rutile, zircon, monazite, leucoxene, sillimanite, and garnet (Mongabay 2018).

3.2 Participatory Rural Appraisal Technique

The study employed participatory rural appraisal (PRA) as a methodology. PRA is a tool which provides a way of enabling

Fig. 5 Location map of the study area. *Sources* Esri India, 2023. *Source* Authors

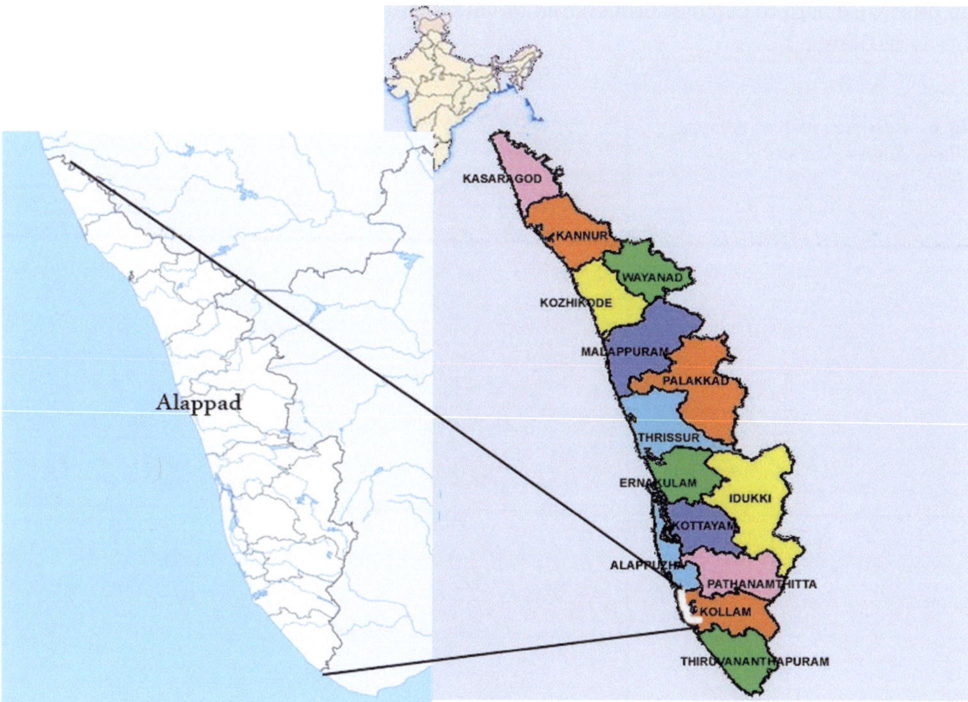

local (rural and urban) people to analyze their living conditions, share the outcomes, and plan their activities (Chandra 2010). The Participatory Rural Appraisal (PRA) methodology was chosen over the other methods because it shifts from a closed to an open system, facilitating the capture of the ground realities of the people and uses a comparative approach rather than a measurement-based one and enables a group approach as opposed to an individual one. PRA has the potential to encourage the active participation of people in the development process, enabling them to take control of their development. It is also known for its ability to capture trends and changes quickly and easily, even in the absence of baseline data. PRA is a group activity that provides in-depth information and analysis in a shorter time, making it a useful tool for researchers and practitioners alike.

3.2.1 Resource Map

According to (Sontakki et al. 2019), a resource map serves as a valuable tool for comprehending a village's spatial forms. This map portrays a detailed overview of various landforms, agricultural land, fishing resources, erosional lands, vegetation, minerals, transportation, means of transport, and other pertinent factors. The map's efficacy is evidenced in Fig. 6, where the extensive coastline of Alappad is highlighted by a red arrow, signifying its vastness. Information on these physical properties was gathered by the research team in agreement with the villagers while walking through the village. It also helped to pinpoint locations where coastal erosion was noticeable and the various measures underway to arrest the erosion. However, as it was drawn in the field the resource map is not to scale but contains details depicting local knowledge which are otherwise hard to capture through traditional quantitative survey designs.

3.2.2 Transect Walk

Transects are systematic walks taken with community members or key informants through an area. They can follow a loop, a watercourse, a straight line, or an S-shape to fit in with the local topography. The transect walk through the community highlighted issues related to environmental challenges in terms of erosion, its risks and vulnerabilities and suggested workable solutions. The transect walk was attended by participants from the village group and the research team, this brought clarity and understanding to soil type, vegetation, livelihood, opportunities, climate-related problems, and possible interventions. The result of this transect walk of Alappad is shown in Fig. 7.

3.3 Methods

The descriptive design was adapted where socio-economic conditions, and effects of erosion on the livelihoods of the community needed to be described.

Hence unstructured interviews were conducted.

3.3.1 Informal Discussions

Informal discussions were done among fisher folks in the Alappad community where fishermen came out with their experience with hazards and what had caused the erosion. All six group discussions were done. They were selected based on convenience and in the spirit of willingness to share their experience.

3.3.2 Unstructured Interviews

Unstructured interviews were conducted with twelve people comprising both government officials and fishermen to share

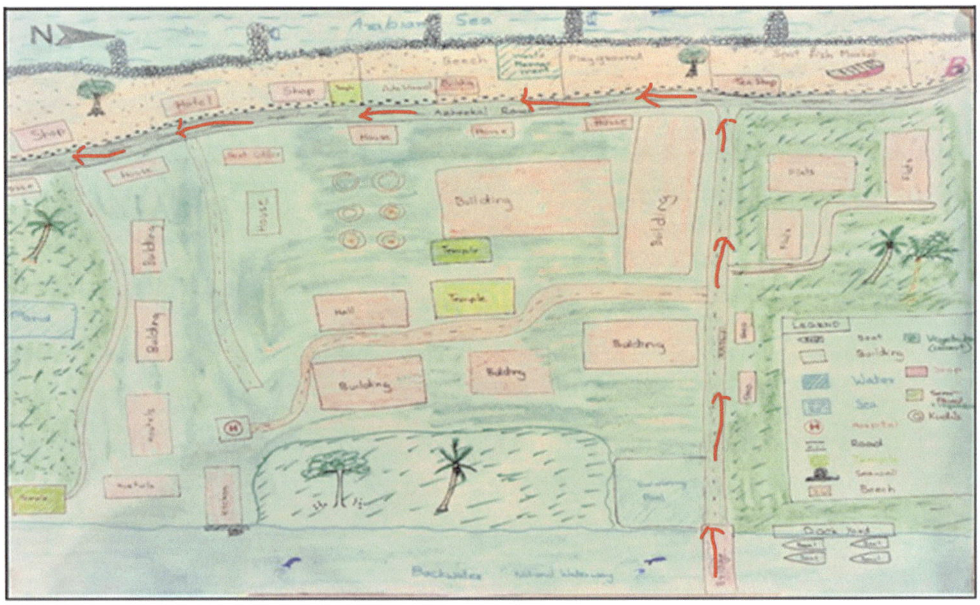

Fig. 6 Resource map of Allapad village. *Source* Authors

Fig. 7 Transect walk. *Source* Authors

Transect Walk/ Route										
Resource	Shop	Coconut Vegetation	Sand Mining	Temple	Rick Shop	Local fishing	Beaches	Waste Collection Point	Playground	Chai/Tea Shop
Soil	Sandy	Sandy	Sandy	Sandy	Sandy	Sandy	Sandy	Sandy	Sandy	Sandy
Vegetation	Grass	Coconut	Casuarina/ Bandam			Coconut	Coconut	Grass	Grass/ Coconut	Coconut/Casuarina
Livelihood	Primary source of income	Coconut farming	Illegal source of income		Skill Training	Income from sale of fishes		Income from waste collection		Small scale business
Opportunities	Expansion of shops	Local coconut food production	Legalising mining activities	Spiritual tourism	Livelihood diversification		Beach tourism	Waste Management Plant	Game tourism	Business expansion
Common Challenges/ Concerns	Livelihood in decline	Coastal erosion	Coastal erosion/		Air pollution	Bycatch/ Coastal erosion/ Livelihood in decline	Coastal erosion/ Livelihood in decline			Water contamination

Fig. 8 Excerpts of the informal discussions and unstructured interview sessions. *Source* Authors

their personal views on coastal erosion and the possible mitigation. For example, Fig. 8 gives excerpts on how the interview was done. During the interview, ten to fifteen minutes were allotted for every interviewee.

The target groups of fishermen selected were the Elderly—as their experience will be able to provide a snapshot of the historical environment, the Youth—as their feelings may be able to provide a snapshot of the future of Alappad, and SHGs—as these women provide one of the greatest opportunities for capacity building in Alappad, and Government Officials—as they will be able to speak about the government support provided for the study.

Table 1 Demographics of respondents. *Source* Authors

Age	Frequency	Percent
30yrs	6	33
50–60 years	7	39
60–70 years	5	28
Total	18	100
Monthly income	Frequency	Percent
10–20 k	11	92
20–25 k	1	8
Total	12	100
Ownership of houses	Frequency	Percent
Own	10	83
Rent	2	17
Total	12	100

4 Findings and Discussions

4.1 Social Demographics

The fishing industry in Alappad is a male-dominated trade and for that matter, all interviews were done with the male respondents. From Table 1, the highest (50%) of the respondents were between the age group 50–60 years. 92% Of the respondents had a monthly income of 10000–20000 rupees representing 92%. 83% Of the respondents live in their own houses, the reason being that they have been living in the same location for generations.

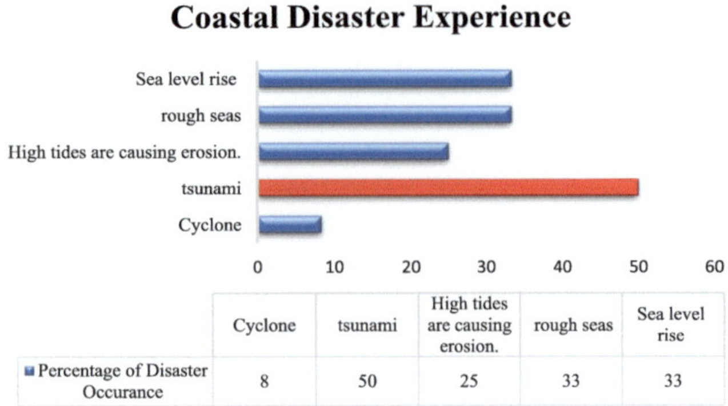

Fig. 9 Coastal disaster experience of respondents. *Source* Authors

Many of the respondents owned houses simply because they had stayed in those habitats all their lives and had had very good experiences of the actions of disasters.

Figure 9 shows the coastal disaster experiences the fishing community/members have faced over time in their fishing occupation. The respondents were asked what are the environmental hazards (e.g., floods, cyclones, tsunamis etc.) common to your community? 50% Mentioned that it was the tsunami that affected them most. There was a tie of 33% as respondents said that sea level rise and cyclones causing erosion were the next common disasters in their community. During the informal discussions, respondents were asked about how the hazards had affected them and they had this to say that the Cyclone which occurred in 2019 had impacted housing significantly and reduced fish concentration in the sea to their estimation. These hazards have led to coastal erosion and have brought myriad physical, economic, and psychological stressors upon the family.

Many respondents shared that they are forced to borrow money from money lenders due to an increase in economic pressure and therefore they are in debt.

Government officials and stakeholders' interviews also revealed that cyclones, a tsunami, and sea level rise were the most common environmental threats they had experienced thus far.

4.2 Causes of Coastal Erosion in Alappad

Anthropogenic Activities such as dredging, dumping of waste, sand mining, and coral mining have contributed to coastal erosion causing sediment deficit, and modification of water depth leading to longshore drift and altered wave refraction.

Erosion has also been sparked by fishing harbors and dams constructed in the catchment area of rivers and ports reducing the flow of sediments from river estuaries.

Coastal erosion broadly emanates from variations in temperature and precipitation which impacts the flora and fauna in the sea. This situation hampers the fishing industry's output and has severe consequences for the livelihoods of fishing communities as it directly affects their primary source of income and sustenance.

4.3 Effects of Coastal Erosion on Livelihoods

Respondents mentioned that the impact of coastal erosion was not only on the fisheries sector and affecting the catch but also on tourism. As a result, it impacts the fishermen's economic situation, as some may rely on tourists as a source of supplementary revenue.

Again, climate change has exacerbated the problem of damaged fishing equipment. This leads to borrowing at high-interest rates from lenders to fix the damaged equipment and other needs, which eventually puts them into more debt. The fishing community in Alappad heavily relies on the coastal ecosystem for their day-to-day livelihood activities including fishing fish processing and related trades. However, with the increasing coastline erosion, these activities are greatly hampered leading to a significant decline in the community's overall well-being. Respondents further stressed that sand mining activity had exacerbated the issue of erosion and has caused a reduction in the fish population affecting the livelihoods of fishermen. Coastal erosion disrupts traditional fishing practices making it difficult for fishermen to access fishing grounds. As the shoreline recedes the distance between the community and the fishing grounds increases resulting in longer travel times and higher fuel costs.

Moreover, the erosion of coastal habitats such as mangroves and coral reefs diminishes fish populations and alters their migration patterns further reducing the catch and affecting the income of the fishing community. The spider diagram in Fig. 10 shows common challenges and concerns that the respondents shared. Sand mining, less government support, and a decline in livelihoods were the most common challenges of great concern (Table 2).

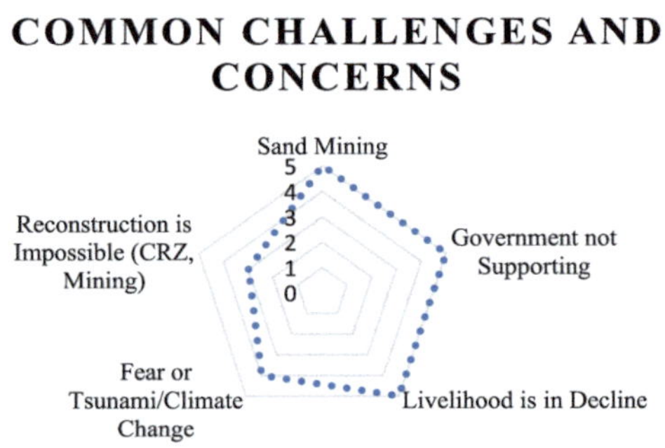

Fig. 10 Common challenges and concerns. *Source* Authors

Table 2 Responses by-government officials. *Source* Authors

Changes observed	Large-scale coastal erosion	Sand mining and climate change have changed the landscape entirely	
Effectiveness of seawall and other mitigation methods	The construction technique has been largely ineffective	Need to instate more seawall groins—lengthened and strengthened with tetrapod construction	Focus on vulnerable sections first
	Sand instability is a problem for revegetation of the coastline	Nature-based solutions directly on the coast have been ineffective for various reasons	
Other factors resulting in coastal erosion	Sand mining (especially beach washing)		
Local mining regulations	Not supportive of mining—it is exploiting the people and the land	There are no local regulations in place—this is state government-controlled	Mining is not sustainable in Alappad, it is ruining the land
Future of Alappad	The future of Alappad is teetering on the edge		

Table 3 Mitigation methods derived from qualitative trends. *Source* Authors

	Additional seawall groins	Improved seawall construction	Revegetation	Improved sand mining regulations	Total score	Mean	Rank
Additional seawall groins	27	57	43	43	170	14	**2**
Improved seawall construction	27	43	27	43	140	12	**4**
Revegetation	57	57	27	57	310	26	**1**
Improved sand mining regulations	27	27	27	73	154	13	**3**

Based on unstructured interviews with government officials, it was observed that significant changes were happening along the coastline. These changes included large-scale coastal erosion and sand instability, which were causing problems for the revegetation of the coastline. The major cause of these environmental challenges was identified as sand mining, which was found to be a real agent of coastal erosion. Unfortunately, the officials also noted that there were no local regulations in place to control these activities. The State government was responsible for these regulations, making it difficult for local communities to take any effective mitigation actions. Everyone was concerned about the future of Allapad due to these issues. They were of the view that mining is not a sustainable option for keeping the land. The study conducted a rudimentary weighted evaluation of the various mitigation methods, and the resulting scores are presented in Table 3. The evaluation criteria used was the Garrett Ranking method. Respondents were asked to prioritize various mitigation strategies that would work for them by scoring which of them was very important one which was not important. After that, the Garrett score of ranking was used to establish the ranking scores in Table 3 to determine which mitigation strategy was most important.

$$\text{Percent Position} = \frac{100\left(R_{ij} - 0.5\right)}{N_j}$$

where Rij = Rank given for the ith variable by jth respondents.

Nj = Number of variables ranked by jth respondents.

With the help of Garrett's Table, the percent position estimated is converted into scores by referring to the table given by Garret and Woodworth (1969). Then for each factor, the scores of each individual are added and then the total value of scores and mean values of scores are calculated. The factors having the highest mean value are considered to be the most important factors. Below is the tabular representation of the mitigation strategies suggested during the unstructured interviews.

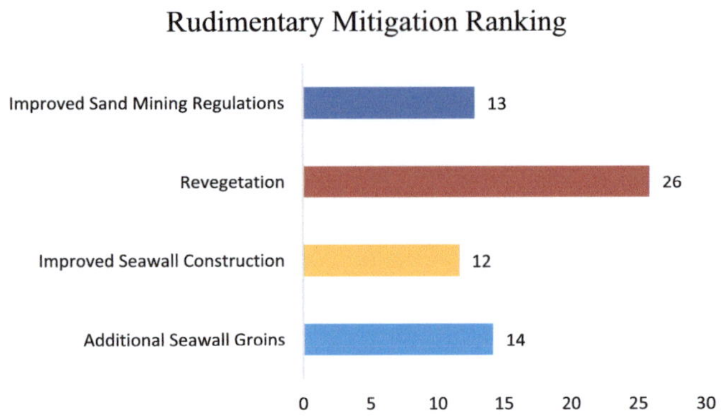

Fig. 11 Responses of mitigation methods. *Source* Authors

It was found that revegetation received the highest mean score, indicating that it may be the most effective mitigation method from the perspective of the community members. Additionally, the study participants proposed the implementation of more coherent mining regulations to control the rate of mining and prevent future occurrences of erosion and land degradation. These high scores present strong evidence that, from a sustainability perspective, further research should be carried out and sustainable mitigation strategies that address each of these areas should be proposed.

The mean mitigation scores are displayed in Fig. 11 as a clustered bar chart, with revegetation earning the highest score of 26. It is crucial to remember that the decrease in fish supply has resulted in a large income drop for stakeholders and locals, seriously endangering their ability to maintain a sustainable economy. The challenges faced by the impacted communities are made worse by this issue and the growing expense of life.

Our conversations with responders made it clear that, in addition to the impending threat of tsunamis, climate change has had a significant impact on land erosion. Land may completely vanish if sea levels were to rise by just one meter. The land's degradation has also been exacerbated by the detrimental environmental effects of sand mining, and the Coastal Regulation Zone (CRZ) regulations have made it more difficult for the inhabitants to reconstruct their homes. Daily routines have been further upended by changes in fishing schedules, forcing some family members to leave on multi-day fishing trips as early as 2 a.m. in order to make ends meet.

The responses offered a number of ways to address these issues, such as strengthening already-existing seawalls, revegetating, stopping sand mining, and building new seawall groins to promote sand deposition. These actions are necessary to reduce coastal erosion and ensure that future generations in Alappad have access to sustainable means of subsistence.

5 Conclusion

Based on the research findings, coastal erosion has significantly impacted the financial stability of fishermen, leading many to seek additional income through tourism. The situation has been exacerbated by climate change, causing damage and loss of fishing equipment, increasing their debt burden, and hindering their ability to afford essential expenses such as equipment repairs. Consequently, fishermen have resorted to taking out high-interest loans from moneylenders to address these challenges.

The study also discovered that the community's general well-being has suffered as a result of the disruption of fish processing and related trade activities brought on by deteriorating coastline erosion. Respondents noted that the erosion issue has been made worse by sand mining, which has decreased fish populations and put fishermen's livelihoods at additional risk and coastal erosion has also interfered with conventional fishing methods. The distance to the fishing regions grows as the shoreline recedes, adding to the expense and duration of their fishing journey. Furthermore, fish populations have declined and migratory patterns have changed as a result of the degradation of coastal ecosystems like mangroves and coral reefs, which has a negative impact on the community's revenue and catch.

Regarding potential future remedies, respondents suggested putting an end to sand mining, increasing the number of seawall groins to facilitate sand deposition, revegetating the area, and fortifying the seawalls. The community also voiced support for building seawalls using tetrapods. The land is still in danger of being lost entirely if government action is not taken and a clear plan is not put in place to solve the problem. The study's overall findings highlight the critical need for swift action to lessen the negative effects of coastal erosion on the fishing community's means of subsistence.

5.1 Coastal Rehabilitation and Erosion Mitigation Proposal

The proposed solution in the study suggests a comprehensive, phased approach to address the issue of coastal erosion in Alappad, Kerala, India. The approach will combine nature-based solutions and traditional structural mitigation techniques with eco-friendly solutions to safeguard the coastline and restore local industries. In addition, the study highlights the various initiatives undertaken by the Alappad Panchayat to support the community, including initiatives in agriculture, education, healthcare, and infrastructure development.

5.2 Limitations

It is worth noting that the scope of this study was limited to the Alappad village and a small number of respondents were taken into consideration. The research focused on evaluating the socio-economic and demographic factors of the participants, in addition to examining the impacts of coastal erosion on their livelihoods. This limitation was mainly due to the constraints of time and financial resources, which inevitably restricted the extent of the study. Considering the effect of coastal erosion, future studies need to be conducted further with a larger sample size and a longitudinal panel study to examine whether the schemes, programs, technology, and nature-based solutions that were implemented have been working successfully over the period.

Writers Contribution

Sheikh OF: Conceptualization, Methodology, Investigation, writing.

Ekuban CAA: Writing Original draft, preparation, and Analysis.

David B: Data Collection and Investigation.

Sudha A: Conceptualization, Supervision, Reviewing and Editing.

Acknowledgements The authors would like to thank the E4LIFE International Ph.D. Fellowship Programme offered by Amrita Vishwa Vidyapeetham, Kerala, India. We extend our gratitude to the Amrita Live-in-Labs® academic programme for providing all the necessary support and facilitating this study's conduct.

Conflict of Interest The authors declare that there is no competing interest.

Funding Declaration No funding support was received.

Ethics Approval Not applicable.

References

Angnuureng, B.D., Adade, R., Chuku, E.O., Dzantor, S., Brempong, E.K., Mattah, P.A.D.: Effects of coastal protection structures in controlling erosion and livelihoods. Heliyon **9**(10), Article e20633 (2023). https://doi.org/10.1016/j.heliyon.2023.e20633

Bai, J., Chen, Y., Hou, J., Zhao, J., Ma, Y., Zhai, J.: Analysis on the hotspot characteristics of bird diversity distribution along the continental coastline of China. Front. Mar. Sci. **9**, Article 1007442 (2022). https://doi.org/10.3389/fmars.2022.1007442

Barua, P., Rahman, S.H., Barua, S., Ismail, M.M.R.: Climate change vulnerability and responses of fisherfolk communities in the south-eastern coast of Bangladesh. Water Conserv. Manag. **4**(1), 20–31 (2020a)

Barua, P., Rahman, S.H., Mitra, A., Zaman, S.: Review on coastal erosion, displacement and resettlement strategies of South Asian countries. Glob. Environ. Eng. **7**, 52–72 (2020). https://doi.org/10.15377/2410-3624.2020.07.4

Basheer Ahammed, K.K., Pandey, A.C.: Shoreline morphology changes along the Eastern Coast of India, Andhra Pradesh by using geospatial technology. J. Coast. Conserv. **23**(2), 331–353 (2019)

Basheer Ahammed, K.K., Pandey, A.C.: Assessment and prediction of shoreline change using multi-temporal satellite data and Geostatistics: a case study on the eastern coast of India. J. Water Clim. Change **13**(3), 1477–1493 (2022). https://doi.org/10.2166/wcc.2022.270

Chandra, G.: Participatory rural appraisal. ResearchGate. https://www.researchgate.net/publication/265127750 (2010)

Dalezios, N.R., Eslamian, S., Ostad-Ali-Askari, K., Rabbani, S., Saeidi-Rizi, A.: Sediments. In: Encyclopedia of Earth Sciences Series, pp. 1–2. Springer Science and Business Media B.V. (2017). https://doi.org/10.1007/978-3-319-12127-7_254-1

Economic Review.: (2023) Government of Kerala Volume One. Centre for Development of Imaging Technology. www.cdit.org (2023)

Haneefa, M.: (2023). Fisherfolk and Football: An Ethnographic Tale of Loss in a Changing Climate. Scite. https://scite.ai/reports/ (2023). https://doi.org/10.1111/1467-8322.12839

Irshad, S.M.: Disaster Management and Livelihood of Fishermen: A Study on the Selected Areas of Kerala (n.d.)

Matovu, B., Brouwer, F., Bleischwitz, R., Aljanabi, F., Alkoyak-Yildiz, M.: Resource nexus perspectives in the Blue Economy of India: the case of sand mining in Kerala. Environ. Sci. Policy 151 (2024). https://doi.org/10.1016/j.envsci.2023.103617

Miyab, M., Sallu, S., Hubacek, K., Paavola, J.: Vulnerability of fishery-based livelihoods to the impacts of climate variability and change: insights from coastal Bangladesh. Reg. Environ. Change **14**(1), 281–294 (2013). https://doi.org/10.1007/s10113-013-0487-6

Monalisha, M., Panda, G.K.:. Coastal erosion and shoreline change in Ganjam coast along East Coast of India. J. Earth Sci. Clim. **9**, Article 467 (2018)

Mongabay-India.: Villages Vanish in this Coastal District of Kerala as they Succumb to Sand Mining. https://india.mongabay.com/2018/10/villages-vanish-in-this-coastal-district-of-kerala-as-they-succumb-to-sand-mining/ (2018)

Onda, K., Sinha, P., Gaughan, A., Stevens, F., Kaza, N.: Missing millions: undercounting urbanization in India. Popul. Environ. **41**(2), 126–150 (2019). https://doi.org/10.1007/s11111-019-00329-2

Paul, S., Chowdhury, S.: Investigation of the character and impact of tropical cyclone Yaas: a study over the coastal district of West Bengal, India. Saf. Extrem. Environ. **3**(3), 219–235 (2021). https://doi.org/10.1007/s42797-021-00044-y

Pillai, A., Prasad, G.: Land Use Pattern as an Indicator of Sustainability: A Case Study. ResearchGate. https://www.researchgate.net/publication/350442478 (2020)

Power, H.E., Pomeroy, A., Kinsela, M.A., Murray, T.: Research priorities for coastal geoscience and engineering: a collaborative exercise in priority setting from Australia. Front. Mar. Sci. https://scite.ai/reports/ (2021). https://doi.org/10.3389/fmars.2021.645797

Pradeep Kumar, B., George, R., Baby Reena, S.: Fisheries sector in Kerala: emerging issues and policy responses. Shanlax Int. J. Econ. **10**(3), 1–8 (2022). https://doi.org/10.34293/economics.v10i3.4777

Prieto, F.: The complexity of studying coasts: from forms and processes to management. Cuadernos De Investigación Geográfica **48**(2), 219–255 (2022). https://doi.org/10.18172/cig.5451

Punniyarajah, K.: An assessment on effects of coastal erosion on coastal environment: a case study in the coastal belt between Kalu River Mouth and Bolgoda River Mouth, Sri Lanka. In: Jana, N.C., Singh, R.B. (eds.) Climate, Environment and Disaster in Developing Countries, pp. 387–407. Springer (2022). https://doi.org/10.1007/978-981-16-6966-8_20

Santha, S.D.: Adaptation to Coastal Hazards: The Livelihood Struggles of a Fishing Community in Kerala, India. https://scite.ai/reports/ (2014). https://doi.org/10.1111/disa.12094

Satish, A., Geetha, P.: Assessment of shoreline change detection in Alappad region Kerala using geospatial technology. In: 2021 4th International Conference on Electrical, Computer and Communication Technologies (ICECCT 2021) (2021). https://doi.org/10.1109/ICECCT52121.2021.9616727

Sharon, C.: Planning for black sand mining affected Alappad coastal region. Int. J. Sci. Res. (n.d.). https://doi.org/10.21275/SR21713123229

Sontakki, B., Venkatesan, P., Rao, J.: Participatory Rural Appraisal (PRA): Tools and Techniques. https://www.researchgate.net/publication/336230869 (2019)

Subramoniam, V., Suresh, V.: A study on the waste disposal practices and its impact in Alappad Panchayat, Kerala. Int. J. Soc. Sci. Manag. **2**, Article 12289 (2015). https://doi.org/10.3126/ijssm.v2i2.12289

Swathi.: Vulnerability of Coastal Areas and Livelihood Challenges of Fisher Communities: A Case Study in Kerala. http://www.ijpam.eu (2018)

Varea, R., Kant, R., Farrelly, T.: Qi no tu I baba ni qwali (living down by the river): impacts of flooding and mining on ecosystems and livelihoods. Front. Mar. Sci. 9 (2022). https://doi.org/10.3389/fmars.2022.954062

Vieira, B., Pinho, J., Barros, J.: Assessment of different coastal defence structures to promote wave energy dissipation and sediments retention. In: Advances in Coastal and Ocean Engineering, pp. 33–45 (2021). https://doi.org/10.1007/978-3-030-76465-4_4

Wang, W., Cao, C.: Evaluation of NOAA-20 VIIRS reflective solar bands early on-orbit performance using daily deep convective clouds recent improvements. IEEE J. Sel. Top. Appl. Earth Obs. Remote. Sens. **13**, 3975–3985 (2020). https://doi.org/10.1109/jstars.2020.3007863

Digital Monitoring Systems to Mitigate the Impact of Climate Change in Cultural Heritage Protected Areas: The Case Study of Wadi Rum UNESCO Site

Rania Aburamadan, Victoria Cotella, and Luigi Petti

Abstract

World Heritage sites are increasingly vulnerable to climate change. Extreme weather, rising temperatures, high humidity and sea-level affect historical structures and archaeological sites accelerating deterioration, threatening authenticity, integrity and putting cultural buildings at risk of total loss. A comprehensive strategy, particularly through systematic documentation and monitoring, enabling early interventions are urgently necessary. Wadi Rum UNESCO-inscribed area, situated in southern Jordan is of extraordinary significance. The rock art, inscriptions, and archaeological evidence can be considered exceptional testimony to the cultural traditions of its early inhabitants promoting tourism activities and consequently, being relevant for local economy. Unfortunately, this rock art and its numerous inscriptions are threatened by extreme weather conditions, which cause them to fade and make some petroglyphs difficult to distinguish and even disappear. The present paper proposes a novel 'DUNE' framework for the wide-ranging documentation and monitoring of the Cultural and Natural Heritage under hazard in the UNESCO area of Wadi Rum, Jordan. It highlights collaboration with local stakeholders, enabling early intervention in case of decay of inscriptions and petroglyphics due to harsher climate conditions and ensuring its safeguarded for future generations. Furthermore, the framework proposes the use of advanced digital technologies to create a sustainable and replicable approach. The methodology consists of three key phases: (i) Collaborative data documentation; (ii) Real-time monitoring and visualisation; and (iii) Knowledge dissemination and community engagement. While DUNE Framework offers a promising solution, the discussions address some challenges related to its implementation and long-term effectiveness.

Keywords

Climate change · Endangered cultural heritage · Archaeological landscapes · Monitoring · IoT systems

1 Introduction

Jordan, a land steeped in history and natural wonders, stands at the forefront of a pressing environmental challenge: climate change. The region, once known for its stable weather, now experiences extreme weather conditions such as high temperatures, irregular rainfall, and frequent droughts. On the other hand, while Cultural Heritage play an important role in the tourism and recreational industries (e.g., support economic growth and employment); enhance public education and lifelong learning, shape socio-cultural capital through community identity, traditions, and sense of place, climate change poses serious threats to the protection, preservation, and transmission of these non-renewable heritage and resources to future generations (Fatorić and Seekamp 2017).

Jordanian Cultural Heritage, encompassing ancient archaeological sites, historical monuments, and unique cultural practices, is at risk due to the adverse impacts of climate change. Rising temperatures (Fayez 2020), changing precipitation patterns (Al Balasmeh et al. 2019), droughts (Aladaileh et al. 2019) and increased frequency of extreme

R. Aburamadan
Department of Architecture, Applied Science Private University, Amman, Jordan
e-mail: R_aburamadan@asu.edu.jo

V. Cotella (✉)
Department of Architecture, University of Naples Federico II, Naples, Italy
e-mail: victoriaandrea.cotella@unina.it

L. Petti
Department of Civil Engineering, University of Salerno, Fisciano, Italy
e-mail: petti@unisa.it

P. Mohammad et al. (eds.), *Geographic Approaches to Climate Change and Mitigation: Urban and Rural Perspectives (Volume 1)*, Advances in Science, Technology & Innovation, https://doi.org/10.1007/978-3-031-92119-3_12

weather events threaten the preservation and integrity of these invaluable cultural assets. In response, Jordan has established a network of protected areas. As defined by the Jordanian Antiquities Law n° 21 of 1988 (and its amendment in 2004), these areas go beyond just conserving biodiversity and natural resources. They encompass the environment and Cultural Heritage within their boundaries, recognizing the interconnectedness of these elements and the unique challenges posed by climate change.

Wadi Rum Protected Area (WRPA) is located in the southern part of Jordan close to the border with Saudi Arabia (Fig. 1), around 290 km south of Amman and 60 km northeast of the coastal city of Aqaba. Following the criterion (iii) of the World Heritage Dataset (International Union for Conservation 2017) "*The WRPA bears a unique testimony to the practice of rock art and inscriptions that has been on-going for millennia. The combination of 25,000 petroglyphs with 20,000 inscriptions and their continuity over a period of at least 12,000 years sets Wadi Rum apart from other rock art and/or inscription sites*". The rock art remains in its original setting, mostly unaltered except for the effects of weathering, which have led to its fading because of rain and wind erosion, making some petroglyphs hard to distinguish. In fact, extreme weather conditions, including extreme and sudden increases and drops in temperatures, are putting enormous pressure on this heritage and leading it to disappear (Fig. 1).

Furthermore, the impact of increased tourism on the deterioration of the rock at WRPA (Wadi Rum Protected Area) is evident (Shatnawi and Taleb Obaidat 2022). Damage and fragmentation of rocks may be accelerated by increased tourism activities via physical contact, such as climbing. The preservation and conservation of these exceptional environments need practical research to help understand weathering effects. These rock signs, testifying to humankind's evolution, are exposed to weather conditions, which in the last decades have increasingly become affected by extreme occurrences. Climate change isn't just intensifying weather events in desert regions like Jordan; it's also dramatically increasing their frequency. This alarming rise in extreme weather, from scorching temperatures to erratic rainfall, accelerates rock decay, causing surface changes and even deep cracks. The consequence is a rapid escalation in the risk of losing this irreplaceable cultural heritage.

The lack of systematic monitoring is causing some of these unique documents to slowly fade and eventually disappear. In fact, the loss of Cultural assets due to erosion and cracks could be prevented if timely monitored and tackled. Therefore, the urgency of a comprehensive strategy cannot be overstated.

The present paper is based on the following research question: "*How can IoT technology be effectively integrated into a participatory framework to monitor and protect not only tangible but also intangible Cultural Heritage in the face of climate change?*". In these regards, the following objectives are pursued:

- Develop a comprehensive framework that includes both tangible and intangible cultural heritage elements in the WRPA, such as traditional practices, and oral histories.
- Quantify the rate of deterioration of tangible heritage sites using IoT sensors that measure environmental factors and physical changes.
- Collaborate with local communities and relevant organizations to ensure the feasibility and sustainability of the IoT-based monitoring system.

The research makes a major contribution by proposing an integrative framework for monitoring protected areas including community engagement, considering the WRPA as a specific case of study. This framework is based on a participatory approach and utilizes IoT systems, laying the groundwork for future practical applications. The following sections of the manuscript provide a literature review that presents several protected areas facing climate hazards. It also explores the technical applications of IoT systems in

Fig. 1 Images of the Alameleh and Khazali Canyon inscriptions in the Wadi Rum Protected Area. *Source* Author's photo, 2024

large, Cultural Heritage areas. Additionally, the current risk situation of the Wadi Rum Protected Area is analysed. The finals sections discuss the outlining issues in the proposed framework and conclusions and future work are presented.

2 Literature Review

This section presents an overview of several protected areas around the world that are facing challenges related to climate change hazards. Additionally, it describes the current state of the art regarding the application of IoT systems for monitoring cultural heritage outside. Subsequently, the concept of "Heritage at Risk" is highlighted by examining the specific hazards faced by the Wadi Rum Protected Area and how they are faced. This review highlights the urgent need for an efficient monitoring system implementation.

2.1 An Overview of the Protected Areas Under Climate Change

Given the accelerating pace of climate change, a sensitive focus is being placed on the resilience of protected areas—critical landscapes for safeguarding biodiversity and the ecosystem services they provide. This global phenomenon manifests disproportionately, with some protected areas experiencing more severe impacts than others. A recent wide literature review carried out by (Sesana et al. 2021) addressing climate change impacts on Cultural Heritage, highlights the relevance of changes in temperature, precipitation, and wind in the complex integrity (Giglio et al. 2024). The available study highlights the following relevant impacts for the case of archaeological sites:

- An increase in rain and humidity, along with higher temperatures, can lead to various ways of damage to materials like corrosion, biological breakdown, warping and splitting of materials that absorb moisture, and the formation of salt crystals that cause efflorescence and subflorescence (Jain et al. 2022).
- The increase of wind, especially when combined with sand, salt, and atmospheric pollutants can result in surface erosion, heightened water infiltration, structural impairment, and the possibility of structures collapsing (Wang et al. 2022).
- An increase in temperature could lead to more instances of freezing and thawing, or greater fluctuations in temperature throughout the day, which would impact the frequency of thermoclastism leading to increased physical damage to stone and ceramic materials due to weathering (Ren et al. 2023).
- An increase in temperature and humidity will increase biological degradation as it can create conditions favourable to mould growth and insect activity (Chau et al. 2023).

Moreover, it is necessary to also consider the different effects that both fast and extreme changes could have with the main goal to avoid major damage and above all disasters. Beyond Wadi Rum Protected Area, numerous protected areas worldwide grapple with similar challenges due to climate change. The Great Barrier Reef Marine Park in Australia (Benedetti-Cecchi et al. 2024), the world's largest coral reef system, faces threats from rising ocean temperatures and acidification, both linked to climate change, which damage coral health and bleach the reef.

High in the Andes, the Machu Picchu Historic Sanctuary in Peru experiences melting glaciers and increased landslide risks due to changing weather patterns, potentially damaging the archaeological structures in addition to the earthquakes (Rodríguez-Pascua et al. 2020). In the United States, Everglades National Park, a unique wetland ecosystem (Leatherman 2022), is threatened by rising sea levels caused by climate change. Saltwater intrusion disrupts the delicate balance of freshwater and saltwater plants and animals, impacting the entire food chain. Similarly, the Serengeti National Park in Tanzania (Kilungu et al. 2017), famous for its annual wildlife migration, experiences changes in precipitation and rising temperatures, disrupting migration patterns and affecting plant growth, impacting food sources for various species.

Finally, Venice Lagoon in Italy, a historic city built on canals, faces rising sea levels and more frequent flooding events due to climate change (Rubinetti et al. 2022). This not only puts cultural heritage sites like St. Mark's Square at risk but also threatens the very foundations of the city itself. These examples highlight the widespread impact of climate change on protected areas and the urgent need for innovative solutions to monitor and safeguard these invaluable cultural and natural resources.

2.2 Applying IoT Technologies in Monitoring Large Heritage Protected Areas

Facing increasing threats from climate change and human activity, safeguarding Cultural Heritage sites has become a major concern. To address this challenge, researchers are increasingly turning to the potential of Internet of Things (IoT) technologies. Several current research efforts are exploring innovative applications of IoT for cultural heritage monitoring.

(Gribaudo et al. 2017) Proposes an IoT-based monitoring system specifically designed for extensive Cultural Heritage areas facing large crowds. Using a network of diverse sensors, the system aims to anticipate potential damage caused by crowds in UNESCO-protected sites like Matera, Italy. The technique is based on the use of Markovian agent models to study the parameters and the dynamics of a selected scenario. Through analysing crowd movement and environmental data, the system can predict threats and help authorities take preventative measures to safeguard these irreplaceable cultural treasures.

(Lerario and Varasano 2020) Presents a Wireless Sensor Network (WSN) for monitoring environmental conditions within a historical church in Matera, Italy. Employing a standardized protocol stack and adhering to the OneM2M standard for machine-to-machine communication, the WSN transmits sensor data securely to a cloud platform via a concentrator node and an NB-IoT 5G network gateway. This method eliminates the need for cabling within the structure and enables remote monitoring of the church's environment. (Mitros et al. 2022) Present the development of Smart IoT devices (SmartTags) that monitor specific environmental conditions in the vicinity of Cultural Heritage buildings. The Smart Tags have shown limitations in their data processing capabilities due to their limited resources, making them unsuitable for complex computations due to its constrained resources (e.g., memory, power consumption).

(Merello et al. 2014) Proposes using sensor networks to monitor microclimatic conditions (temperature and humidity) in open-air archaeological sites like Ariadne's house in Pompeii. By analysing this sensor data statistically, the researchers aim to identify irregular climate patterns that could pose a threat to the Cultural Heritage and update preventative conservation efforts. (Lecari et al. 2021) Examines the viability of employing stationery and SLAM-based terrestrial laser scanning, close-range photogrammetry, automated surface change detection, GIS (Geographic Information System), and WebGL (Web Graphics Library) visualisation methods to enhance the preservation of cultural resources in California. The methodology is tested in response to recent environmental threats from wildfire and earthquakes in the historic town of Bodie, an iconic Gold Rush-era boom town located on the California and Nevada border.

(Colace et al. 2024) Suggests an innovative monitoring approach for complex archaeological parks. This method aims to automatically identify and classify potential critical issues, and subsequently communicate them to relevant stakeholders. The approach is being tested at the Archaeological Park of Pompeii (Italy), which utilizes high-definition aerial drone imagery captured bi-weekly to generate high-resolution orthophotos of the park and its artifacts. However, current methods are time-consuming and hinder real-time maintenance. Therefore, the research focuses on automating the process by using machine learning techniques. Specifically, they use the YoloV7 library to automatically detect and classify problems within the images, followed by a mechanism for notifying park workers.

2.3 Heritage at Risk: The Wadi Rum Protected Area

The Wadi Rum Protected Area (WRPA) was declared in 1998 "*to protect and preserve a representative area of the Eastern Desert (HISMA) land type, together with indigenous flora and fauna, and to preserve a large section of the scenic mountains of Rum*" (Figs. 2 and 3). The lack of effective governance and the presence of unskilled users living in the area underscore the importance of researchers and experts in guiding local communities towards sustainable practices for conservation.

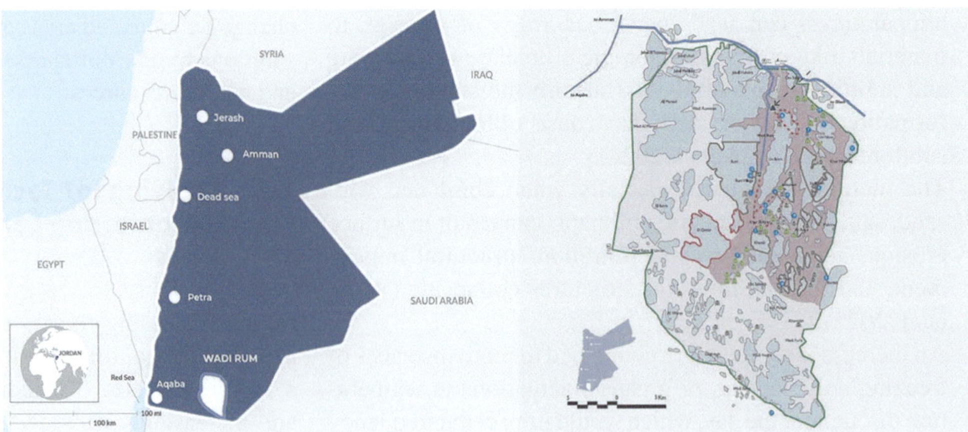

Fig. 2 The map on the left shows Jordan with the WRPA highlighted in the south. The map on the right provides a zoomed-in view of the WRPA highlighting in green the buffer zone. *Source* Author's elaboration

Fig. 3 Wadi Rum Protected Area during the initial field observation campaign. *Source* Author's photo, 2024

In this sense, one of the main initiatives to protect the area was the Wadi Rum Protected Area Management Plan, developed by ICOMOS and USAID in 2014 to guide and set priority to the many, diverse activities for Wadi Rum Protected Area (WRPA's) management in the period 2015–2019 (ICOMOS–IUCN 2014).

The plan includes various activities, such as monitoring petroglyphs and archaeological sites in WRPA, waste management, and tourism management. Still, there are several threats identified by UNESCO, including visitor pressure, tourism infrastructure, encroachment by the village of Rum, groundwater exploitation, and firewood collection. The wider tourism and planning policies for the property, its buffer zone and wider setting also should prioritise the protection of its Outstanding Universal Value.

Research in the field highlights some interest as: (Shatnawi and Taleb Obaidat 2022) present research using satellite images acquired in the period from 1987 to 2018, aimed to discover any influence of erosion on the geological structure of (WRPA) and any potential of floods, using the topographic wetland-index analysis. Supervised classification was used to construct land-cover/land-use maps of WRPA. The findings revealed a shift in Wadi deposits, but no change in mountain ranges, indicating that satellite photos might anticipate soil erosion. On the other hand, (Gharib 2020) proposed a methodology based on Geographic Information System (GIS) such as ArcGIS Online, to apply the Rock Art Stability Index (RASI) in order to assess a large sample of panels based on their current state of decay. Statistical analysis of this data aims to identify decay processes most harmful to the art.

Currently, the official WRPA website (WRPA 2014) declares to have conservation projects underway in the protected area to try and safeguard its unique landscape and ecology and to raise awareness among local people and visitors like: (i) Ecological surveys have been carried out and are being used to monitor the populations of endangered animals like the Gray Wolf and Nubian Ibex; (ii) Ranger patrols to enforce the protected area by-laws and reduce illegal hunting; (iii) one main network for tourist vehicles is created and thereby reduce the widespread damage to desert habitats caused by "free-for-all" driving; (iv) a large and impressive visitor centre has been opened to create a single gateway to the protected area that will enable all tourism activities to be well managed and coordinated and (v) several education and outreach programmes are being implemented to promote the importance of the protected area and encourage local communities and visitors to participate in its conservation.

However, until today, there is not a monitoring network well established to control the current state of the rocks and petroglyphs and develop strategies of intervention in order to protect it against climate hazards.

3 Methodology: DUNE Framework

The paper examines the precarious situation confronting preservation approaches in historical sites (Sect. 2.2), attributed to a combination of natural and human-induced factors and raises the need for a preservation approach that considers both tangible and intangible Heritage.

In these regards, DUNE Framework is proposed as a novel methodology for tangible and intangible heritage documentation and monitoring that leverages the power of the Internet of Things (IoT) in externally protected Cultural Heritage areas. It is important to highlight that before formulating the DUNE theoretical framework a direct field observation campaign was carried out in February 2024. This on-site investigation at the Alameleh and Khazali Canyon provided crucial data that would condition the development of the framework. Key observations included: (i) The absence of an internet network in the area; (ii) The specific location of the Cultural Heritage—hieroglyphs within the Alameleh and Khazali Canyon; (iii) The unique environmental conditions, particularly the lack of direct sunlight leading to high humidity; (iv) The concerning direct contact between the hieroglyphs and a high number of tourists, with no apparent protective measures.

The proposed DUNE framework is divided into three interconnected phases (Fig. 4): (i) Qualitative data documentation (Intangible Heritage) (ii) Real-time monitoring and visualisation (Tangible Heritage); and iii) Knowledge dissemination and community engagement (Intangible and Tangible Heritage).

In phase I and II, a comprehensive documentation process is proposed, centred around qualitative and quantitative data collection methods. This involves, on the one hand, the collection of primary data through semi-structured questionnaires and interviews and, on the other hand, the establishment of a scalable monitoring network to assess the state of Wadi

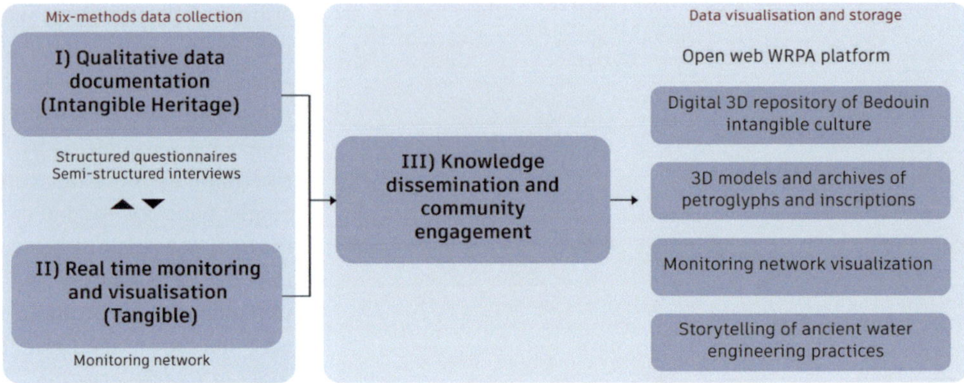

Fig. 4 DUNE methodological process highlighting the three main phases. *Source* Author's elaboration

Rum, with the active participation of the responsible entities and NGOs in the region. On the other hand, phase III involves knowledge dissemination and community engagement through data visualisation in an open web WRPA platform where the roles of various stakeholders can be enhanced, and results disseminated on an international scale.

3.1 Qualitative Data Documentation

Primary data collection approach is the first step in structuring the framework for data collection. Contact with Bedouin communities and women allows researchers to better understand their experiences, perspectives and contributions to their environment.

In this stage, semi-structured questionnaires and interviews are planned in order to have a more flexible and in-depth exploration of the topics covered such as ancient water management practices, local ecological knowledge, and the challenges posed by climate change. The goal is to empower local communities and fill a gap in the representation of the Bedouin Intangible Heritage.

Finally, in order to feed the storytelling platform, collecting stories of resilience, adaptation and initiatives to protect their cultural heritage and natural resources from participants will be crucial.

3.2 Real-Time Monitoring and Visualisation

DUNE framework is proposed as a multi-pronged approach to documenting and safeguarding the cultural and natural heritage of Wadi Rum (Fig. 5). It exploits cutting-edge technology to establish a comprehensive monitoring network for both environmental and archaeological preservation using a layered approach: Ground-based sensors and aerial drones equipped with photogrammetry and LiDAR technology will provide high-resolution, real-time data on the site's condition, specifically 3D models and archives of petroglyphs and inscriptions, in Alameleh and Khazali Canyon will be developed. Additionally, satellite interferometry will offer broader insights into subtle changes. This wide-ranging data stream will be fed into a central processing unit, enabling continuous analysis and proactive maintenance to prevent emergencies.

Furthermore, DUNE will document and monitor the cultural and historic heritage of WRPA by: Piloting an up-scalable monitoring network to assess the evolutions of the WR site conditions with the aim to support proactive maintenance process and avoid emergency conditions. This includes non-invasive inspection techniques; data acquisition by terrestrial and aerial photogrammetry by using DSLR cameras and UAV (Unnamed Aerial Vehicles); data acquisition by LiDAR (Light Detection and Ranging) technology and data from satellite interferometric analysis.

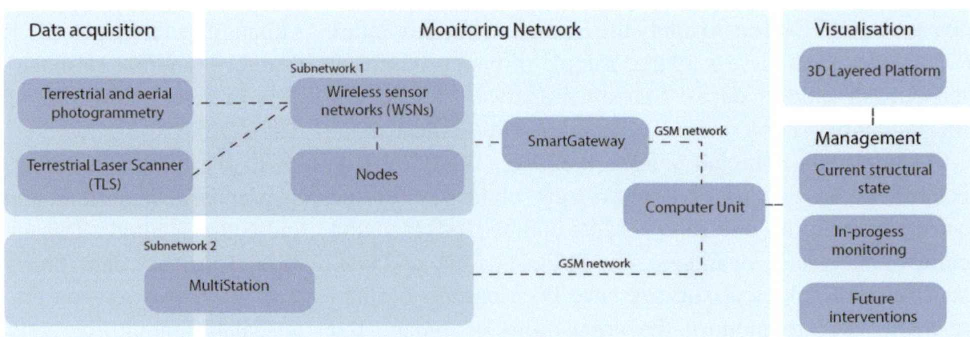

Fig. 5 An overview of the DUNE framework to WRPA monitoring and visualisation. *Source* Author's elaboration

The monitoring network is divided into two subnetworks: (1) Sensors and nodes; and (2) Multistation type MS60 LEICA. From a functional point of view, as far as the first subnetwork is concerned, the data read by the sensor is acquired by the relative node (to which it is connected via cable), which wirelessly transfers the given to the acquirer (smart gateway), which at the same time, via the GSM network forwards them to the central computer unit, which collects and processes them through the dedicated software platform. The second subnetwork (MultiStation) is connected to a Leica ComBox60 type modem via the network GSM that forwards the data to the central computer. Geological and structural assessment of site conditions is planned to be developed with the aim of assessing critical conditions and safety against climate change pressure through digital technologies.

The system also enables visualisation, which offers the possibility of simultaneously "seeing" and analysing heterogeneous events, significantly increasing the ability to monitor and investigate the various situations and climate risks. The information of interest to be represented will be acquired by means of the use of special IoT Adapters. The proposed project will include a 3D front end of the platform. The 3D platform is organized according to "Layers", each of which represents a subsystem to be represented, and which represents functionalities and heterogeneous information of a selected element/feature.

Through the selection of the Layer item, it is possible to activate animations and information within the 3D map. This feature will show the immediate visual evidence of the critical issues in progress. Hence, allowing better site management; (1) immediately perceive the state of the entire infrastructure, (2) perceive the actions in progress to manage any problems, (3) possibility of intervening personally to address the governance and management of the problems highlighted.

3.3 Knowledge Dissemination and Community Engagement

The methodology concludes in knowledge dissemination and community engagement by an open web platform that disseminates research findings and training materials on the utilized technologies T this platform, along with targeted training programs, aims to equip local professionals to manage the monitoring system and expand its reach, ensuring the project's long-term sustainability and empowering the local population.

Upon completing the data documentation phase, DUNE endeavours to establish a digital repository showcasing Bedouin intangible cultural practices by storytelling practices and 3D models of their local heritage, encompassing aspects such as pottery and jewellery. This initiative will be developed in collaboration with local communities. Using digital technologies such as video recordings, interviews, and 3D replicas, DUNE aims to preserve these intangible assets while simultaneously fostering local economic growth and supporting local entrepreneurship. Digitalizing data takes a part of data presentation methodology through creating WR platform which makes the data deliverable publicly.

Furthermore, the project aims to shift the narrative by portraying Bedouin communities, particularly women, as active participants in heritage preservation and environmental stewardship. This includes highlighting their valuable knowledge in traditional water management and their efforts to combat local flora and fauna extinction. A dedicated web-based platform will be created to share research findings and project activities with a wider international audience. This platform will not only feature training webinars on the utilized technologies but will also house a collection of videos, interviews, and other materials that promote the intangible heritage of Wadi Rum, making it accessible to a global audience.

4 Discussions

While the reviewed studies in Sect. 2.1 showcase diverse applications of IoT for Cultural Heritage management, a crucial takeaway is the absence of a universal solution. Each heritage site possesses unique characteristics, necessitating tailored approaches that consider specific needs and limitations. Many heritage sites, particularly those in remote areas, face challenges like limited connectivity: (i) Rural locations may have weak or non-existent internet infrastructure, hindering real-time data transmission; lack of on-site technical expertise, (ii) The absence of readily available technical support can pose difficulties in system deployment, maintenance, and troubleshooting; and restricted power availability and, (iii) Reliable and sustainable power sources might be scarce in certain locations, impacting sensor operation and data collection.

In the context of Wadi Rum Protected Area (WRPA), these challenges are particularly relevant. Here, a gap exists in comprehensive and continuous monitoring activities. Furthermore, measures at the government level are crucial to protect cultural heritage from direct tourist contact. Unrestricted access can lead to degradation through physical touch or accidental damage. This highlights the need for a framework like DUNE, which is adaptable and addresses the specific limitations often encountered in protected areas.

The Wadi Rum Protected Area (WRPA) exemplifies these challenges perfectly. Here, comprehensive and continuous monitoring activities are lacking. This gap is further intensified by the absence of strong measures at the government level to protect Cultural Heritage from direct tourist contact. Unrestricted access can lead to degradation through physical touch or accidental damage. This highlights the need for an adaptable framework like DUNE, allowing it to address the specific limitations often found in protected areas.

DUNE Framework builds upon and surpasses existing research on IoT-based Cultural Heritage monitoring in several keyways. While previous works like Gribaudo et al. (2017) and Merello et al. (2014) utilize sensor networks for monitoring, DUNE goes beyond basic environmental data collection. It integrates real-time threat assessment through a robust sensor network and high-precision MultiStation, enabling proactive management of climate change pressures. Furthermore, DUNE acknowledges the limitations of existing frameworks that focus solely on technology or local knowledge. It fosters a collaborative approach like Lecari et al. (2021) by integrating both elements, empowering local communities through culturally sensitive training and knowledge dissemination. This fosters a sense of ownership and ensures the project's long-term sustainability. Additionally, DUNE surpasses the sensor-monitor limitations of SmartTags (Mitros et al. 2022) by offering a comprehensive data visualisation platform. This user-friendly 3D interface translates complex data into a format accessible to a wider audience, facilitating informed decision-making for site management and targeted conservation efforts. In essence, DUNE builds upon the strengths of existing research while addressing their limitations, offering a more comprehensive, collaborative, and sustainable approach to safeguarding cultural heritage in the face of climate change.

4.1 Outstanding Issues

However, implementing the DUNE Framework in the Wadi Rum Protected Area (WRPA) requires addressing several outstanding issues. First, securing the necessary approvals for aerial laser scanning and potentially deploying specific sensors might require additional planning and negotiation with relevant authorities like the Royal Army. In second stage, ensuring the project's long-term sustainability hinges on effective community training and capacity building. This involves developing culturally sensitive training materials and fostering a sense of ownership among local communities regarding the monitoring system's operation and maintenance.

Finally, establishing robust data security protocols is crucial to protect sensitive information collected about the site and local communities. Addressing these outstanding issues will be essential for the successful implementation of the DUNE Framework in WRPA paving the way for a comprehensive and sustainable approach to safeguarding Cultural Heritage in the face of climate change.

5 Conclusion

The present paper introduced the DUNE Framework, a novel and adaptable methodology for documenting and monitoring cultural heritage in external protected areas vulnerable to climate change. This framework starts from a "one-size-fits-all" approach, acknowledging the specific needs and limitations of each site.

DUNE Framework goes beyond existing approaches by integrating a robust sensor network and high-precision MultiStation for real-time environmental monitoring, enabling proactive management of climate change threats, and offering a significant advantage. Unlike frameworks focused solely on technology or local knowledge, DUNE adopts a collaborative approach. It integrates both elements through culturally sensitive training, empowering local communities and ensuring long-term project sustainability.

Beyond the Wadi Rum Protected Area, the DUNE Framework holds immense promise for safeguarding cultural heritage sites worldwide, particularly those facing similar challenges from climate change and limited resources. While implementing the DUNE Framework requires addressing outstanding issues like securing approvals, community training, and data security, these challenges can be overcome through careful planning and collaboration. Future research can focus on optimizing the framework for different Cultural Heritage sites and exploring its integration with advanced data analysis techniques for even more comprehensive threat assessment and risk management.

In conclusion, the DUNE Framework represents a significant step forward in Cultural Heritage preservation. By offering a comprehensive, collaborative, and sustainable approach, DUNE empowers local communities and equips stakeholders with the tools necessary to safeguard irreplaceable cultural resources in the face of climate change.

References

Al Balasmeh, O., Babbar, R., Karmaker, T.: Trend analysis and ARIMA modeling for forecasting precipitation pattern in Wadi Shueib catchment area in Jordan. Trend Anal. ARIMA Model. Forecast. Precip. Pattern Wadi Shueib Catchment Area Jordan **12**(27) 2019. https://doi.org/10.1007/s12517-018-4205-z

Aladaileh, H., Al-Qinna, M., Barta, K., Al-Karablieh, K., Rakonczai, J.: An investigation into the spatial and temporal variability of the meteorological drought in Jordan. Climate **7**(6) (2019). https://doi.org/10.3390/cli7060082

Benedetti-Cecchi, L., Bates, A.E., Strona, G., Bulleri, B., Horta e Costa, B., et al.: Marine protected areas promote stability of reef fish communities under climate warming. Nat. Commun. **15**(1822) (2024). https://doi.org/10.1038/s41467-024-44976-y

Chau, K., Fleck, R., Irga, P., Torpy, F., Wilkinson, S., Castel, A.: Hempcrete as a substrate for fungal growth under high humidity and variable temperature conditions. Constr. Build. Mater. **398**(5) 2023. https://doi.org/10.1016/j.conbuildmat.2023.132373

Colace, F., De Santo, M., Gaeta, R., Loffredo, R., Petti, L.: FAUNO: a machine learning-based methodology for monitoring and predictive maintenance of structures in archaeological parks through image analysis. In: HCI International 2024 Conference. Springer (2024). (In printing)

Fatorić, S., Seekamp, E.: Are cultural heritage and resources threatened by climate change? A systematic literature review. Clim. Change **142**, 227–254 (2017). https://doi.org/10.1007/s10584-017-1929-9

Fayez, A.: 21st Century climate change projections of precipitation and temperature in Jordan. Procedia Manuf 197–204 (2020). https://doi.org/10.1016/j.promfg.2020.02.222

Gharib, Z.: Using the rock art stability index to facilitate management of rock art in Wadi Rum, Jordan. In: Inquiry@Queen's Undergraduate Research Conference Proceedings (2020). https://doi.org/10.24908/iqurcp.14016

Giglio, F., Frontera, P., Malara, A., Armocida, F.: Materials and climate change: A set of indices as the benchmark for climate vulnerability and risk assessment for tangible cultural heritage in Europe. Sustainability **16**(5) (2024). https://doi.org/10.3390/su16052067

Gribaudo, M., Iacono, M., Levis, A.: "An IoT-based monitoring approach for cultural heritage sites: the Matera case. Concurr. Comput. Pract. Exp. **29**(11) (2017). https://doi.org/10.1002/cpe.4153

ICOMOS–IUCN.: Report on the joint World Hhritage Centre-ICOMOS-IUCN reactive monitoring mission to Wadi Rum Protected Area, Jordan. https://whc.unesco.org/document/129738 (2014)

International Union for Conservation.: Wadi Rum Protected Area. http://world-heritage-datasheets.unep-wcmc.org/datasheet/output/site/wadi-rum-protected-area/ (2017). Last accessed 2024

Jain, P., Castellanos-Acuna, D., Coogan, S., Abatzoglou, J., Flannigan, M.: Observed increases in extreme fire weather driven by atmospheric humidity and temperature. Nat. Clim. Change **12**, 63–70 (2022). https://doi.org/10.1038/s41558-021-01224-1

Kilungu, H., Leemans, R., Munishi, P., Amelung, B.: Climate change threatens major tourist attractions and tourism in Serengeti National Park, Tanzania. In: Climate Change Adaptation in Africa, pp. 375–392. Springer (2017). https://doi.org/10.1007/978-3-319-49520-0_23

Leatherman, S.: Management of invasive snakes in coastal environments: a baseline assessment of the Burmese python invasion in the Florida Everglades. Mar. Pollut. Bull. 113996 (2022). https://doi.org/10.1016/j.marpolbul.2022.113996

Lecari, N., Jaffke, D., Campiani, D., Guillem, D., McAvoy, S., Jiménez Delgado, G., Bevk Neeb, A.: Building cultural heritage resilience through remote sensing: an integrated approach using multi-temporal site monitoring, datafication, and Web-GL visualization. Remote. Sens. **13**(20) (2021). https://doi.org/10.3390/rs13204130

Lerario, A., Varasano, A.: An IoT smart infrastructure for S. Domenico Church in Matera's "Sassi": a multiscale perspective to built heritage conservation. Sustainability **12**(16) (2020). https://doi.org/10.3390/su12166553

Merello, P., García-Diego, F., Zarzo, M.: Diagnosis of abnormal patterns in multivariate microclimate monitoring: a case study of an open-air archaeological site in Pompeii (Italy). Sensors **488**, 14–25 (2014). https://doi.org/10.1016/j.scitotenv.2014.04.068

Mitros, N., Krommyda, M., Amditis, A.: Smart tags: IoT sensors for monitoring the micro-climate of cultural heritage monuments. Appl. Sci. **12**(5) (2022). https://doi.org/10.3390/app12052315

Ren, J., Wang, Y., Wang, T., Hu, J., Wei, K., Guo, Y.: Numerical analysis of the effect of groundwater seepage on the active freezing and forced thawing temperature fields of a new tube–screen freezing method. Sustainability **15**(12) (2023). https://doi.org/10.3390/su15129367.

Rodríguez-Pascua, M.A., Benavente Escobar, C., Rosell Guevara, L., Audin, L., Walker, R., García, B., Aguirre, E.: Did earthquakes strike Machu Picchu? J. Seism. 883–895 (2020). https://doi.org/10.1007/s10950-019-09877-4

Rubinetti, S., Taricco, C., Zanchettin, D., Arnone, E., Bizzarri, I., Rubino, A.: Interannual-to-multidecadal sea-level changes in the Venice lagoon and their impact on flood frequency. Clim. Change (2022). https://doi.org/10.1007/s10584-022-03448-2

Sesana, E., Gagnon, A.S., Ciantelli, C., Cassar, J.A., Hughes, J.H.: Climate change impacts on cultural heritage: a literature review. WIREs Clim Change **12**(4) (2021). https://doi.org/10.1002/wcc.710

Shatnawi, N., Taleb Obaidat, M.: Remote-sensing approach to assess erosion and floods in Wadi Rum Protected Area. Jordan J. Civ. Eng. **16**(4) (2022).

Wang, W., Zhang, J., Li, Z., Wang, C., Wang, D.: Study on transports of atmospheric pollutants under global wind by dynamical system approach. J. Environ. Account. Manag. **10**(4), 345–360 (2022). https://doi.org/10.5890/JEAM.2022.12.002

WRPA.: Wadi Rum Protected Area. http://wadirum.jo/protecting-wadi-rum/conservation-projects/ (2014). Last accessed 2024

Meteorology and Disaster Prevention in Urban and Rural Environments

Leveraging High-Quality Seismic Datasets to Earthquake Prediction Occurrence: A Review

Ari Fadli, Tri Kuntoro Priyambodo, Agfianto Eko Putra, and Wiwit Suryanto

Abstract

Earthquakes are natural disasters caused by the movement of the Earth's tectonic plates due to the release of large amounts of energy. Earthquakes have claimed many lives and resulted in billions of dollars in economic losses. In recent years, studies in the field of seismology regarding the availability of seismic datasets to mitigate these impacts have experienced rapid development. We conducted a literature review on a collection of seismic datasets. High-quality seismic datasets are essential for producing earthquake prediction systems with high accuracy. We provide insight into a type of seismic dataset that contains seismic wave recordings from three components: east–west, north–south, and vertical, from each monitoring station, along with its meta information and the parameters contained therein. The information is structured and arranged chronologically to support earthquake prediction systems. Next, we compare the capabilities of seismic datasets in event detection, phase identification, and onset time picking for P waves and S waves. The results of this research can serve as a reference for researchers in the field of seismology in determining the appropriate seismic datasets for developing earthquake prediction systems using machine learning (ML) and deep learning (DL) models.

Keywords

Earthquake prediction · Seismic dataset · Event identification · Phase detection · Seismic measurements · Seismology

1 Introduction

Earthquakes are natural disasters caused by the movement of the Earth's tectonic plates due to the release of relatively large amounts of energy. Their destructive power characterizes them, and they often occur without warning (Aanuoluwa et al. 2016; Florido et al. 2016; Moustra et al. 2011). These seismic events result in numerous fatalities and substantial economic losses, as demonstrated by several notable earthquakes worldwide. For instance, the 2011 Tohoku Earthquake in Japan, which measured 9.1 on the Richter scale, claimed the lives of 20,000 people and incurred economic losses of 360 billion U.S. dollars. Similarly, the 2015 Nepal earthquake, with a magnitude of 7.8, resulted in 8,964 fatalities and financial losses totaling 10 billion U.S. dollars. Furthermore, the 2023 earthquake in Turkey and Syria, registering 7.5 on the Richter scale, caused 45,000 casualties and economic losses of 100 billion U.S. dollars (Maharani 2023).

Meanwhile, Indonesia also grapples with the devastating impact of earthquakes. In 2016, the country experienced 18 destructive earthquakes, followed by 19 in 2017 and 23 in 2018 (Siswadi 2018). Notable among these seismic events was the Sumenep earthquake on October 10, 2018, with a magnitude of 6.4, resulting in the loss of three lives and severe damage to numerous houses. Additionally, the Donggala-Palu earthquake on September 28, 2018, measuring 7.5 on the Richter scale, triggered a tsunami and liquefaction, leading to

A. Fadli · T. K. Priyambodo (✉) · A. E. Putra
Department of Computer Science and Electronics, Universitas Gadjah Mada, Yogyakarta, Indonesia
e-mail: mastri@ugm.ac.id

A. E. Putra
e-mail: agfi@ugm.ac.id

W. Suryanto
Department of Physics, Universitas Gadjah Mada, Yogyakarta, Indonesia
e-mail: ws@ugm.ac.id

A. Fadli
Department of Electrical Engineering, Universitas Jenderal Soedirman, Purwokerto, Indonesia
e-mail: arifadli@mail.ugm.ac.id

P. Mohammad et al. (eds.), *Geographic Approaches to Climate Change and Mitigation: Urban and Rural Perspectives (Volume 1)*, Advances in Science, Technology & Innovation, https://doi.org/10.1007/978-3-031-92119-3_13

over 2,000 fatalities and 1,000 people being declared missing, along with the destruction of thousands of homes. Similarly, the Lombok earthquake on August 5, 2018, with a magnitude of 7, claimed 259 lives and left more than 1,200 individuals injured while causing significant damage to public infrastructure (BBC News 2018; Gumelar 2018). These incidents underscore the urgent need for effective disaster preparedness and mitigation measures in Indonesia and globally.

Understanding the nature of earthquakes is crucial for comprehending the devastation they cause. As mentioned earlier, earthquakes result from the movement of the Earth's tectonic plates, releasing vast amounts of energy (Aanuoluwa et al. 2016; Florido et al. 2016; Moustra et al. 2011). While the destructive power of earthquakes is evident, it is equally important to recognize their diverse origins. Approximately 90% of earthquakes stem from natural tectonic shifts, with the remainder attributed to volcanic activities and human-induced factors (Cremen and Galasso 2020; Rathnaweera et al. 2020). These seismic events often exhibit significant force, necessitating accurate monitoring and analysis. Seismographs, equipped with sophisticated sensors, play a pivotal role in recording ground motion and seismic waves, providing essential data for understanding earthquake behavior and mitigating their impact.

The crucial role of seismographs in recording seismic events extends to their ability to capture the complex nature of seismic waves. Three-component seismic waves, including east–west, north–south, and vertical components, are recorded at each seismic station. These waves, originating from the Earth's crust, propagate with distinct characteristics. The P wave, characterized by its higher velocity compared to the S wave, is typically the first recorded vibration, evident in the BHZ (vertical) component. Conversely, with its perpendicular vibration direction to the propagation, the S wave is more distinctly visible on the horizontal BHE and BHN components. Notably, the amplitude and period of the P wave are typically smaller than those of the S wave.

P waves, or primary waves, are longitudinal waves that can propagate through solid, liquid, and gaseous materials. Because they can propagate through various media, P waves provide information about the density and elasticity of materials within the Earth. On the other hand, S waves, or secondary waves, are transverse waves that can only propagate through solid materials. S waves cannot pass through liquid or gaseous materials. When S waves pass through materials, they cause the materials to vibrate perpendicular to the direction of the waves, providing information about the elasticity and structure of the Earth's layers. By studying the characteristics of P waves and S waves detected from earthquakes, seismologists can identify the nature of the Earth's materials, determine the Earth's internal structure, and even understand the mechanisms by which earthquakes occur. Therefore, understanding P waves and S waves is critical in modeling Earth's structure and researching earthquakes (Lowrie 2007; Sykes 2019).

In addition to seismic wave data, earthquake metadata provides vital information for understanding and predicting seismic events. This metadata includes critical parameters such as the earthquake's time, magnitude, epicenter location, depth, and intensity. Magnitude, in particular, is a fundamental measure of an earthquake's strength, quantifying the energy released during the event. Developed by K. Wadati and C. Richter in the 1930s, earthquake magnitude remains a cornerstone of seismic analysis (Lay and Wallace 1995). With the proliferation of seismic sensor networks worldwide, there has been exponential growth in seismic datasets and metadata availability. These datasets play a pivotal role in enabling seismologists to analyze wave phase arrival times, pinpoint earthquake locations, and generate ground motion maps, enhancing our understanding of earthquake dynamics and bolstering efforts in earthquake prediction.

The abundance of seismic datasets necessitates advanced tools for processing and extracting insights to further our understanding of earthquake occurrences. Machine learning (ML) and deep learning (DL) methods emerge as pivotal solutions to address this demand, leveraging their capability to model complex datasets effectively. In seismological research, ML and DL methods have demonstrated superior performance to traditional techniques in recognizing seismic phase arrival times and determining associated physical quantities (Choi et al. 2022; Zhu et al. 2019). Moreover, in recent years, with the increase in the number of globally distributed seismic stations and the density of seismic networks, a large amount of seismic data continues to accumulate. As a data-based scientific discipline, the application of ML and DL methods in seismology has developed as an early warning method for earthquakes (Kuran et al. 2023; Li et al. 2023). This underscores the vital role of machine learning in translating seismic datasets into actionable insights for earthquake preparedness and response efforts. As machine learning techniques evolve, seismic data processing becomes more efficient, accurate, and automated. Numerous studies, particularly in earthquake detection and seismic phase identification, focus on harnessing machine learning and deep learning methods for seismic data processing (Perol et al. 2018; Wang et al. 2019; Xiao et al. 2021; Zhou et al. 2019). This research aims to conduct a comprehensive literature review of available seismic datasets and their use in earthquake prediction research using ML and DL. This endeavor can provide valuable information and insights for seismologists, researchers, and the public interested in earthquake research.

The rest of this paper is organized as follows. Section 1 Introduction: We provide an overview of the global impact of earthquakes, explaining seismic wave characteristics, metadata, and seismic dataset parameters. Section 2 presents the methodology used to assess the utilization of several seismic datasets in previously conducted studies. Section 3 Results and Discussion explains the detailed review of the use of seismic datasets. Section 4 concludes with a summary of the findings from the review of several datasets, offering an overview of how researchers determine the selection of seismic datasets.

2 Methodology

A systematic and bibliometric literature review, which has garnered the interest of researchers from various disciplines, is a method of gathering and summarizing prior research (Tranfield et al. 2003). This review technique describes research findings that target specific concerns and identifies patterns in research efforts (Mariano et al. 2015). Bibliometric literature evaluation can assist in quantifying and highlighting research trends and directions on emerging topics while indicating obstacles and areas for future research. This approach aligns with Snyder's (Snyder 2019) argument that an in-depth review, such as a bibliometric or systematic literature review, effectively provides evidence that can inform practice and policy by synthesizing a collection of studies on a similar topic. Consequently, this study performed a bibliometric analysis and systematic literature review (SLR) to identify the type, characteristics, metadata, and parameters of seismic datasets, as well as to provide an accurate depiction of trends in the use of machine learning and deep learning in earthquake prediction based on these datasets. This systematic review of seismic datasets focuses on top academic research databases, such as Scopus and Web of Science.

Bibliometric analysis utilizes various seismic datasets with private and public access, encompassing local, regional, and global scopes. The study relies on keywords such as "Earthquake Prediction," "Seismic Dataset," "Event Identification," "Phase Detection," "Seismic Measurements," "Time of Seismic Arrival," "P Wave Detection," and "S Wave Detection," incorporating Boolean combinations of these terms. The scope of this paper is limited to a review of seismic dataset types and characteristics, complete with metadata and parameters, and their application in earthquake prediction systems.

This article outlines the composition of seismic datasets comprising seismic wave recordings across three components. These datasets capture earthquake events in the form of seismograms, depicting ground motion vectors through a three-component instrument known as a seismograph. These components include recordings of north–south waves, east–west waves, and vertical ground movements. Additionally, the datasets feature structured data encompassing event timing, geographical coordinates (longitude and latitude), earthquake magnitude, and depth. These seismic datasets are available in diverse formats, such as CSV, HDF5, and ASCII.

This paper thoroughly analyzes publicly available seismic datasets, focusing on event detection, phase identification, and onset time picking. Before comparing these datasets, we explore their types, metadata, and the earthquake parameters they contain. This preliminary exploration is essential for understanding their structure and content, setting the stage for evaluating their efficacy in earthquake prediction systems leveraging machine learning and deep learning algorithms.

This comparison is essential as we evaluate the suitability of seismic datasets in earthquake prediction systems based on machine learning and deep learning. Our evaluation is limited to assessing the compatibility of these datasets for the following aspects:

1. Event Detection: We measure the positive rate detected on the seismic dataset.
2. Phase Identification: Seismic phase identification classifies the type of seismic wave received at a station based on the waveform recorded by a seismometer.
3. Onset Time Picking: This technique detects the presence of P and S waves in a dataset. Since these two waves indicate the occurrence of an earthquake, with the P wave being the first to arrive, research is necessary to develop automated methods for detecting the onset of the P wave.

Through these assessments, we aim to provide insights into the effectiveness of various seismic datasets for enhancing earthquake prediction systems using machine learning and deep learning algorithms.

3 Result and Discussion

3.1 Types of Seismic Dataset

The availability of high-quality data sets will undoubtedly play an essential role in the success of any research. Choosing the proper data set will help validate the analysis properly. Table 1 provides a list of datasets used in earthquake prediction research. Using data sets available in open sources on the internet is good news for every researcher. However, sometimes researchers have to clean up the dataset to be able to support the proposed research method.

The dataset used in classification and grouping research must be balanced so that it does not tend to become a single class. Therefore, several studies have added sensitivity and specificity to make the research results objective (Ali et al. 2012). Over the past few years, various parties have attempted to collect and create earthquake data sets. Some data sets are publicly accessible, but some have private access.

Table 1 Types of seismic datasets

Dataset	Type	Description
DiTing	Private	China E.Q. Networks Center (Zhao et al. 2022)
LENDB	Public	LENDB—Local EQ detection (Magrini et al. 2020)
NCEDC	Public	Berkeley Seismological Laboratory (Adagunodo et al. 2018)
INSTANCE	Public	The Italian Seismic Dataset (Michelini et al. 2021)
STEAD	Public	Stanford Earthquake Dataset (Mousavi et al. 2019a, b)
SCEDC	Public	Southern California EQ Data Center (Azam et al. 2014)
USGS	Public	United States Geological Survey (Amar et al. 2015)
GEOFON	Public	GFZ Seismological Data Archive (Woollam et al. 2021)
BMKG	Private	Badan Meteorologi, Klimatologi dan Geofisika (Suma'inna and Mahmudi 2015)
ETHZ	Private	Swiss Seismological Service (Woollam et al. 2021)
JMA	Private	Japan Metrological Agency (Bilal et al. 2022; Sahana and Choudhary 2014)
PHIVOLCS	Private	Philippine Institute of Volcanology and Seismology (Mejia et al. 2014)
MET Malaysia	Private	Malaysian Meteorological Department (Marhain et al. 2021)

3.2 Meta Information Seismic Dataset

Since 1981, the Southern California Seismic Network (SCSN), Caltech-USGS TriNet, Southern California SAR satellite imagery, and several regional crust and mantle refraction profiles have recorded seismic waves. The Southern California Earthquake Data Center (SCEDC) has cataloged nearly 3,000 data channels originating from 375 stations. On average, the center processes and archives approximately 20,000 earthquakes yearly. These recordings and data channels contribute significantly to earthquake research and monitoring efforts in the Southern California region (Ross et al. 2018).

SCEDC's mission is to provide easily accessible, organized, and high-quality earthquake catalogs to benefit seismological research. The central archive of seismological data for Southern California at the Southern California Earthquake Data Center (SCEDC) operates within the Seismology Laboratory at Caltech. The earthquake catalog (1932-present) and wave archive (1977-present) are among the most comprehensive and integrated seismic data libraries in the United States. SCEDC, accessible via the web portal https://scedc.caltech.edu/data/, consists of a 4.8 million time series recorded by nearly 700 recipients and stores more than 270,000 earthquake events in southern California (Meier et al. 2019; Ross et al. 2018).

Furthermore, the seismic dataset is called DiTing. DiTing is a seismic dataset containing incident reports from 2013–2020 at the China Earthquake Networks Center. DiTing records data using broadband and short-period seismometers. This dataset consists of 2,734,748 traces of three-component waveforms originating from 787,010 regional seismic events and includes 641,025 P-wave first-motion polarity labels. The seismic wave has a sampling frequency of 50 Hz and a duration of 180 s. Additionally, the recorded seismic data has three conditions, namely: (1) distance to the epicenter <330 km; (2) P and S phases are manually labeled; and (3) the signal-to-noise ratio ensures that the waveform does not contain a "NaN" or "inf" value (Zhao et al. 2022). DiTing dataset meta information is stored in a file in CSV format, as shown in Table 2.

LENDB is a seismic dataset for local earthquakes, comprising 1,487 global receivers. It contains 1,244,942 seismogram components, categorized as Noise (AN) and Earthquake (E.Q.), with 615,847 and 629,095 components, respectively. The dataset is publicly available in HDF5 binary format at https://doi.org/10.5281/zenodo.3648232. The labels within the LENDB dataset include Earthquake (E.Q.) and Noise (AN). Noise (AN) metadata is provided in Table 3, while Table 4 displays the Earthquake information (E.Q.) dataset (Magrini et al. 2020).

The INSTANCE Dataset includes nearly 1.2 million traces of three-component (3C) waveforms from approximately 50,000 earthquake events and over 130,000 3C noise waveforms. It records data from January 2005 to January 2020, based on the Italian Seismic Bulletin from the Istituto Nazionale di Geofisica e Vulcanologia. INSTANCE is accessible at https://doi.org/10.13127/instance and includes metadata with over 100 parameters, providing comprehensive earthquake source information, recording stations, and trace features (Michelini et al. 2021).

The INSTANCE Dataset selects a list of earthquakes based on the Italian Seismic Bulletin from the Istituto Nazionale di Geofisica e Vulcanologia between January 2005 and January

Table 2 Diting seismic datasets meta information

No.	Terms	Description
1	Part	Relate to the appropriate HDF5 file number
2	Key	Relate to the index of the waveform data, which includes the seismic event number and the station number
3	ev_id	Relate to the desensitized serial number of the earthquake associated with the waveform file
4	ev_mag	Relate to the magnitude of a seismic earthquake
5	mag_type	Relate to the relevant type of seismic magnitude
6	p_pick	Relate to the arrival time of P waves;
7	p_sharpness	Relate to the sharpness of the P-wave initial motion
8	p_motion	Relate to the P-wave first-motion polarity
9	s_pick	Relate to the arrival time of S waves
10	net_id	Relate to the serial number of the seismic network
11	sta_id	Relate to the stations' desensitized serial number
12	dis	Relate to the epicentral distance in the waveform file
13	st_mag	Relate to the single station magnitude in the waveform file
14	Baz	Relate to the back azimuth in the waveform file
15	Z_P_amplitude_snr	SNR Amplitude Z Channel
16	Z_P_power_snr	SNR Power Z Channel
17	N_S_amplitude_snr	SNR Amplitude N Channel
18	N_S_power_snr	SNR Power N Channel
19	E_S_amplitude_snr	SNR Amplitude E Channel
20	E_S_power_snr	SNR Power E Channel
21	P_residual	Relate to the residual of the manually labeled P-wave arrival time against the theoretical arrival time

Table 3 LENDB noise datasets meta information

No.	Terms	Description
1	Net	Network code
2	Sta	Station code
3	Stla	Station latitude, in degrees, positive north
4	Stlo	Station longitude, in degrees, positive east
5	Stel	Station elevation, in meters
6	Start time	Start time waveform
7	End time	End time waveform

2020. The Italian National Seismic Network (network code IV) records the INSTANCE dataset on the H.H., HE, and H.N. channels. Earthquake event data are recorded in the magnitude range of $0 \leq M \leq 6.5$, with distances ranging from 0 to 600 km, though most events fall within the range of $2 \leq M \leq 3$ with a distance of 250 km. The INSTANCE Dataset comprises recordings from more than 600 stations across Italy (Michelini et al. 2021).

Meanwhile, the STEAD Dataset records two categories of signals: seismic and non-seismic. The earthquake signal category contains approximately 1,050,000 three-component seismograms, each with a one-minute duration, originating from 450,000 earthquake events between January 1984 and August 2018. Globally, 2,613 receivers (seismometers) within the STEAD dataset record earthquake events. The non-earthquake signal category contains seismic noise with 100,000 samples. Additionally, the STEAD Dataset consists of approximately 1.2 million time series, equivalent to over 19,000 h of recorded seismic signals, with a frequency of 100 Hz (Mousavi et al. 2019a, b).

The metadata accompanying the STEAD dataset is derived from multiple reputable sources, including the International Seismological Center, the National Earthquake Information Center, the Northern California Seismic Network, the

Table 4 LENDB seismic datasets meta information

No.	Terms	Description
1	Mag	Magnitude
2	Evla	Latitude of the epicenter, in degrees, positive north
3	Evlo	Longitude of the epicenter, in positive east degrees
4	Evdp	Depth of the hypocenter to sea level in meters
5	Time	Station elevation, in meters
6	Tist	Epicentral distance, in meters
7	Time	Time of origin of events
8	Az	The azimuth of events to stations, in degrees
9	Baz	Stations to azimuths of events in degrees

Southern California Seismic Network, the Pacific Northwest Seismic Network, the New Madrid Seismic Network, the Incorporated Research Institutions for Seismology (IRIS), the Advanced National Seismic System Composite Catalog, and the Global Seismograph Network (GSN). This diverse array of sources ensures the dataset's comprehensive coverage and reliability.

Furthermore, the ETHZ dataset provides locally and regionally recorded seismicity data across Switzerland and its border areas. With 57 metadata variables available, the dataset focuses on events with magnitudes greater than 1.5 from 2013 to 2020, comprising 231 events containing 36,743 waveform examples (Woollam et al. 2021). Seismic data recordings obtained from public networks encompass various sources, including the Swiss Seismological Service (SED), CERN, the Swiss Seismological Service (SED) at ETH Zurich, and the AlpArray Seismic Network (X.T.). Additionally, the GEOFON seismic network, established in 1993 by the German Geo Forschungs Zentrum (GFZ), comprises over 800 seismic stations globally. The GEOFON seismic dataset encompasses 2,270 events and 275,000 waveforms recorded between 2009 and 2013, contributing significantly to seismic research and monitoring efforts on a global scale (Woollam et al. 2021).

Moreover, the Iquique seismic dataset is a benchmark dataset featuring locally recorded seismic arrivals across northern Chile. Initially utilized to train the deep learning picker, this dataset comprises 13,400 waveform examples, including 13,327 manual P-phase picks and 11,361 manual S-phase picks. With all waveform units presented in raw counts, the dataset incorporates 23 metadata variables, offering valuable insights into seismic activity and waveform characteristics (Woollam et al. 2021).

On the other hand, the BMKG dataset represents a comprehensive collection of earthquake occurrences throughout Indonesia. Categorized into three types of earthquake data—M 5.0+ Earthquakes, Felt Earthquakes, and Potential Tsunami Earthquakes—this dataset provides crucial information for seismic monitoring and disaster management efforts in the region. Together, these datasets contribute significantly to seismic research and hazard mitigation strategies, facilitating a deeper understanding of earthquake dynamics and their potential impacts (BMKG Indonesia 2023).

The availability of the Northern California Earthquake Data Center (NCEDC) dataset, with its extensive archive covering northern and central California, underscores the critical role of comprehensive data repositories in seismic research and monitoring. With more than 80 terabytes of continuous time series data and over 890,000 recorded events since 1984, the NCEDC dataset provides valuable parameters, moment tensors, and first-motion mechanism information essential for understanding seismic activity. Moreover, its support for web services in FDSN and IRIS ensures accessibility and usability for researchers and practitioners alike. Accessible via http://service.ncedc.org, the NCEDC dataset is a cornerstone in advancing our understanding of earthquake dynamics and enhancing seismic hazard mitigation efforts.

3.3 Earthquake Parameter

Based on the comparison of seismic datasets in the preceding section, we present a multi-parameter analysis from the literature review. Table 5 illustrates the comparison results, focusing on seismic datasets such as INSTANCE, STEAD, LENDB, and DiTing. Parameters assessed include earthquake and noise seismogram components, epicenter distance, waveform length, earthquake and noise metadata, sampling frequency, magnitude range, and coverage area (Michelini et al. 2021).

Table 5 Seismic dataset parameter comparison

Parameter	INSTANCE	STEAD
Seismogram (Event)	~1.15 M	~1.05 M
Seismogram (Noise)	~0.13 M	~0.1 M
Metadata (Event)	115	35
Metadata (Noise)	46	8
Sampling Rate (Hz)	100	100
Region Coverage	Italy	World
The epicentral distance (km)	<620	<350
Magnitude range	0–6.5	0–7.9
Waveform length	120 s	60 s

The DiTing dataset stands out with the highest number of earthquake seismogram components—approximately three times more than the other datasets—and notably lacks noise. Given the vast amount of seismic data available, machine learning (ML) algorithms play a crucial role in deciphering earthquake causality and facilitating predictions. As an evolving learning technique, ML exhibits substantial potential and forms an integral part of the analytical process for handling large datasets (Zhou et al. 2017). Meanwhile, deep learning models (DL) require extensive datasets for training to prevent "overfitting," given their complex architecture and numerous parameters (Zhao et al. 2022).

The INSTANCE and DiTing seismic datasets are tailored to specific locations, where seismic activity in one area can correlate with activity in another. Additionally, seismic events in a particular locale often exhibit distinctive patterns in the time domain (Wang et al. 2017). Our investigation delves into the seismic vulnerability of the Maluku region, analyzing potential earthquake hazards by examining seismic parameters such as the a-value and b-value derived from statistical data on tectonic earthquakes within the region over a specified period (Apriliani and Prastowo 2021).

In earthquake prediction, the NEIC Earthquake Catalog employs three distinct convolutional neural network models comprising 1.3 million seismic phase arrivals. These models predict the onset time of arrival, phase type, and distance (Yeck and Patton 2020). However, it is worth noting that past research primarily focuses on temporal correlations in earthquake prediction models, neglecting spatial relationships between dependent and independent variables (Wang et al. 2017). Recognizing that earthquakes manifest from both spatial and temporal associations, predictions for specific locations rely on historical data from the area in question as well as broader geographical contexts (Wang et al. 2017). Waveforms recorded at stations over 100 km from the epicenter often exhibit relatively low signal-to-noise ratios (SNR), posing challenges in phase picking (Huang et al. 2020). Consequently, SNR values serve as critical indicators when selecting datasets for earthquake prediction, offering insights into the clarity of the P and S phases (Huang et al. 2020).

Seismic datasets like SCEDC and DiTing provide crucial information on arrival times and first-motion polarity, which are vital for seismic phase identification algorithms and polarity determination. The extended 120-s time window in the INSTANCE dataset accommodates a wide range of seismic events, capturing phenomena from shallow to profound depths (Magrini et al. 2020). Conversely, the DiTing Dataset boasts an even longer window length of 180 s, incorporating noise and seismic events (Zhao et al. 2022).

These seismic datasets are foundational for developing innovative earthquake disaster warning systems leveraging deep learning and machine learning techniques. With public access to these datasets, seismologists can explore new methods in earthquake prediction systems. Therefore, standardizing data and metadata formats is paramount. Standardization facilitates comparative analysis across different datasets using the same ML models, enabling researchers to derive insights into event detection, phase identification, and onset time picking (Münchmeyer et al. 2022). Moreover, researchers utilize these datasets to swiftly distinguish between volcanic and tectonic earthquakes and accurately characterize earthquake sources, distances, and depths.

3.4 Seismic Dataset Performa Comparison

Building on the insights from the preceding discussion on seismic datasets, this section presents a comparative analysis of dataset performance, incorporating machine learning and deep learning algorithms alongside findings from existing literature. In recent years, machine learning has become a predominant tool for processing seismic data, enabling tasks such as event detection, phase identification, location mapping, magnitude estimation, and source mechanism determination. Notably, the adoption of deep learning techniques has demonstrated significant potential in improving earthquake localization, effectively reducing location errors when compared to traditional algorithms (Choi et al. 2022).

Previous studies have explored diverse deep learning algorithms, including EQTransformer (Mousavi et al. 2020), CNN RNN Earthquake Detector (CRED) (Mousavi et al. 2019a, b), Generalized Phase Detection (GPD) (Ross et al. 2018), Deep Phase Pick (DPP) (Soto and Schurr 2021), PhaseNet (P.N.) (Zhu and Beroza 2018), BasicPhaseAE (BPAE) (Woollam et al. 2019). These algorithms have been applied across various datasets, encompassing six events within local-to-regional distances (<100 epicentral distance): ETHZ, INSTANCE, LenDB, Iquique, SCEDC, and STEAD. Two datasets featuring teleseismic events (>100 epicentral distance) are incorporated: the NEIC and GEOFON datasets.

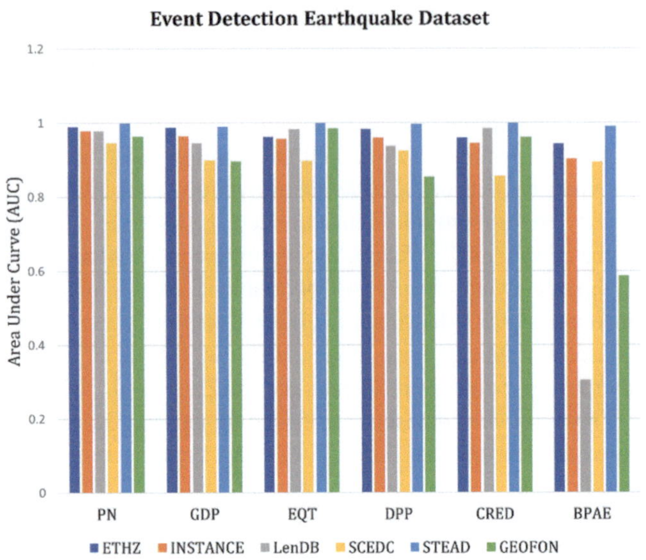

Fig. 1 AUC score event detection (by authors)

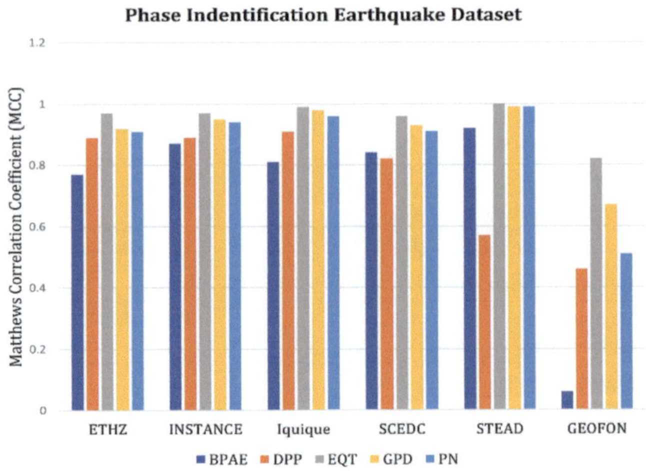

Fig. 2 MCC score for phase identification (by authors)

The seismic detection criteria aim to discern seismic events and high seismic activity regions (Sharma et al. 2010). Notably, our study highlights the superior performance of the STEAD seismic dataset in accurately capturing each earthquake event, as depicted in Fig. 1. Leveraging deep learning algorithms such as P.N., GPD, EQT, DPP, CRED, and BPAE, earthquake events can be detected with remarkable accuracy values of 0.997, 0.999, 1, 0.997, 1, and 0.991, respectively (Münchmeyer et al. 2022).

Furthermore, the ETHZ and INSTANCE seismic datasets demonstrate strong performance. These datasets exhibit event detection averages of 0.97 and 0.95, respectively, indicating high-quality labeling for both seismic and non-seismic signals recorded by seismic instruments. This quality of labeling will help accelerate advancements in applying machine learning and deep learning algorithms to address earthquake disaster challenges. Additionally, such well-labeled seismic datasets will significantly enhance the training, validation, and performance of predictive models.

The next criterion is phase identification, which evaluates performance using the Matthews Correlation Coefficient (MCC). The MCC is a widely recognized metric for assessing classification accuracy (Zhu 2020). B.W. Matthews originally introduced this metric to evaluate the performance of protein secondary structure prediction (Matthews 1975). Then, it became a widely used performance measure in biomedical research (Song et al. 2006).

The Matthews Correlation Coefficient (MCC) average values of the ETHZ, INSTANCE, Iquique, SCDEC, STEAD, GEOFON, and NEIC datasets are 0.89, 0.92, 0.93, 0.89, 0.89, 0.50, and 0.81, respectively (Münchmeyer et al. 2022). The Matthews Correlation Coefficient (MCC) provides essential insight into the effectiveness of models in identifying phases in seismic data. A higher MCC value indicates better model performance in distinguishing between different phases. In comparisons between datasets, higher MCC values suggest that models are more successful in identifying phases within those datasets. Therefore, datasets with higher MCC values demonstrate superior phase identification quality compared to those with lower MCC values, as shown in Fig. 2.

The next criterion examines the residual time of the P wave and S wave. Residual time refers to the difference between the pick time produced by the machine learning model (ML-pick time) and the reference pick time provided in the dataset. Root Mean Squared Error (RMSE) and Mean Absolute Error (MAE) are used to assess this difference, as both metrics indicate the level of prediction error in the models across the various seismic datasets under evaluation.

Figure 3 presents the Root Mean Squared Error (RMSE) value for the P wave residual time. This analysis provides information about the error rate of the prediction model across the seismic datasets being evaluated. By utilizing various automatic picking algorithms, including EQTransformer, Generalized Phase Detection (GPD), Deep Phase Pick (DPP), PhaseNet (P.N.), BasicPhaseAE (BPAE), and Baer-Kradolfer Picker (B.K.) (Baer and Kradolfer 1987), we determined the residual time of the P wave. Based on these calculations, we recorded the average RMSE value of the P wave residual time for each dataset as follows: ETHZ (0.70 s), INSTANCE (0.92 s), Iquique (0.69 s), SCEDC (0.83 s), STEAD (0.38 s), GEOFON (1.42 s), and NEIC (1.39 s) (Münchmeyer et al. 2022).

Based on these findings, the low RMSE value in predicting the P wave residual time indicates a high level of accuracy in the information that is critical for seismic analysis. Detecting the precise timing of P wave residuals is crucial for mapping earthquake locations and further understanding the Earth's internal structure. The presence of P waves in

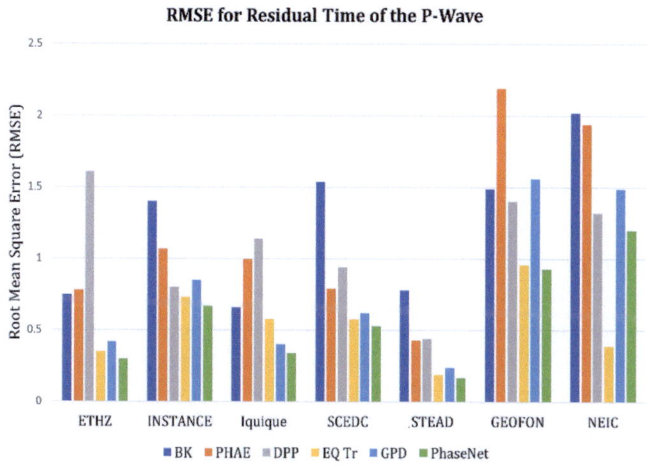

Fig. 3 RMSE for a residual time of the P wave (by authors)

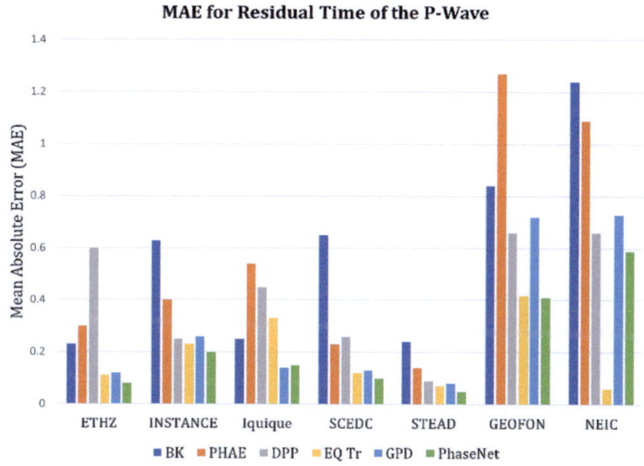

Fig. 5 MAE for a residual time of the P wave (by authors)

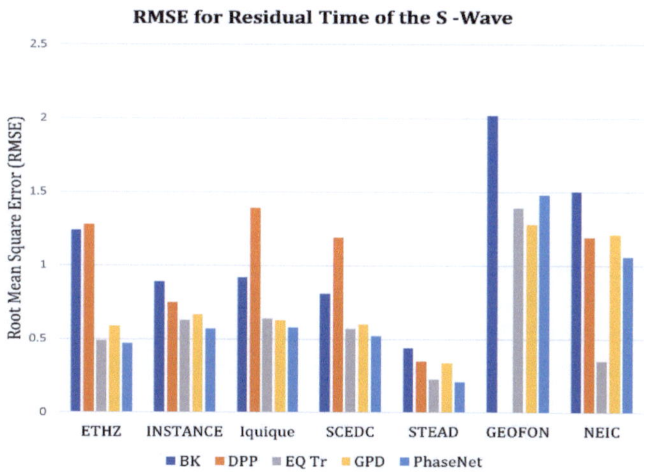

Fig. 4 RMSE for residual time of the S wave (by authors)

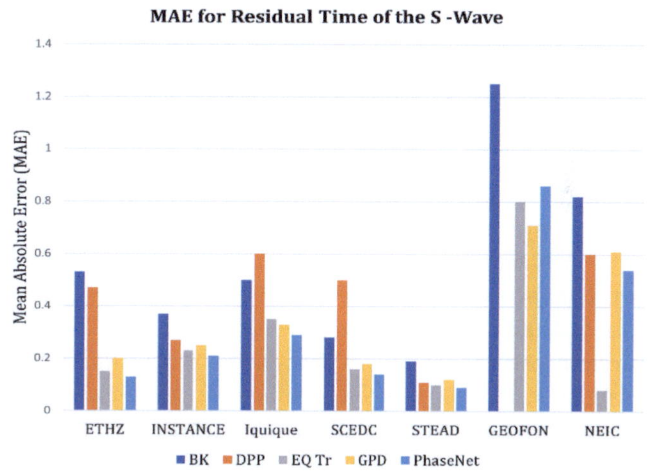

Fig. 6 MAE for a residual time of the S wave (by authors)

seismic data significantly impacts overall earthquake prediction and analysis. Therefore, the analysis results illustrate the quality of P wave residual time predictions and contribute to the understanding and mitigation of earthquake risk.

Moreover, Fig. 4 illustrates the S wave residual time values for each dataset using several automatic retrieval algorithms, including EQTransformer, Generalized Phase Detection (GPD), Deep Phase Pick (DPP), PhaseNet (P.N.), and BasicPhaseAE (BPAE). Based on these calculations, we determine the average S-wave residual time RMSE value for each dataset as follows: ETHZ (0.81 s), INSTANCE (0.70 s), Iquique (0.83 s), SCEDC (0.73 s), STEAD (0.31 s), GEOFON (1.54 s), and NEIC (1.06 s) (Münchmeyer et al. 2022).

Figure 5 shows the Mean Absolute Error (MAE) value of the residual time of the P wave. This analysis provides information about how close the model predictions are to the actual values of the data. The lower the MAE value, the better the model's performance in predicting the data. The average MAE values obtained for each dataset, in sequential order, are as follows: ETHZ (0.24 s), INSTANCE (0.33 s), Iquique (0.31 s), SCEDC (0.25 s), STEAD (0.11 s), GEOFON (0.72 s), and NEIC (0.73 s).

Meanwhile, Fig. 6 illustrates the Mean Absolute Error (MAE) value of the residual time of the S wave, with the average MAE values obtained for each dataset, in sequential order, as follows: ETHZ (0.27 s), Iquique (0.41 s), SCEDC (0.25 s), STEAD (0.12 s), GEOFON (0.91 s), and NEIC (0.53 s). The mean MAE values for the GEOFON and NEIC teleseismic datasets indicate a poorer signal-to-noise ratio (Münchmeyer et al. 2022). Moreover, the signal-to-noise ratio effectively reflects the clarity of the P and S wave labels. In contrast to the STEAD and INSTANCE datasets, the DiTing dataset contains substantial data with a low signal-to-noise ratio. The insufficient data with a low signal-to-noise ratio in the STEAD and INSTANCE datasets have limited effectiveness in certain areas. Hence, incorporating the DiTing dataset into the training process of machine learning algorithms would enhance detection performance for data with low signal-to-noise ratios.

Furthermore, in other studies, we apply machine learning to the LENDB Dataset to distinguish earthquakes from noise and conduct tests beyond the training data. The results exhibit 96.7, 95.3, and 93.2% accuracy in the training, validation, and test sets. These findings demonstrate that the dataset offers comprehensive information supporting the generalization of algorithms used in earthquake prediction.

Moreover, researchers utilized the STEAD and NCEDC datasets with the PhaseNet and EQTransformer network architectures to predict the Yangbi EQ & Maduo Earthquake. The PhaseNet Network Architecture yielded Precision of 0.171, Recall of 0.829, and F1-Score of 0.283, while EQTransformer produced Precision of 0.613, Recall of 0.391, and F1-Score of 0.478 (Jiang et al. 2021). Similarly, researchers applied the same method to the following datasets: ETHZ, INSTANCE, Iquique, SCEDC, STEAD, GEOFON, and NEIC, resulting in MAE and RMSE values as follows: MAE values were 0.308, 0.278, 0.426, 0.264, 0.126, 0.756, and 0.546, respectively, while RMSE values were 0.834, 0.734, 0.846, 0.568, 0.328, 1.286, and 1.092, respectively (Münchmeyer et al. 2022).

In analyzing the presented seismic dataset, it is essential to distinguish explicitly between high-quality and low-quality seismic datasets. Precise event detection, accurate phase identification, and distributions of P-wave and S-wave time residuals characterize high-quality earthquake seismic datasets. The distribution of time residuals can reflect the extent to which the model can predict picking times correctly based on the given data if the dataset has good quality, such as accurate annotation, sufficient data, and precise signal conditions. Low Mean Absolute Error (MAE) and low Root Mean Squared Error (RMSE) reflect that the model tends to make predictions close to the actual values with reasonable accuracy when applied to high-quality datasets. Conversely, if the distribution of time residuals in the dataset is more variable or asymmetrical, this may indicate a higher error rate in the model predictions. In this context, the Mean Absolute Error (MAE) and Root Mean Squared Error (RMSE) values tend to be larger, indicating that the model has difficulty predicting values that are close to the actual value with reasonable accuracy when faced with datasets that have poor characteristics or are referred to as low-quality datasets.

4 Conclusion

Datasets like INSTANCE, STEAD, LENDB, and DiTing offer valuable insights into various seismic parameters, including earthquake components, epicenter distances, and waveform characteristics. The DiTing dataset, distinguished by its extensive earthquake seismogram components and absence of noise, underscores the importance of robust data for machine learning and deep learning algorithms in earthquake analysis and prediction. Furthermore, the localized nature of datasets like INSTANCE and DiTing enables focused studies on specific regions, such as the Maluku region, aiding in the assessment of earthquake vulnerabilities and hazards.

Moreover, seismic datasets like SCEDC and DiTing provide essential information for seismic phase identification algorithms and polarity determination, enhancing the accuracy of seismic analysis. Extended time windows in datasets like INSTANCE and DiTing accommodate diverse seismic events, further enriching seismic research endeavors. Additionally, the standardization of data and metadata formats across datasets facilitates comparative analysis and ensures consistency in earthquake prediction models.

Based on the evaluation of the AUC Score for Event Detection, MCC for Phase Identification, as well as MAE and RMSE for the residual times of P wave and S wave, it is evident that datasets such as STEAD, INSTANCE, and ETHZ exhibit outstanding performance across multiple metrics. The AUC Scores for Event Detection for STEAD, INSTANCE, and ETHZ datasets are notably high, with values of 0.997, 0.999, and 0.997, respectively, indicating their precise identification of seismic events. Regarding Phase Identification, INSTANCE, Iquique, and ETHZ datasets stand out with MCC values of 0.92, 0.93, and 0.89, showcasing their ability to accurately distinguish between various seismic phases. Furthermore, the STEAD dataset accurately estimates the residual times of both P wave and S wave, with MAE values of 0.11 s and 0.12 s, and RMSE values of 0.38 s and 0.31 s, respectively. These findings underscore the superior performance of STEAD, INSTANCE, and ETHZ datasets in earthquake detection, phase identification, and precise estimation of seismic wave residual times, particularly for P waves.

Acknowledgements LPDP funded this research through the Beasiswa Pendidikan Indonesia in the Computer Science Doctoral Study Program, Department of Computer Science and Electronics, Faculty of Mathematics and Natural Sciences, Universitas Gadjah Mada. I would like to thank my supervisor and co-supervisor for their patient guidance, encouragement, and advice throughout my time as their student. I was fortunate to have supervisors and co-supervisors who cared about my situation and who responded to my questions and inquiries in a timely manner.

References

Aanuoluwa, A., Ayobami, S. L., Adagunodo, T. A., & Sunmonu, L. A. (2016). Earthquake: a terrifying of all natural phenomena. In: Journal of Advances in Biological. Retrieved from http://www.asdpub.com/index.php/jabbr

Adagunodo, T.A., Lüning, S., Adeleke, A.M., Omidiora, J.O., Aizebeokhai, A.P., Oyeyemi, K.D., Hammed, O.S.: Evaluation of $0 \leq M \leq 8$ earthquake data sets in African–Asian region during 1966–2015. Data Brief **17**, 588–603 (2018). https://doi.org/10.1016/J.DIB.2018.01.049

Ali, W., Shamsuddin, S.M., Ismail, A.S.: Intelligent Naïve Bayes-based approaches for web proxy caching. Knowl.-Based Syst. **31**, 162–175 (2012). https://doi.org/10.1016/j.knosys.2012.02.015

Amar, E., Khattab, T., Zada, F.: Intelligent Earthquake Prediction System Based on Neural Network (2015). https://doi.org/10.5281/ZENODO.1337607

Apriliani, T., Prastowo, T.: Penentuan Parameter Seismik A-Value dan B-Value untuk analisis Potensi Bencana Gempa di Wilayah Maluku. Inovasi Fisika Indonesia **10**(1), 11–20 (2021). https://doi.org/10.26740/ifi.v10n1.p11-20

Azam, F., Sharif, M., Yasmin, M., Mohsin, S.: Artificial Intelligence Based Techniques for Earthquake Prediction: A Review. https://www.researchgate.net/publication/267039190_ARTIFICIAL_INTELLIGENCE_BASED_TECHNIQUES_FOR_EARTHQUAKE_PREDICTION_A_REVIEW (2014)

Baer, M., Kradolfer, U.: An automatic phase picker for local and teleseismic events. Bull. Seismol. Soc. Am. **77**, 1437–1445 (1987)

BBC News.: Gempa Lombok: Korban meninggal dunia mencapai 436 orang, kerugian tembus Rp5 triliun. https://www.bbc.com/indonesia/indonesia-45165817 (2018)

Bilal, M.A., Ji, Y., Wang, Y., Akhter, M.P., Yaqub, M.: An early warning system for earthquake prediction from seismic data using batch normalized graph convolutional neural network with attention mechanism (BNGCNNATT). Sensors **22**(17) (2022). https://doi.org/10.3390/S22176482

BMKG Indonesia.: Gempabumi. https://www.bmkg.go.id/gempabumi (2023)

Choi, Y., Bae, S., Song, Y., Seol, S.J., Byun, J.: P- and S-wave arrival picking and epicentral distance estimation of earthquakes using convolutional neural networks. In: EGU General Assembly (2022). https://doi.org/10.5194/egusphere-egu22-11098

Cremen, G., Galasso, C.: Earthquake early warning: recent advances and perspectives. Earth Sci. Rev. **205**, 103184 (2020). https://doi.org/10.1016/j.earscirev.2020.103184

Florido, E., Aznarte, J.L., Esteban, A.M., Álvarez, F.M.: Earthquake magnitude prediction based on artificial neural networks: a survey. CRORR J. Regul. **7**(2) (2016)

Gumelar, G.: BNPB Sebut Total Kerugian Gempa Lombok Capai Rp12 Triliun. https://www.cnnindonesia.com/ekonomi/20180910124912-532-329123/bnpb-sebut-total-kerugian-gempa-lombok-capai-rp12-triliun (2018)

Huang, Y., Zhang, S., Lv, Y., Li, Y., Zhang, Y., Liu, M.: Earthquake detection in the Jiangsu region, China using graphics-processing-unit-based Match Locate and rapid earthquake association and location. Earthq. Sci. **33**(1), 23–33 (2020). https://doi.org/10.29382/eqs-2020-0023-03

Kuran, F., Tanırcan, G., Pashaei, E.: Performance evaluation of machine learning techniques in predicting cumulative absolute velocity. Soil Dyn. Earthq. Eng. **174**, 108175 (2023). https://doi.org/10.1016/j.soildyn.2023.108175

Lay, T., Wallace, T.C.: Modern Global Seismology, vol. 58. Academic Press, San Diego (1995)

Li, C., Ji, D., Zhai, C., Ma, Y., Xie, L.: Vertical ground motion model for the NGA-West2 database using deep learning method. Soil Dyn. Earthq. Eng. **165**, 107713 (2023). https://doi.org/10.1016/j.soildyn.2022.107713

Lowrie, W.: Fundamentals_of_Geophysics, 2nd edn. Cambridge University Press (2007)

Magrini, F., Jozinović, D., Cammarano, F., Michelini, A., Boschi, L.: Local earthquakes detection: a benchmark dataset of 3-component seismograms built on a global scale. Artif. Intell. Geosci. **1**, 1–10 (2020). https://doi.org/10.1016/j.aiig.2020.04.001

Maharani, E.: Infografis Kerusakan Ekonomi Akibat Gempa Besar di Dunia. Republika. https://visual.republika.co.id/berita/rrsyo1335/infografis-kerusakan-ekonomi-akibat-gempa-besar-di-dunia (2023)

Marhain, S., Ahmed, A.N., Murti, M.A., Kumar, P., El-Shafie, A.: Investigating the application of artificial intelligence for earthquake prediction in Terengganu. Nat. Hazards **108**(1), 977–999 (2021). https://doi.org/10.1007/s11069-021-04716-7

Mariano, E.B., Sobreiro, V.A., do Nascimento Rebelatto, D.A.: Human development and data envelopment analysis: a structured literature review. Omega **54**, 33–49 (2015). https://doi.org/10.1016/j.omega.2015.01.002

Matthews, B.W.: Comparison of the predicted and observed secondary structure of T4 phage lysozyme. Biochim. Biophys. Acta (BBA)-Protein Struct. **405**(2), 442–451 (1975). https://doi.org/10.1016/0005-2795(75)90109-9

Meier, M., Ross, Z.E., Ramachandran, A., Balakrishna, A., Nair, S., Kundzicz, P., Li, Z., Andrews, J., Hauksson, E., Yue, Y.: Reliable real-time seismic signal/noise discrimination with machine learning. J. Geophys. Res. Solid Earth **124**(1), 788–800 (2019). https://doi.org/10.1029/2018JB016661

Mejia, J., Rojas, K., Valeza, N., Villagracia, A.R.: Earthquake Prediction through Kannan-Mathematical-Model Analysis and Dobrovolsky-Based Clustering Technique. DLSU Research Congress. https://www.researchgate.net/publication/293330773_Earthquake_Prediction_through_Kannan-Mathematical-Model_Analysis_and_Dobrovolsky-Based_Clustering_Technique (2014)

Michelini, A., Cianetti, S., Gaviano, S., Giunchi, C., Jozinović, D., Lauciani, V.: (2021). INSTANCE the Italian seismic dataset for machine learning. In: Seismic Waveforms And Associated Metadata Published 2021 in Istituto Nazionale Di Geofisica e Vulcanologia (INGV) (2021)

Mousavi, S.M., Zhu, W., Sheng, Y., Beroza, G.C.: CRED: a deep residual network of convolutional and recurrent units for earthquake signal detection. Sci. Rep. **9**(1), 10267 (2019b). https://doi.org/10.1038/s41598-019-45748-1

Mousavi, S.M., Ellsworth, W.L., Zhu, W., Chuang, L.Y., Beroza, G.C.: Earthquake transformer—an attentive deep-learning model for simultaneous earthquake detection and phase picking. Nat. Commun. **11**(1), 3952 (2020). https://doi.org/10.1038/s41467-020-17591-w

Mousavi, S.M., Sheng, Y., Zhu, W., Gregory C.B.: STanford EArthquake Dataset (STEAD): a global data set of seismic signals for AI. IEEE Access **7**, 179464–179476 (2019). https://doi.org/10.1109/ACCESS.2019.2947848

Moustra, M., Avraamides, M., Christodoulou, C.: Artificial neural networks for earthquake prediction using time series magnitude data or seismic electric signals. Expert Syst. Appl. **38**(12), 15032–15039 (2011). https://doi.org/10.1016/j.eswa.2011.05.043

Münchmeyer, J., Woollam, J., Rietbrock, A., Tilmann, F., Lange, D., Bornstein, T., Diehl, T., Giunchi, C., Haslinger, F., Jozinović, D., Michelini, A., Saul, J., Soto, H.: Which picker fits my data? A quantitative evaluation of deep learning based seismic pickers. J. Geophys. Res. Solid Earth **127**(1) (2022). https://doi.org/10.1029/2021JB023499

Perol, T., Gharbi, M., Denolle, M.: Convolutional neural network for earthquake detection and location. Sci. Adv. **4**(2) (2018). https://doi.org/10.1126/sciadv.1700578

Rathnaweera, T.D., Wu, W., Ji, Y., Gamage, R.P.: Understanding injection-induced seismicity in enhanced geothermal systems: from the coupled thermo-hydro-mechanical-chemical process to anthropogenic earthquake prediction. Earth Sci. Rev. **205**, 103182 (2020). https://doi.org/10.1016/j.earscirev.2020.103182

Ross, Z.E., Meier, M., Hauksson, E.: P wave arrival picking and first-motion polarity determination with deep learning. J. Geophys. Res.

Solid Earth **123**(6), 5120–5129 (2018). https://doi.org/10.1029/2017JB015251

Sahana, B.C., Choudhary, M.M.: Comparative analysis of performance of earthquake early warning system based on wavelet based techniques. Int. J. Eng. Res. & Technol. (IJERT) **3**(6) (2014)

Sharma, B.K., Kumar, A., Murthy, V.M.: Evaluation of seismic events detection algorithms. J. Geol. Soc. India **75**(3), 533–538 (2010). https://doi.org/10.1007/s12594-010-0042-8

Siswadi, A.: 23 Gempa Merusak di Indonesia Sepanjang 2018. https://tekno.tempo.co/read/1159731/23-gempa-merusak-di-indonesia-sepanjang-2018 (2018)

Snyder, H.: Literature review as a research methodology: an overview and guidelines. J. Bus. Res. **104**, 333–339 (2019). https://doi.org/10.1016/j.jbusres.2019.07.039

Song, J., Burrage, K., Yuan, Z., Huber, T.: Prediction of cis/trans isomerization in proteins using PSI-BLAST profiles and secondary structure information. BMC Bioinform. **7**(1), 124 (2006). https://doi.org/10.1186/1471-2105-7-124

Soto, H., Schurr, B.: DeepPhasePick: a method for detecting and picking seismic phases from local earthquakes based on highly optimized convolutional and recurrent deep neural networks. Geophys. J. Int. (2021). https://doi.org/10.1093/gji/ggab266

Suma'inna, & Mahmudi.: Earthquake periodicity analysis in Indonesia using wavelet. Far East J. Math. Sci. (FJMS) **98**(2), 221–234 (2015). https://doi.org/10.17654/FJMSSep2015_221_234

Sykes, L.R.: Plate_Tectonics_and_Great_Earthquakes. Columbia University Press (2019)

Tranfield, D., Denyer, D., Smart, P.: Towards a methodology for developing evidence-informed management knowledge by means of systematic review. Br. J. Manag. **14**(3), 207–222 (2003). https://doi.org/10.1111/1467-8551.00375

Wang, Q., Guo, Y., Yu, L., Li, P.: Earthquake prediction based on spatio-temporal data mining: an LSTM network approach. IEEE Trans. Emerg. Top. Comput. **8**(1), 148–158 (2017). https://doi.org/10.1109/TETC.2017.2699169

Wang, J., Xiao, Z., Liu, C., Zhao, D., Yao, Z.: Deep learning for picking seismic arrival times. J. Geophys. Res. Solid Earth **124**(7), 6612–6624 (2019). https://doi.org/10.1029/2019JB017536

Woollam, J., Rietbrock, A., Bueno, A., De Angelis, S.: Convolutional neural network for seismic phase classification, performance demonstration over a local seismic network. Seismol. Res. Lett. **90**(2A), 491–502 (2019). https://doi.org/10.1785/0220180312

Woollam, J., Münchmeyer, J., Tilmann, F., Rietbrock, A., Lange, D., Bornstein, T., Diehl, T., Giunchi, C., Haslinger, F., Jozinović, D., Michelini, A., Saul, J., Soto, H.: SeisBench–A Toolbox for Machine Learning in Seismology (2021)

Xiao, Z., Wang, J., Liu, C., Li, J., Zhao, L., Yao, Z.: Siamese earthquake transformer: a pair-input deep-learning model for earthquake detection and phase picking on a seismic array. J. Geophys. Res. Solid Earth **126**(5) (2021). https://doi.org/10.1029/2020JB021444

Yeck, W.L., Patton, J.: Waveform Data and Metadata used to National Earthquake Information Center Deep-Learning Models, USGS (2020). https://doi.org/10.5066/P9OHF4WL

Zhao, M., Xiao, Z.W., Chen, S., Fang, L.H.: Diting: a large-scale chinese seismic benchmark dataset for artificial intelligence in seismology. Earthq. Sci. 35 (2022)

Zhou, Y., Yue, H., Kong, Q., Zhou, S.: Hybrid event detection and phase-picking algorithm using convolutional and recurrent neural networks. Seismol. Res. Lett. **90**(3), 1079–1087 (2019). https://doi.org/10.1785/0220180319

Zhou, L., Pan, S., Wang, J., Vasilakos, A.V: Machine learning on big data: opportunities and challenges. Neurocomputing **237**, 350–361 (2017). https://doi.org/10.1016/j.neucom.2017.01.026

Zhu, Q.: On the performance of Matthews Correlation Coefficient (MCC) for imbalanced dataset. Pattern Recogn. Lett. **136**, 71–80 (2020). https://doi.org/10.1016/j.patrec.2020.03.030

Zhu, W., Beroza, G.C.: PhaseNet: a deep-neural-network-based seismic arrival time picking method. Geophys. J. Int. (2018). https://doi.org/10.1093/gji/ggy423

Zhu, L., Peng, Z., McClellan, J., Li, C., Yao, D., Li, Z., Fang, L.: Deep learning for seismic phase detection and picking in the aftershock zone of 2008 M7.9 Wenchuan earthquake. Phys. Earth Planet. Inter. **293**, 106261 (2019). https://doi.org/10.1016/j.pepi.2019.05.004

A Framework of the Remote Sensing Database for Hydrometeorological Hazards Adaptation in Urban City of Malaysia

Mohd.Rizaludin Mahmud, Nor Wahida Awang, Siti Solehah Azmi, and Visvamalar Vijayakumaran

Abstract

Many tropical urban cities faced significant impacts due to hydrometeorological hazards, particularly during the extreme rainfall and rainfall deficits. Having an efficient geodatabase consisting of significant high temporal and large hydrometeorological dataset is one of the keys to successful hazard adaptation, mitigation, and prevention. However, acquiring and managing such large raster data for that database is a huge challenging task because many of them were categorized as developing countries and facing the data conflict situations (DCS). The DCS is either the ground data is (i) unavailable, (ii) scarce, (iii) may involve unavailable ground data, scarce data, data interoperability and sharing conflicts, or (iv) unknown data quality. With the availability of publicly accessed remote sensing raster data, there is an opportunity to utilize them as alternative solutions. However, those data required specific downscaling and localization as they come in various spatial, temporal, and radiometric characteristics. It is theorized that storing all the variables in a single consistent vector-based grid GIS layers upon the completion of downscaling could be a useful alternative in anticipating the above-mentioned conflicts. Therefore, this paper initiates the development of the framework to create a spatial vector grid database consisting of multiple hydrometeorological variables derived from the open-source remote sensing data. The framework place significant emphasis on the (1) hardware and software requirements, (2) methodology on the deriving an appropriate environmental variables, indicators, and their validation, and (3) methodology for creating the spatial database based on a vector grid approach. The indicators and variables that will be incorporated are (1) rainfall, (2) solar radiation, (3) Standardized Precipitation Index, (4) topography, (5) impervious surfaces, (6) vegetation index, and (7) land surface temperature. The successful realization of the framework would enabled spatial-based hazards adaptation initiatives at the city scale, including near-real-time forecasting or nowcasting, early warning, post-disaster assessment, and future risk assessment.

Keywords

Flash flood · Heat wave · Tropical cities · Open-source remote sensing · Disaster

M. Mahmud (✉) · S. S. Azmi · V. Vijayakumaran
Department of Geoinformation, Faculty of Built Environment and Surveying, Universiti Teknologi Malaysia, Johor Bahru, Malaysia
e-mail: rizaludin@utm.my

S. S. Azmi
e-mail: sitisolehah@graduate.utm.my

V. Vijayakumaran
e-mail: visvamalar@graduate.utm.my

M. Mahmud
Geoscience and Digital Earth Centre, Research Institute for Sustainable, Universiti Teknologi Malaysia, Johor Bahru, Malaysia

N. W. Awang
Faculty of Applied Sciences, Universiti Teknologi MARA Sarawak, Kota Samarahan, Malaysia
e-mail: norwahida@uitm.edu.my

1 Introduction

170 Tropical cities globally face ongoing risks from hydrometeorological hazards, leading to property loss, human impact, and economic repercussions (Nohrstedt et al. 2022; Albrecht 2017; Skidmore and Lim 2020). In the tropics, hydrometeorological hazards primarily occur due to two main factors: too much rain and too little rain. Heavy rain can cause significant traffic jams, reduce visibility, and activate sudden floods (Putra et al. 2022; Pramono and Savitri 2018). In the

P. Mohammad et al. (eds.), *Geographic Approaches to Climate Change and Mitigation: Urban and Rural Perspectives (Volume 1)*, Advances in Science, Technology & Innovation, https://doi.org/10.1007/978-3-031-92119-3_14

tropics, a significant lack of rainfall causes drought and heat waves, particularly during the daytime because of the constant solar radiation all year (Zhao et al. 2019; Kajtar et al. 2022). In order to tackle this problem, town municipalities, governing bodies, ministries, and international associations have begun efforts to enhance response to hydrometeorological hazards or disasters. Regrettably, numerous tropical urban areas globally are classified as developing or least-developed nations, encountering data conflict situations (DCS) (Sayed and Salah 2023). Consequently, DCS has impeded the implementation of urban hydrometeorological hazard adaptation initiatives, mainly due to inadequate data supply.

DCS refers to a situation in which there is a lack of sufficient data, particularly in terms of in-situ or ground data, to carry out proper analysis or implement the framework. Various significant factors contribute to this, such as a lack of ground measurements, incomplete data, uncertain data quality, conflicts regarding data sharing, and data retrieval conflicts (Mahmud et al. 2018). DCS may be the result of a variety of factors, either alone or in combination. The lack of technical knowledge, expertise, experience, financial capacity, efficient planning and execution, and adequate physical and infrastructure capacity are reasons why least-developed and developing countries struggle with DCS. The countries encountered similar challenges in preparing for DCS, such as inadequate financial and human resources, as noted by Osti et al. (2008), Rosmadi et al. (2023), Kovacs et al. (2017), and Seibert (2024), which included restricted funding for system operations and lack of skilled personnel. Hence, utilizing alternative data such as remote sensing to reduce reliance on ground measurement could be a beneficial approach to tackle the DCS problem.

A lot of environmental and hydrological remote sensing measurement data fall under the classification of publicly accessible data. Using them could greatly lower data acquisition expenses. Currently, numerous compressed images of meteorological variables from both ground and satellite sources are provided at frequent time intervals, ranging from 10 min to an hour. For instance, the HIMAWARI satellite provides brightness temperature images every 30 min for nearly all inhabited areas around the world (JMA 2015). Nevertheless, immediate utilization of this data is impractical due to the necessity of carrying out many important pre-processing and localization procedures. This involves changing RGB images into real radiometric values, transforming reflectance into useful variables, and calculating desired indices like vegetation indices. Moreover, it is important to validate remote sensing data for errors, whether systematic or unsystematic, as a quality control measure (Loew et al. 2017; Benkirane et al. 2022). The upcoming significant task is to align large raster data from various pixel dimensions into a unified database.

Integrating remote sensing data from different spatial, spectral, temporal, and product levels is essential for effective spatial analysis needed for the framework. The key variables in tropical hydrometeorology, such as rainfall and cloud data, are extremely variable, and forecasting methods may necessitate a large temporal database (Flaming et al. 2020; Misgwa et al. 2024). Handling large datasets with temporal information can lead to the generation of numerous data layers. For example, a precipitation radar recording hourly rainfall rates would result in 24 layers per day and around 720 layers per month. The current method of resizing, downsizing spatially, and layering raster data has disadvantages like intricate processing of raster data which includes resizing pixels, aligning pixels, reconstructing raster images, and displaying data (Silva-Coira et al. 2020; Chen and Knoblock 2016). Moreover, significant data storage space is necessary for processing raster data. Converting localization and big raster remote sensing data into vector grids is a strategic solution to create a database and address related issues. The idea is to gather and arrange data from localized remote sensing sources into a unified grid size, holding pixel-based details from different sensors in attributes within a vector pixel-based layer. In order to accomplish this, it is necessary to have a suitable framework in place to offer direction for carrying out the procedure.

Prior to the aforementioned concern, this study intends to create a suitable structure for constructing a spatial database with the use of remotely collected data to aid in hydrometeorological hazard adaptation efforts in tropical urban areas. The proposed plan will focus on the tropical city of Johor Bahru in Malaysia and will cover: (i) the range of hydrometeorological dangers in a tropical city, (ii) necessary remote sensing data and specifications, (iii) strategies for developing a remote sensing database, and (iv) verifying the accuracy of the database. Three major hydrometeorological risks in Johor Bahru city include monsoonal flooding, flash flooding, and heatwaves. Different open-source remote sensing data, spanning from hourly to bi-annual temporal scales, are incorporated into the framework for developing an effective remote sensing database. The goal is to transform all raster data into a unified vector grid with various attributes to address hydrometeorological risks effectively. The model suggested in this paper could act as a relevant blueprint for utilizing remote sensing information as a substitute spatial data aid in developing a tropical city that is resilient to natural disasters.

2 The Concept of Vector Grid Database Using Raster Remote Sensing Data

The goal of this framework is to develop a versatile, user-friendly, and efficient system for analyzing or conducting additional analysis on the extensive remote

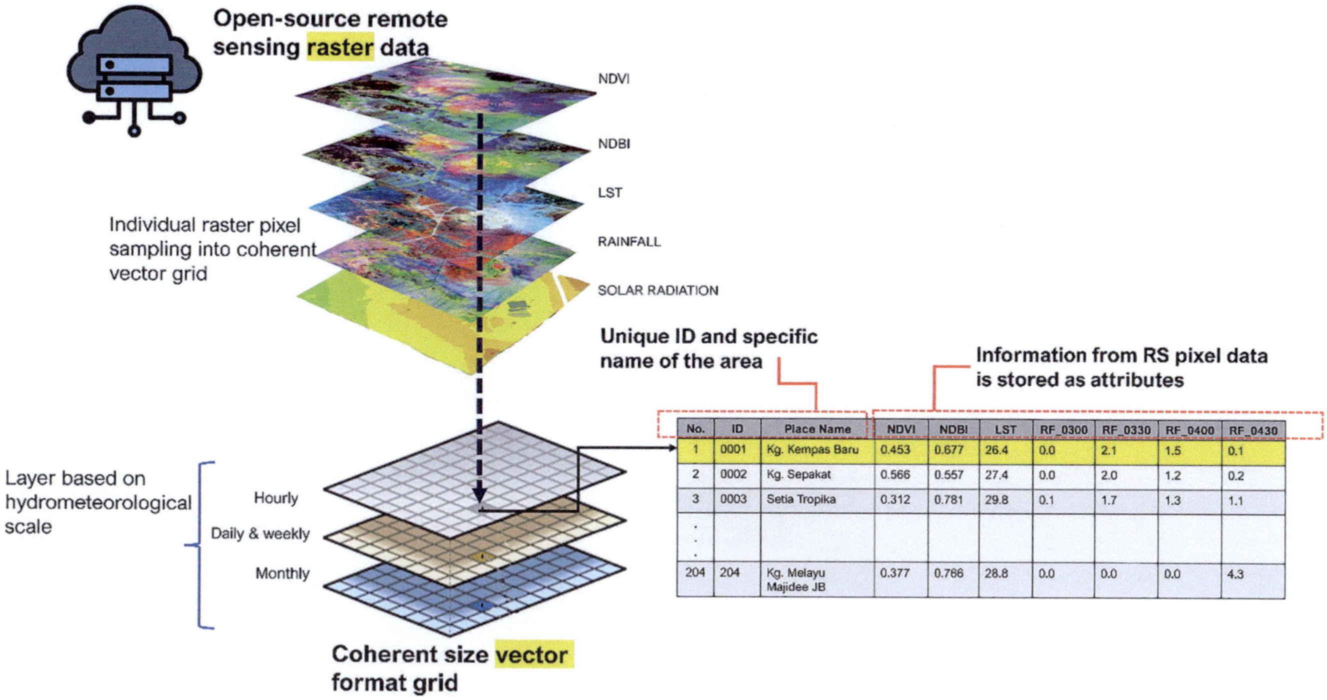

Fig. 1 Concept of the vector grid database from open-source remote sensing data

sensing geodatabase. The primary difficulties in managing various remote sensing data analyses include variations in pixel dimensions, extensive raster data sizes, and the task of handling large data raster (Cruces et al. 2019; Tsutsumida et al. 2019). In order to address these difficulties, it is advised to consolidate all data from various remote sensing sources into a unified vector grid layer. All remote sensing data will be saved as attributes to each grid in this database structure, as depicted in Fig. 1.

Having a single consistent vector grid for specific areas has several benefits such as easy updates, unique identification, efficient data retrieval, compact storage, and effective data handling. This method is utilized, such as in the case of quick and effective 3D rendering (Wang and Jin 2022). By assigning specific names, IDs, and other pertinent details such as postal codes to every grid, it would possess a distinguishable identity and be beneficial for user applications. Retrieving and showing multiple information effectively is also possible using vector data format since it can include multiple attributes. Hence, additional examinations, particularly those that incorporate numerous factors, can be smoothly carried out. This idea is applied in geographic information system GIS by connecting each area and object with appropriate spatial data (Wyszomirski and Gotlib 2020).

Many studies suggest that employing vector grids in database operations can address various challenges associated with handling large raster remote sensing data (e.g., Li et al. 2017; Ye et al. 2018). The updating process also becomes somewhat simpler since aligning pixels from various remote sensing data is difficult and involves time-consuming steps. Aligning large raster remote sensing data will require an increasing amount of time and money as time goes on.

This framework proposes the vector layer grid creation to align with the temporal scale of hydrometeorological events. This is due to the fact that having all the data stored in a single vector grid could result in attributes becoming too lengthy, particularly for frequent measurements such as hourly or daily ones. Figure 1 recommends organizing the vector layer based on temporal scale, starting from yearly and moving down to monthly, weekly, daily, and hourly. The rest of the low temporal resolution data, like vegetation and topography, can be duplicated easily for every vector grid. This method will aid in the updating process and simplify data management.

3 The Framework of the Remote Sensing Database for Urban Hydrometeorological Hazards

The framework consisted of three main components: hardware, software, and database creation. Each one will be detailed in the following subsection. The experimental site chosen is the Johor Bahru City Council, also known

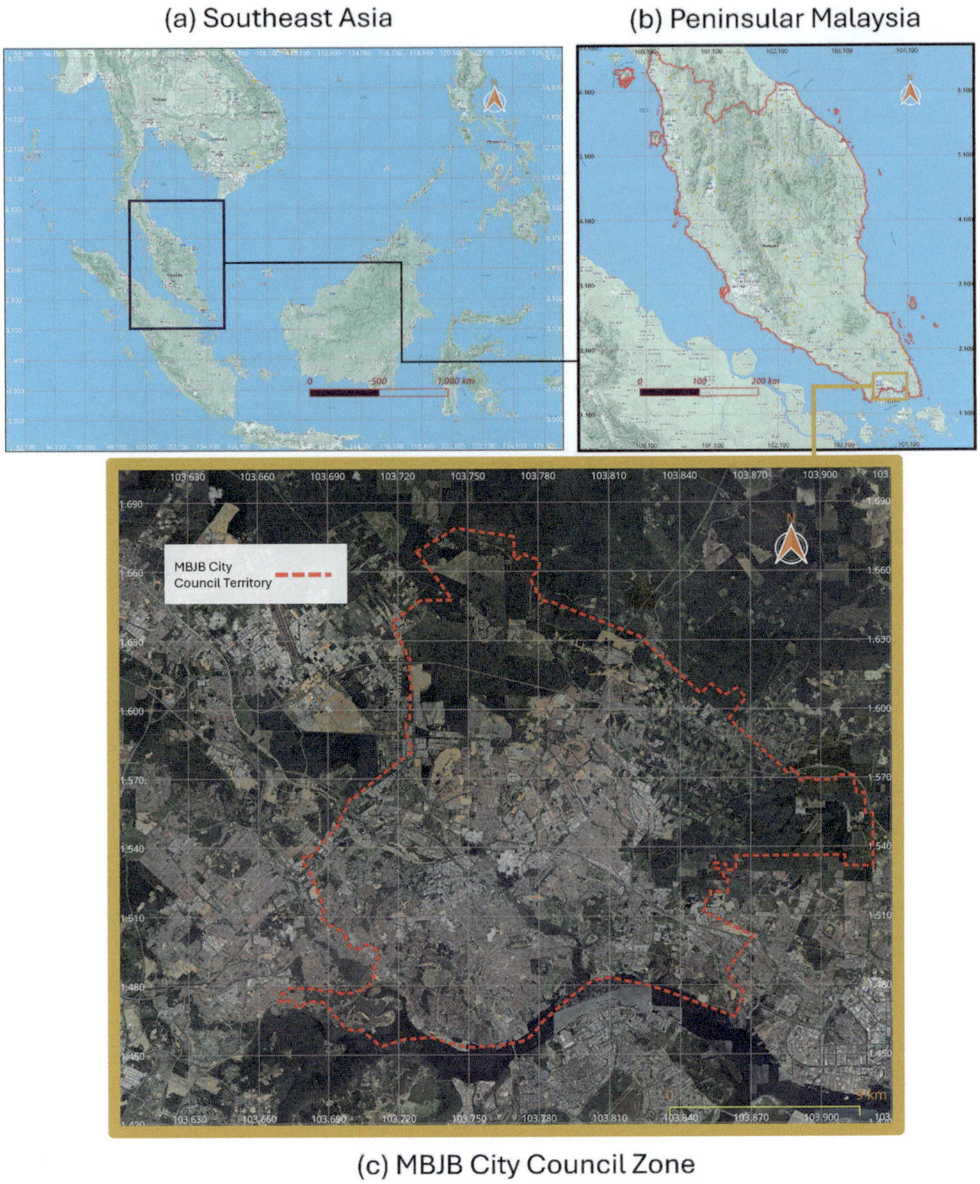

Fig. 2 Johor Bahru City Council territory area

as Majlis Bandaraya Johor Bahru (MBJB), situated in the southernmost area of Peninsular Malaysia. It holds the rank of third largest city in Malaysia, covering an expanse of 391 km^2, primarily consisting of built-up sectors comprising of industries (25%), commercial and administrative zones (24%), urban residences (33%), and green spaces (18%) (MBJB 2021). In 2021, the estimated population is around 858,100 individuals. Johor Bahru has a humid tropical climate with steady rainfall all year round, receiving 2400 mm of rainfall annually and having daily temperatures fluctuating between 26 and 33 °C. Situated in the southern part of Peninsular Malaysia (Fig. 2), Johor Bahru is surrounded by the Singapore Straits, Malacca Straits, and the South China Sea and regularly experiences rainfall, particularly convective precipitation. Notable rainy seasons typically happen in the monsoon season (November to January) or the periods between monsoons (Mar.–Apr. and Sep.–Oct.). As a result, hydrometeorological hazards such as riverine floods or flash floods are frequently seen in urban areas with heavy rainfall. In periods of low rainfall, particularly lasting 3–4 days, daytime temperatures may rise up to 35 °C, leading to minor heatwaves. Torrential rain showers can also cause significant traffic jams, vehicle collisions, and challenging driving circumstances. Table 1 provides an overview of the hydrometeorological characteristics. The focus of this city council's urban disaster adaptation is on preventing, reducing, and providing early warning and reporting for the hazards mentioned. Table 2 displays the variables, origins of open-source data, and additional details for the spatial database.

Table 1 Summary of the hydrometeorological characteristics in Johor Bahru District

Parameters	Description/value
Rainfall (mm)	Monthly average: 120–350 Annual average: 2200–3100
Temperature (Celcius)	Min.: 20 Avg. daily: 25 Max.: 35
No. of rainy days (average per annum)	~100
Land cover	Industries (25%) Commercial and administrative (24%) Urban settlements (33%) Green areas (18%)
Solar radiation (W/m^2)	Avg.: 510 Max.: 1120
Acreage (km^2)	391

3.1 Hardware Specification

The hardware required for the remote sensing database is primarily influenced by the following factors: (1) size of the urban city, (2) temporal resolution of the data, (3) number of variables, and (4) GIS software systems. The size of the urban city influences the number of spectral images used to represent the entire city area. In the context of the Johor Bahru City area, one single Sentinel image (290 × 290 km) is sufficient to cover the entire area. However, the storage capacity might vary depending on the size of the city area. Common cities in Asia, for instance, are around 100–700km^2, except for several main metropolitan areas where it could reach 1000–3000km^2. A single high temporal remote sensing data like HIMAWARI or precipitation radar is relatively small, around 200–400 KB, and the processed data will be double in size due to format changes, mostly to GeoTIFF format. Thus, a one-year precipitation radar still images could reach 4–5 GB in storage.

The next important element in the database hardware is the specification for the GIS software system, including processor, memory, and storage. The most common system, for instance, ArcGIS, requires the largest storage with 16 GB. Other GIS systems like QGIS and MapInfo only require 1.3 and 2.3 GB of storage, respectively. Regarding other aspects of the hardware, it is recommended that it should at least comply with the common workstation standards, with 16 GB of memory, 36 MB cache, 24 cores, 32 threads, and a processor speed of 2.00–5.60 GHz or higher. Table 3 shows the recommended hardware specifications for the remote sensing database based on the assumptions above. It might vary as the parameters of the database can be slightly different from one city to another.

3.2 Software Specification

For the software requirements, it is highly recommended to utilize a GIS-based environment since the core of the database is to support raster and vector data. The software system should possess the following fundamental features: (1) vector and raster database management system (VR-DBMS), (2) map creation, visualization, and display, (3) graphical user interface (GUI), (4) spatial analysis function, (5) spatial data interoperability function. It is highly convenient to opt for established GIS software systems such as ESRI ArcGIS, MapInfo, QGIS, and Hexagon Geomedia since they are fully equipped with many of the required features. An open-source GIS platform like QGIS has the ability to store a large number of data and project files. Nevertheless, developing your own GIS system for the database is also plausible as it gives more flexibility and could reduce maintenance and updating costs in the future. Table 4 provides a description of the elements required in the software system.

3.3 Database Development

There are four major phases in database development: data retrieval and preparation (Phase 1), data validation (Phase 2), database construction (Phase 3), and information retrieval and display (Phase 4). The schematic diagram of the process is provided in Fig. 3. A detailed elaboration is given later in this subsection.

Phase 1: Data Retrieval and Preparation

In this first phase, two major processes were involved: (1) data retrieval and (2) data preparation. In data retrieval, all the selected remote sensing data will be downloaded

Table 2 Specification of the remote sensing data and variables for the database. (/) Elements, variables, or characteristics that are considered in the database

No.	Open-source remote sensing data	Data source	Spatial variables	Hydrometeorological hazards				Spatial resolution	Temporal resolution (initial)				
				Flash flood	Monsoon flood	Heavy rainfall	Heat wave		Hourly	Daily	Monthly	Bi-Annual	Annual
1	Ground precipitation weather data	Malaysian Meteorological Dept	Rainfall	/	/	/	/	~3 km	/				
2	Himawari Satellite	Malaysian Meteorological Dept. @Japanese Aerospace Exploration Agency	Cloud top temperature				/	5 km	/				
3	Himawari satellite and ground precipitation weather data	Malaysian Meteorological Dept	Solar radiation				/	2 km	/				
4	Sentinel/ Landsat/ Feng Yun	Copernicus Access Hub, United States Geological Survey, and Feng Yun Satellite Data Center	Vegetation indices				/	20–30 m				/	
5			Surface imperviousness	/	/							/	
6			Urban areas	/	/							/	
7			Land surface temperature				/	60–100 m				/	
8	Shuttle radar topographic mission	NASA Earthdata	Elevation	/	/			30 m				/	
9			Slope	/	/							/	
10			Aspect	/	/							/	

Table 3 Hardware requirement for the proposed database

No.	Items	Specification
1	Storage	5 Years maximum storage, 140 GB–2 TB; for maximum of 50 km city size
2	Memory/RAM	Minimum of 8 GB, maximum 32 GB
3	Graphics memory	A minimum of 6 GB is recommended, with 16 GB or more being optimal
4	CPU	Minimum 1 GHz 64-bit (×64) processor; above 2 GHz recommended for top performance
5	Graphics card	GPU from minimum of 20–40 GB
6	SQL server	Microsoft® SQL Server® 2016 or FoundationDB, NoSQL database or Red Hat Enterprise Linux (RHEL), SUSE Linux Enterprise Server (SLES), and Ubuntu

Table 4 Software requirement for the proposed database

No.	Item	Purpose
1	Vector and raster database management system (VR-DBMS)	– Store and organize the vector and raster data – Read various types of common raster and vector data – Request and access data from the database – Performance monitoring, security, and backup and recovery
2	Map creation, visualization, and display	– Creating maps from vector or raster data – Visualize the vector and raster data – Cartography and related display function
3	Graphical User Interface (GUI)	– Enable users to interact with electronic devices using graphical icons and visual indicators – Allow users to operate functions based on user interactions
4	Spatial analysis function	– Operate algebraic process – Vector data processing – Raster data processing
5	Spatial data inter-operability	– Support various data readings from various format – Multiple data format interchange – Multiple data aggregation and merge

from publicly accessed open-source sites (Table 2). For the low temporal resolution data, or in this context bi-annual, manual retrieval from the direct provider directory is recommended. Meanwhile, for the high temporal resolution data, for instance, the ground precipitation radar data and HIMAWARI satellite images that come in 10-min intervals and half-hourly intervals, the automated storage and retrieval system is suggested.

Generally, there will be two types of raster data formats: (1) with georeferenced format and consists of actual radiometry information of the variables (reflectance from optical satellite images, etc.) and (2) non-georeferenced and compressed images, which mainly include the ground precipitation radar data and HIMAWARI satellite images. It contains picture-compressed information representing the information in color code instead of actual radiometry information. For those belonging to category 2, conversion of the compressed image and georeferenced needs to be carried out as outlined in pre-processing 1.

In data preparation, all the remote sensing data parameters and information are extracted. For parameters that require computation such as NDVI and NDBI, the corresponding raster layer will be calculated and derived. Those variables were computed based on the standard model as widely known except for surface imperviousness. Selection of the appropriate model should be done by considering the suitability of the algorithm with the localized condition as well as the availability of the open-source data. Common guidelines can be found in Weng (2012) or Slonecker et al. (2001) as many surface impervious models have been developed until the present. Among those that we recommend are one from Sun et al. (2023) and Su et al. (2022).

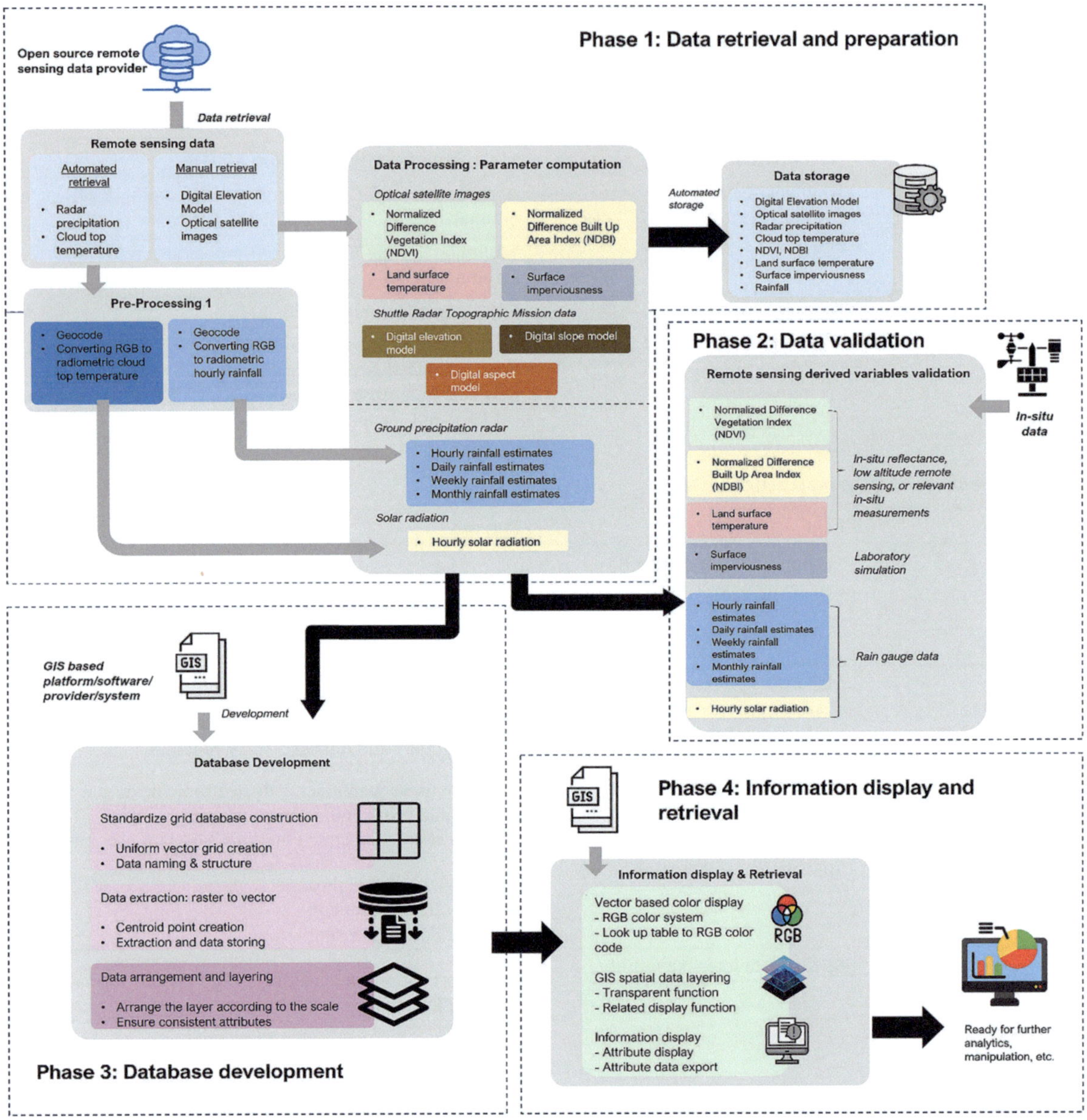

Fig. 3 Overall framework of the database development. Phase 1: Data retrieval and preparation, Phase 2: Data validation, Phase 3: Database development, Phase 4: Information display and retrieval

Another significant computation is the computation of various scales of rainfall layers from the hourly estimated precipitation radar data. These multi-scale products serve different initiatives in the framework. The hourly rainfall could be useful for forecasting and early warning of heavy rainfall or flood if it can be performed in real time. Meanwhile, the daily and weekly data can be useful for drought analysis or heatwave analysis. For the hourly solar radiation, a localized empirical model is adapted from Mahmud (2024) by using input from the cloud top temperature and precipitation radar data. All the computed remote sensing information will be stored as raster layers, subjected for archiving and other purposes.

Phase 2: Data Validation

The second phase of the framework is validation, which serves as quality control of the database inputs. Every single variable retrieved or derived from the open-source remote sensing data will be validated against the reference data. There are three categories of validation, each representing a different nature of data. For low temporal data, including vegetation indices, land surface temperature, and built-up index, validation can be done via in-situ-based derived indices or measurements. This in-situ data can be acquired from field-based measurements or low-altitude remote sensing that carries identical spectral information, such as multi-spectral drones or thermal drones.

This is because it is assumed that low-altitude scanning could provide a more accurate representation and estimation of the indices or variables due to negligible atmospheric, geometric, and radiometric errors.

Evaluating the rainfall and solar radiation variables is also conducted directly against the rain gauge data. Statistical indicators, including Pearson correlation and root mean square error, can be used. Because this variable is continuous measurement, evaluating based on time series data is strongly recommended. Sufficient rain gauge samples are required to perform this verification on selected pixels. Validating the surface imperviousness could be slightly challenging as many studies do not perform in-situ procedures. Nonetheless, we suggest that the surface imperviousness can be validated via in-situ laboratory simulation of measuring water retention, adopting the sandbox experiment but conducted on the ground with different surfaces as subjects.

All the validation reports should be compiled and used as references to indicate the standards of the database accuracies. The threshold for acceptable accuracies was adapted from numerous studies as shown in Table 2. The information obtained from the validation can later be used as an adjustment value for future data, especially during updating processes. Decent remote sensing data accuracies are important to ensure the credible results upon further hazard analyses.

Phase 3: Database Development

The core process in this phase is to unify all the raster information into a single, consistent, and coherent size vector grid. There are three major processes in this phase: (1) creating the standardized grid, (2) extracting information from raster remote sensing data to vector, and (3) arranging the attributes and layering. In the first process, a rectangular vector grid is created. In this framework context of Johor Bahru city, the chosen size is 50 × 50 m. As we highlighted earlier, three types of vector grids are created, each representing a different temporal scale: hourly rainfall, daily and weekly, and monthly. The basic information for each grid to be manually inserted is name and ID. The grid is constructed following the vector of the city, which in our case example is Johor Bahru City Council, or known as Majlis Bandaraya Johor Bahru.

The next step of the process is to extract all the information from raster remote sensing data into the grid. This is conducted by sampling the values from each pixel into the corresponding attributes. A single remote sensing data pixel will represent one column attribute in the vector grid (Fig. 4). Eventually, the centroid of each grid is created as some GIS software sample raster values through a point data basis. Once the data is extracted, it will be integrated into the vector grid based on ID matching (Fig. 5). This process is done for every single layer of remote sensing data. In the case of the remote sensing data having smaller spatial resolution than the grid, it will somehow degeneralize the data, a compensation in creating a coherent database.

Because the development takes place in a spatial or GIS-based environment, the vector grid layer and attributes are recommended to be arranged accordingly based on the temporal scale of the rainfall variables; ranking from hourly (top) to monthly (bottom). This is for convenient updating and visualization since hourly data were supposed to undergo frequent updates. Regarding the attributes, the low temporal variables or land-based variables such as NDVI, NDBI, were stored on the left side while the high temporal variables toward the right. The illustration is provided in Fig. 4.

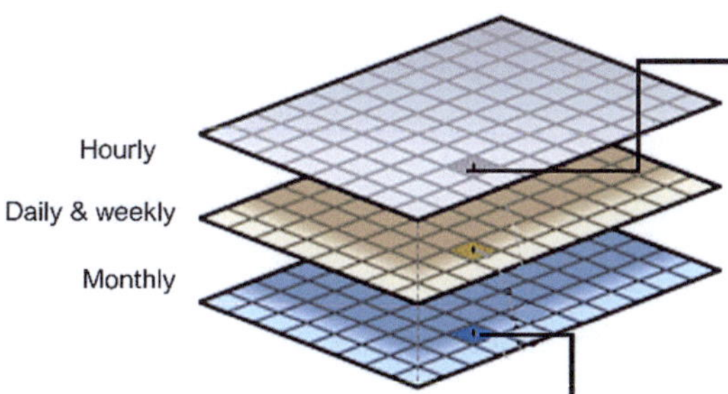

No.	ID	Place Name	NDVI	NDBI	LST	RF_0300	RF_0330	RF_0400	RF_0430
1	0001	Kg. Kempas Baru	0.453	0.677	26.4	0.0	2.1	1.5	0.1
2	0002	Kg. Sepakat	0.566	0.557	27.4	0.0	2.0	1.2	0.2
3	0003	Setia Tropika	0.312	0.781	29.8	0.1	1.7	1.3	1.1
. . .									
204	204	Kg. Melayu Majidee JB	0.377	0.766	28.8	0.0	0.0	0.0	4.3

No.	ID	Place Name	NDVI	NDBI	LST	RF01_23	RF02_23	RF03_23	RF04_23
1	0001	Kg. Kempas Baru	0.453	0.677	26.4	154	110	130	128
2	0002	Kg. Sepakat	0.566	0.557	27.4	120	122	123	145
3	0003	Setia Tropika	0.312	0.781	29.8	110	112	99	96
. . .									
204	204	Kg. Melayu Majidee JB	0.377	0.766	28.8	100	130	142	121

Fig. 4 Attribute structure of the vector grid database with examples

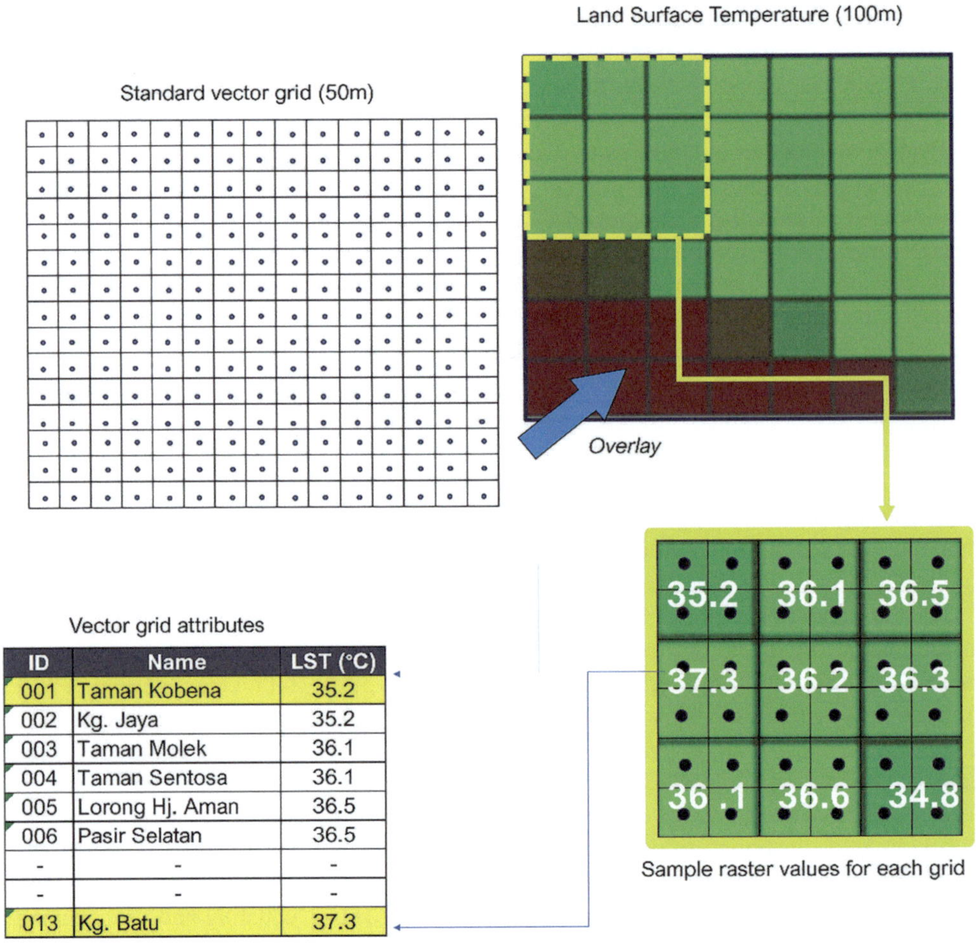

Fig. 5 Conceptual diagram depicting the process of sampling raster values into vector grid values

Phase 4: Information Display and Retrieval

In this final phase, the system should be able to visualize, display, and perform basic data retrieval on attributes. Fundamentally, the system should be able to perform three basic functions: (1) vector data display, (2) GIS layering, and (3) information or attribute-based display. The vector data display would operate based on linking the attribute value of each grid with the desired color code such as RGB, CMYK, or HIS. In addition, other supporting functions such as the information retrieval tab and transparent function were also equipped to assist effective visualization. Most concurrent GIS software systems such as ArcGIS, QGIS, and MapInfo are capable of performing all these tasks. In the case of the in-house GIS-based development system, all the required functions must be developed using the appropriate platform. The most common platforms can be JavaScript, SQL, HTML, or VB.Net.

4 Discussion

The developed framework has significant implications for other tropical cities that have experienced similar conditions of DCS. It can be used as useful guidelines for replicating similar databases based on open-source remote sensing data. Nonetheless, a localization process is recommended to ensure the framework fits the nature to support hydrometeorological hazard initiatives in a specific country. Primary localization factors include (1) the size of the grid vs. the size of the city, (2) critical data or variables and their details scale (hourly, daily, etc.), and (3) validation strategy. These factors can be observed in various examples of downscaling or localizing remote sensing or satellite data (e.g., Yan et al. 2021; Ullah et al. 2020). The size of the grid should align with the purpose; for instance, in Johor Bahru city, a 50 m grid was chosen due to the severity of flash floods at the micro-scale (~<1 km).

Choosing a smaller grid would also impact the processing burden and storage systems. The validation strategy largely depends on the availability of in-situ data. It is common for less developed countries to have insufficient data, which can affect the quality of modeling, simulation, or data processing.

Furthermore, if the developed framework can be executed in real-time with automated processing, it could be highly significant for near-real-time forecasting and early warning systems for floods, heat waves, or extreme rainfall occurrences. This is because precipitation radar data is operationally supplied at 10–30 min intervals. Many early warning frameworks or real-time forecasting systems have struggled to find appropriate data supplies that cover both spatial and temporal scales. Having access to real-time rainfall data would enable more accurate and complex environmental or hydrometeorological indicators to be computed, visualized, and streamed in near real-time, such as the Standardized Precipitation Index and Drought Monitoring Index. This could be extremely beneficial for cities with large populations and serve a wide range of users. Such preventive initiatives toward hydrometeorological extremes can be developed based on the database, including early warning systems for floods or UV index alerts indicating the risk of urban heat.

The proposed framework could also address issues related to raster data management. Using vector grids could help overcome storage problems (Hein and Blankenbach 2021). Many hydrometeorological variables come in high temporal scales (e.g., 10 min, 30 min, hourly, etc.), such as rainfall, winds, and solar radiation, resulting in a large amount of raster data. Opting for consistent and coherent vector grid storage would eliminate spatial raster data structure errors that occur during raster data creation, especially during interpolation and resampling. Additionally, utilizing vector grids enables seamless and faster analytics, such as time series analysis. Constructing time series from raster data can be complex, especially when data comes from various sources with diverse spatial resolutions and dimensions.

Another significant implication of this framework, when executed in automated mode, is that it can be a highly operational, practical, and cost-effective spatial database suitable for the high spatio-temporal dynamics of hydrometeorology in the tropics. It can reduce dependency on ground data, manpower, and physical infrastructure requirements. Moreover, it can anticipate bureaucratic factors in the data ecosystem due to less human and governance interference. These benefits could be key in addressing the data challenges faced by many developing regions (Sayed and Salah 2023).

A major potential drawback of this approach is dealing with very high temporal data storage (10–20 min intervals), especially if there are many variables. Storing such data appropriately in a consistent grid ID structure requires creating numerous attributes. While this situation is anticipated in existing DBMS approaches (Sharma et al. 2023; Iqbal 2020), where unique ID profiling and indexing are applied, it can still lead to complexities in the attribute data structure. An alternative to overcome this limitation is to create pseudogrid layers to represent each sub-variable. This approach could allow the vector data structure to consume fewer attributes, reduce complexity, and maintain a consistent grid vector structure.

5 Conclusion

This paper presents a framework for the development of a better and more efficient spatial database using open-source remote sensing data for hydrometeorological hazards in tropical cities by adapting the vector grid database concept. By using Johor Bahru City Council, Malaysia, as an experimental site, the initiated framework has overcome various issues pertinent to supporting hydrometeorological hazard or urban disaster adaptation initiatives, particularly the data conflict situation (DCS) and big raster database management and operation. The developed framework would be able to anticipate the issue of (1) unavailable ground or in-situ measurement, (2) sparse ground measurement, (3) data missing or incomplete, (4) ambiguous data quality, (5) data sharing conflict, and (6) data retrieval conflict. Secondly, the approach of opting for a coherent vector-based grid instead of maintaining the original raster data format could significantly reduce the burden of processing, storage, and pixel alignment complexity. Moreover, more complex and detailed analysis can be executed with a consistent grid and allows effective spatial visualization with minimum display burden requirements. Finally, executing the framework in real time, especially through automated processing, could be highly operational in performing near-real-time forecasting and early warning initiatives in a cost-effective and quick fashion.

Acknowledgements The authors would like to thank all the data providers, the local city council of Majlis Bandaraya Johor Bahru for their cooperation, and those who contributed to the production of this manuscript. The work is part of the research grant initiated by Universiti Teknologi Malaysia titled ‘Projecting the near future of worst micro-climate scenarios in urbanized tropics city using big-geospatial data’ (Q.J130000.2652.17J09).

References

Albrecht, F.: Natural hazard events and social capital: the social impact of natural disasters. Disasters **42**(2), 336–360 (2017)

Benkirane, M., Laftouhi, N., Khabba, S., Portillo, Á.: Hydro statistical assessment of TRMM and GPM precipitation products against ground precipitation over a mediterranean mountainous watershed (in the moroccan high atlas). Appl. Sci. **12**(16), 8309 (2022)

Chen, C., Knoblock, C.: Conflation of geospatial data. Encylopedia GIS 1–10 (2016)

Cruces, N., Seco, D., Guitérrez, G.: A Compact Representation of Raster Time Series, 2019 Data Compression Conference (DCC), Snowbird, UT, USA, pp. 103–111 (2019). https://doi.org/10.1109/DCC.2019.00018

Flamig, Z., Vergara, H., Gourley, J.: The ensemble framework for flash flood forecasting (ef5) v1.2: description and case study. Geosci. Model. Dev. **13**(10), 4943–4958 (2020)

Hein, N., & Blankenbach, J.: Evaluation of a NoSQL database for storing big geospatial raster data. GI_Forum. Munster, 76–84 (2021). https://doi.org/10.1553/giscience2021_01_s76

Iqbal, F.: The Method to Handle Big Data in Different DBMS (2020). https://doi.org/10.13140/RG.2.2.32510.41280

JMA.: Himawari-8/9. Himawari Standard Data User's Guide. Version 1.2. 20 May, 2015. Japan Meteorological Agency (2015)

Kajtar, J., Hernaman, V., Holbrook, N., Petrelli, P.: Tropical western and central pacific marine heatwave data calculated from gridded sea surface temperature observations and cmip6. Data Brief **40**, 107694 (2022)

Kovacs, Y., Doussin, N., Gaussens, M.: Flood risk and cities in developing countries. Technical Reports. No. 35, SEPIA Conseils (2017)

Li, C., Wu, Z., Yin, J.: Research on oracle-based integrative storage and management of spatial data. In: Proceedings of the 2017 International Conference on Information System and Data Mining (2017). https://doi.org/10.1145/3077584.3077587

Loew, A., et al.: Validation practices for satellite-based Earth observation data across communities. Rev. Geophys. **55**, 779–817 (2017). https://doi.org/10.1002/2017RG000562

Mahmud, M.R., Hashim, M., Matsuyama, H., Numata, S., Hosaka, T.: Spatial downscaling of satellite precipitation data in humid tropics using a site-specific seasonal coefficient. Water **2018**(10), 409 (2018)

Mahmud, M.R.: Mapping the Near Real Time of Solar Radiation in Humid Tropics Using Open-Source Remote Sensing Data (2024). (Manuscript submitted for publication)

MBJB.: Strategic Plan of MBJB. Majlis Bandaraya Johor Bahru (2021)

Msigwa, A., Makinde, A.S., Mwaijengo, G.N.: Flash flood early warning systems: a critical review of their effectiveness in reducing disaster risks. In: Maftei, C., Muntean, R., Vaseashta, A. (eds.) Modeling and Monitoring Extreme Hydrometeorological Events, pp. 70–89. IGI Global (2024). https://doi.org/10.4018/978-1-6684-8771-6.ch00

Nohrstedt, D., Hileman, J., Mazzoleni, M., Baldassarre, G., Parker, C.: Exploring disaster impacts on adaptation actions in 549 cities worldwide. Nat. Commun. **13**(1) (2022)

Osti, R., Tanaka, S., Tokioka, T.: Flood hazard mapping in developing countries: problems and prospects. Disaster Prev. Manag. Int. J. **17**(1), 104–113 (2008)

Pramono, I., Savitri, E.: Flash flood in Arau watershed, West Sumatera: a mitigation study. Matec Web Conf. **229**, 03002 (2018). https://doi.org/10.1051/matecconf/201822903002

Putra, D., Sofwany, B., Zulfinanda, H., Kamaruddin, I.: Flash flood risk and damage assessment in Batu, East Java. Jurnal Teknosains **12**(1), 72 (2022)

Rosmadi, H.S., Ahmed, M.F., Mokhtar, M.B., Lim, C.K.: Reviewing challenges of flood risk management in Malaysia. Water **15**(13), 2390 (2023)

Sayed, M., Salah, W.: Secondary data-associated challenges for environmental assessment in developing countries; case of Egypt. MSA Eng. J. **2**(2), 1225–1244 (2023)

Seibert, J.: Getting your money's worth: testing the value of data for hydrological model calibration. Hydrol. Process. **38**(2) (2024)

Sharma, S., Singh, S., Sharma, S., Negi, N.: Big data in cloud computing: an overview. Int. J. Res. Appl. Sci. Eng. Technol. **11**(4), 241–245 (2023)

Silva-Coira, F., Paramá, J., Ladra, S., López, J., Gutiérrez, G.: Efficient processing of raster and vector data. PLoS ONE **15**(1), e0226943 (2020)

Skidmore, M., Lim, J.: Natural disasters and their impact on cities. Obo Urban Stud. (2020). https://doi.org/10.1093/obo/9780190922481-0014

Slonecker, E.T., Jennings, D.B., Garofalo, D.: Remote sensing of impervious surfaces: a review. Remote Sens. Rev. **20**(3), 227–255 (2001). https://doi.org/10.1080/02757250109532436

Su, S., Tian, J., Dong, X., Tian, Q., Wang, N., Xi, Y.: An impervious surface spectral index on multispectral imagery using visible and near-infrared bands. Remote Sens. **14**, 3391 (2022). https://doi.org/10.3390/rs14143391

Sun, G., Li, Z., Zhang, A., Wang, X., Yan, K., Jia, X., Liu, Q., Li, J.: A 10-m resolution impervious surface area map for the greater Mekong subregion from remote sensing images. Sci. Data **10**(607), 2023 (2023)

Tsutsumida, N., Rodriguez-Veiga, P., Harris, P., Balzter, H., Comber, A.: Investigating spatial error structures in continuous raster data. Int. J. Appl. Earth Obs. Geoinf. **74**, 259–268 (2019)

Ullah, S., Zuo, Z., Zhang, F., Zheng, J., Huang, S., Lin, Y., Yan, L.: GPM-based multitemporal weighted precipitation analysis using GPM_IMERGDF product and ASTER DEM in EDBF algorithm. Remote Sens. **12**(19), 3162 (2020)

Wang, G., Jin, C.: A vector data model for efficiently rendering large vector maps on global 3D terrain surfaces. ISPRS Int. J. Geo Inf. **11**(4), 234 (2022)

Weng, Q.: Remote sensing of impervious surfaces in the urban areas: requirements, methods, and trends. Remote Sens. Environ. **117**(2012), 34–49 (2012)

Wyszomirski, M., Gotlib, D.: A unified database solution to process BIM and GIS data. Appl. Sci. **10**(23), 8518 (2020)

Yan, X., Chen, H., Tian, B., Sheng, S., Wang, J., Kim, J.: A downscaling–merging scheme for improving daily spatial precipitation estimates based on random forest and cokriging. Remote Sens. **13**(11), 2040 (2021)

Ye, S., Liu, D., Yao, X., Tang, H., Xiong, Q., Wen, Z., Song, C.: RDCRMG: a raster dataset clean & reconstitution multi-grid architecture for remote sensing monitoring of vegetation dryness. Remote Sens. **10**(9), 1376 (2018)

Zhao, Q., Li, S., Coelho, M., Saldiva, P., Hu, K., Huxley, R.R., Abramson, M., Guo, Y.: The association between heatwaves and risk of hospitalization in Brazil: a nationwide time series study between 2000 and 2015. PLoS Med. **16**(2), e1002753 (2019)

Flash Floods in Poland, Current Status and Prospects

Daniel Wojciech Słyś and Przemysław Ogarek

Abstract

In recent years flash floods have become a growing problem that affects almost all areas of Europe and the world. Urban and highly developed areas are especially vulnerable to the dangers of rapid rainwater runoff. Infrastructure damage, high financial losses, and the lives and health of residents are all a consequence of their contributions. Many large floods have occurred in Polish cities in recent years. In this paper, flash floods in Poland are examined based on selected cases. The identification of causes, challenges, and prospects for similar events in the future is done in the context of ongoing climate change. Quantitative analysis using the Standardized Precipitation Index (SPI) highlights the severity of the situation, with SPI for Poland projected to range from 5.18 to 5.4 under the RCP4.5 climate change scenario and from 5.19 to 5.82 under the RCP8.5 scenario between 2021 and 2100, reflecting significant shift in precipitation patterns, which are primary drivers of flash floods in the region.

Keywords

Flood forecasting · Urban flooding · Climate change · Rainwater · Water management

D. W. Słyś (✉)
Department of Infrastructure and Water Management, Rzeszow University of Technology, Rzeszow, Poland
e-mail: daniels@prz.edu.pl

P. Ogarek
Ignacy Lukasiewicz Doctoral School of Rzeszow University of Technology, Rzeszow, Poland
e-mail: p.ogarek@prz.edu.pl

1 Introduction

Climate change is currently one of the most significant challenges facing humanity. The rise in average temperatures, shifts in precipitation patterns, and increased frequency of extreme weather events have profoundly impacted various aspects of life (Żywiec et al. 2021; Nassah et al. 2022). One of the key areas where climate change exerts considerable influence is the functioning of cities and their infrastructure (Starzec et al. 2023; Gallazzi et al. 2021; Fan et al. 2023).

Changes in the amount and distribution of rainfall, along with the occurrence of sudden flash floods, are becoming increasingly pressing issues that require comprehensive analysis and understanding. These phenomena also pose significant financial, technical, and social challenges (Pregnolato et al. 2017).

Although flash floods typically affect relatively small areas, their impacts can vary depending on local conditions, sometimes leading to significant losses of public and private property, job losses, transportation disruptions, and infrastructure damage (Tan et al. 2024). It is noteworthy that between 1980 and 2022, extreme flood events caused close to 650 billion euros in damages in European Union member countries, with nearly 60 billion euros in 2021 and over 50 billion euros in 2022 alone (European Environment Agency 2023).

In Poland, as in other regions, increasingly intense rainfall is accompanied by extended dry and snow-free periods, contributing to the more frequent occurrence of droughts (Sobieraj et al. 2022; Słyś and Stec 2014). This trend results in reduced crop yields, worsening the financial situation of farms, and decreases the availability of drinking water for the population (Ruau et al. 2024).

P. Mohammad et al. (eds.), *Geographic Approaches to Climate Change and Mitigation: Urban and Rural Perspectives (Volume 1)*, Advances in Science, Technology & Innovation, https://doi.org/10.1007/978-3-031-92119-3_15

2 Definition of the "Flash Flood" Phenomenon

A flash flood is a type of flood characterized by a very short lead time from the onset of its cause, making it difficult to predict and violent in nature. The occurrence mechanism is typically associated with short-duration but intense rainfall caused by a storm or a series of storms occurring in quick succession. Flash floods can also result from other phenomena, such as failures of hydraulic structures, including dams and levees, or the flow of mudslides into rivers and lakes (Abdelkareem and Mansour 2023).

In Poland, flash floods are primarily triggered by rainfall from Cumulonimbus clouds within homogeneous air masses and in zones of cold fronts, as well as from Nimbostratus clouds associated with warm fronts and low-pressure systems (Franczak 2014; Bryndal 2014a, b).

The likelihood of a flash flood caused by intense rainfall is determined by a number of factors, some of which are illustrated in Fig. 1.

Flash floods most commonly affect small catchment areas, typically less than 1000 km^2. A distinctive feature of flash floods is their rapid onset and progression, combined with their unpredictability, which is particularly critical for protecting urban areas and ensuring the safety of the population (Marchi et al. 2016).

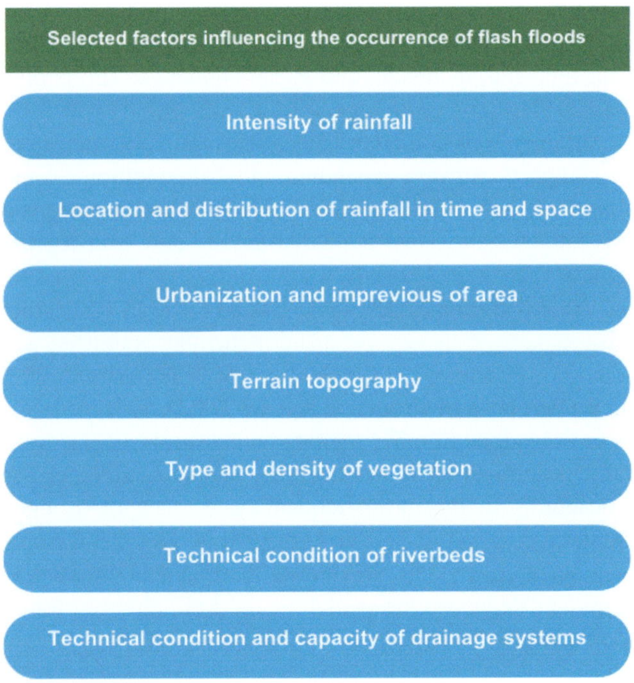

Fig. 1 Factors influencing the occurrence of flash floods

3 Selected Examples of Flash Floods

3.1 Flash Flood—Gdansk 2001

Between 1992 and 2016, the Gdansk area recorded 23 instances of sudden urban floods. The flood that occurred on July 9, 2001, was particularly severe in its effects. As a result of this event, 304 people required immediate evacuation, and four individuals lost their lives due to flood-related stress. During the violent storm, 127.7 mm of rain fell in just 8 h. The average monthly rainfall for July in Gdansk ranges from 70 to 80 mm. The flood caused enormous damage to the city's infrastructure, estimated at approximately 50 million euros, excluding the losses suffered by residents related to property loss, destruction, and health impacts (Plan Adaptacji Miasta Gdańska…).

3.2 Flash Flood—Gdansk 2016

After the catastrophic flash flood in 2001, dubbed the "flood of the century" in Gdansk, a series of flood safety improvement investments were completed by 2016. These included increasing the number of urban retention basins from 26 to 49 and expanding their capacity from 136,944 to 678,826 m^3. The capacity of open channels transporting water from the city to the sea was also increased (Raport Prezydenta Miasta…). Approximately 90 million euros were spent on this purpose over a 15-year period. A major test of the city's flood preparedness occurred on July 14, 2016. On that day, rainfall reached 160 mm, surpassing the amount recorded during the "flood of the century" (Raport Prezydenta Miasta…). As a result of the flood, two people lost their lives. Nevertheless, the city's significantly improved flood infrastructure mitigated the financial impacts. The losses were estimated at 2.5 million euros. Since then, Gdansk has implemented a series of additional infrastructure investments, becoming one of the best-managed Polish cities in terms of water and resource management.

3.3 Flash Flood—Katowice 2019

On July 27, 2019, a static supercell storm developed over Katowice. According to data from the Polish Institute of Meteorology and Water Management, 74.2 mm of precipitation was recorded locally within one hour, with the total reaching 83.8 mm. The average monthly rainfall in July from 1991 to 2020 ranged between 100 and 120 mm (Rekordowa ulewa…).

In a single day, 409 fire brigade interventions were carried out. These activities included evacuating customers from stores and shopping centers. Parking lots, tunnels, and underground garages were submerged. The flood barrier of the Rawa River was breached, and significant sections of nearby roads were washed away by the water.

3.4 Flash Flood—Gorzow Wielkopolski 2022

A flash flood also occurred in the city of Gorzow Wielkopolski on September 8, 2022. It was triggered by storms along a convergence line in the wind field. Due to the stationary nature of the storm, approximately 130 mm of rain fell throughout the day, with local amounts reaching up to 220 mm. Within one hour, 77.6 mm of rainfall was recorded. The average monthly rainfall for July ranges from 70 to 80 mm (Powódź błyskawiczna…).

This extreme precipitation completely overwhelmed the drainage system, rendering it ineffective. Many streets in the city center were flooded, landslides occurred, and urban infrastructure was damaged, including railway tracks (Powódź błyskawiczna…).

3.5 Flash Flood—Rzeszow 2020

On June 22, 2020, a peak of torrential rain and storms occurred in the eastern part of Poland. The zone of most intense precipitation stretched across almost the entire eastern part of the country. By 23:00, the highest rainfall totals were recorded in the center and around Rzeszów in the Subcarpathian region. Locally, total rainfall ranged from 60 to 80 mm. A new daily rainfall record for Rzeszów was set at 67.5 mm, surpassing the previous record of 65.2 mm from 2010. The average monthly rainfall for June ranges from 90 to 100 mm (Potężna ulewa…).

As a result of the rainfall, there was a rapid rise in the levels of local streams, leading to flooding due to the overloaded drainage systems' inability to discharge the waters.

3.6 Flash Flood—Poznan 2021

On June 22, 2021, a storm resulted in 64 mm of rainfall within 30 min, the third highest in the 70-year history of measurements. The average monthly rainfall for June ranges from 80 to 90 mm.

Urban drainage systems proved inadequate for efficiently managing such large volumes of rainwater in a short period. In city districts with lower terrain, which served as natural accumulation areas for the torrential rainwater, the water level reached about 90 cm. Many streets in the city center and railway tracks were flooded (Poznań pływał…).

3.7 Flash Floods—Malopolska Province 2021, 2022, 2023, 2024

The area of Poland located south of Cracow, extending toward the Tatra Mountains, the highest in Poland, which form their foothills, is a region where flash floods occur very frequently. The conditions in this area are particularly conducive to the formation of violent surface runoff, leading to flash flood events that affect many villages and towns almost every year. Examples include flash floods in 2020, 2022, 2023, and 2024 (Nawałnice w Małopolsce…).

4 A General Overview of Flash Floods in Poland

Flash floods occur throughout Poland, with the most susceptible areas being in the south of the country, where mountainous and sub-mountainous regions exist, as well as coastal areas near Gdańsk. In both cases, the steep gradients of the terrain generate rapid and violent surface runoff.

Analyzing the climatic conditions in Poland, it is evident that intense precipitation events (rainfall exceeding 30 mm/day) predominantly occur in June (in the mountainous and sub-mountainous areas of Lesser Poland), July (in the mountainous and sub-mountainous areas of Lesser Poland, Subcarpathia, and the Sudeten Foothills), and August (in the mountainous and sub-mountainous areas of Lesser Poland and the Sudeten Foothills).

Locations of areas with the highest probability of rainfall exceeding 30 mm/day over successive ten-day periods in the summer months are shown in Fig. 2.

The analysis of flash flood events within cities across Poland shows that, in the vast majority of cases, these events coincide with the areas most likely to experience rainfall exceeding 30 mm/day. Figure 3 illustrates the occurrence of the most significant flash floods in Poland from 2018 to 2023.

5 The Impact of Climate Change on Flash Floods in Poland

Several factors contribute to the increase in the number of flash floods, both through direct human impact and environmental changes, such as alterations in land use, deforestation, and the encroachment of buildings towards watercourses, as well as indirectly through anthropogenic effects on climatic conditions, including climate change driven by greenhouse gas emissions.

Given the complex and multifaceted nature of human interactions with the climate, the question of the extent to which climate change contributes to the occurrence of flash floods

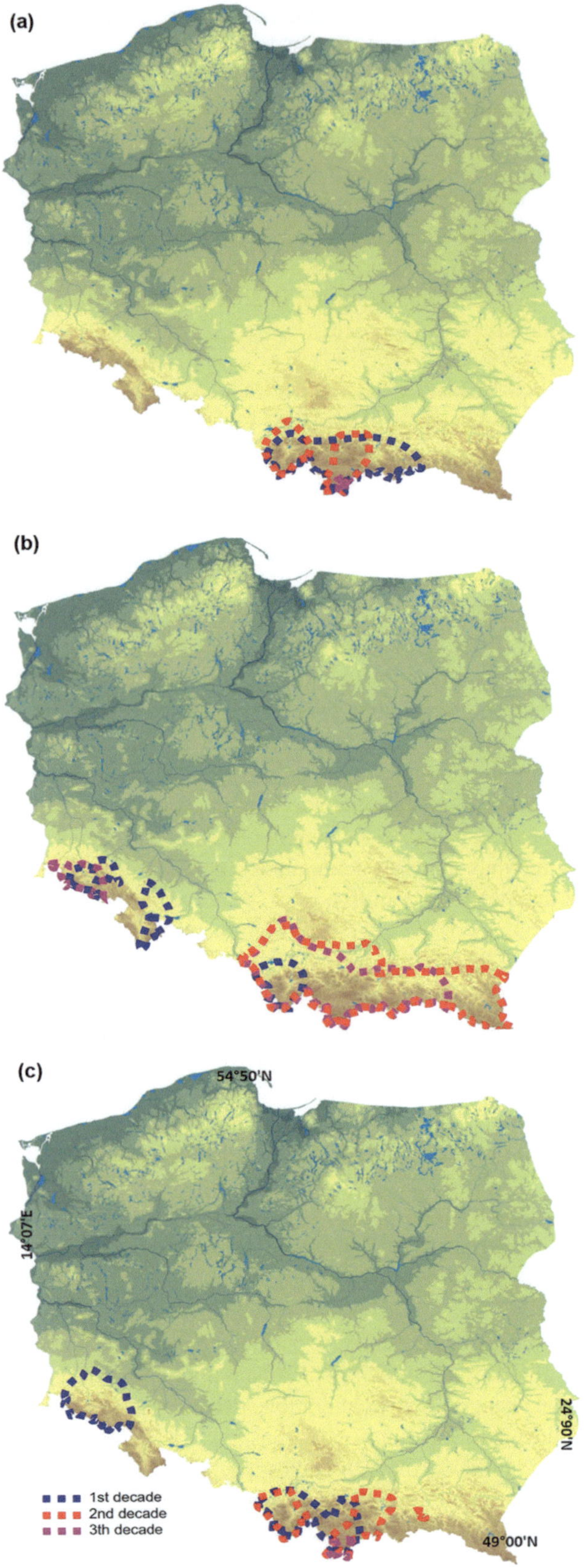

Fig. 2 Locations of areas with the most probable rainfall exceeding 30 mm/day over successive ten-day periods in the months of: **a** June, **b** July, **c** August (author's own study based on: https://imgw.isok.gov.pl/mapy-klimatologiczne/intensywne-opady-atmosferyczne.html)

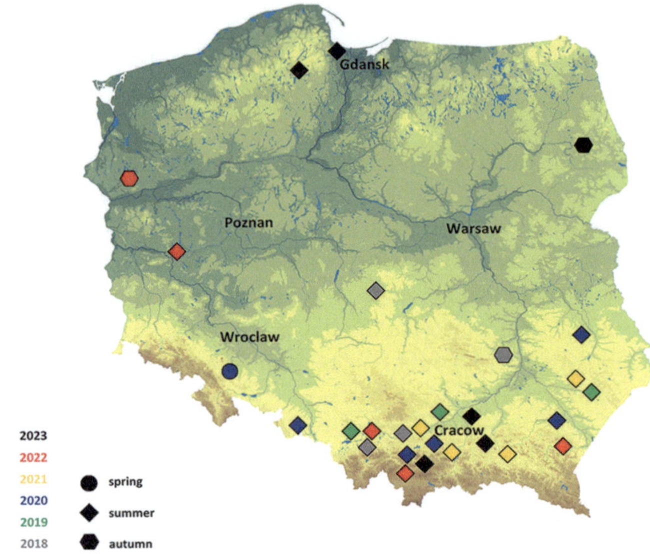

Fig. 3 Occurrence of the most significant flash floods in Poland in 2018–2023 based on geocoded data from local media sources (own study)

versus the role of direct human activities remains highly relevant. It is certain that both factors are closely linked and may act synergistically, though the scope of their impacts can differ significantly. Factors related to direct human impact on water management in catchment areas typically have a local character and significant influence. In contrast, factors indirectly resulting from human activities, manifesting as climate change, have a global character, multifaceted impact, and are more difficult to clearly identify. Figure 4 illustrates selected factors that contribute to the increased frequency of flash floods.

The increase in flood risk in Poland in the coming years will primarily be driven by changes in the frequency and structure of precipitation, the reduction of solid precipitation such as snow, an increase in transpiration levels, and the rise in the Baltic Sea level (Hieronymus et al. 2018).

Statistical analysis of historical precipitation reveals significant variability across multi-year, monthly, and annual intervals. Intense precipitation exhibits noteworthy spatial divergence, primarily depending on the physiogeographic conditions of the area and the time of year. Ongoing climate alterations are driving changes in hydrological processes and their intensity, with a noticeable shift in atmospheric precipitation patterns from the winter months to the summer period.

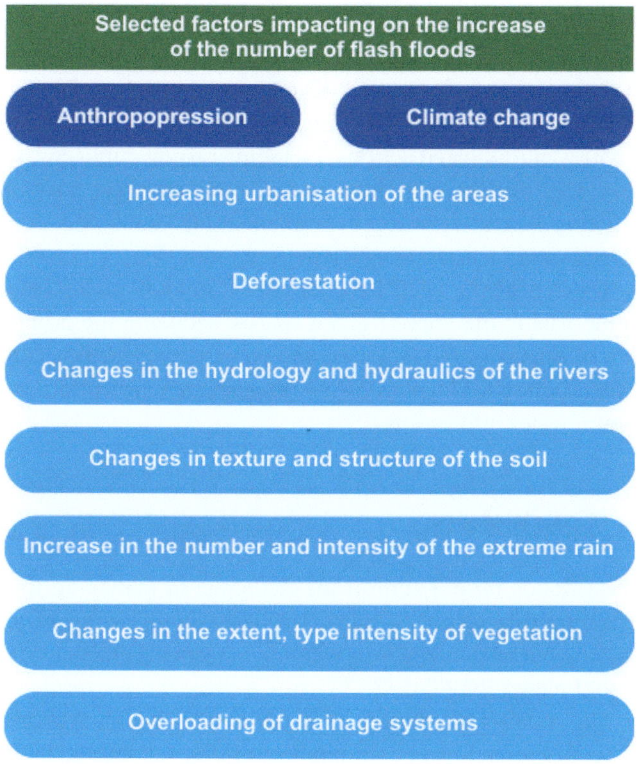

Fig. 4 Factors influencing the increase of the number of flash floods

Winter precipitation is also undergoing transformation, with snowfall increasingly being replaced by rainfall due to rising average air temperatures. This results in diminished snow cover extent and thickness, reduced retention, heightened surface runoff dynamics, and decreased groundwater aquifer recharge.

The rise in temperature, coupled with changes in precipitation type, exacerbates the risk associated with autumn and winter rainfall events.

Climate change scenarios indicate a pervasive impact across Poland, albeit with varying degrees of intensity. Regions particularly susceptible to flash floods are situated in the southern part of the country, encompassing foothills and mountainous areas.

The assessment of climate change's influence on flash flood occurrences can be indirectly conducted using predictive climate change models.

One of the parameters that can be considered is the Standardized Precipitation Index (SPI), defined by Eq. (1). In the context of Poland's climate, SPI is particularly suitable due to its focus on precipitation patterns, which are the primary drivers of flash floods in this region. Unlike the Standardized Precipitation Evapotranspiration Index (SPEI) (Mokhtar et al. 2021), which includes evapotranspiration and is more suitable for arid and semi-arid climates, SPI is especially effective in reflecting changes in the number of days with precipitation exceeding 1 mm relative to the total number of precipitation days within a year. Additionally, the availability of reliable evapotranspiration data, especially on a fine spatial and temporal scale, is limited in Poland compared to precipitation data, further supporting the use of SPI in this context.

$$SPI_n = \frac{\sum_{i=1}^{precipitation\_days\_in\_year} PR_{in}}{Number\ of\ days\ with\ PR > 1\text{mm}} \tag{1}$$

where

SPI_n—Standardized precipitation index for a specific year n,

PR_{in}—Total daily precipitation on day i, in year n.

According to available predictive models for Poland, an increase in the Standardized Precipitation Index (SPI) is expected in the coming decades, indicating a rise in high-intensity precipitation (Klimada).

The averaged SPI for the entire territory of Poland is projected to range from 5.18 to 5.4 between the years 2021–2100 for the climate change scenario RCP4.5, and from 5.19 to 5.82 for the scenario RCP8.5 (Klimada).

Figure 5 illustrates the simulated changes in the Standardized Precipitation Index (SPI) for selected Polish cities with populations above 400,000 under the RCP4.5 and RCP8.5 climate change scenarios. Meanwhile, Fig. 6 presents the simulated changes in SPI for selected locations particularly vulnerable to flash floods under the RCP4.5 scenario.

6 Challenges

The implementation of the European Union Directive 2007/60/EC on the assessment and management of flood risks into Polish practice has resulted in the development of flood hazard maps covering the entire country. These maps provide information on the extent of floodwaters with specified probabilities of occurrence and the depth of water inundation. Additionally, for cities with populations exceeding 100,000, the maps also indicate flow directions and velocities. Furthermore, Poland has developed flood risk management plans that include a list of necessary investments to reduce flood risk.

It is important to note, however, that flood hazard maps do not encompass the protection of areas and populations from flash flood events. This is partly because flash floods are often not closely associated with the proximity of rivers or seas and typically affect small areas (Bryndal 2014a, b).

The challenges in developing such tools and utilizing them as effective risk management instruments are influenced, among other factors, by the dynamic growth of urbanized areas and the accompanying changes in their development and topography.

Nonetheless, in recent years, there has been growing interest among cities in creating maps of areas prone to flash

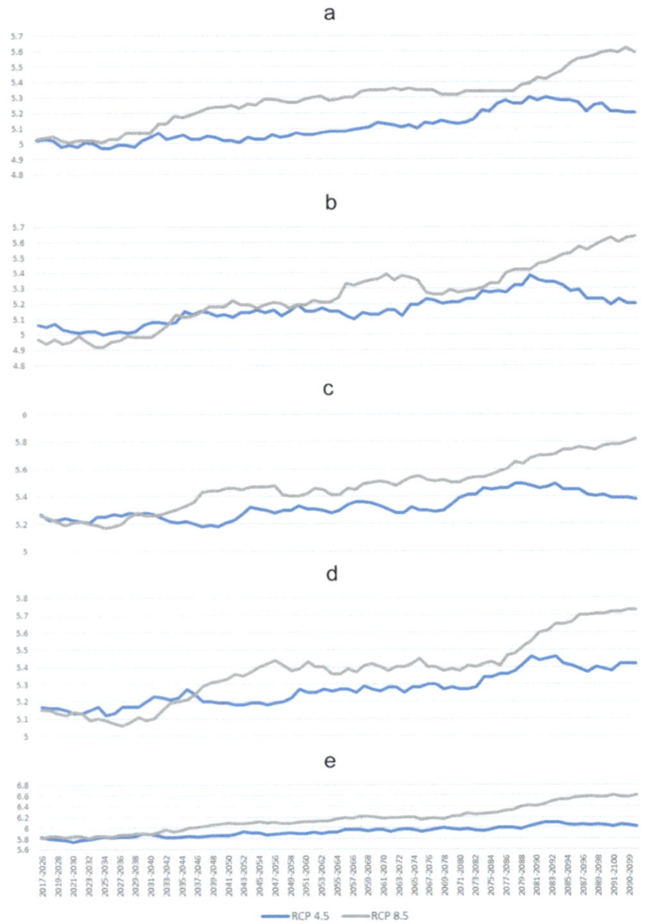

Fig. 5 The change in the SPI index for the climate change scenarios RCP4.5 and RCP8.5, **a**—Gdansk, **b**—Poznan, **c**—Warsaw, **d**—Wrloclaw, **e**—Cracow (author's own study based on: https://klimada2.ios.gov.pl/wskaznik-intensywnosci-opadu)

floods. Pilot projects for such maps are currently being implemented in selected large Polish cities in the southern part of the country, such as Cracow, Katowice, and Bielsko-Biala.

It should be noted that mitigating the risk of flash floods requires multifaceted actions, including raising public awareness and readiness to take immediate measures, such as assisting rescue services in implementing dams and flood barriers and facilitating efficient evacuation.

There is also a need for changes in urban planning, such as increasing the area of green spaces, reducing the proportion of impermeable surfaces, and constructing facilities and infrastructure for local rainwater management, retention, and groundwater infiltration.

7 Conclusions

Flash floods pose a significant challenge in managing their risk, requiring not only a long-term vision for preparing infrastructure but also changes in water management practices and spatial planning policies.

The small-scale nature of flash floods, which typically affect individual rural communities or cities, combined with the intense rainfall that causes rapid inundation and the mechanisms of flood wave formation, makes flood risk management a complex and somewhat unpredictable issue, underscoring the importance of preventive actions.

A key challenge in mitigating the impact of flash floods on population and property is reducing exposure to the sudden onset of flash flooding. At the current stage of flood risk management tool development, these challenges must be addressed through well-informed policies implemented regionally by local officials and decision-makers.

Fig. 6 The change in the SPI index for the climate change scenario RCP4.5 (author's own study based on: https://klimada2.ios.gov.pl/wskaznik-intensywnosci-opadu)

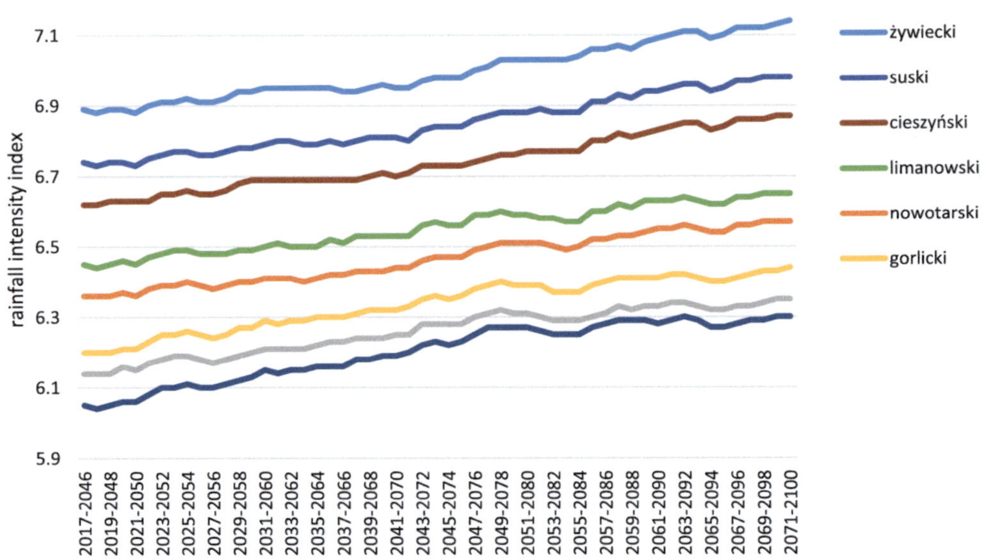

It is important to note that while flood risk maps available for the entire country only indirectly contribute to reducing exposure to flash floods, they are very helpful in the case of "standard" floods.

As noted in this publication, an increase in the number of flash floods in Poland is expected in the coming years, driven by changes in land use, intensification of surface runoff, and climatic changes, particularly the increased frequency of extreme precipitation events.

Limiting the number of unpredictable flash floods requires significant financial investments in flood protection measures, which may not always yield the expected results. Therefore, in many cases, a more targeted approach is to reduce the exposure of residents and property to flash floods through public education, changes in water management practices in watersheds, and proper urban planning in cities and villages.

References

Abdelkareem, M., Mansour, A.M.: Risk assessment and management of vulnerable areas to flash flood hazards in arid regions using remote sensing and GIS-based knowledge-driven techniques. Nat. Hazards **117**, 2269–2295 (2023). https://doi.org/10.1007/s13593-023-00942-4

Bryndal, T.: Flash flood in small Carpathian catchments–selected aspects of flood risk management. Annales Universitatis Pedagogicae Cracoviensis. Studia Geographica (7), 69–80. https://bibliotekanauki.pl/articles/471355 (2014)

Bryndal T.: The role of flood hazard and flood risk maps in reducing of effects of flash flooding in municipalities. In: Ciupa, T., Suligowski, R., (eds.) Water in the City, pp. 29–37. I G U J K, Kielce (2014)

Directive 2007/60/EC of the European Parliament and of the Council of 23 October 2007 on the assessment and management of flood risks. Document 32007L0060. https://eur-lex.europa.eu/eli/dir/2007/60/oj

European Environment Agency: Economic losses from weather- and climate-related extremes in Europe. https://www.eea.europa.eu/en/analysis/indicators/economic-losses-from-climate-related

Fan, X., Zhang, X., Yu, et al.: Assessment of the impacts of climat change on water supply system pipe failures. Sci. Rep. **13**(1) (2023). https://doi.org/10.1038/s41598-023-33548-7

Franczak P. (2014). The role of major floods in shaping the life of the population in the catchment area of the upper Skawa River since the 15th century. In: Krąż, P. (ed.) Contemporary Problems and Research Directions in Geography, pp. 117–129. I G i G P U J, Kraków (2014)

Gallazzi, A., Molinari, D., Ballio, F., et al.: Development of a flood damage model for urban drainage networks. In: Proceedings of the FLOODrisk 2020—4th European Conference on Flood Risk Management, Online Conference, 22–24 June 2021. https://repozitorium.omikk.bme.hu/items/4795d049-7b21-45d2-8b4c-812eb80a9ae3 (2021)

Hieronymus, M., Dieterich, C., Andersson, H., et al.: The effects of mean sea level rise and strengthened winds on extreme sea levels in the Baltic Sea. Theor. Appl. Mech. Lett. **8**(6), 366–371 (2018). https://doi.org/10.1016/j.taml.2018.06.008

Hydroportal. https://wody.isok.gov.pl/imap_kzgw/?gpmap=gpMZP

Klimada 2.0. https://klimada2.ios.gov.pl/wskaznik-intensywnosci-opadu

Marchi, L., Cavalli, M., Amponsah, W., et al.: Upper limits of flash flood stream power in Europe. Geomorphology **272**, 68–77 (2016). https://doi.org/10.1016/j.geomorph.2015.11.005

Mokhtar, A.., Mohammadnabi, J., Hongming, H., et al.: Estimation of SPEI meteorological drought using. IEEE Access **9**, 65503–65523. https://ieeexplore.ieee.org/document/9408611 (2023)

Nassah, H., Daghor, L., Chatoui, H. et al.: Climate change impact on agricultural production in the Sahel region. Nutr. Hum. Health Eff. Environ. Impacts 3–11 (2022). https://doi.org/10.1007/978-3-030-93971-7_1

Nawałnice w Małopolsce. Powódź błyskawiczna w Rabce-Zdroju. (Storms in Małopolska. Flash flood in Rabka-Zdrój). https://www.rmf24.pl/regiony/zakopane/news-nawalnice-w-malopolsce-powodz-blyskawiczna-w-rabce-zdroju,nId,6907702#crp_state=1

Plan adaptacji Miasta Gdańska do zmian klimatu do roku 2030. (City of Gdańsk Climate Change Adaptation Plan until 2030). https://baw.bip.gdansk.pl/api/file/GetZipxAttachment/216/1169238/false/preview

Potężna ulewa w Rzeszowie. Liczne zalania i podtopienia na Podkarpaciu. (Heavy rainfall in Rzeszów. Numerous floods and inundations in the Podkarpacie region). https://dobrapogoda24.pl/artykul/ulewa-zalania-podtopienia-rzeszow-22-czerwca-2020-podkarpacie

Powódź błyskawiczna w Gorzowie Wielkopolskim. Do tego grad. (Flash flood in Gorzów Wielkopolski. Accompanied by hail). https://www.polsatnews.pl/wiadomosc/2022-09-08/ogromna-ulewa-nad-gorzowem-wielkopolskim

Poznań pływał po największej czerwcowej ulewie w historii. Burze poczyniły szkody nie tylko tam. (Poznań was flooded after the largest June storm in history. The storms caused damage not only there). https://www.twojapogoda.pl/wiadomosc/2021-06-22/poznan-i-torun-plywaja-po-wielkich-ulewach-niebawem-czeka-to-kolejne-miasta-ktore-zdjecia

Pregnolato, M., Ford, A., Wilkinson, S.M. et al.: The impact of flooding on road transport: a depth-disruption function. Transp. Res. Part D Transp. Environ. **55**, 67–81 (2017). https://doi.org/10.1016/j.trd.2017.06.020

Raport Prezydenta Miasta Gdańska. Gdańsk po nawałnicy 14/15 lipca 2016. (Report of the Mayor of Gdańsk. Gdańsk after the storm of July 14/15, 2016). https://download.cloudgdansk.pl/gdansk-pl/d/20160776631/1.pdf

Rekordowa ulewa w Katowicach. Ponad 200 razy interweniowali strażacy. (Record rainfall in Katowice. Firefighters intervened over 200 times). https://dobrapogoda24.pl/artykul/ulewa-potop-katowice-bytom

Ruau, C., Naipal, V., Gagnaire, N. et al.: Soil erosion has mixed effects on the environmental impacts of wheat production in a large, semi-arid Mediterranean agricultural basin. Agron. Sustain. Dev. **44**(6) (2024). https://doi.org/10.1007/s13593-023-00942-4

Słyś, D., Stec, A.: Optimization of the hydraulic system of the storage reservoir hydraulically unloading the sewage network. Ecol. Chem. Eng. **21**, 215–228 (2014). https://doi.org/10.2478/eces-2014-0017

Sobieraj, J., Bryx, M., Metelski, D.: Stormwater management in the City of Warsaw: a review and evaluation of technical solutions and strategies to improve the capacity of the combined sewer system. Water 14 (2022). https://doi.org/10.3390/w14132109

Starzec, M., Kordana-Obuch, S. Słyś, D.: Assessment of the feasibility of implementing a flash flood early warning system in a small catchment area. Sustainability **15**(10) (2023). https://doi.org/10.3390/su15108316

Tan, X.Z., Li, Y., Wu, X.X., et al.: Identification of the key driving factors of flash flood based on different feature selection techniques coupled with random forest method. J. Hydrol. Reg. Stud. **51**, 2214–5818 (2024). https://doi.org/10.1016/j.ejrh.2023.101624

Żywiec, J., Boryczko, K., Kowalski, D.: Analysis of the negative daily temperatures influence on the failure rate of the water supply network. Resources **10**(9), 89 (2021). https://doi.org/10.3390/resources10090089

Centrifuge Testing for Assessing Rain-Induced Slope Instability

Samuel A. Espinosa F. and M. Hesham El Naggar

Abstract

The overarching objective of this research is to assess slope instability hazards related to the heightened and more frequent rainfall patterns attributed to climate change. Additionally, the study aims to develop predictive models and early warning systems while providing guidelines for establishing safe zones on slopes. Centrifuge testing is employed in this research to replicate slope responses under varying rainfall intensities and durations. The experimental approach seeks to yield crucial data, including the development of the Intensity-Duration (ID) threshold curve and the degree of saturation versus void ratio threshold curve (Sr–e). Both curves play pivotal roles in evaluating the vulnerability of slopes to rainfall-induced instability. The ID threshold curve offers insights into the correlation between rainfall intensity and duration, delineating thresholds at which slopes become susceptible to instability. Concurrently, the degree of saturation versus void ratio threshold curve provides essential information on the variation of soil's void ratio, offering additional dimensions to our assessment. The anticipated outcomes from these centrifuge tests are poised to demonstrate the profound influence of climate change-induced alterations in rainfall patterns on slope stability. Thus, the research aims to provide valuable insights to evaluate the hazards associated with climate-induced slope instability, which will eventually inform comprehensive and targeted policies and strategies for slope management. The emphasis lies on the development of predictive models and early warning systems. These elements collectively form the foundation for ensuring the resilience of infrastructure and, in turn, the safety of communities amid the evolving challenges posed by climate patterns.

Keywords

Early warning system · Climate change · Slope stability · Rainfall · Centrifuge

S. A. Espinosa F. · M. H. El Naggar (✉)
Geotechnical Research Center, Department of Civil and Environmental Engineering, University of Western Ontario, London, Canada
e-mail: naggar@uwo.ca

S. A. Espinosa F.
e-mail: sespinos@uwo.ca

1 Introduction

Climate change presents a formidable challenge to the stability of slopes, posing a threat to the resilience of infrastructure and the safety of communities. As climate patterns undergo unprecedented shifts, the heightened frequency of rainfall events emerges as a significant contributor to slope instability hazard (Leroueil 2001). This research endeavors to contribute to the understanding and mitigation of slope instability hazards associated with the escalating and more frequent rainfall patterns attributed to climate change. The primary goal is to advance our comprehension of slope instability mechanisms and develop a predictive model for the stability of slopes during heavy rain events.

As with similar studies in the field, this research adopts a comprehensive framework that integrates various facets of slope assessment (Jayakody et al. 2023; Matziaris et al. 2015; Eab et al. 2014; Ling et al. 2012; Tamate et al. 2012). A key cornerstone of this framework is the utilization of centrifuge testing, a sophisticated experimental technique designed to replicate slope responses under diverse rainfall intensities and durations (Ling et al. 2009). Through this approach, the research seeks to generate essential data, specifically focusing on the development of the Intensity-Duration (ID) threshold curve and the degree of saturation vs void ratio threshold curve (Sr–e). These curves play pivotal roles in evaluating the vulnerability of slopes to climate-induced instability (Wang et al. 2021).

P. Mohammad et al. (eds.), *Geographic Approaches to Climate Change and Mitigation: Urban and Rural Perspectives (Volume 1)*, Advances in Science, Technology & Innovation, https://doi.org/10.1007/978-3-031-92119-3_16

Anticipated outcomes from the centrifuge tests will shed light on the profound influence of climate change-induced alterations in rainfall patterns on slope stability. The research aims to deliver valuable insights for evaluating the hazards associated with climate-induced slope instability. These insights, in turn, will contribute to the development of comprehensive and targeted policies and strategies for effective slope management.

2 Centrifuge Modeling

Centrifuge testing offers the opportunity to mimic elevated stress levels and directly observe recurrent failures. Geotechnical centrifuge testing can yield valuable insights into slope behavior across various scenarios, including rainfall, as highlighted by Schofield (1978). By subjecting slope models to centrifugal forces, we can accurately reproduce the influence of gravity and the pore pressure induced by rainfall, both pivotal factors in slope stability assessments. Moreover, centrifuge testing facilitates the examination of transient phenomena, such as the dissipation of pore pressure over time. This capability enhances our comprehension of how slopes react to rainfall events and the subsequent alterations in pore pressure distribution. Such changes play a crucial role in evaluating slope stability amid rainfall conditions.

2.1 Centrifuge Facility and Model Box

The centrifuge testing will take place at the Western Geotechnical Drum Centrifuge, located within the Geotechnical Research Centre (GRC) at the University of Western Ontario, Canada. This centrifuge apparatus comprises a 360-degree drum channel measuring 0.4 m in depth and 0.7 m in width, with a diameter of 2.2 m, capable of accommodating up to 3.5 tonnes of soil. The slope models will be constructed within a sturdy enclosure featuring internal dimensions of 50 × 25 × 39 cm (length × width × height), predominantly constructed from aluminum alloy material (refer to Fig. 1).

The scaling factors required to replicate soil behavior within the sturdy enclosure, enabling the correlation of model-scale parameters with their prototype-scale counterparts, are detailed in Table 1.

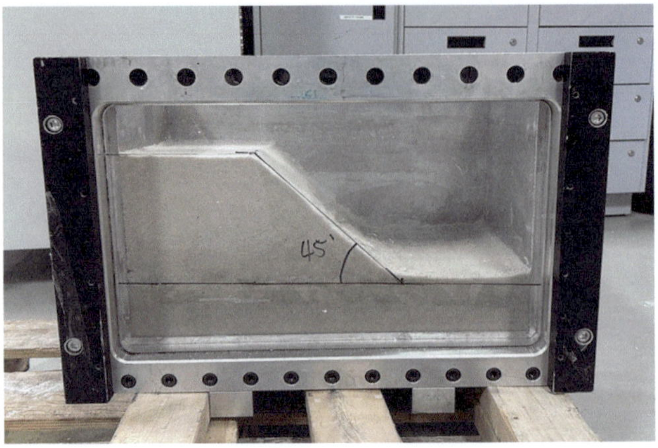

Fig. 1 Centrifuge strongbox with a model slope

Table 1 Scale factors

Quantity	1*g* (prototype)	*ng* (model)
Length	1	n
Stress	1	1
Pore water pressure	1	1
Hydraulic gradient	1	n
Rainfall duration	1	$1/n^2$
Rainfall intensity	1	$n^{0.5}$

2.2 Physical Properties of the Slope Soil

The model soil utilized in this research is identified as silica sand. To ensure consistency and reliability, ASTM standardized tests for soil analysis and experimentation are adhered to. In particular, tests were conducted to establish the soil's fundamental physical properties. Analysis of particle size distribution, as illustrated in Fig. 2a, demonstrates a range of sand particles spanning from 0.075 to 0.42 mm, with a mean grain diameter (D50) measuring 0.35 mm. Employing the Unified Soil Classification System (USCS), the soil is designated as poorly graded sand (SP). Subsequent compaction tests were performed on the silica sand to ascertain critical parameters, including the maximum dry unit weight (17.21 kN/m^3), optimum water content (14.2%), and saturated water content (20%) (see Fig. 2b).

In this investigation, the shear strength of unsaturated soil is evaluated employing the extended Mohr–Coulomb criterion, represented as:

$$\tau = c' + (\sigma - u_a)\tan\emptyset' + (u_a - u_w)\tan\emptyset^b \qquad (1)$$

where σ, u_a, and u_w are the total stress, pore air pressure and pore water pressure, respectively, and $(u_a - u_w)$ denotes the matric suction.

The hydraulic conductivity of the silica sand was assessed using the constant head permeability test, resulting in a coefficient of saturated hydraulic conductivity of 5.5×10^{-4} cm/s. Direct shear tests were conducted to determine the internal friction angle, yielding a value of 32°. These tests were performed under four different normal stress conditions: 27.2 kPa, 109 kPa, 204.3 kPa, and 313.3 kPa, corresponding to depths of 2 m, 7.5 m, 15 m, and 22.5 m, respectively. The

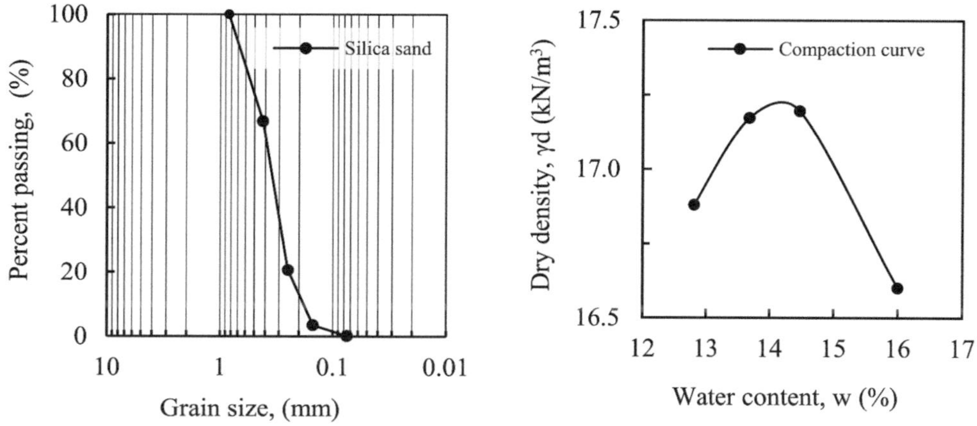

(a) Grain size distribution **(b)** Compaction curve

Fig. 2 Soil properties for silica sand: **a** grain size distribution curve and **b** compaction curve

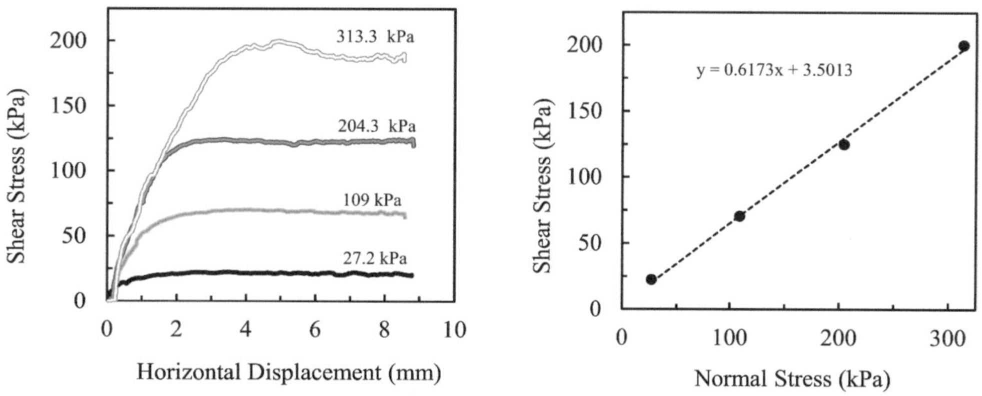

(a) Plot for shear stress against shear displacement for dense dry silica sand **(b)** Shear strength parameter for a dry silica sand

Fig. 3 Direct shear test results: **a** horizontal displacement against shear stress and **b** normal stress against shear stress

relationship between horizontal displacement and shear stress is depicted in Fig. 3a, while the correlation between normal stress and shear stress is illustrated in Fig. 3b. These figures offer valuable insights into the shear behavior of the silica sand under varying normal stress conditions, facilitating the analysis of slope stability and deformation characteristics.

The findings pertaining to basics soil properties are summarized in Table 2.

Table 2 Soil properties

Parameter	Value
Specific gravity, G_s	2.65
D_{60}, D_{30}, D_{10} (mm)	0.4, 0.28, 0.20
Sat. hydraulic conductivity (cm/s)	5.5×10^{-2}
Relative density, D_r (%)	78.6
e_{max}, e_{min}	1.023, 0.622
Friction angle, ϕ' (°) and cohesion, c (kPa)	32, 0

The soil water characteristics curve (SWCC) is derived using the formulation proposed by Van Genuchten (1980). According to Van Genunchten, the volumetric water content θ_w is expressed as:

$$\theta_w = \theta_{res} + \frac{\theta_{sat} - \theta_{res}}{\left[1 + (a'\varphi)^n\right]^m} \tag{2}$$

In this context, the parameters *a*, *n*, and *m* are curve fitting parameters that control the shape of the volumetric water content function. Additionally, θ_{sat} represents the saturated volumetric water content, while θ_{res} denotes the residual volumetric water content. Figure 4a illustrates the drying branch of the SWCC for the silica sand, utilizing the fitting parameter *a*, *n*, and *m*. The hydraulic conductivity curve is generated by the Fredlund, Xing and Huang model (Fredlund et al. 1994) incorporated in the computer program SEEP/W (GEO-SLOPE International Ltd. 2022). Figure 4b

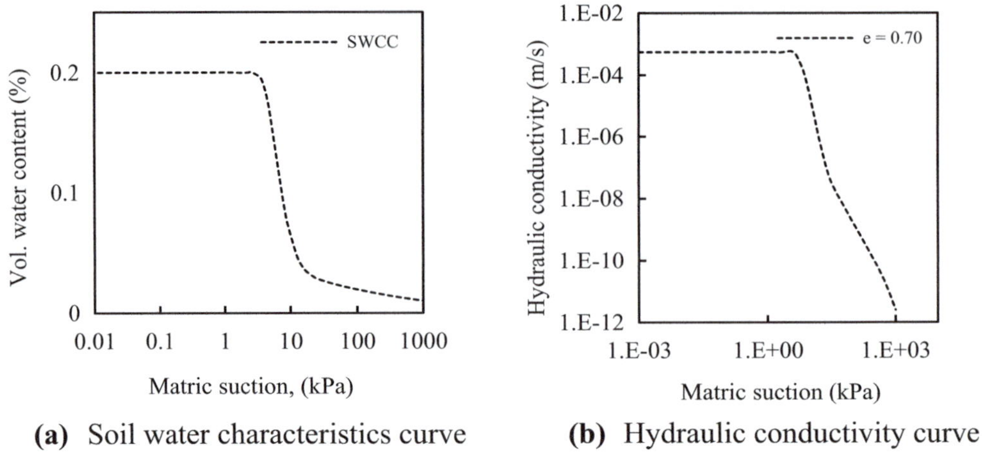

Fig. 4 Hydraulic properties of the silica sand: **a** SWCC based on Van Genuchten (1980) and **b** hydraulic conductivity based on Fredlund, Xing and Huang

Table 3 Centrifuge tests program

Test no	Slope angle, β (°)	Slope height, H_m(cm)	Acceleration, *n* (g)	Slope height, H_p(m)	Rainfall Intensity, I_p (mm/hr)
1	45	17	89	15.13	10
2	45	17	89	15.13	35
3	45	17	178	30.26	60
4	45	17	178	30.26	70

The subscripts "m" and "p" indicate the model and prototype scale, respectively

illustrates the hydraulic conductivity curve for the silica sand and shows that the hydraulic conductivity decreases with the matric suction.

3 Centrifuge Test Program

Four homogeneous slopes are being prepared for the centrifuge experiments, each represents unique initial conditions and rainfall intensities. The initial slope conditions encompass parameters such as the slope angle β, initial void ratio e_0 and initial water content w_0. All slopes will be compacted to a dense state with $e_0 = 0.707$, with an initial water content of $w_0 = 13\%$ ($S_r = 0.65$). Each slope will be subjected to varying rainfall intensities, specifically. $I_p = 10$, 35, 60 and 70 mm/hr, representing moderate (2.5–10 mm/hr), heavy (10–50 mm/hr) and extreme (>50 mm/hr) conditions at the prototype scale as detailed in Table 3. The rainfall will persist until the initial failure occurs, the rainfall duration, accumulated rainfall and failure time will be recorded.

3.1 Modeling Result

The configuration of the test model will be analyzed using SLOPE/W software. The result of this analysis will be compared with the centrifuge test results of the same model conducted without rainfall. Such a comparative analysis aims

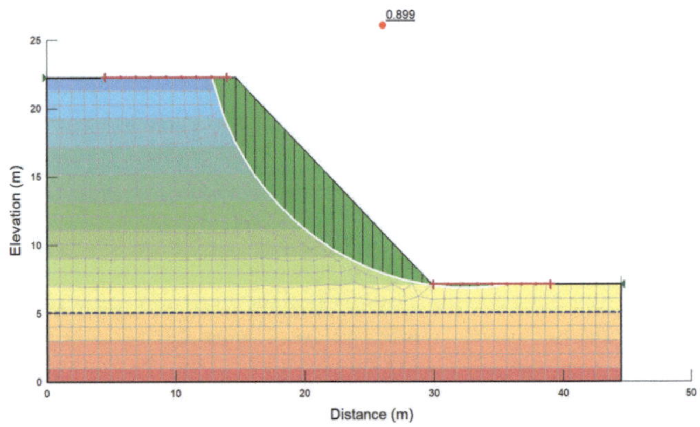

Fig. 5 Stability analysis of slope using SLOPE/W

to elucidate the stability of the slope at the prototype scale and compare it with the stability observed at the model scale inside the centrifuge.

Subsequently, the same procedure will be repeated with rainfall conditions to assess its impact on slope stability for both prototype and model scales. This comparison will provide valuable insights into how rainfall influences slope stability across different scales, aiding in the assessment of climate-induced slope instability and the validation of centrifuge testing as a reliable experimental method for studying such phenomena (Fig. 5).

The numerical model and the interpretation of its results will account for the scaling effects and the representation of real-world conditions. Despite simulating at the prototype scale, numerical models inherently involve simplifications and assumptions, which may influence the accuracy of the results compared to physical experiments. Therefore, the interpretation of the FS value obtained from the numerical model should consider the limitations and uncertainties associated with the physical and numerical models. While numerical modeling allows control over various parameters, it may not fully capture all the complex and uncertain real-world conditions. On the other hand, centrifuge testing provides a physical representation of the slope behavior under scaled conditions, allowing for direct observation of failure mechanisms and behavior.

4 Conclusion

The manuscript outlines a comprehensive framework for investigating climate-induced slope instability using centrifuge testing, delineating the planned experimental setup and methodologies. Although the physical tests remain pending, the document provides an intricate exposition of the proposed soil properties and experimental conditions essential for the study's execution.

Through meticulous soil characterization encompassing particle size distribution, compaction tests, hydraulic conductivity determination, and direct shear testing, the groundwork has been laid for forthcoming experiments. Furthermore, the derivation of the Soil Water Characteristic Curve (SWCC) based on the Van Genuchten formulation offers critical insights into soil–water interactions, setting the stage for in-depth slope stability analyses.

The structured construction of four homogeneous slopes, subjected to varying initial conditions and rainfall intensities, presents a systematic approach to evaluating slope behavior under climate-induced scenarios. By meticulously recording essential parameters such as rainfall duration, accumulated precipitation, and failure time, a comprehensive understanding of slope responses to diverse rainfall scenarios is sought.

While the actual experimental outcomes are yet to be realized, the methodologies delineated herein provide a robust foundation for future research endeavors in geotechnical engineering and slope stability analysis. The anticipated findings are poised to inform the development of proactive strategies aimed at mitigating slope instability risks in regions vulnerable to climate-induced hazards, thereby enhancing the resilience of infrastructure and land use planning practices.

References

Eab, K.H., Takahashi, A., Likitlersuang, S.: Centrifuge modelling of root-reinforced soil slope subjected to rainfall infiltration. Geotech. Lett. **4**(July–September), 211–216 (2014). https://doi.org/10.1680/geolett.14.00029

Fredlund, D.G., Xing, A., Huang, S.: Predicting the permeability function for unsaturated soils using the soil-water characteristic curve. Can. Geotech. J. **31** (1994)

GEO-SLOPE International Ltd.: GEO-SLOPE International Ltd. (2022)

Jayakody, S.H.S., Uzuoka, R., Ueda, K.: Centrifuge modelling of unsaturated slopes subjected to the integrated effect of groundwater and rainfall infiltration. E3S Web Conf. 382 (2023). https://doi.org/10.1051/e3sconf/202338210003

Leroueil, S.: Natural slopes and cuts: movement and failure mechanisms. Géotechnique **51**(3), 197–243 (2001)

Ling, H., Ling, H. I., Asce, M.: Centrifuge Model Simulations of Rainfall-Induced Slope Instability (2012). https://doi.org/10.1061/(ASCE)GT.1943

Ling, H.I., Wu, M.-H., Leshchinsky, D., Leshchinsky, B.: Centrifuge Modeling of Slope Instability (2009). https://doi.org/10.1061/ASCEGT.1943-5606.0000024

Matziaris, V., Marshall, A. M., Heron, C.M., & Yu, H.S.: Centrifuge model study of thresholds for rainfall-induced landslides in sandy slopes. IOP Conf. Ser. Earth Environ. Sci. **26**(1) (2015). https://doi.org/10.1088/1755-1315/26/1/012032

Schofield, A.N.: Use of Centrifugal Model Testing to Assess Slope Stability (1978)

Tamate, S., Suemasa, N., Katada, T.: Simulation of precipitation on centrifuge models of slopes. Int. J. Phys. Model. Geotech. **12**(3), 89–101 (2012). https://doi.org/10.1680/ijpmg.11.00009

Van Genuchten, M.T.: A closed-form equation for predicting the hydraulic conductivity of unsaturated soils. Soil Sci. Soc. Am. J. **44**(5), 892–898 (1980)

Wang, S., Idinger, G., Wu, W.: Centrifuge modelling of rainfall-induced slope failure in variably saturated soil. Acta Geotech. **16**(9), 2899–2916 (2021). https://doi.org/10.1007/s11440-021-01169-x

European Union Strategies for Climate Change Mitigation and the Development of Smart Cities

Debora Scala and Ángel Ignacio Aguilar Cuesta

Abstract

This study examines the European Union's (EU) strategies to mitigate climate change and develop smart cities, with a particular focus on the integration of environmental policies into sustainable urban environments. This study examines the relationship between climate change and the smart city model in the European context, emphasising the crucial need for cities to implement preventive strategies to reduce harmful emissions and mitigate climate change, with the objective of enhancing community resilience. The strategic vision of the smart city initiative is designed to provide an effective response to the complexities posed by such environmental issues. The EU provides support for smart city ecosystems through the implementation of innovative funding and policy initiatives, with the objective of achieving long-term sustainability and an enhanced quality of life. Notable initiatives include Horizon Europe, a research and innovation funding programme, and the Smart City Marketplace, which fosters collaboration between cities, industry, and researchers. The European Commission's European Mission of Smart and Climate-Neutral Cities (2021–2027) has the objective of guiding 100 EU cities towards climate neutrality by 2030, in alignment with the United Nations' Sustainable Development Goals, Agenda 2030. The Turin 2030 Action Plan serves as a case study, exemplifying the promotion of sustainable and resilient urban development through the implementation of citizen participation, cultural promotion, and strategies for the establishment of a more accessible, circular, and green city model.

Keywords

Climate mitigation · Environmental policy · Smart cities · Urban development · Urban sustainability

1 Introduction

The implementation of effective strategies to address climate change will necessitate significant and far-reaching transformations within urban areas. The rapid growth of urban populations, coupled with a range of challenges including traffic congestion, deterioration in air quality, outdated infrastructure, and rising energy demands, has positioned cities as central actors in efforts to combat climate change.

As reported by the United Nations Population Fund, a significant milestone was reached in 2008 when over half of the global population, or 3.3 billion individuals, were residing in urban areas. Projections indicate that this proportion will increase to 85% by 2050 (Fig. 1).

In Europe, 75% of the population already resides in urban areas, which presents considerable challenges in regard to environmental and social sustainability (OECD 2012). The significance of urban areas as a global phenomenon is further evidenced by the proliferation of megacities with over 20 million inhabitants in Asia, Latin America, and Africa (UN 2008).

As indicated by the United Nations, the majority of global resource consumption occurs in urban areas, underscoring both their economic significance and their detrimental environmental impact. Urban regions are responsible for approximately 60–80% of worldwide energy consumption and contribute considerably to greenhouse gas emissions (UN 2008). Consequently, it is imperative that cities implement long-term sustainability-oriented strategies to effectively address evolving circumstances and the necessity for regeneration (Bulkeley and Betsill 2005).

D. Scala (✉)
University of Castilla-La Mancha, Ciudad Real, Spain
e-mail: debora.scala@alu.uclm.es

Á. I. A. Cuesta
Autonomous University of Madrid, Madrid, Spain
e-mail: angel.aguilar@uam.es

P. Mohammad et al. (eds.), *Geographic Approaches to Climate Change and Mitigation: Urban and Rural Perspectives (Volume 1)*,
Advances in Science, Technology & Innovation, https://doi.org/10.1007/978-3-031-92119-3_17

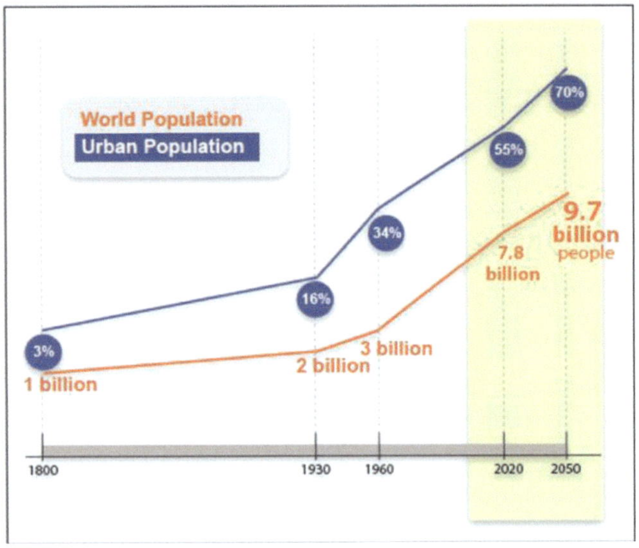

Fig. 1 World population (past–present–future projections). *Source* Food and Agriculture Organization of the United Nations—FAO. *FAO Framework for the Urban Food Agenda*; FAO: Rome, Italy, 2019 adapted by Schiavo, Fabienne T., and Cláudio F. de Magalhães (2022)

The role of urban areas in both the generation and reduction of global warming effects represents a significant challenge for our planet. In the context of Europe, which is characterised by high levels of urbanisation and cities that act as hubs of economic and social life, the commitment to combating climate change has become imperative. Concurrently, technological advancements have ushered in the era of smart cities, which, due to digital innovation, are redefining the concept of urban sustainability by increasing efficiency and improving the quality of urban living.

Over the past two decades, the concept of the smart city has become increasingly prominent within the academic field and in international policy programmes (Albino et al. 2015). In order to gain a full understanding of this concept, it is essential to comprehend why urban environments are regarded as indispensable elements of future advancement. It is evident that urban areas occupy a pivotal role in global social and economic spheres, exerting a substantial influence on the ecosystem (Bibri et al. 2020). Consequently, current discussions and policies in Europe consistently focus on the crucial role of sustainability in urban design and development, in response to the significant challenges posed by the rapid expansion of urbanisation and the unsustainability of existing urban configurations (Bibri and Krogstie 2017).

The objective of this study is to delineate the trajectory that the European Union has pursued in addressing the pivotal challenge of climate change, with a comprehensive examination of the nexus between environmental policies and the evolution of smart cities. Such cities may be regarded as pioneering and proactive projects, situated at the vanguard of efforts to mitigate the effects of climate change and offering innovative solutions for a more sustainable future.

This article examines the interrelationship between the concept of a smart city and climate change mitigation, taking into account factors such as urbanisation, environmental pressures, and technological innovation. The initial section presents the theoretical underpinnings of sustainable development and smart cities, delineating a historical context. Subsequently, the role of the European Union in the development of these smart cities through significant initiatives and development programmes is discussed. The third part of the text examines the specific efforts of some European cities, namely Rotterdam, Lyon, Barcelona, and Turin, which is a case study of the 100 Mission City project. Finally, the specific initiatives and programmes undertaken by Turin are analysed in detail.

2 The Theoretical Framework

2.1 Towards Sustainable Urban Development

In the light of the numerous international reports that have focused attention on urban development over the years, it can be argued that policies and actions worldwide in various countries (Fig. 2) are oriented towards sustainable urban development (UN 2020).

The concept of sustainable development can be traced back to the seminal "Limits to Growth" report, which was drafted by the Club of Rome in 1972. Over a decade later, the UN issued a critique of nations that, in pursuit of maximum economic and social welfare, had disregarded the indiscriminate exploitation of limited natural resources and the environment. Furthermore, the report drew attention to the disparity in development between countries in the global north and those in the global south. Whilst the report did not propose specific solutions, it conveyed a compelling message that established an intergenerational compact, defining sustainable development as development that meets the needs of the present without compromising the ability of future generations to meet their own needs (UN 1987).

At the Earth Summit, held in Rio de Janeiro in 1992, representatives from 179 countries signed the Agenda 21 document, which was regarded as the most ambitious plan to promote a new approach to global development. The designation "Agenda 21" was selected to signify the

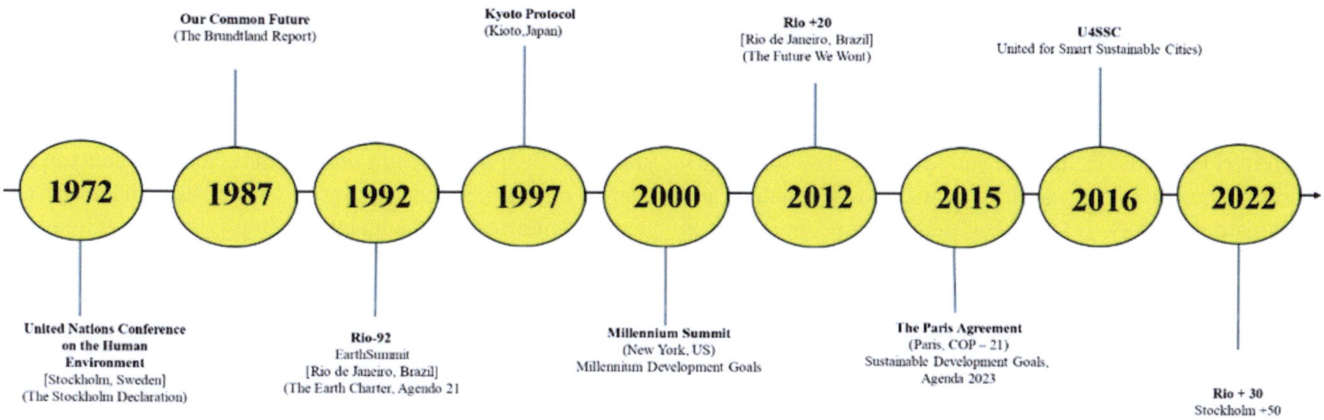

Fig. 2 World summits and documents concerning the future we desire. *Source* United Nations Environment Programme (UNEP). Environmental Moments: A UN75 Timeline. Adapted by Schiavo, Fabienne T., and Cláudio F. de Magalhães (2022)

pledge towards a sustainable development paradigm for the twenty-first century. This model was devised as a planning instrument with the objective of establishing sustainable communities in diverse geographical settings. The concept is structured around three fundamental tenets: environmental conservation, social equity, and economic efficiency (UN 2020).

Subsequently, at the United Nations Millennium Summit held in New York in 2000, representatives from 191 countries signed an agreement with the objective of establishing a future based on peace, justice, and sustainability. The agreement comprised eight principal objectives, which were further divided into 18 specific targets. The success of the agreement was to be measured through 48 indicators (UN 2020).

The Sustainable Development Goals (SDGs) superseded the previous Millennium Development Goals (MDGs), with the advent of a novel global agenda devised to facilitate sustainable development. The Sustainable Development Goals (SDGs) were formally introduced in 2015 at the United Nations Summit. The SDGs comprise 17 goals and are intended to provide a framework for global policies until 2030. The overarching objective of the SDGs is to eradicate poverty, safeguard the environment, and promote well-being and stability for all. SDG 11 is particularly concerned with the creation of more inclusive, safe and sustainable cities and communities.

In order to facilitate the implementation of the 2023 Agenda and Sustainable Development Goal 11, the United Nations initiated the "United for Smart Sustainable Cities" (U4SSC) initiative in 2016. This initiative advocates the utilisation of digital technologies with the objective of enhancing the sustainability and innovation of urban centres (ITU 2023).

2.2 Smart City

The transformation of urban environments into sustainable and intelligent ones represents one of the most significant challenges of our era, both in terms of the scale of the task and the urgency with which it must be addressed. Cities across the globe are confronted with the challenge of identifying and implementing novel strategies to mitigate the issues and capitalise on the opportunities presented by the demographic growth of metropolitan areas. This necessity arises from the imperative to confront pivotal issues such as pollution, congestion, unemployment, and socio-economic disparities (Anthopoulos 2017; Dameri 2013; Lee et al. 2014; Nam and Pardo 2011). In response to these issues, urban planners are orchestrating and promoting socially innovative initiatives (Edwards-Schachter et al. 2012) through the implementation of the advanced concept of a smart city. This model represents an evolutionary paradigm for urban development, characterised by an interdisciplinary approach and multifunctional objectives (Dameri 2013; Monfaredzadeh and Berardi 2015; Stratigea et al. 2015).

It is important to note that the concept of a smart city is subject to constant evolution and does not adhere to a singular, standardised definition. The definition of a smart city is dependent on the source of reference and the entity conducting the study or evaluation. Moreover, the concept of a smart city is in a state of constant evolution and is characterised by a broad scope, encompassing a multitude of interpretations. In the academic literature, a variety of descriptions of smart cities can be found, each highlighting different aspects, ranging from socio-economic to technological elements. In their study, Caragliu et al. (2011) cite certain characteristics derived from Hollands (2008), who adopted an economic approach to identify them. In the present context, the concept of a "smart city" is defined as "an urban environment that applies technology to enhance the benefits and reduce the deficiencies of urbanisation for its citizens" (IMD 2023).

The term is defined as the utilisation of infrastructure, innovation, and technology, in addition to an intelligent, active, and participatory society. It is evident that the conceptualisation of a smart city cannot be achieved without the input of the individuals who reside and work within it.

To gain a deeper insight into the concept of a smart city, we may turn to the comprehensive definition put forth by Caragliu. In his view, a city can be regarded as smart when "investments in human and social capital and in traditional and modern infrastructure promote sustainable economic growth and a high quality of life, with wise management of natural resources, through participatory governance". The scholar employs a holistic methodology, emphasising the significance of an integrated perspective that encompasses the multifaceted dimensions of urbanity. Nam and Pardo (2011) categorised and optimised the fundamental elements of the smart city concept into three principal categories, which represent:

- technology
- individuals
- institutions.

Nevertheless, an additional concept, namely sustainability, serves to enhance the comprehensive perspective of the smart city. This has become a pivotal topic in contemporary discourse, with a focus on addressing social and environmental concerns. In recent decades, the conservation of non-renewable resources has emerged as a priority on the international political and economic agendas. The concept of the sustainable city emerged concurrently with the recognition of cities as the primary agents of sustainable development, acting as both users and suppliers of products and services (Batagan 2011).

The integration of sustainability within the framework of smart cities is a topic that has been extensively researched and discussed in numerous articles and books. Prominent examples include works by Ahvenniemi et al. (2017), Araya, Dameri. Kramers et al. (2014) observed that certain definitions of a "smart city" incorporate specific sustainability objectives. The concept of sustainable smart cities (SSC) is the subject of study and is considered an aggregate concept. In order for an entity to be classified as a sustainable smart city, three fundamental elements must be present: the concept of the "City, Smart, and Sustainable" (Kramers et al. 2014) is an aggregate concept that encompasses the design of SSC to address urban challenges. The overarching objective of these SSC is to enhance the quality of life for citizens through a harmonious equilibrium between the natural environment, social structures, and economic prosperity (Höjer and Wangel 2014).

The definition put forth by Caragliu et al. (2009) provides a comprehensive overview of the factors and objectives that underpin the phenomenon of the smart city. The concept of a "smart city" is defined as a city that has invested in human and social capital, coupled with the development of traditional and modern communication infrastructures, in order to facilitate sustainable economic growth and a high quality of life, whilst ensuring the prudent management of natural resources and the implementation of participatory governance. This definition synthesises a range of concepts, emphasising technology, human resources, governance, and sustainability as pivotal elements in the creation of a smart city.

2.2.1 The Six Characteristics of a Smart City

In order to gain a deeper understanding of the holistic vision of smart cities, an approach that is widely recognised by both the academic community and European policy frameworks, it is essential to examine the characteristics of a smart city as outlined by Giffinger. Giffinger et al. have identified six characteristics that are characteristic of a smart city (see Table 1). These attributes identify the areas of urban life where the smart city model can be applied to enhance existing infrastructures by leveraging data and ICT to provide efficient services to citizens. The table below outlines the various areas of application of the characteristics of smart cities in urban environments and in addressing challenges.

Table 1 Characteristics of smart cities

Smart economy	*Smart people*
Entrepreneurship Flexibility Internationally integrated Productivity Transformative	Affinity for lifelong learning Creativity Levels of educational achievement Participation in public life Social and ethnic plurality
Smart governance	*Smart mobility*
Participatory decision-making Provision of public and social services	Availability of ICT Infrastructure Innovative and safe transportation system Local and international accessibility
Smart environment	*Smart living*
Environmental protection Levels of pollution Sustainable resource management	Educational facilities Health facilities Housing quality Individual security Provision of cultural facilities Social cohesion Tourist attractiveness

Source Giffinger et al. (2007)

This approach is used to classify cities in a more straightforward manner. It is not necessary for a city to encompass all six dimensions in order for it to be considered intelligent; it is sufficient for it to have one of them, to a greater or lesser extent.

In the context of climate change mitigation, two particularly salient dimensions emerge. The initial dimension is that of the "Smart Environment", which is subdivided into four principal aspects: (1) the appeal of the natural environment, (2) pollution control, (3) environmental protection, and (4) sustainable resource management. The indicators employed for these aspects include hours of sunlight, the percentage of green spaces, the level of atmospheric pollution, efforts for environmental conservation, and the efficient use of resources such as water and energy. The aforementioned smart city indicators frequently coincide with climate change mitigation measures. Indeed, they are conducive to improvements in air quality, green infrastructure, water and waste management, coastal protection, and energy efficiency (García-Fernández and Peek 2020).

The second dimension is "smart mobility", which benefits from the use of information technologies to make traffic flows more efficient. This results in a reduction in the time spent by motorists searching for parking and the promotion of the development of suitable infrastructure for electric vehicles. The primary focus of smart mobility is the reduction in greenhouse gas emissions, thus aligning with climate mitigation strategies. In contrast, the smart environment adopts a broader approach, focused on adaptation to climate change.

3 Evolution of the European Union's Strategies for Smart Cities

The increasing global urbanisation, coupled with the interrelated issues of global warming, water shortage, ecological deterioration, economic transformation, and social marginalisation affecting the entire planet, necessitates a comprehensive examination of the future of our cities in Europe.

It is estimated that urban areas in Europe, which house over two-thirds of the European economy, play a pivotal role in fostering creativity and innovation throughout the Union (EC). Consequently, urban development constitutes a central element of EU regional policy.

Since the early 1990s, the European Union has positioned itself as a key player in environmental protection, adopting an increasingly proactive approach. This commitment has manifested in the form of a series of incisive and targeted measures, the principal objective of which is to integrate the principles of sustainable development into European environmental policy. These initiatives have had tangible effects, particularly in the context of urban centres, and are not merely theoretical or detached from everyday reality.

The European Union's approach to urban environmental challenges is well documented and can be traced through a series of strategic documents and international meetings. Amongst these, the 1990 Green Paper represents one of the EU's inaugural efforts to confront urban environmental challenges. Although the term "smart city" had not yet entered the lexicon, the document established the foundations for a more sustainable approach to urban development and initiated a comprehensive discourse on the means by which cities could become more sustainable. In the 1990s, the EU acknowledged the necessity for a more precise and unified approach to urban and environmental concerns. This resulted in the formation of an Expert Group on the Urban Environment, which was tasked with examining how environmental objectives could be integrated into future urban and spatial planning strategies. Additionally, the Expert Group was charged with assisting the European Commission in developing an urban environmental dimension within EU environmental policy. Concurrently, the Treaty on the European Union (1992, revised in 2002) incorporated environmental considerations into its fundamental legislation, acknowledging the significance of balanced and sustainable development. Of international importance was the United Nations World Summit on the Environment in Rio in 1992, which marked a significant shift in awareness regarding the sustainability of territories and of cities. This new orientation led to the signing of the Aalborg Charter in Denmark in 1994, which defines a sustainable city as one that effectively integrates the three pillars of sustainable development—social, environmental, and economic—with the aim of enhancing the quality of life.

Subsequently, a series of influential United Nations Habitat II and III conferences (1996, 2006) emphasised the urgency of addressing urban sustainability. These meetings provided important platforms for dialogue and international cooperation, culminating in a series of recommendations calling for immediate and strategic action for the future of cities. These recommendations highlighted the crucial role of cities not only at the local level but also in a global perspective.

This was followed by the adoption of planning instruments, such as the European Union's Urban Environment Strategy (2006), and important documents, such as the Leipzig Charter (2007), which sought to reach a consensus amongst ministers responsible for the objective was to advance urban development and territorial cohesion, as well as the development of sustainable cities in Europe. This was pursued by seeking to enhance policies related to sustainable urban development, with a particular focus on addressing climate and environmental changes. This resulted in the publication of a new Leipzig Charter in 2020, which updated the previous version. The Leipzig Charter, adopted in 2007 and updated in 2020, serves as a foundational reference point for the definition of European urban policies. These documents, in conjunction with the increasing focus on the development of smart cities, provide a framework for European cities, offering guidance on innovative strategies to address the challenges of the present and the future. The initial 2007 Charter advocated an integrated methodology to urban planning, emphasising sustainability, urban density, and quality of life. It provided a model for the development of cities capable of responding effectively to environmental, social, and economic challenges in a manner that is sustainable, dense, and conducive to a high quality of life. The 2020 update extended this vision by introducing the concepts of urban resilience and adaptation to climate change, reflecting a more holistic approach that considers the physical environment together with social and economic resilience. Concurrently, the European Union has endorsed the advancement of smart cities, advocating the deployment of sophisticated technologies to enhance urban efficacy and quality of life. This methodology, which emphasises the utilisation of data and technology to optimise urban services, mobility, environmental impact reduction, and active citizen participation, is wholly consistent with the tenets of the Leipzig Charter. The smart city initiative prioritises the use of data and technology to optimise urban services, mobility, environmental impact reduction, and active citizen participation to provide a better quality of life and make cities a more pleasant place to live. The Leipzig Charter, in both its original and updated versions, provides a solid theoretical and practical basis for sustainable urban planning, which integrates harmoniously with the EU's vision of smart cities. This integration guides European cities towards a more efficient and resilient future that meets the needs of the twenty-first century.

The European Union's vision of urban services and technology has evolved to encompass a more expansive perspective, underscoring the potential for transformative change through the development of smart cities. Such solutions not only address the ever-evolving urban challenges but are also perceived as catalysts for significant societal change. In May 2016, a significant turning point was reached in the field of urban development when the European Union ministers responsible for urban affairs reached an agreement to adopt an Urban Agenda for European Member States. This significant document seeks to advance an innovative model of urban development, thereby facilitating the implementation of two major global initiatives: the United Nations 2030 Agenda for Sustainable Development and the New Urban Agenda outlined by the Habitat III Secretariat. These initiatives have a common core theme, namely the pivotal role of sustainable urban development in fostering economic prosperity, social and territorial cohesion, and global well-being. A fundamental aspect of this approach is the acknowledgement of the development of smart cities as a crucial element, where the implementation of cutting-edge technologies can enhance urban safety, inclusivity, resilience, and sustainability. In 2019, the European Union adopted the Green Deal, a comprehensive and ambitious integrated programme of reforms, investments, and research. The Green Deal represents the most ambitious package of measures ever conceived by the European Union. It aims to make Europe the first climate-neutral continent by 2050, radically transforming the economy and society in a greener and more sustainable direction.

The European Green Deal, as set forth by the European Commission, signifies a significant qualitative advancement in the global effort to combat climate change and environmental degradation. The Green Deal is not merely a set of environmental directives; it is a comprehensive strategy that integrates environmental transition with economic innovation and social justice. This ensures that the green transition is equitable and inclusive for all citizens and regions of Europe. The Green Deal has consolidated the EU's global environmental leadership by establishing rigorous standards for reducing greenhouse gas emissions, promoting renewable energy, enhancing energy efficiency, and recycling materials. The package of legislation and initiatives has twofold objectives: firstly, to mitigate the effects of climate change; and secondly, to promote a sustainable and resilient economy. This will be achieved by stimulating green investment, technological innovation, and job creation.

The advancement of urban development in accordance with the smart city model constitutes a pivotal element in the pursuit of sustainable and innovative urban growth. Cognizant of the pivotal role that smart cities will play in the future of Europe, the European Union has initiated a series of initiatives and programmes aimed at fostering the adoption of technologies in urban settings as part of its broader mission towards urban sustainability.

4 EU Programmes and Funding

The policies adopted by the European institutions over the last two decades demonstrate a progression in environmental thinking (Circle of Diplomatic Studies 2020). It is increasingly acknowledged that technology and innovation can offer efficacious solutions to environmental concerns and the challenge of climate change. The concepts of sustainability and quality of life are at the core of contemporary urban development agendas, with international initiatives such as those of the European Union and the United Nations exerting considerable influence in this domain. The concept of a smart city is closely aligned with these objectives, which include the efficient utilisation of resources, the enhancement of urban living standards, and the establishment of a sustainable urban environment, as well as the mitigation of climate change.

As urban populations continue to expand, the need for innovative solutions to address challenges such as pollution, traffic congestion, and resource management is becoming increasingly apparent. It is in this context that the European

Union has identified the importance of providing support for the development of smart cities. It is evident that programmes such as Horizon 2020, Horizon Europe, and the Digital Europe Programme play a pivotal role in this initiative, providing much-needed funding and support to facilitate the sustainable and innovative development of European cities. In addition to fostering innovation, these programmes are structured to be participatory, equitable, and inclusive. As articulated in the European Green Deal, the objective is to "protect, conserve, and enhance the EU's natural capital and safeguard the health and well-being of citizens" (European Commission 2019a, b, p. 2), thereby ensuring that the environmental transition is equitable and inclusive.

A concrete example of this approach is Horizon Europe, the European Union's research and development funding programme, which plays a crucial role in supporting sustainable research and development projects. With an allocated budget of approximately €95.5 billion, Horizon Europe represents a pivotal instrument for advancing urban sustainability. The funds are allocated to projects that advance the scientific understanding of climate change, develop solutions to reduce greenhouse gas emissions, and facilitate adaptation to climate change. Horizon Europe plays a pivotal role in effecting a transformation of the built environment and land use, promoting a regenerative approach to design. The Horizon Europe programme is structured around three principal pillars. The three pillars of the programme are "Excellent Science", "Global Challenges and European Industrial Competitiveness", and "Innovative Europe". These pillars are supported by a programme designed to enhance the research and innovation capabilities of the Member States. With a substantial budget, Horizon Europe has the potential to generate substantial economic, social, and scientific benefits.

The programme aims to stimulate technological innovation, develop new green technologies, and create resilient urban infrastructures. A key and ambitious goal is to make the EU the first continent to achieve climate neutrality by 2050, with a strong emphasis on climate targets (Asvis 2020).

In comparison with the preceding Horizon 2020 programme (2014–2020), Horizon Europe (2021–2027) introduces an innovative approach based on missions to address urgent societal challenges. On 29 September 2021, the European Commission (2021) initiated five missions under the Horizon Europe 2021–2027 programme. This reorientation is intended to achieve a greater social impact by focusing on the mission and on citizen engagement. As outlined in the Mazzucato Report (European Commission, Directorate-General for Research and Innovation, Mazzucato 2019), the EU missions are devised with the objective of delivering tangible solutions to pivotal challenges, with ambitious goals and concrete results by 2030. This is to be achieved by prioritising research and innovation, along with new forms of governance and collaboration.

Of particular interest is the Mission for Smart and Climate-Neutral Cities. This mission acknowledges the pressing need to address climate change at the urban level and the pivotal role of cities in achieving net-zero emissions through the utilisation of smart data and technologies. The objective is to achieve climate neutrality in at least 100 European cities by 2030, with the aspiration that these solutions will serve as models applicable to all European cities by 2050.

The EU's strategy for smart cities and climate neutrality, as embodied in Horizon Europe, represents a significant advancement in the global effort to combat climate change. These initiatives seek to transform European cities into sustainable, efficient, and equitable environments through an innovative and inclusive approach, in alignment with the ambitious goals set forth in the Green Deal.

A recent study by Mazzucato defined five criteria for missions: they should be bold, inspiring, and of broad social relevance; they should have a clear direction; they should envisage ambitious but realistic actions; they should be multidisciplinary and cross-sectoral; and they should promote bottom-up solutions. As the authors themselves have emphasised, the success of these missions is contingent upon public engagement. It is imperative that the missions are perceived as solutions to challenges of great societal importance, thereby ensuring their longevity and success beyond political cycles. Furthermore, it is crucial that citizens understand the direct benefits of European initiatives.

5 The European Urban Agenda and the Italian Urban Agenda

Since the end of the last century, Europe has been engaged in a concerted effort to enhance the quality of life in urban areas. This has entailed a shift away from traditional sector-specific policies towards a more integrated and holistic approach to urban development. At the core of these endeavours lies the Urban Agenda for the European Union, which was inaugurated in 2016 with the Amsterdam Pact. This represents a substantial stride towards an integrated and coordinated approach to urban development.

The document "Urban Agenda for the EU" represents the culmination of a significant evolution in European urban policy, which has developed through a series of key milestones, including the Lille Programme of 2000, the Leipzig Charter of 2007, and the Riga Declaration of 2015.

The Amsterdam Pact establishes the European Urban Agenda, which is based on three essential pillars for cities: optimising funding, enriching knowledge, and refining regulation. These tenets are operationalised through twelve thematic areas, which encompass a range of urban challenges and opportunities. These include the integration of migrants and refugees, air quality, the fight against urban poverty, housing policies, and the formation of partnerships between European cities, their networks, urban programmes such as Urbact and UIA, and other relevant stakeholders to address the circular economy, climate adaptation, and the digital transition.

These are therefore epochal challenges that align the European and international urban agendas. Concurrently, the endorsement of the Amsterdam Covenant occurs within a global context wherein the United Nations (UN) itself adopted its own Urban Agenda at the Habitat III Conference in Quito in 2016. This agenda, which is aligned with the UN's broader 2030 Agenda for Sustainable Development, emphasises the critical importance of cities in achieving the 17 Sustainable Development Goals. Goal 11, in particular, focuses on transforming cities into inclusive, safe, resilient, and sustainable places. It is an ambitious goal that includes promoting adequate and accessible housing, inclusive and sustainable urbanisation, safe and green transport systems, reducing the environmental impact of cities, and adapting to climate change.

6 Mission City: Climate-Neutral and Smart Cities

The European Union has set ambitious targets for reducing greenhouse gas emissions. The European Green Deal sets a target of a 55% reduction by 2030 and the ultimate goal of achieving climate neutrality, or zero CO_2 emissions, by 2050 (European Council 2019). In order to accelerate progress, the EU has initiated the "100 Climate-Neutral and Smart Cities by 2030—for and by the citizens" mission, also referred to as "Mission Cities", as part of the broader Climate-Neutral and Smart Cities Mission (European Commission 2020). The objective of the Horizon Europe programme is to select and promote 100 examples of EU cities that aspire to become carbon neutral by 2030, thereby establishing themselves as centres of innovation and role models. To achieve this ambitious goal, it is necessary to implement not only ecological but also cultural transformations, which will require the involvement of a range of stakeholders, including public authorities, businesses, researchers and citizens (Rathnayaka et al. 2011).

A total of 377 cities participated in the 2021 call, including 40 Italian cities. Amongst these, Turin was selected together with Bologna, Bergamo, Florence, Milan, Padua, Parma, Prato, and Rome.

In order to become designated as a "Mission City", a city must demonstrate its compliance with the "EU's climate change mitigation and adaptation goals, as well as national and/or local objectives concerning energy, air quality, and climate".

Each city has developed its own smart strategies in a variety of ways, including the creation of lists of initiatives and the production of official reports. Some cities are rapidly advancing in the development of smart city projects, exceeding the policies that govern their development. The definition of a universal framework for smart cities is complicated due to the unique characteristics of each city, which require customisation based on local priorities and visions. Amongst the exemplary cities in this field are Rotterdam, Lyon, and Barcelona, selected amongst the 100 Mission Cities precisely for their innovative approach to climate change mitigation.

Despite the Dutch long-standing fight against sea level rise, the city of Rotterdam is confronted with the challenges of heat waves, droughts, and intense storm events. In order to address this latter phenomenon, the Dutch city has created an innovative water plaza, Bentem Square, in the ZOHO district, which is becoming Rotterdam's inaugural climate-proof district. The square is equipped with a rainfall prediction system that protects the city from heavy rains and the overflow of the sewer system, whilst simultaneously collecting water for future use and serving as a sports and recreational area (García-Fernández and Peek 2020; Elsner et al. 2019). In the ZOHO district, further eco-friendly solutions are evident, including a rainwater collection system in large barrels that facilitates water reuse. Furthermore, the "Polderroof" is a green roof constructed over a parking lot that serves to mitigate the heat island effect whilst also facilitating the efficient management of excess rainwater from nearby buildings.

In order to enhance flood control, the city of Rotterdam has integrated information and communication technologies (ICT) into its systems. This includes the utilisation of sophisticated forecasting techniques, the deployment of sensors to monitor the real-time conditions of critical infrastructure such as dams, and the utilisation of satellite imagery. A particularly innovative element is the "Flood Control 2015" programme, which aims to facilitate more expedient and precise decision-making processes. The programme also incorporates a novel, engaging element in the form of a flood management simulation game, designed to educate and prepare the next generation to safeguard the city from the impact of rising sea levels.

The city of Lyon is pursuing a comprehensive strategy to adapt to the challenges posed by climate change. Recent forecasts indicate that the city will be particularly affected by global warming. In response, the city government is implementing a series of preventive measures, including the increase of vegetation and the deployment of green technologies. A notable illustration of this initiative is the Lyon Confluence project. The project has already seen the installation of 7,000 m^2 of solar panels, with the long-term objective of reaching 20,000 m^2. This intervention has facilitated the establishment of a positive energy urban district, comprising residential, commercial, and retail premises that harness solar energy, a thermal power plant operating on vegetable oil, and a cooling system utilising water from the Saône River. The buildings are designed to consume 50–60% less energy than would be the case under conventional standards and are equipped with autonomous energy production systems. Furthermore, 275 housing units have been upgraded through eco-renovation, including insulation and modernisation of heating systems.

Since 2020, the city has been engaged in the redevelopment of urban spaces for youth through the "school streets" initiative. This initiative has the objective of making the surroundings of schools safer, greener, and more welcoming. This initiative, also known as "streets for youth", has already resulted in over 60 interventions in primary and secondary schools. These include participatory design with students, traffic calming measures, the introduction of green areas, and the conversion of urban spaces for pedestrian use. A total of 22 sites have been fully pedestrianised, thereby introducing urban greenery to counteract the urban heat island effect. In other instances, these spaces have been transformed into shared zones utilising traffic calming measures, including a 20 km/h speed limit and the creation of one-way streets. A further noteworthy urban transformation is currently taking place in the Danton district, a commercial area that has historically been oriented towards automobile use. The project has been inspired by the "superblocks" concept from Barcelona, resulting in the implementation of green solutions and traffic calming measures, including a 20 km/h speed limit, with the objective of reducing crossing traffic. The process commenced with the establishment of a school street on Rue Léon Jouhaux, which was subsequently transformed into a shared space designed for pedestrians and cyclists. New trees have been planted and existing gardens expanded with the objective of enhancing the environment for residents and increasing its resilience to climate change. This has been achieved by utilising permeable surfaces and vegetation to mitigate the effects of heat islands. In a similar vein, Boulevard Garibaldi, a historic eight-lane road dedicated to motor traffic, is undergoing a significant transformation in the vicinity of the Danton district.

Barcelona is dedicated to confronting long-term urban challenges, including climate change and the scarcity of natural resources. The city's objective is to foster a more collaborative and sustainable economy, with the aim of reducing social inequalities and maintaining its position as a leader in innovation. Whilst climate action is not explicitly categorised, the strategy focuses on four principal areas: governance, city services, the digital socio-economic fabric, and citizen engagement. The city's principal challenges include the management of urban flooding, the mitigation of heatwaves, the allocation of water resources, and the protection of coastal areas from erosion (Sinaeepourfard et al. 2016). In order to address these challenges, Barcelona has implemented a number of solutions, including the creation of a network of stormwater tanks that are controlled by ICT systems. This has resulted in a significant reduction in operational costs and the risk of flooding. Furthermore, the city has developed the Sentilo platform, an open-source software that collects data from a wide network of environmental sensors to monitor air quality, humidity, and temperature. Finally, Barcelona has installed Districlima, an innovative urban heating and cooling system powered by municipal waste incineration, which has been demonstrated to reduce electricity usage by 35% and emissions by 50% in comparison to conventional systems (March and Ribera-Fumaz 2016).

7 Turin as a Smart City

This paper will focus on the city of Turin, the capital of the Piedmont region in north-west Italy, as a case study in sustainable and intelligent urban planning. Turin was selected from a pool of 100 Mission Cities, a designation given to cities around the world that are engaged in innovative and forward-thinking urban development.

Since the 1970s, the city, like other cities with economies centred on a single company or industry, has had to navigate the turbulent waters of the crisis of Fordism. These changes have had a significant impact on FIAT, the city's primary economic driver, and consequently on the social structure of Turin. Despite the fact that the city has not yet completed the transition to a post-Fordist model and continues to have a strong link with industry, efforts have been made to diversify its economy. Initiatives have been implemented with the objective of promoting sectors such as information technology, higher education, cultural industries, and tourism. A pivotal moment in this metamorphosis was the hosting of the 2006 Winter Olympics, which was perceived as a potential catalyst for the growth of tourism and culture. Nevertheless, the initial optimism was confronted with challenges, including elevated local debt and the repercussions of the global economic crisis. This resulted in an uptick in unemployment, which climbed from 6.1% in 2004 to 9.8% in 2012, with a peak of 34% amongst the younger demographic.

7.1 The Process of Transforming Turin into a Smart City

The concept of the "smart city" has recently emerged as a pivotal topic in the discourse surrounding Turin's urban transformation. The economic crisis has prompted a shift in political priorities, with a greater focus on economic sustainability and urban development through innovative technologies, at the expense of environmental sustainability. However, as Crivello (2015) has observed, the adoption of this vision in Turin has often followed a more European political line and has been limited to renaming existing initiatives, which has led to questions being raised about their effectiveness and sustainability.

The case of Turin illustrates the challenges and opportunities that a traditional industrial city may encounter on its path towards a more intelligent and sustainable future. The city has participated in numerous national and international urban sustainability projects, driven by the opportunities for development and growth offered by the EU. Additionally, it has developed its own smart city programme.

In 2009, Turin took a significant step towards sustainability and urban intelligence. At that time, the Mayor of Turin, together with the leaders of 3,000 other European cities, joined the Covenant of Mayors in Brussels, an initiative supported by the European Commission. The goal of this Covenant was ambitious: to reduce CO_2 emissions by 20% by 2020, engaging cities in a collective movement towards sustainability.

In a continuation of this trajectory, on 13 September 2010, the City Council of Turin formally adopted the "Turin Action Plan for Energy (TAPE)". This plan represented an even more ambitious commitment by the city to reduce CO_2 emissions by 40%, which was twice the original target. The strategy adopted by Turin was not an isolated action, but rather part of a wider context of interventions and projects that had been underway since the 1990s. During this period, the city had already begun to respond to the challenges posed by the decline of the Fordist industrial model by moving towards a more sustainable and integrated urban approach (Crivello 2015).

The Turin Action Plan for Energy (TAPE) offered a comprehensive vision and a set of practices that had been developed over the years, representing a turning point in urban management. This plan was part of a larger project that required a profound change not only in policies and technologies, but also in the culture and mentality of the city and its inhabitants. It outlined a path towards a more responsible and environmentally conscious urban transformation.

The decision to join the Covenant of Mayors and to draft the TAPE document constituted a fundamental shift in Turin's approach to the concepts of sustainability and urban innovation. Nevertheless, it was through Turin's active involvement in various European Union calls for proposals that the city began to assume a pivotal role in the smart city landscape. In the summer of 2011, the city put forth three significant proposals: the "EeCoFit" project concentrated on the energy efficiency of buildings and infrastructure, whilst the "Energy Smart Cities" initiative was dedicated to the utilisation of innovative technologies for energy storage. The "E.Plan" project, on the other hand, was focused on the development of databases for sustainable urban planning in the energy sector.

These proposals demonstrated that the pursuit of urban intelligence in Turin was significantly shaped and directed by the directives and prospects presented at the institutional level by the European Union.

Despite the lack of initial success, with none of the projects presented in 2011 receiving funding from the EU, the city of Turin remained optimistic. In February 2013, the city reached a significant milestone by responding to a national call, receiving a substantial sum of 183 million euros to fund 12 projects under the national programme "Cities and Smart Communities and Social Innovation", promoted by the Ministry of Education, University and Research. This success not only brought tangible resources for the city's development, but also reinforced Turin's confidence in its capacity to obtain European funding in the future.

In the light of this promising outcome, Turin has reinvigorated its efforts to become a prominent hub for urban innovation and sustainability at the European level. To this end, the city has renewed its focus on Europe and is preparing to participate in new EU calls and tenders. This proactive approach exemplifies Turin's dedication to advancing its position as a leading European city in these two crucial domains.

8 Turin's European Projects

The financial support of the European Union has enabled Turin to implement a series of ambitious projects that have reinforced its status as a prominent city in Italy and Europe. This article will examine some of the most significant projects funded by the EU, focusing on those that have been completed and analysing their positive impact on the city and its inhabitants (Table 2).

The analysis of the table reveals that the majority of projects undertaken in the city of Turin are aligned with the Horizon 2020 programme. The most ambitious projects can be examined, followed by a description of the ongoing ones (City of Turin, https://www.torinoeuprojects.it/tutti-i-progetti/).

The WeGovNow project, which received funding from the European Union's Horizon 2020 research and innovation programme, sought to enhance and revitalise urban areas by actively engaging citizens in decision-making processes. The WeGovNow platform and other technologies were employed to facilitate citizen participation in the co-design of services and the implementation of public policies. In Turin, the project was implemented in two principal appli-

Table 2 Completed European projects

Project	European programme project	Duration	Allocated	Objectives
WeGovNow	Horizon 2020	2016–2019	223.847,00 €	Development and use of technological platforms for citizen engagement in decision-making processes concerning the improvement, rehabilitation, and regeneration of the areas where they live
ProGIreg	Horizon 2020	2018–2023	€ 896.500,00	Utilise nature for urban regeneration of post-industrial areas
Innovato-R	Urbact	2018–2021	€ 177.105,00	Improve the performance of public administration. Counteract the contraction of resources, both human and financial
5G-EVE	Horizon 2020	35 months	€ 136.825,00	Develop the spread of 5G, with the support of TIM Polytechnic of Turin
ToNite	Urban Innovative Actions. Urban Security	2020–2023	3.235.843,50	Enhance the liability and safety during evening hours in areas adjacent to the Dora River
PACTESUR	Fondo per la sicurezza interna dell'Unione europea Polizia	2019–2021 Extension 2022	€ 391.786,92	Strengthen the capacities of cities and local actors in protecting and securing urban public spaces
TInnGO	Horizon 2020	2019–2021	€ 80.125,00	Create a European Observatory for Gender Innovation in the Transport sector
SUITS	Horizon 2020	2016–2021	€ 129.000,00	Strengthen the capacities of local authorities and transport stakeholders and transfer learning to smaller-sized cities, making them more effective and resilient to change in the implementation of sustainable transport measures
ROCK	Horizon 2020	2017–2020	€ 417.286,81	Enable "replicator cities" to learn from "model cities" about the policies, processes, and actions that have enabled the regeneration of historic centres
MONICA	Horizon 2020	2017–2020	€ 374.500,00	Develop IoT technologies that help cities address challenges of sound, noise, and security during large outdoor cultural events in the city
MOLOC	INTERREG EUROPE	2017–2021	€ 264.000	Adapt its morphology, inherited from a process of urbanisation to current challenges related to climate change
CWC—CITY WATER CIRCLES	INTERREG CENTRAL EUROPE	2018–2021	€ 270.391,60	Reform urban water infrastructure systems by applying a circular economy approach
CO-CITY	UIA—Urban Innovative Actions	2017–2020	€ 4.125.891,44	Create project communities as a sustainable solution to break the vicious cycle of socio-spatial polarisation in disadvantaged neighbourhood
SUPER	REC	2021–2023	€ 239.671,44	Strengthen the role of local authorities in combating racism and xenophobia
VALUES	Fondo Asilo Migrazione e Integrazione (AMIF)	2019–2021	€ 24.229,08	Promote the exchange of knowledge, experiences, and best practices at the European level amongst cities and volunteer networks on the best ways to successfully integrate migrants
CO3	Horizon 2020	2019–2022	€ 156.875,00	Assess the benefits and risks of blockchain, augmented reality, geolocated social networks, within co-creation, co-production, and co-management processes of public services and common goods, and on creating a new model of interaction between public administration and citizens

(continued)

Table 2 (continued)

Project	European programme project	Duration	Allocated	Objectives
5G-TOURS	Horizon 2020	2019–2022	€ 348.250,00	Test and utilise 5G technologies to provide useful, efficient, and reliable services to citizens and tourists
YOUNG IN&UP	FAMI	2019–2022	€ 12.196,00	Stimulate reflection on good practices for the inclusion and reception of young people and minors from non-EU countries
HARMONY	Horizon 2020	2019–2023	€ 127.500,00	Develop a new generation of tools for planning multimodal transportation
STEVE	Horizon 2020	2017–2021	€ 220.006,25	Develop light electric vehicles to test new e-mobility services
GOEASY	Horizon 2020	2017–2021	€ 137.000,00	Build a new generation of mass-market-based location-based services (LBS) to engage, stimulate, and reward citizens for sustainable behaviours
SETA	Horizon 2020	2016–2019	€ 220.000,00	Create technological and methodological solutions to utilise such large amounts of data to change the way mobility is organised, monitored, and planned in metropolitan areas
CESBA MED	MED	2016–2019	€ 366.605,00	Develop innovative tools for assessment and policy-making support, to implement more effective energy efficiency plans that consider public buildings in the context of the urban districts where they are located
SOLEZ	CENTRAL EUROPE	2016–2019	€ 1.933.734,00	Find solutions for low-carbon mobility
CO4CITIES	UIA-URBACT	2017–2021	€ 206,360.00	Adapt and transfer the methodological structure of the UIA CO-CITY project to three other European cities
ASIS	Horizon 2020	2019–2021	€ 97.756,04	Initiate, develop, and promote a new vision of innovation in the Alpine space to increase its capacity for innovation to new needs, economic, and social challenges faced by the Alpine regions
SEED	EaSI	2021–2023	€ 900.000,00	Create four Competence Centres for Social Innovation
APPRAISE	Horizon 2020	2021–2023	265.000,00 €	New security solutions that will enhance the safety of cities and public spaces in the EU
Pop-AI	Horizon 2020	2021- 2023	€ 98.500,00	Promote trust in AI for the security sector through increased awareness and ongoing social engagement
RESPONDET	COSME-SEM	2022–2023	€ 18.992,00	Promote local and regional development to strengthen the social economy by making common resource management the main element of social sustainability in the Green transition process
HANDSHAKE	Horizon 2020	2018–2022	232.188,00 €	Designed to make cycling a more appealing daily means of transportation

cation scenarios. The initial application scenario involved utilising the WeGovNow platform for electronic voting with a considerable number of participants, thereby demonstrating the efficacy and dependability of the system on a large scale. The second case study involved 73 associations in a co-design process for Parco Dora, an area of approximately 500 m^2. This process aimed to facilitate dialogue between different local communities, including territorial stakeholders, residents, citizens' committees, sports organisations, associations, and park users. The co-design process, overseen by the Civic Design Table, involved the Departments of Environment, Urban Regeneration, Sport and Safety, and Social Policies, in addition to the Polytechnic, the University of Turin, and the Fondazione Brodolini. Almost 50 citizens participated in each workshop, with over 100 new users registering on the platform for this particular use case. At the conclusion of the participatory process, seven initiatives were uploaded to the platform, supported, and voted for by approximately 100 citizens. The winning initiative reflects the desire of the sports associations residing in the area to establish a hub for street sports.

The ProGIreg project, which was funded by the European Commission under the Horizon 2020 programme, sought to utilise nature-based solutions (NBS) for urban regeneration in post-industrial areas in Europe and China. The project encompassed four pilot cities, including Turin, and four replicator cities that closely monitored the progress of the Living Labs. In Turin, the Mirafiori Sud district served as the site of the ProGIreg Living Lab, where a number of nature-based schemes (NBS) were subjected to experimental testing and subsequently implemented in specific areas of the city. The implementation of NBS included the creation of regenerated fertile soil for green spaces, community urban gardens, aquaponic systems, green walls and roofs, accessible green corridors, and local environmental compensation processes. Additionally, the project promoted the enhancement of pollinator biodiversity. Awareness-raising and training activities were also conducted with the objective of involving citizens in the creation, monitoring and promotion of pollinator-friendly spaces and the promotion of sustainable agricultural practices.

The SUITS project made a significant contribution to the advancement of sustainable mobility and the enhancement of urban capacity to effectively address transport-related challenges. In Turin, the project concentrated on the logistics of goods in the city centre, with the objective of reorganising the loading and unloading of goods and improving the loading and unloading areas in car parks. In order to promote the use of low-emission vehicles in the limited traffic area, the city collaborated with transport operators. As part of the SUITS capacity-building programme, Turin benefited from enhanced technical expertise and the opportunity to engage in workshops and webinars. Additionally, the city facilitated the dissemination of knowledge regarding sustainable urban mobility plans (PUMS) within the metropolitan area, whilst concurrently endeavouring to alter the prevailing organisational mindset within the municipality. The MONICA project, which received funding from the Horizon 2020 programme, sought to enhance the management of large outdoor cultural events in European cities, addressing challenges related to sound, noise, and safety. The deployment of Internet of Things (IoT) technologies in six major cities, including Turin, involved the participation of over 100,000 users. At the core of the project was the establishment of a comprehensive IoT ecosystem, capable of overseeing the management of numerous devices, sensors, networks, and data. The development of specialised applications included the creation of designated sound zones for outdoor concerts, with the objective of reducing noise levels and enhancing crowd management and safety.

The MONICA project demonstrated the efficacy of IoT technologies in addressing urban challenges and enhancing the quality of life and safety in urban areas during major events, thereby establishing Turin as a noteworthy exemplar of sustainable urban innovation. In Turin, the project was implemented in two distinct contexts: the Kappa Futur Festival, which attracted approximately 20,000 attendees, and the nightlife in the San Salvario district, an area undergoing rapid transformation due to its vibrant nightlife and events. The challenges faced included crowd management, security, and noise control. In response, concrete actions were taken in Turin, including the infrastructure of key areas with improved data networks, the integration of acoustic monitoring platforms, the deployment of next-generation sensors for crowd counting and sound environment characterisation, active community engagement, data analysis to identify predictive patterns, and the organisation of a hackathon to develop new business models transferable to similar contexts.

The Handshake project, which involved 13 European cities, including Turin, had the objective of making cycling a more attractive daily mode of transport. The project concentrated on the various stages of introducing new cycling policies, encompassing the planning, implementation, and monitoring phases. The participating cities, including Amsterdam, Copenhagen, and Munich, which have all been designated as cycling capitals, collaborated on over 60 measures in 23 types of solutions, with the support of six technical partners. As a consequence of EU and Italian government investment, the number of people cycling in Turin is increasing. Notwithstanding the city's traditional car culture, there is a discernible shift towards a more balanced modal mix. The Handshake project facilitated the exchange of knowledge and expertise on cycling between the participating cities. In Turin, this involved the installation of cycle counting sensors, the introduction of new cycle lanes, the creation of temporary pedestrian areas, the expansion of the cycle network through the Cycle Route

Decree, the implementation of new shared mobility parking areas, and the introduction of new infrastructure such as the Parkin'Station. Training and mentoring were also provided, with workshops and practical activities led by experts from Amsterdam.

In the field of sustainable urban mobility, the STEAVE project led to the development of electric quadricycles, which were initially made available to municipal employees. These compact electric vehicles were developed in collaboration between the Polytechnic and a local enterprise.

The Turin Strategic Plan, including the SOLEZ Action Plan, places an emphasis on integrating low-carbon policies into the city's mobility planning in order to achieve the objectives set out in the Sustainable Urban Mobility Plan. Initiatives have been developed since 1995, with a particular focus on improved traffic management, parking regulations, and sustainable transport. The plan's strategic objectives are designed to ensure technical, social, environmental, and economic sustainability in mobility, reflecting a commitment to reducing emissions, promoting public transport, and improving urban life.

Two pilot actions serve to exemplify the innovative approach that underpins the plan.

The pilot project, entitled "Living Lab for a MaaS Paradigm in Turin", entailed the assessment of the Mobility as a Service (MaaS) paradigm through the utilisation of a technological platform, accessible via a mobile application. The pilot project concentrated on route planning, booking, payment, and data collection for various modes of transport, with the objective of promoting sustainable mobility within a limited ecosystem of users.

The second pilot project involved the introduction of dedicated lanes for freight mobility in congested areas. This pilot project tested multiuser lanes for freight transport with the aim of reducing conflicts and negative impacts on traffic. The objective was to improve logistics efficiency by allowing specific commercial vehicles to share the reserved lanes for buses, without affecting the public transport system and reducing congestion for all road users.

In addition to the projects enumerated in the aforementioned table and briefly discussed above, there are currently 21 ongoing projects in Turin. These projects are designed to achieve the following objectives (Table 3).

In the contemporary urban context of Turin, there is a notable acceleration in the implementation of development and innovation policies. This process is characterised by a holistic and multidisciplinary approach, in which the involvement of academic institutions, particularly the Polytechnic University of Turin, renowned for its excellence in engineering and technology, makes a significant contribution to the design, development, and implementation of innovative projects.

It is also noteworthy that the involvement of all stakeholders, including public bodies, private companies, the academic community, and citizens, has in recent years begun to create a fertile ecosystem for innovation. This synergy between different actors facilitates the transfer of knowledge and expertise that is essential for the adoption of effective and sustainable policies. Whilst there is still a long way to go to achieve climate neutrality by 2030, this collaborative environment provides a solid foundation for progress.

9 The "Turin 2030" Action Plan

The Turin 2030 Action Plan (City of Turin 2009) represents a comprehensive strategy for the transformation of Turin into a sustainable and resilient city. The Turin 2030 Action Plan represents a synthesis of projects and initiatives designed to facilitate an innovative renewal of the city. The plan encompasses the redevelopment of historic neighbourhoods, the implementation of sustainable transport, and the promotion of advanced technologies for a more liveable city with a reduced environmental impact. Furthermore, the Turin 2030 Action Plan (2019) is fully aligned with the United Nations' Agenda 2030 and its 17 Sustainable Development Goals. Indeed, the Action Plan compares each action with the UN goals, highlighting the challenges they address.

Despite facing challenges such as the preservation of its historical heritage and the need to modernise the city, Turin has already achieved significant successes in its journey towards 2030. The transformation of industrial areas into green spaces and the introduction of smart mobility systems are clear indications of the city's efficacious approach. There is a notable dedication to the involvement of active citizens in the decision-making processes, with a particular emphasis on the promotion of culture, innovation, and economic development. The necessity for a more accessible, circular, healthy, and green city is underscored, with a pronounced commitment to social solidarity and rights (City of Turin2009). The development strategies encompass the enhancement of digital technologies, the implementation of citizen co-design projects, and the enactment of policies designed to foster a liveable and attractive urban environment. The priority for Turin and the other selected cities is to develop so-called Climate City Contracts (CCC), which are agreements that, although not legally binding, will formalise a clear and well-defined political commitment towards the Commission and national and regional authorities, including citizens, research institutions, and the private sector. The CCCs will be signed by the mayors and representatives of the Commission. The aforementioned agreements are comprised of three distinct components: strategic commitments, actions, and investments. The strategic commitments have been formulated through a joint

Table 3 Ongoing projects

Project	Programme	Duration	Allocated	Objectives
5g4lives	Connecting Europe facility	36 months	€ 396.750, 00	Utilise 5G and drones to enhance public safety and environmental health
Tips4ped	CEF 5G smart communities	42 months	€ 305.000,00	Implement the first PED (Positive Energy District) and demonstrate the potential to increase renewable energy capacity in densely populated urban areas
Pactesur2	Fondo interno per la sicurezza	24 months	€ 233.604,53	Deploy 15 augmented reality viewers to improve the protection of public spaces
Eui c2c	City-to-city exchanges—European urban initiative capacity building	5 months	€ 6.786,00	Share Turin's experience in the recovery of former industrial areas, including through environmental rehabilitation, with the Metropolitan City of Lille
Let'sGOv	Horizon Europe	24 months	€ 120.375,00	Reduce emissions related to the energy system
Sefit	Fondo sociale europeo plu	24 months	€ 40.940,02	Enhance collaboration between public institutions and private actors to grow the innovation ecosystem
Femact-cities	Femac	30 months	€ 66.618,75	Promote gender equality in leadership and decision-making, recruitment and career progression, organisational well-being, and gender-based violence
Woodcircles	Horizon Europe	48 months	€ 175.00,00	Reduce the production of wood waste, improve the recovery of such materials in demolition sites, and acquire new design methods based on circular economy principles
Cofarm4cities	INTERREG CENTRAL EUROPE	36 months	€ 216.898,00	Transform the Falchera district in terms of social, environmental, and economic sustainability, and involve local residents in participatory decision-making with a focus on social inclusion
Trialsnet	Horizon Europe	36 months	€ 188.750,00	Demonstrate the deployment of XR, 5G/6G solutions in cultural/recreational venues to enhance visitors' experiences of cultural sites and public heritage from an entertainment perspective
Climaborough	Horizon Europe	48 months	€ 292.500,00	Work as a leading city in circular economy and waste management
Convince	Horizon Europe	42 months	€ 149.500,00	Improve the capabilities of robots to perform complex actions safely and appropriately within unstructured environments
Emeritus	Horizon Europe	36 months	€ 312.250,00	Develop an investigative protocol for effective investigation of environmental crime, leveraging the creation of a geo-intelligence platform
Desire	Horizon Europe	24 months	€ 72.500,00	Regenerate the territory of Falchera, starting from Cascina Falchera and involving other key areas of the zone like its lakes
Sme4green	Cosme-lgd (Local green deals)	41 months	€ 41.516,00	Launch the implementation of two local green deals in Turin, mobilising local stakeholders towards a new, sustainable, and inclusive recovery
Icarus	Horizon 2020	24 months	€ 119.225,00	Foster production, social, and technological innovation to strengthen the strategic approach to urban security
Incit ev	Horizon 2020	54 months	€ 228.000,00	Install infrastructure for electric vehicle charging, develop associated technologies and business models

(continued)

Table 3 (continued)

Project	Programme	Duration	Allocated	Objectives
Fusilli	Horizon 2020	47 months	€ 556.875,00	Implement integrated and systemic food policies that support the transition to sustainable food systems in urban, peri-urban, and nearby rural areas
Conexus	Horizon 2020	50 months	€ 175.287,50	Develop a shared understanding of NBS, and develop and integrate knowledge of cultural, social, environmental, and economic forces
Decido	Horizon 2020	36 months	€ 65.625,00	Improve civil protection policies for flood risk in the Meisino area

process involving local, regional, and national stakeholders, with the objective of defining a strategy to achieve climate neutrality as soon as possible. In contrast, the action plan outlines the specific interventions that will be implemented in order to achieve the mission's goal by 2030. These interventions are coordinated with an investment plan that serves as a business plan, guiding the interventions and assessing the associated costs and impacts. Upon the signing of the CCCs, the European Commission will undertake an assessment of the documents with a view to awarding a "Mission Label". This will facilitate access to EU funds earmarked for the energy and environmental transition. By March 2024, Turin is required to define its intervention strategy and finalise its CCCs, including contributions from local actors involved in the co-design of short- and medium-term actions. Furthermore, a monitoring phase is to be implemented to measure progress towards the objectives, considering both emission reductions and social and economic impacts.

The objective of achieving climate neutrality by 2030 necessitates a substantial commitment, which Turin aspires to attain through the formation of strategic alliances and the involvement of pivotal stakeholders across all sectors, including citizens, businesses, and the research community. In order to achieve this goal, the city has initiated a dialogue with local institutions and strategic partners, including the Polytechnic University of Turin, the University of Turin, and numerous other entities. Furthermore, Turin has established a working group dedicated to the third sector for the Challenge of 100 Climate-Neutral and Smart Cities, emphasising the significance of an integrated approach to the various dimensions of sustainable development, including social, environmental, economic, and governance.

The planned interventions are to be implemented in a number of sectors, including the residential and tertiary sectors, the public sector, the transportation and urban mobility sector, the expansion of the district heating network and local production of renewable electricity, and urban afforestation initiatives. It is anticipated that these measures will result in an estimated 65.98% reduction in CO_2 emissions, calculated from 1991 to 2030 (PAESC).

10 Discussion

It has been observed that the dimensions of the smart environment and climate change mitigation have similar objectives, both of which are aimed at addressing the same vulnerabilities. Notwithstanding the aforementioned similarities, it is imperative to refrain from conflating the two concepts in the context of urban development. To illustrate, a city may encourage the installation of green roofs without necessarily utilising ICT. However, the deployment of sensors to collect environmental data does not inherently ensure enhanced resilience to climate impacts unless these data are effectively integrated into decision-making processes.

In the future, the implementation of technologies designed to enhance resilience to climate change has the potential to reinforce the protective measures already in place in urban areas. However, the market for smart environment solutions is still in its infancy, with numerous technologies still in the early stages of development. Mitigation and adaptation to climate change are crucial and modern cities should be at the forefront of this field of innovation.

In order to ensure their continued relevance in the future, city administrations must prioritise the improvement of the quality of life for their residents. This should include a commitment to integrating the natural environment in order to guarantee urban sustainability and resilience. However, given the multifaceted nature of the concept of a smart city, the degree of emphasis placed on the smart environment can vary considerably.

Consequently, the decision to adopt smart initiatives aimed at environmental goals is one that is made by the cities themselves, often driven by the urgent need to address specific climate vulnerabilities. Despite the existence of numerous initiatives, there is still a lack of a solid base of effective and shared experiences that can provide guidance so that the climate change mitigation strategies adopted by smart cities can become a significant contribution. This highlights the need for further research to better understand the dynamics and create a more comprehensive mapping of smart initiatives.

11 Conclusions

Turin has become a paradigmatic case study in urban transformation and resilience. It has emerged from the ruins of its industrial legacy to espouse the concept of a smart city. Despite the numerous challenges encountered along the way, Turin has demonstrated remarkable resilience and has risen from the ashes of its industrial past. This resurgence can be attributed to the implementation of innovative policies and the crucial support received from the Italian government and subsequently from European Union funds. The urban regeneration strategy has placed a particular emphasis on the redevelopment of the suburbs, with a focus on those located to the north and south of the city. Whilst these efforts have already yielded tangible results, it remains evident that the path to complete renewal necessitates a consistent commitment and a long-term vision.

At the core of this renewal lies a profound aspiration to actively engage citizens, a tenet that not only aligns with the directives of the European Union but also encapsulates the very essence of Turin's cultural ethos. Initiatives such as the Turin Living Lab, although not the primary focus of this article, exemplify this spirit, underscoring the vital role of social innovation and active involvement in forging a more robust and cohesive community.

The European Commission's recommendation to engage citizens and small associations has resulted in the implementation of numerous projects. Active participation is evident in urban redevelopment initiatives, such as the Tonite project, which encompasses the Aurora and Vanchiglia districts and local schools. The CO4CITIES project facilitated collaboration between citizens and public administration, underscoring the pivotal role of local community centres in fortifying the community and developing a novel collaborative approach. These initiatives underscore the significance of citizens' involvement in fostering sustainable cultural change.

In contrast to a top-down approach to urban development, Turin has adopted a model that recognises and values the interconnected ecosystem of individuals, communities, businesses, and industries. It is this synergy, this continuous exchange of ideas and resources, that drives creativity and social innovation, enabling the city to reduce its environmental footprint and enhance the quality of life of its citizens whilst simultaneously improving its global competitiveness. The transformation of Turin into a smart city is not merely the consequence of infrastructural developments; rather, it is the result of a strategic combination of new technologies and human capital. This synergy is of critical importance for the development of projects and initiatives that are not only aimed at reducing the city's environmental footprint, but also at improving its global competitiveness and the quality of life of its citizens, as highlighted by Michelucci et al. (2016). Such endeavours situate Turin at the epicentre of regional development, catalysing innovation and local collaboration, a pivotal observation made by Battaglia and Tremblay (2011) and Tanda et al. (2017).

However, as Breuer et al. (2014) and Walravens (2015) remind us, whilst city administrators implement top-down strategic plans, the real engine of smart city innovation lies in the city's interconnected ecosystem. People, communities, businesses, and industries collaborate and work together to foster creativity and social innovation, a concept supported by the research of Cosgrave and Edwards-Schachter.

In order to achieve its objectives, Turin could adopt a strategy similar to that of other European cities which have demonstrated a high level of commitment to environmental sustainability, with the aim of mitigating climate change. Cities such as Copenhagen, Amsterdam, and Stockholm are at the vanguard of this field of endeavour. Copenhagen, with its ambitious target of becoming carbon neutral by 2025, provides an illustrative example of the efficacy of a well-structured cycling network and policies that incentivise an eco-friendly lifestyle (AlZu'bi et al. 2023). Turin, with its relatively flat terrain, is ideally situated to develop an extensive network of bike paths. In this regard, notable advancements have been made, with the expansion of bike lanes being a particularly noteworthy achievement. However, there is still a considerable scope for further development. The city of Amsterdam, renowned for its sustainability strategies, has introduced extensive infrastructure for electric vehicles and incentives for eco-friendly housing, thereby demonstrating how technological innovation can be seamlessly integrated into daily urban life. The city of Stockholm, aiming to phase out the use of fossil fuels by 2040, exemplifies how a city can leverage advanced technologies for heat recycling and actively promote green spaces, thereby enhancing the quality of life for its citizens (AlZu'bi et al. 2023).

Similarly, Turin could enhance energy efficiency not only by improving the insulation and energy performance of private homes and public buildings, but also by promoting the adoption of solar panels. The incorporation of these innovative practices would serve to reinforce Turin's trajectory towards sustainable development. The implementation of an integrated approach that considers sustainable mobility, energy efficiency, and active community participation has the potential to effect a profound transformation of the urban environment, thereby establishing the city as a model of resilience and innovation. This journey not only addresses the challenges of the present but also paves the way for future opportunities, thus positioning Turin as a benchmark in the mapping of smart and sustainable cities.

Consequently, the advancement of social innovation and creativity has been identified as a pivotal objective for enhancing the quality of life, competitiveness, and sustainability of urban areas. This assertion is supported by the findings of Battaglia and Tremblay (2011), Cosgrave et al. (2014), and Tanda and De Marco. From this perspective, the case of Turin represents a model for integrating top-down strategies and valuing grassroots innovation, which points the way forward for cities of the future.

It is only through this approach that it will be possible to respond to current challenges and anticipate future ones. In this way, the city will become not only a place to live, but also a dynamic, inclusive, and sustainable environment.

The city's journey towards climate neutrality by 2030 remains to be seen.

References

Ahvenniemi, H., Huovila, A., Pinto-Seppä, I., Airaksinen, M.: What are the differences between sustainable and smart cities? Cities **60**, 234–245 (2017). https://doi.org/10.1016/j.cities.2016.09.009

Albino, V., Berardi, U., Dangelico, R.M.: Smart cities: definitions, dimensions, performance, and initiatives. J. Urban Technol. **22**(1), 3–21 (2015). https://doi.org/10.1080/10630732.2014.942092

AlZu'bi, S., Hendawi, S., Elbes, M., Kanan, T., Mughaid, A.: Towards a sustainable future: exploring key features, challenges, and global examples of building smart cities. In: Proceedings of the 2023 International Conference on Information Technology (ICIT), pp. 284–291 (2023). https://doi.org/10.1109/ICIT58056.2023.10225987

Anthopoulos, L.: Smart utopia VS smart reality: learning by experience from 10 smart city cases. Cities **63**, 128–148 (2017). https://doi.org/10.1016/j.cities.2016.10.005

ASVIS.: The Urban Agenda for Sustainable Development: Objectives and Proposals. https://asvis.it/public/asvis2/files/Rapporto_ASviS/Rapporto_ASviS_2020/Report_ASviS_2020_FINAL8ott.pdf (2020)

Batagan, L.: Smart cities and sustainability models. Inform. Econ. **15**(3), 80–87 (2011)

Battaglia, A., Tremblay, D.-G.: 22@ And the innovation district in Barcelona and Montreal: a process of clustering development between urban regeneration and economic competitiveness. Urban Stud. Res. (2011). https://doi.org/10.1155/2011/568159

Bibri, S.E., Krogstie, J.: Smart sustainable cities of the future: an extensive interdisciplinary literature review. Sustain. Cities Soc. **31**, 183–212 (2017). https://doi.org/10.1016/j.scs.2017.02.016

Bibri, S.E., Krogstie, J., Kärrholm, M.: Compact city planning and development: emerging practices and strategies for achieving the goals of sustainability. Dev. Built Environ. **4**, 100021 (2020). https://doi.org/10.1016/j.dibe.2020.100021

Breuer, J., Walravens, N., Ballon, P.: Beyond defining the smart city: meeting top-down and bottom-up approaches in the middle. TeMA J. Land Use Mobil. Environ. (2014). https://doi.org/10.6092/1970-9870/2475

Bulkeley, H., Betsill, M.: Rethinking sustainable cities: multilevel governance and the 'Urban' politics of climate change. Environ. Polit. **14**(1), 42–63 (2005). https://doi.org/10.1080/0964401042000310178

Caragliu, A., Del Bo, C., Nijkamp, P.: Smart cities in Europe. J. Urban Technol. **18**(2), 65–82 (2009). https://doi.org/10.1080/10630732.2011.601117

Caragliu, A., Del Bo, C., Nijkamp, P.: Smart cities in Europe. J. Urban Technol. **18**(2), 65–82 (2011). https://doi.org/10.1080/10630732.2011.601117

Circle of Diplomatic Studies.: The Challenge of Climate Change and Biodiversity Shifts: Their Geopolitical Implications (2020)

Città di Turin.: Piano d'azione per l'Energia Sostenibile. http://www.comune.Turin.it/ambiente/bm%7Edoc/tape-2.pdf (2009)

City of Turin.: European Projects. https://www.torinoeuprojects.it/tutti-i-progetti/ (n.d.)

Comisión Europea, Dirección General de Investigación e Innovación, Mazzucato, M.: Governing missions in the European Union. Publications Office (2019). https://doi.org/10.2777/618697

Cosgrave, E., Tryfonas, T., & Crick, T.: The smart city from a public value perspective. In: ICT for Sustainability, ICT4S 2014, pp. 369–377 (2014). https://doi.org/10.2991/ict4s-14.2014.45

Crivello, S.: Urban policy mobilities: the case of Turin as a smart city. Eur. Plan. Stud. **23**(5), 909–921 (2015). https://doi.org/10.1080/09654313.2014.891568

Dameri, R.: Searching for smart city definition: a comprehensive proposal. Int. J. Comput. & Technol. **11**, 2544 (2013). https://doi.org/10.24297/ijct.v11i5.1142

Edwards-Schachter, M.E., Matti, C.E., Alcántara, E.: Fostering quality of life through social innovation: a living lab methodology study case. Rev. Policy Res. **29**(6), 672–692 (2012). https://doi.org/10.1111/j.1541-1338.2012.00588.x

Elsner, I., Monstadt, J., Raven, R.: Decarbonising Rotterdam? Energy transitions and the alignment of urban and infrastructural temporalities. City **23**(4–5), 646–657 (2019). https://doi.org/10.1080/13604813.2019.1689735

European Commission.: Sustainable Urban Development. https://ec.europa.eu/regional_policy/policy/themes/urban-development_en (n.d.)

European Commission.: Governing Missions: Governing Missions in the European Union by Mariana Mazzucato. https://research-and-innovation.ec.europa.eu/document/download/d960bad0-ffce-4e7d-a38e-517650304065_en?filename=ec_rtd_mazzucato-report-issue2_072019.pdf (2019a)

European Commission.: The European Green Deal. https://ec.europa.eu/info/strategy/priorities-2019-2024/european-green-deal_en (2019b)

European Commission.: 100 Climate-Neutral Cities by 2030—By and for the Citizens. https://ec.europa.eu/commission/presscorner/detail/en/ip_20_2531 (2020)

European Commission.: Fit for 55—Delivering the EU's 2030 Climate Target on the Way to Climate Neutrality. https://ec.europa.eu/commission/presscorner/detail/en/IP_21_3541. (2021)

European Council.: European Council Conclusions, 12 December 2019. https://www.consilium.europa.eu/it/press/press-releases/2019/12/12/european-council-conclusions-12-december-2019/ (2019)

García-Fernández, C., Peek, D.: Smart and sustainable? Positioning adaptation to climate change in the European smart city. Smart Cities **3**(2), 511–526 (2020). https://doi.org/10.3390/smartcities3020027

Giffinger, R., Fertner, C., Kramar, H., Kalasek, R., Milanović, N., Meijers, E.: Smart cities—ranking of European medium-sized cities. Centre of Regional Science, Vienna University of Technology. http://www.smart-cities.eu/download/smart_cities_final_report.pdf (2007)

Höjer, M., Wangel, J.: Smart sustainable cities: definition and challenges. In: Advances in Intelligent Systems and Computing, vol. 310, pp. 333–349 (2014). Springer. https://doi.org/10.1007/978-3-319-09228-7_20

Hollands, R.: Will the real smart city please stand up? City **12**, 303–320 (2008). https://doi.org/10.1080/13604810802479126

IMD.: Smart City Index 2023. IMD World Competitiveness Center. https://www.imd.org/smart-city-observatory/smart-city-index/ (2023)

International Telecommunication Union (ITU).: U4SSC. https://u4ssc.itu.int (2023)

Kramers, A., Höjer, M., Lövehagen, N., Wangel, J.: Smart sustainable cities: exploring ICT solutions for reduced energy use in cities. Environ. Model Softw. **56**, 52–62 (2014). https://doi.org/10.1016/j.envsoft.2013.12.019

Lee, J.H., Hancock, M.G., Hu, M.-C.: Towards an effective framework for building smart cities: lessons from Seoul and San Francisco. Technol. Forecast. Soc. Chang. **89**, 80–99 (2014). https://doi.org/10.1016/j.techfore.2013.08.033

March, H., Ribera-Fumaz, R.: Smart contradictions: the politics of making Barcelona a self-sufficient city. Eur. Urban Reg. Stud. **23**, 816–830 (2016)

Michelucci, F.V., De Marco, A., Tanda, A.: Defining the role of the smart-city manager: an analysis of responsibilities and skills. J. Urban Technol. **23**(3), 23–42 (2016). https://doi.org/10.1080/10630732.2016.1164439

Monfaredzadeh, T., Berardi, U.: Beneath the smart city: dichotomy between sustainability and competitiveness. Int. J. Sustain. Build. Technol. Urban Dev. **6**(3), 140–156 (2015). https://doi.org/10.1080/2093761X.2015.1057875

Nam, T., Pardo, T.: Conceptualizing smart city with dimensions of technology, people, and institutions. In: Proceedings of the 12th Annual International Digital Government Research Conference: Digital Government Innovation in Challenging Times, pp. 282–291 (2011). https://doi.org/10.1145/2037556.2037602

Organisation for Economic Co-operation and Development (OECD).: Compact Cities Policies: A Comparative Assessment. OECD Publishing (2012)

Rathnayaka, D., Potdar, V., Hussain, O., Dillon, T.: Identifying prosumer's energy sharing behaviours for forming optimal prosumer-communities. In: Proceedings of the IEEE Consumer Communications and Networking Conference, pp. 230–234 (2011). https://doi.org/10.1109/CCNC.2011.5766651

Schiavo, F.T., de Magalhães, C.F.: Smart sustainable cities: the essentials for managers' and leaders' initiatives within the complex context of differing definitions and assessments. Smart Cities **5**(3), 994–1024 (2022). https://doi.org/10.3390/smartcities5030050

Sinaeepourfard, A., Garcia, J., Masip-Bruin, X., Marín-Tordera, E., Cirera, J., Grau, G., Casaus, F.: Estimating smart city sensors data generation: current and future data in the city of Barcelona. In: Proceedings of the 15th IFIP Annual Mediterranean Ad Hoc Networking Workshop, pp. 20–21. Barcelona, Spain (2016).

Stratigea, A., Papadopoulou, C.-A., Panagiotopoulou, M.: Tools and technologies for planning the development of smart cities. J. Urban Technol. 22 (2015). https://doi.org/10.1080/10630732.2015.1018725

Tanda, A., De Marco, A., Rosso, M.: Evaluating the impact of smart city initiatives-The Turin living lab experience. In: Proceedings of the International Conference on Smart Cities and Green ICT Systems, pp. 281–286. Scitepress (2017)

United Nations.: Our common Future: Report of the World Commission on the Environment and Development. Oxford University Press (1987)

United Nations.: Millennium Summit of the United Nations. New York (2000)

United Nations.: Transforming Our World: The 2030 Agenda for Sustainable Development. https://sustainabledevelopment.un.org/post2015/transformingourworld (2015)

United Nations.: New Urban Agenda. file:///C:/Users/Utente/Downloads/New-Urban-Agenda-GA-Adopted-68th-Plenary-N1646655-E.pdf (2016)

United Nations.: The Sustainable Development Goals report 2020. United Nations. https://unstats.un.org/sdgs/report/2020/The-Sustainable-Development-Goals-Report-2020.pdf (2020)

United Natoins.: World Population Prospects. https://www.un.org/development/desa/pd/sites/www.un.org.development.desa.pd/files/files/documents/2020/Jan/un_2008_world_population_prospects-2008_revision_volume-ii.pdf (2008)

Walravens, N.: Qualitative indicators for smart city business models: the case of mobile services and applications. Telecommun. Policy **39**(3), 218–240 (2015). https://doi.org/10.1016/j.telpol.2014.12.011

A Two-Decade Revelation on Climate-Smart Agriculture and Soil Carbon in Brazil: The Bibliometric and Bibliographic Coupling Approaches

Chukwudi Nwaogu, Bridget E. Diagi, Chidinma A. Acholonu, Chinonye V. Ekweogu, Susan I. Ajiere, Cosmas C. Uche, Chinwe G. Onwuagba, Iheoma E. Mbuka-Nwosu, Chidi P. Ugochukwu, Jonathan C. Anyanwu, Enos I. Emereibeole, Dike H. Ogbuagu, Samuel I. Ogbaa, and Mauricio R. Cherubin

Abstract

The nexus and synergy between climate-smart agriculture (CSA) and soil organic carbon (SOC) has enhanced environmental sustainability in the food production sector. This work aimed to close the gap in knowledge between CSA and SOC and their potential in achieving a sustainable environment. Trends in the literature were unveiled on how the concepts were shaped and institutionalized by mapping their intellectual structure. The study provided answers to the where, when, and how of CSA-SOC nexus trends in scientific literature. From Web of science (WOS), the literature search was performed for the last 20 years by applying specific search, document selection criteria, and bibliometric coupling method. After excluding articles that were outside the goal/scope, a total of 478 papers were qualified for bibliometric analysis. Cluster analysis and bibliographic coupling algorithms were conducted using VOSviewer software. The result revealed that outside Brazil, authors and institutions from nineteen countries collaborated with the Brazilian authors. Year 2019 (having 49 articles) recorded the highest. In terms of impacts to the scientific community, Soil and Tillage Research (having Rv: 0.136) and Field Crops Research (Rv: 0.120) had the highest impacts. University of São Paulo and University of Rio Grande Do Sul had the highest number of citations and

C. Nwaogu (✉) · M. R. Cherubin
Department of Soil Science "Luiz de Queiroz" College of Agriculture, University of São Paulo, Piracicaba, Brazil
e-mail: cnwaogu@gmail.com

M. R. Cherubin
e-mail: cherubin@usp.br

C. Nwaogu · B. E. Diagi · C. A. Acholonu · C. V. Ekweogu · C. C. Uche · C. G. Onwuagba · I. E. Mbuka-Nwosu · C. P. Ugochukwu · J. C. Anyanwu · E. I. Emereibeole
Department of Environmental Management, Federal University of Technology, Owerri, Nigeria
e-mail: chidinma.acholonu@futo.edu.ng

C. V. Ekweogu
e-mail: chinonye.ekweogu@futo.edu.ng

C. C. Uche
e-mail: cosmas.uche@futo.edu.ng

C. G. Onwuagba
e-mail: chinwe.onwuagba@futo.edu

I. E. Mbuka-Nwosu
e-mail: iheoma.mbukanwosu@futo.edu.ng

J. C. Anyanwu
e-mail: jonathan.anyanwu@futo.edu.ng

E. I. Emereibeole
e-mail: enos.emereibeole@futo.edu.ng

S. I. Ajiere
Department of Geography and Environmental Management, University of Port-Harcourt, Port-Harcourt, Nigeria

D. H. Ogbuagu
Department of Biology, Federal University of Technology, Owerri, Nigeria

S. I. Ogbaa
Department of Urban Regional Planning, Federal University of Technology, Owerri, Nigeria
e-mail: samuel.ogbaa@futo.edu.ng

M. R. Cherubin
Center for Carbon Research in Tropical Agriculture (CCARBON), University of São Paulo, Piracicaba, Brazil

P. Mohammad et al. (eds.), *Geographic Approaches to Climate Change and Mitigation: Urban and Rural Perspectives (Volume 1)*, Advances in Science, Technology & Innovation, https://doi.org/10.1007/978-3-031-92119-3_18

total link strength with 15.06% and 10.25% papers, respectively. Considering papers that focused on the Sustainable Development Goals (SDGs), SDGs 02, 13, and 15 had the highest with 414 (38%), 340 (31%), and 324 (30%) papers, respectively. Findings from the study might support the Brazilian government in achieving its Nationally Determined Contributions (NDCs) as agreed at the Paris Summit where the country promised to tackle climate change by reducing carbon emissions and increasing C-sequestration through agricultural systems. This finding might contribute to narrow the gap in knowledge by enhancing targeted debate for effective discussion and policies for a better Brazil and society.

Keywords

Climate-smart agriculture · Sustainable development · VOSview · Brazil · Bibliometric

1 Introduction

Climate-smart agriculture (CSA) is a low-carbon farming system with nature-base practices to achieve sustainable food production chain (Shrestha et al. 2020). CSA has the goal of promoting higher environmental safety and greater benefits for the agricultural sector. This sustainable agricultural approach tends to increase SOC stocks, improve soil health, boost biodiversity, provide safer water, and mitigate climate change. CSA potentially contributes to carbon sequestration, which is one of its primary merits. By helping to reduce greenhouse gases emissions especially CO_2, CSA as an integrated agricultural system is one of the most rewarding farming strategies adopted to improve soil global ecosystem functions such as SOC storage, soil biodiversity, and other SDGs (Tamburini et al. 2020). The improvement in the quality, quantity, and biochemical diversity of plant-derived carbon inputs to soils, consequently, promotes the enrichment of soil microbial communities that support the formation and storage of SOC (Zhang et al. 2021), thus leading to a larger accumulation of SOC in the agricultural system with higher crop diversity. This contrasts with intensive use of farm machinery and the addition of chemicals, such as fertilizers, which increase soil compaction, erosion, and salinization, as well as reduce soil organic matter (Smith and Gregory 2013).

CSA is a promising approach to increase food security without much environmental degradation (Jat et al. 2020). By lowering atmospheric CO_2 levels, carbon sequestration in agricultural soils will not only ameliorate global warming but will also enhance soil health, water retention, and ecosystem resilience. Therefore, building sustainable and CSA needs the integration of conservation agricultural methods for human and environmental health.

The role of CSA in climate change resilience through the storage of carbon has been globally studied (Bhattacharyya 2024; Vishnoi and Goel 2024). For instance, a metadata analysis by Ma and Rahut (2024) established that CSA is significant in the attainment of food security, poverty alleviation, and climate resilience in farming communities globally. In addition to mitigating climate change, CSA has also been reported to support rural employment and food safety (Vishnoi and Goel 2024). In Brazil, some recent studies affirmed that CSA do not only help in absorbing CO_2 from the atmosphere but also prevents the emission of greenhouse gases from the agricultural systems (Bieluczyk et al. 2024). Though many studies have reported the potential of CSA in combating climate change challenges and enhancing the achievement of SDGs, yet it is rare to find any study that integrates the studies in Brazil. Brazil is one of the leading countries where CSA has been substantially adopted. Therefore, it is ideal to map the research trajectory on the interactions between CSA and SOC in the country by identifying intellectual research development through the application of a scientometric analytical method.

The application of bibliometric analysis has been effective in investigating the roles of many farming systems on soil quality, food security, and SDG (Cheng et al. 2023). A systematic review with bibliometric analysis produces a clear understanding of the evolutional trends and development in the scientific literature on various topics. This helps in the implementation of sustainable decisions and policies to improve the agroecological system. On this note, this paper aimed to close the gap in knowledge on CSA and SOC and their potential in achieving a sustainable environment by unveiling and mapping the trends in the literature on how the concepts were shaped and institutionalized in the scientific domain. The study further aimed at identifying when, where, and how research on CSA and SOC trends in the recent decades in Brazil by using bibliometric analysis. It is assumed that this knowledge might ignite targeted debate for effective discussion and policies for a sustainable development in Brazil and the global society at large.

2 Material and Methods

The selection criteria for the data collection and the Preferred Reporting Items for Systematic reviews and Meta-Analyses (PRISMA) (Appendix Fig. 1) were applied for this study. Data collection for the bibliometric review was performed on January 24, 2024 by applying a search engine to extract relevant articles on the topic from the Web of Science Core Collection—Clarivate Analytics platform (https://webofknowledge.com/). The following joint search terms or keywords under the "TOPIC" were used: ("climate-smart agriculture") OR ("climate-smart farm*") AND ("soil organic carbon) OR ("soil carbon") AND ("sustainable development goals") AND ("climate change") AND ("food security"), AND ("01-01-2004 to 10-12-2023") (i.e., for a research period of 20 years). These terms were chosen because they covered the study topic and objective. By adopting the above-mentioned terms, the initial search generated a total of 715 results. The articles were first refined to include only the results from research articles, and papers published in English language and related to the topic. A final total of 478 documents were qualified, archived, and used for the analysis.

In the VOSViewer, various types of analysis were executed such as co-authorship, co-occurrences, bibliographic coupling, citation, and co-citation. On the same hand, the unit of analysis applied involved authors, documents (papers), sources (publication journals), affiliate institutions (or organizations), countries, and keywords. The analysis used the full counting method. In most scenarios, the analyses were done by setting achievable thresholds to focus on the key items that attained the set criteria (or thresholds).

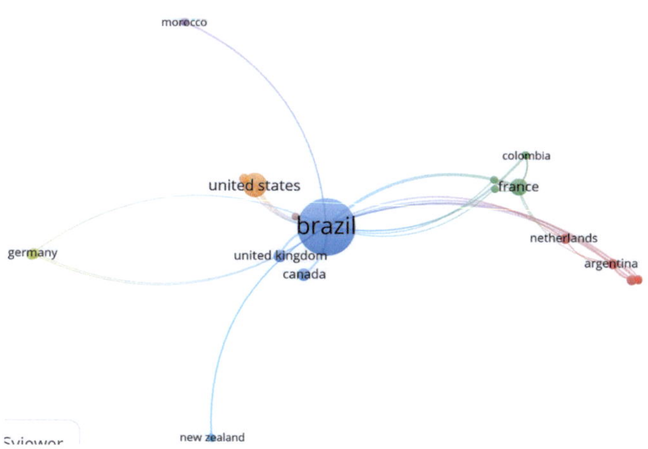

Fig. 1 Countries' publications and collaboration networks on the topic during the study periods. *Note* The circle/ball shapes represent the frequencies of papers while the lines and their thickness represent the link and the link strength (*Source* Authors)

3 Results and Discussions

The result on countries' publications and collaboration networks revealed the involvement of 19 countries with eight clusters (Fig. 1). Cluster 1 (blue cluster) have Brazil, UK, and Canada; cluster 2 (orange cluster) had USA and Spain; cluster 3 (green cluster) had France, India, Pakistan, and Colombia; cluster 5 (red cluster) had the Netherlands, Argentina, Nigeria, Bangladesh, and Mexico; cluster 6 (yellow cluster) had Germany, Sweden, and Switzerland; other clusters have New Zealand and Morocco. Brazil had the highest number of papers on the topic with collaborations within the country followed by USA and France. In consistent with the findings of this work, other authors have reported that Brazil has become one of the top countries globally in the adoption of CSA and other low-carbon farming policies and practices (Cerri et al. 2024). This will support in the achievement of its Nationally Determined Contributions (NDCs) to the 2015 Paris Climate Agreement (Brazil ABC Plan 2021) and remain among the leading producers and exporters of many food, energy, and fiber products (Basso et al. 2024). The government initiatives and implementation of agricultural policies including the ABC supported Brazil in achieving this milestone (Brasil 2022; Brazil ABC Plan 2021; Cerri et al. 2024). Ten years ago, Guilhon-Albuquerque (2014) reported great collaborations between Brazil and USA in research, trade, and other aspects of the economy.

The study and publications on the impacts of CSA on SOC had rapid growth during the two decades of our study (Fig. 2). The early 2000s marked the evolution of this environmental sustainability concept in the agricultural system, and as of 2019, forty-nine papers were published. A substantial decline in the publications was recorded between 2020 and 2021, which rose in 2022 and 2023. The need to provide food for the growing population and mitigate climate change has been a prime concern by governments, technocrats, scientists, and other stakeholders (Iannotti et al. 2024). This therefore was attributed to the increase in the number of articles published on the topic in recent years (Schreiner-McGraw et al. 2024). It is believed that the decline in publications between 2020 and 2021 can be attributed to the global COVID-19 pandemic, which not only resulted in the loss of many lives but also restricted movement and social interactions including research activities. If researchers neither travel nor go to the field or laboratory, it will be difficult to publish; consequently, the number of papers published will decline (Delardas and Giannos 2022). According to Delardas and Giannos (2022), COVID-19 pandemic did not only affect food production but also affected paper production even in the medical fields.

In Fig. 3, the keywords were classified into seven clusters. The study established 115 keywords out of which only 51 keywords had at least five occurrences. The light blue cluster

Fig. 2 Publication chronology (*Source* Authors)

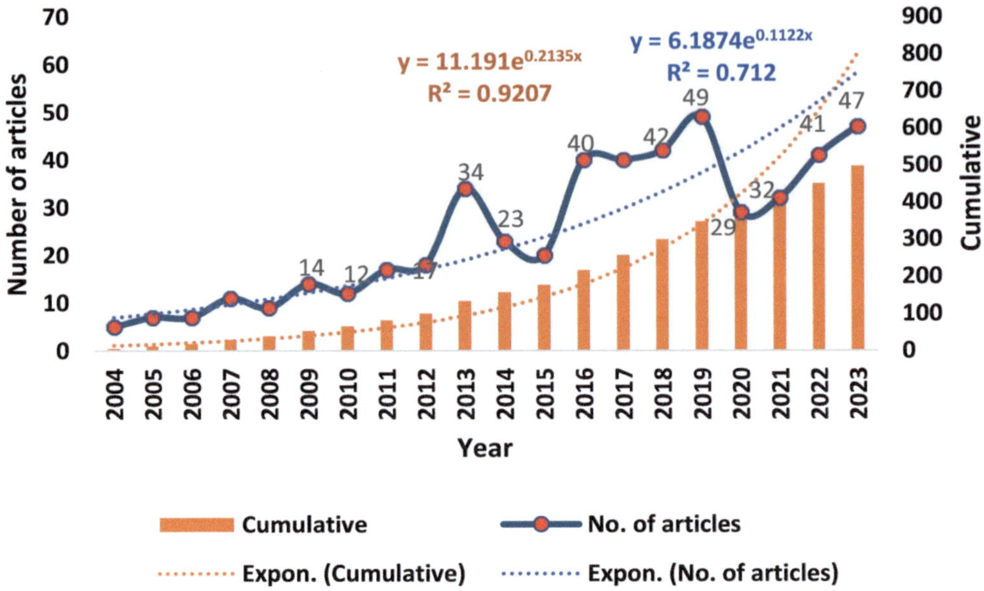

covered soil biology including soil organic matter, soil respiration, geostatistics, caatinga, and soil attributes. The red cluster had more words on soil physical properties such as soil quality, microbial biomass, green manure, nutrient cycling, soybean, and maize which were the focus in the Cerrado biome of Brazil. The navy blue, purple, yellow, and green clusters covered land use and soil management practices including CSA, no-tillage, greenhouse gases, pasture, eucalyptus, crop rotation, organic carbon, soil compaction, as well as soil management and carbon management index. Further, the study observed that most of the keywords were found in papers published between 2010 and 2020 (Fig. 3b). Some previous studies on the impacts of CSA on SOC have covered these keywords on the soil management practices as well as biological, chemical, and physical properties of the soil (Schreiner-McGraw et al. 2024). Meanwhile, these keywords formed the sustainable practices adopted in CSA to reduce carbon emissions, increase soil carbon stocks, mitigate climate change, and enhance soil health (Nwaogu et al. 2024). A recent study in Brazil by Vogado et al. (2024) assessed the total soil carbon covered in different agricultural land uses including CSA (such as integrated crop–livestock system under no-tillage and integrated crop–livestock with recent soil tillage), pasture, and native Cerrado vegetation. They found that the total C stocks in the top 500 cm soil profile were the highest in CSA under no-tillage (127 Mg ha^{-1}). On the other hand, they found that pasture (101 Mg ha^{-1}) and the native Cerrado vegetation (61 Mg ha^{-1}) had the lowest total carbon stocks.

The dataset had more than 80 journals, but only 21 of the journals had at least five articles on the topic and were indicated in five clusters (Fig. 4). The red cluster include Brazilian-based journals such as Revista Brasileira De Ciencia Do Solo, Revista Brasil De Engenharia Agricola E Ambiental, Pesquisa Agropecuaria Brasileira, Revista Caatinga, Revista Ciencia Agronomica, Ciencia E Agrotecnologia, Ciencia Rural, Scientia Agricola, and others. The green cluster had most of the soil biology-oriented journals, namely Soil Biology & Biochemistry, European journal of soil biology, Biology and Fertility of Soils, Global change biology, and others. The yellow, purple, and blue clusters involved soil-, agriculture-, and environmental-related journals such as Geoderma, Geoderma regional, Soil and tillage research, soil science society of America, Catena, and others.

In terms of impacts to the scientific community, Soil and Tillage Research and Field Crops Research accounting for Rv of 0.136 and 0.120, respectively, had the highest impact (Appendix Table 1). On the other hand, Revista Brasileira De Ciencia Do Solo (23.43%) had the highest rate of

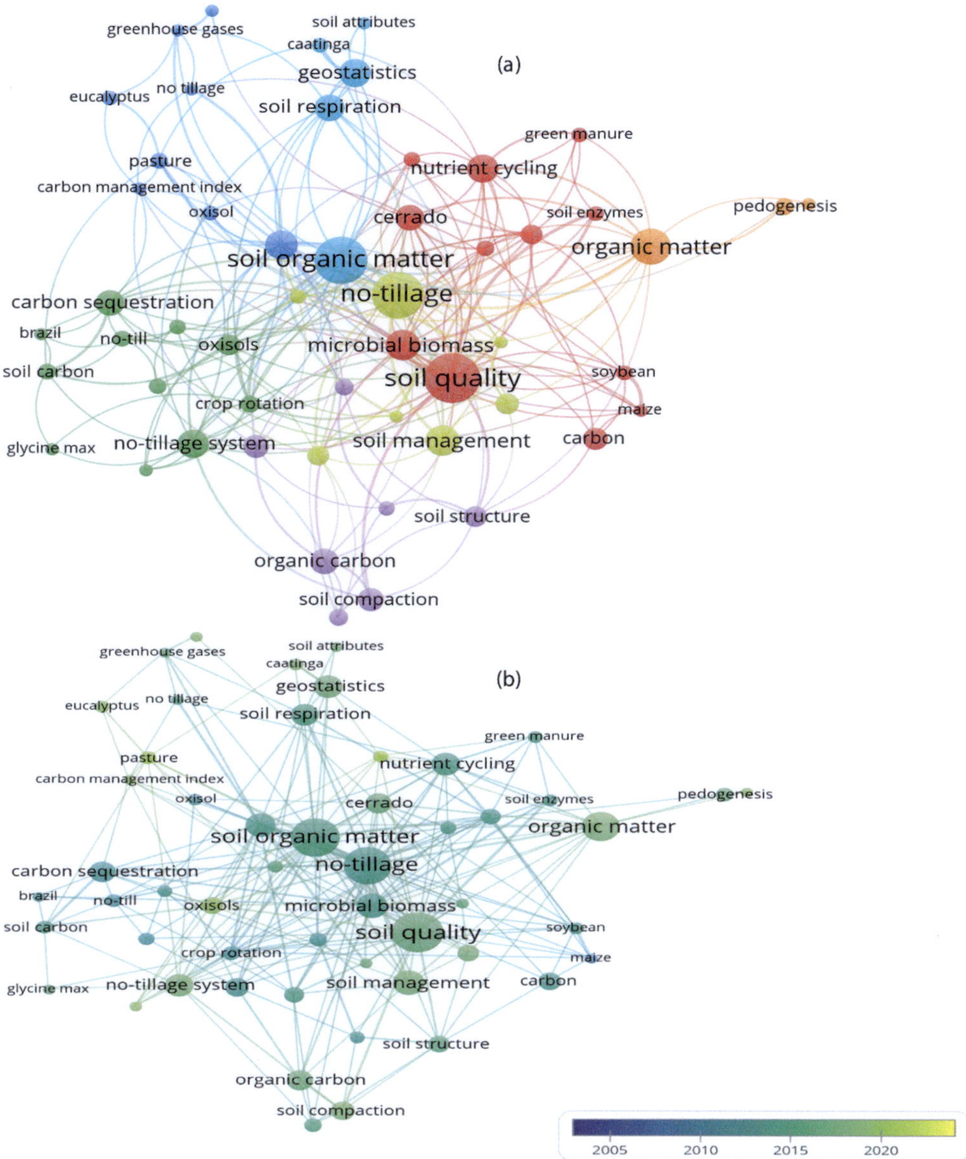

Fig. 3 Keywords **a** thematic areas cluster using network analysis, **b** Temporal-chronological cluster of overlay analysis (*Source* Authors)

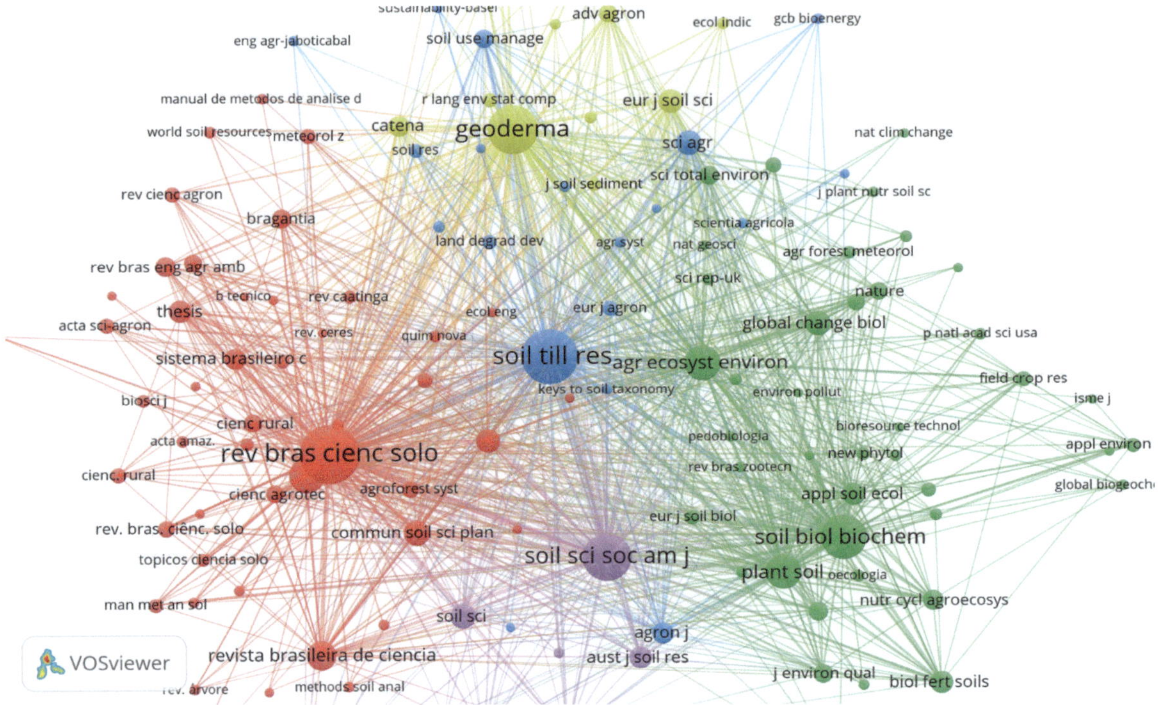

Fig. 4 All journals with at least one paper published on the topic during the study period (*Source* Authors)

publication frequency, followed by Scientia Agricola (8.78%). It was not surprising that the Brazil-based journals recorded the highest number of articles published on the topic because most of the authors might have found it easier publishing in the locally based journals. Many papers have been published on the topic in these top journals (Oliveira et al. 2023).

Thirteen institutions had at least 15 articles published on the topic during the two decades (Appendix Table 2). The University of São Paulo (accounting for 15.06%) had the highest number of papers, followed by Federal University of Rio Grande Do Sul (10.25%). Further, the University of São Paulo and University of Rio Grande Do Sul had the highest number of citations and total link strength (Fig. 5). High publications on the topic as obtainable from these two universities might be attributed to the facts that they are both located in the southeast characterized by Cerrado and Mata Atlantic biomes of Brazil which are popular for sugar cane, soybean, and maize, coffee, and cotton production (Rausch et al. 2019) and have major research funding organizations such as the São Paulo State Research Foundation (FAPESP).

Most of the papers treated some SDG targets with more focus on SDG 02, 13, and 15, and these had the highest number of papers with 414 (38%), 340 (31%), and 324 (30%) papers, respectively (Fig. 6). Several studies have reported the strong impacts of CSA on SOC in mitigating climate change, improving food security and other SDGs in Brazil (Nwaogu and Cherubin 2024; Cerri et al. 2024). In Brazil, it was found that in CSA, soil C–CH_4 influx increased from 0.8 to 1.0 kg ha^{-1} yr^{-1}, when CSA was introduced, and this means that at least 0.2 kg ha^{-1} yr^{-1} was stocked annually in the soil under CSA (Bieluczyk et al. 2024).

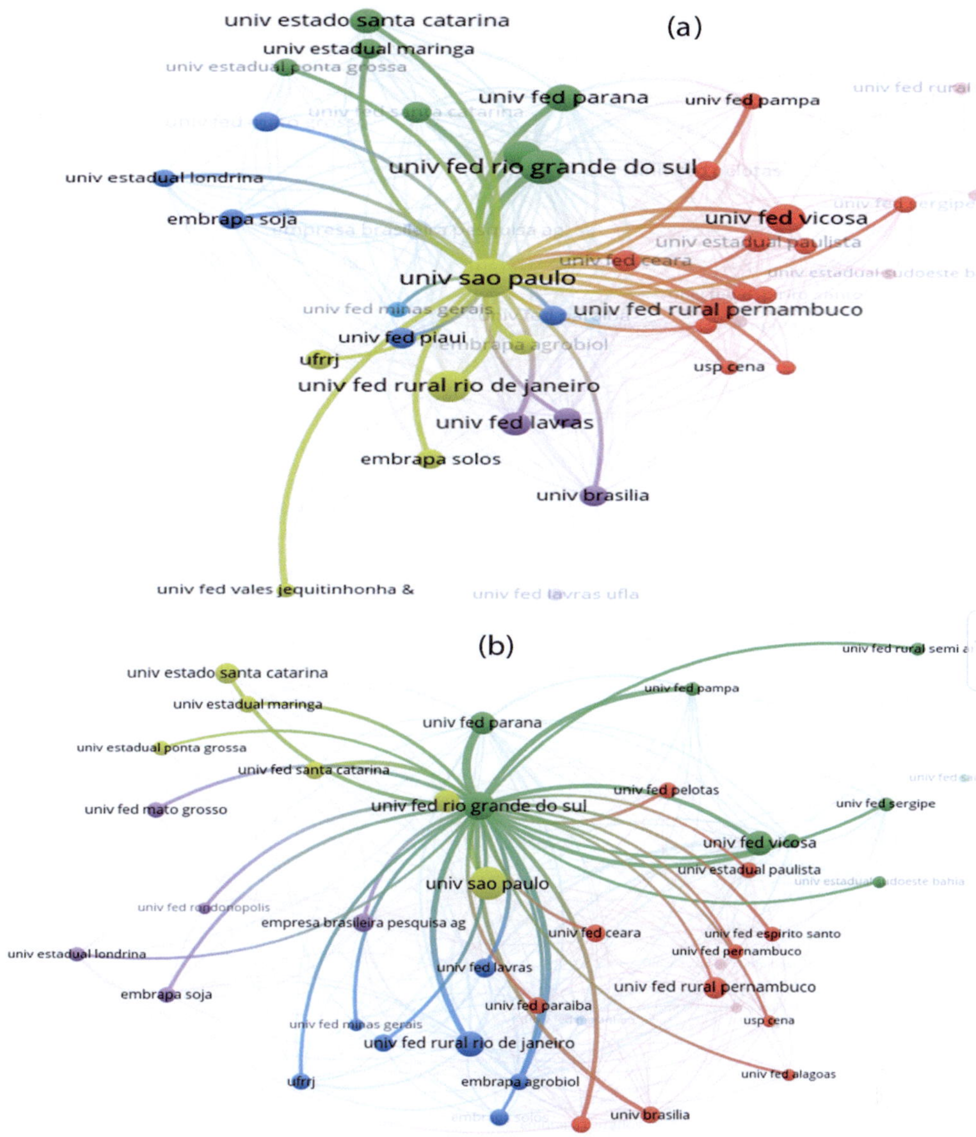

Fig. 5 Institutions citation link strength between **a** University of São Paulo, **b** Federal University of Rio Grande DO Sul and other organizations (*Source* Authors)

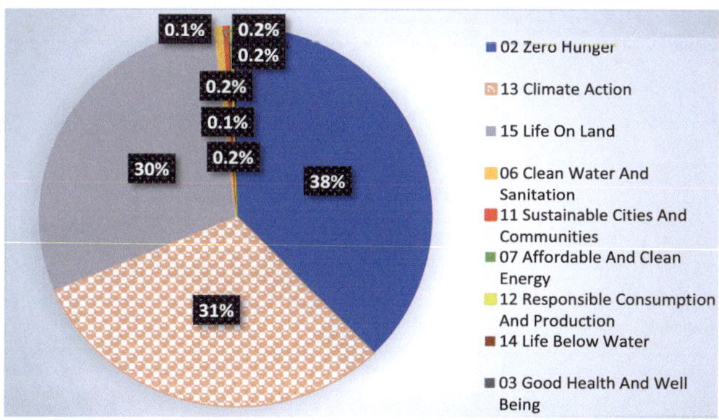

Fig. 6 Proportion of papers that related to SDGs on the studied topic during the two decades (*Source* Authors)

4 Conclusion

The impact of CSA on SOC and the achievement of SGDs have been unveiled in this bibliometric analysis. The result revealed that outside Brazil, authors and institutions from nineteen countries collaborated with Brazilian authors. Year 2019 (accounting for 49 articles) had the highest number of papers. In terms of impacts to the scientific community, Soil and Tillage Research (Rv: 0.136) and Field Crops Research (Rv: 0.120) had the highest impacts. The University of São Paulo and University of Rio Grande Do Sul had the highest number of citations and total link strength with 15.06% and 10.25% papers, respectively. Considering papers that focused on the SDGs, it was observed that SDGs 02, 13, and 15 had the highest with 414 (38%), 340 (31%), and 324 (30%) papers, respectively. The study has addressed questions of where, when, and how CSA-SOC nexus trends and evolved in scientific literature in the recent decades. The gap in knowledge about CSA and SOC and their potential in achieving a sustainable environment has been narrowed by institutionalizing and mapping the intellectual structure on the topic. This finding might contribute to promote targeted debate for effective discussion and policies for a better Brazil and global society. Findings from the study might also support the Brazilian government in achieving its Nationally Determined Contributions (NDCs) agreement at the Paris Summit where the country promised to tackle climate change by reducing carbon emissions and increasing C-sequestration through the agricultural sector. We suggest that further study be conducted on the implications of CSA on other soil properties because SOC has strong correlations with other soil variables.

Acknowledgements The study was funded by São Paulo Research Foundation (FAPESP) (process numbers: 2021/11757-1; 2023/05122-9, and 2021/10573-4). C. Nwaogu and M.R. Cherubin are grateful to FAPESP for the funding.

Appendix

See Fig. A.1 and Tables A.1 and A.2.

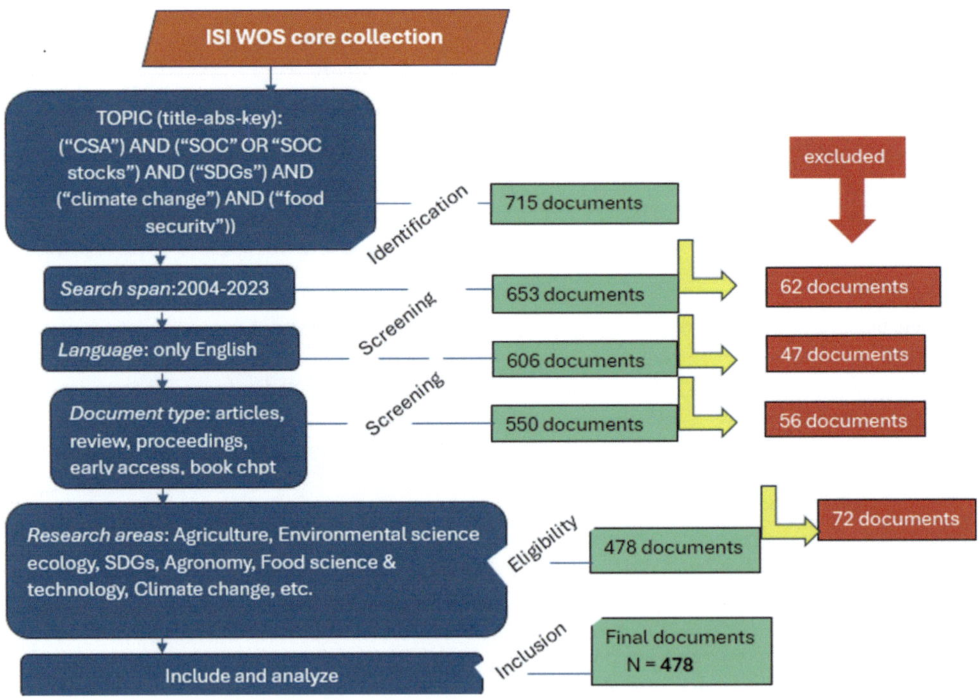

Fig. A.1 PRISMA methodological framework (*Source* Authors)

Table A.1 Top journals with at least five documents (papers) published on the topic

Source	Doc	Cit	TLS	Rv	Freq (%)
Acta Scientiarum-Agronomy	6	37	843	0.013	1.26
Agricultural Systems	5	115	694	0.048	1.05
Applied Soil Ecology	12	307	1116	0.054	2.51
Australian Journal of Soil Research	5	127	982	0.053	1.05
Ciencia E Agrotecnologia	11	100	662	0.019	2.3
Ciencia Rural	8	33	535	0.009	1.67
Field Crops Research	5	287	608	**0.120**	1.05
Geoderma	26	1150	2846	**0.093**	5.44
Geoderma Regional	16	301	1680	0.039	3.35
Nutrient Cycling in Agroecosystems	7	312	651	**0.093**	1.46
Pesquisa Agropecuaria Brasileira	18	195	1254	0.023	3.76
Revista Brasileira De Ciencia Do Solo	112	1008	9261	0.019	23.43
Rev Brasil De Engenharia Agricola E Ambiental	18	55	1149	0.006	3.77
Revista Caatinga	25	72	2270	0.006	5.23
Revista Ciencia Agronomica	16	99	1527	0.013	3.35
Scientia Agricola	42	837	3776	0.042	8.78
Semina-Ciencias Agrarias	24	103	2643	0.009	5.02
Soil & Tillage Research	43	2789	4498	**0.136**	9
Soil Biology & Biochemistry	5	214	295	**0.09**	1.05
Soil Research	12	178	1313	0.031	2.51
Soil Science Society of America Journal	6	233	761	**0.081**	1.26
Others	56				11.7
Total	478				100

Doc = papers or documents, Cit = citations, TLS = total link strength, Rv = Ratio value, Freq = Frequency. Values in bold indicated journals with most scientific impacts to the research community, (*Source* Authors' analysis from data extracted from WoS)

Table A.2 Institutional performances in search on CSA and soil C

Organization	Document	Citation	Tot Links	Ratio value	Freq (%)
Empresa Brasileira Pesquisa Agr	18	119	5538	0.014	3.77
Univ Estado Santa Catarina	22	269	4068	0.026	4.6
Univ Fed Ceara	15	111	3995	0.015	3.14
Univ Fed Lavras	20	280	5472	0.029	4.18
Univ Fed Parana	27	962	7705	0.075	5.65
Univ Fed Piaui	15	168	3897	0.023	3.14
Univ Fed Rio Grande Do Sul	49	1346	14247	0.057	10.25
Univ Fed Rural Pernambuco	25	366	4501	0.031	5.23
Univ Fed Rural Rio De Janeiro	40	849	10952	0.044	8.37
Univ Fed Santa Catarina	16	208	5168	0.027	3.35
Univ Fed Santa Maria	30	503	7995	0.035	6.28
Univ Fed Vicosa	35	367	8757	0.022	7.32
Univ São Paulo	72	1375	16162	0.041	15.06
Others	94				19.66
Total	478				100

Fed = Federal, Doc = documents, Cit = citations, TLS = total length strength, Rv = ratio value, Freq = frequency; (*Source* Authors' analysis from data extracted from WoS)

References

Basso, M.F., Neves, M.F., Grossi-de-Sa, M.F.: Agriculture evolution, sustainability and trends, focusing on Brazilian agribusiness: a review. Front. Sustain. Food Syst. **7**, 1296337 (2024)

Bhattacharyya, T.: Soil carbon footprints and climate-smart soils. Curr. Sci. **126**(5), 548 (2024)

Bieluczyk, W., Cherubin, M.R., Cerri, C.E.P., Siqueira-Neto, M., Abdalla Filho, A.L., Castro, J.I.A., Locatelli, J.L., Tsai, S.M., de Camargo, P.B.: Greenhouse gas fluxes in Brazilian climate-smart agricultural and livestock systems: a systematic and critical overview. J. Clean. Prod. 142782 (2024)

Brasil.: Plan ABC-Agricultura de Baixa Emissão de Carbono. https://www.gov.br/agricultura/pt-br/assuntos/sustentabilidade/plano-abc/plano-abc-agricultura-de-baixa-emissao-de-carbono (2022). Last accessed 03 Apr. 2023

Brazil ABC Plan.: ABC Plus-Brazil's New Climate Change Adaptation and Low Carbon Emission in Agriculture Plan. Brasilia. https://usdabrazil.org.br/ (2021). Last accessed 15 Feb. 2024

Cerri, C.E.P., de Castro Mello, F.F., Renteria, N.B., Cherubin, M.R.: Public Policies and Initiatives to Promote Soil Health and Carbon Sequestration in Brazil, p. 196. Soil Health and Sustainable Agriculture in Brazil (2024)

Cheng, P., Tang, H., Lin, F., Kong, X.: Bibliometrics of the nexus between food security and carbon emissions: hotspots and trends. Environ. Sci. Pollut. Res. **30**(10), 25981–25998 (2023)

Delardas, O., Giannos, P.: How COVID-19 affected the journal impact factor of high impact medical journals: bibliometric analysis. J. Med. Internet Res. **24**(12), e43089 (2022)

Guilhon-Albuquerque, J.A.: Brazil, China, US: a triangular relation? Revista Brasileira De Política Internacional **57**, 108–120 (2014)

Iannotti, L., Kleban, E., Fracassi, P., Oenema, S. & Lutter, C.: Evidence for policies and practices to address global food insecurity. Annu. Rev. Public Health 45 (2024)

Jat, M.L., Chakraborty, D., Ladha, J.K., Rana, D.S., Gathala, M.K., McDonald, A., Gerard, B.: Conservation agriculture for sustainable intensification in South Asia. Nat. Sustain. **3**, 336–343 (2020). https://doi.org/10.1038/s41893-020-0500-2

Ma, W., Rahut, D.B.: Climate-smart agriculture: adoption, impacts, and implications for sustainable development. Mitig. Adapt. Strat. Glob. Change **29**(5), 44 (2024)

Nwaogu, C., Cherubin, M.R.: Integrated agricultural systems: The 21st century nature-based solution for resolving the global FEEES challenges. Adv. Agron. **185**, 1–73 (2024)

Nwaogu, C., Oti, N. N., Enaruvbe, O.G, Cherubin, M.R.: Crop-livestock-forest system as nature-based solutions to combating climate change and achieving SDGs in Brazil. In: Leal Filho, W., et al. (eds.) Handbook of Nature-Based Solutions to Mitigation and Adaptation to Climate Change (2024). https://doi.org/10.1007/978-3-030-98067-2_124-1

Oliveira, D.M.D.S., Tavares, R.L.M., Loss, A., Madari, B.E., Cerri, C.E.P., Alves, B.J.R., Pereira, M.G., Cherubin, M.R.: Climate-smart agriculture and soil C sequestration in Brazilian Cerrado: a systematic review. Rev. Bras. Ciênc. Solo **47**, e0220055 (2023)

Rausch, L., Gibbs, H., Schelly, I., Brandão, A., Morton, D.: Soy expansion in Brazil's Cerrado. Conserv. Lett. **12**, e12671 (2019). https://doi.org/10.1111/conl.12671

Schreiner-McGraw, A.P., Ransom, C.J., Veum, K.S., Wood, J.D., Sudduth, K.A., Abendroth, L.J.: Quantifying the impact of climate smart agricultural practices on soil carbon storage relative to conventional management. Agric. for. Meteorol. **344**, 109812 (2024)

Shrestha, J.I.B.A.N., Subedi, S.U.B.A.S.H., Timsina, K.P., Chaudhary, A., Kandel, M., Tripathi, S.: Conservation agriculture as an approach towards sustainable crop production: a review. Farming Manag. **5**(1), 7–15 (2020)

Smith, P., Gregory, P.J.: Climate change and sustainable food production. Proc. Nutr. Soc. **72**(1), 21–28 (2013)

Tamburini, G., Bommarco, R., Wanger, T.C., Kremen, C., Heijden, M.G.A., van der, M.G., Liebman, M., Hallin, S.: Agricultural diversification promotes multiple ecosystem services without compromising yield. Sci. Adv. **6**, eaba1715 (2020). https://doi.org/10.1126/sciadv.aba1715

Vishnoi, S., Goel, R.K.: Climate smart agriculture for sustainable productivity and healthy landscapes. Environ. Sci. Policy **151**, 103600 (2024)

Vogado, R.F., de Souza, H.A., Sagrilo, E., de Brito, L.D.C.R., Matias, S.S.R., Neto, M.L.T., de Oliveira Junior, J.O.L., de Andrade, H.A.F., Leite, L.F.C.: Soil organic carbon stocks and fractions under integrated systems and pasture in the Cerrado of Northeast Brazil. CATENA **243**, 108196 (2024)

Zhang, K., Maltais-Landry, G., Liao, H.-L.: How soil biota regulate C cycling and soil C pools in diversified crop rotations. Soil Biol. Biochem. **156**, 108219 (2021). https://doi.org/10.1016/j.soilbio.2021.108219

Urban Planning and Climate Change Mitigation

Integrating Passivhaus Standards into Smart City Development in the UK: An Assessment of Energy Efficiency and Policy Implications

Mo Maleki Sadabad, Shahram Tahmasseby, Salwa Salam, and Greg Watts

Abstract

Concerns about building energy conservation are rising, especially in the UK, where residential buildings contribute significantly to CO_2 emissions. The Passivhaus Standard has become a focal point, prioritizing minimal heating energy use, indoor comfort, and superior air quality. However, integrating Passivhaus principles into smart city development faces policy challenges that require alignment, incentives, and regulatory support. While information and communication technology (ICT) is crucial, obstacles remain in incorporating high-efficiency standards into existing urban infrastructure. This highlights the importance of Passivhaus in creating a sustainable housing future. This article explores the evolution of UK policies regarding building energy conservation, presenting a systematic literature review on the integration of Passivhaus within UK smart city development. It highlights energy efficiency, policy implications, and challenges related to urban density and heritage conservation. The study emphasizes the importance of community engagement and stakeholder collaboration for successful Passivhaus implementation, advocating for a context-sensitive approach that includes educating residents and involving them in planning. Additionally, it identifies a promising synergy between smart city technologies and Passivhaus principles, aiming for more livable urban spaces. The research provides actionable management guidance for urban development stakeholders, stressing the need for strategic incorporation of Passivhaus Standards to achieve sustainable urban futures.

Keywords

Smart city · Passive house · Passivhaus · Energy efficiency · Sustainability · UK

1 Introduction

Recent advancements geared toward elevating sustainability and living standards in urban areas through cutting-edge technologies have triggered a wave of urban transformation, giving rise to intelligent cities. According to the United Nations (UN) (2018), over half of the global population currently resides in urban areas, a number projected to rise in the future. Despite occupying a mere 3% of the Earth's land, cities globally contribute significantly to energy consumption, accounting for 60% to 80% of the total and generating 75% of overall carbon emissions (UN 2020). The challenge of managing a sustainable environment and promoting a healthy lifestyle becomes more pronounced for cities grappling with increasing population, demands for infrastructure, heightened resource needs, waste disposal challenges, and the imperative development of essential services (Apanaviciene et al. 2020; Fokaides et al. 2017).

The concept of a "Smart City" first surfaced in the 1990s and has undergone transformative evolution, particularly after 2020 with the advent of the Internet of Things (IoT) (Al Sharif and Pokharel 2022; Tahmasseby 2022). Smart city development involves crafting urban spaces that leverage digital

M. M. Sadabad (✉) · G. Watts
The University of Salford, Manchester, UK
e-mail: M.N.MalekiSadabad1@salford.ac.uk

G. Watts
e-mail: G.N.Watts@salford.ac.uk

S. Tahmasseby
Qatar Transportation and Traffic Safety Center (QTTSC), Department of Civil and Environmental Engineering, College of Engineering, Qatar University, Doha, Qatar
e-mail: stahmasseby@qu.edu.qa

S. Tahmasseby · S. Salam
Qatar Transportation and Traffic Safety Center (QTTSC), College of Engineering, Qatar University, Doha, Qatar
e-mail: ssalam@qu.edu.qa

P. Mohammad et al. (eds.), *Geographic Approaches to Climate Change and Mitigation: Urban and Rural Perspectives (Volume 1)*, Advances in Science, Technology & Innovation, https://doi.org/10.1007/978-3-031-92119-3_19

technologies, data-centric solutions, and innovative infrastructure to improve residents' quality of life, enhance operational efficiency, and foster sustainable progress. This process integrates various technologies, including IoT, artificial intelligence (AI), data analysis, and automation, to streamline city service and asset management (Dameri and Ricciardi 2017; Mishra and Chakraborty 2020; Rotuna et al. 2017; Tahmasseby 2022). Key components of smart city development encompass advanced communication and information systems, efficient transportation networks, eco-friendly energy alternatives, improved public services, and refined urban planning (Hui et al. 2023). These initiatives aim to address diverse urban challenges, including traffic congestion, environmental preservation, resource management, and public security (Fayomi et al. 2019; Hameed 2019; Su et al. 2011). The overarching goal of smart city development is to create an urban environment that is more livable, efficient, and sustainable, meeting the evolving needs of its residents while leveraging technology to drive economic advancement and enhance overall welfare.

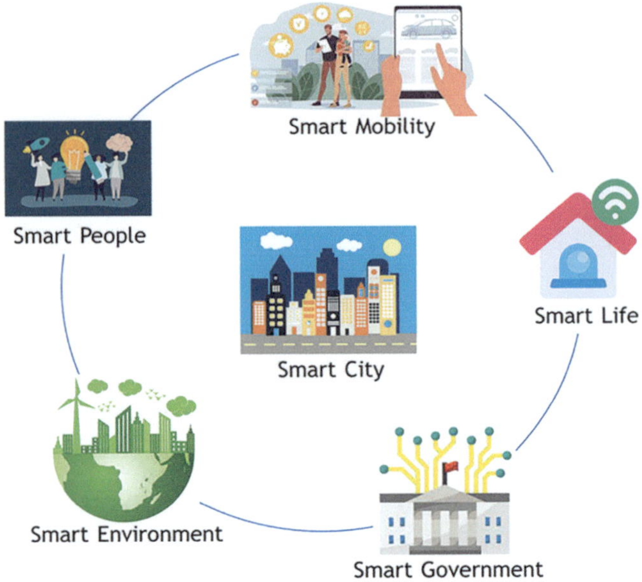

Fig. 1 Representation of smart city (by the authors)

1.1 Harmonizing Smart Buildings and Smart City Development: The Potential of Passivhaus Standards for Sustainable Urban Living

Smart cities are characterized by smart people, smart government, smart mobility, smart environment, and smart lifestyles (Emre and Sarul 2022) as given in Fig. 1. Within this framework, a critical component is the concept of smart buildings, spaces that optimize their structures, systems, services, and management to create efficient environments, showcasing the intricate interplay of these elements. This efficiency, influenced by the utilization of information and communication technologies (ICTs), significantly impacts the operational effectiveness of occupants (Apanaviciene et al. 2020).

The operational efficiency of buildings stands as a linchpin for smart cities, representing a crucial trajectory in shaping the interconnected future environment. One such example is the Passive House or Passivhaus, renowned for its minimal heating energy requirements, commitment to a comfortable indoor environment, and excellent indoor air quality (Liang et al. 2017). Integrating the Passivhaus Standard into smart city development holds the potential to usher in smart living—a more sustainable, comfortable, and energy-efficient urban setting.

The synergy between technology, data, and smart city development optimizes urban services, aligning them with residents' needs. This confluence has the transformative power to make urban spaces not only more sustainable but also more livable. As we explore the intricate relationship between smart buildings and smart city development, it becomes evident that a harmonious integration of these elements is crucial for the evolution of urban landscapes toward enhanced efficiency, sustainability, and quality of life.

The integration of Passivhaus Standards into smart city development within the UK presents a distinctive opportunity to tackle pressing urban challenges. Despite its potential, there is a noticeable gap in comprehensive studies that thoroughly explore this integration, particularly within the context of the UK's specific urban planning and policy landscape. This paper aims to bridge this gap by exploring how Passivhaus Standards can be effectively incorporated into smart city development in the UK and assessing their impact on energy efficiency and sustainable living (Gregory 2023).

Moreover, this study gains significant relevance against the backdrop of recent UK policies dedicated to bolstering urban sustainability and curbing carbon footprints. The UK's commitment to achieving the 2050 Net Zero target calls for innovative approaches in urban development, wherein smart city concepts and energy-efficient building standards play important roles. The anticipated findings of this research are poised to furnish valuable insights for urban planners, policymakers, and various stakeholders, offering guidance for shaping future strategies in urban development (Department for Business, Energy & Industrial Strategy 2021). This research stands not only as a contribution to academic discourse but also as a practical resource informing the dynamic landscape of sustainable urban planning in the UK.

The examination of smart city development within the UK context provides a distinctive perspective for comprehending the intricate interplay between urban expansion, technological progress, and sustainable methodologies. Leading cities such as London, Manchester, and Bristol have taken the lead in deploying smart technologies, spanning from smart grids to intelligent transportation systems. These cities serve as trailblazers, setting a precedent for the integration of energy-efficient building practices, aligning with the principles advocated by the Passivhaus Standards.

While the documented potential of these standards to significantly curtail energy consumption and carbon emissions in residential and commercial buildings is noteworthy, their application within the broader framework of smart cities remains an evolving and dynamic area of study. The ongoing exploration of this intersection promises insights that extend beyond individual buildings, offering a holistic understanding of how energy-efficient structures contribute to the overall fabric of intelligent and sustainable urban landscapes (Mueller 2017).

This study aims to explore the integration of Passivhaus Standards into the development of smart cities within the UK, with a particular focus on energy efficiency and the resulting policy implications. The primary objectives are:

(1) Examine the Current State of UK Urban Development and Smart City Initiatives: Evaluate how urban planning strategies, technological advancements, and sustainability practices currently align with smart city goals in major UK cities, such as London, Manchester, and Bristol. (2) Assess the Feasibility of Incorporating Passivhaus Standards in Smart Cities: Investigate the technical, social, and economic feasibility of incorporating Passivhaus principles, particularly focusing on the challenges posed by urban density, heritage conservation, and local climate conditions. (3) Identify the Synergy Between Smart City Technologies and Passivhaus Standards: Explore how smart city technologies, such as ICT infrastructure and data-driven solutions, can enhance the implementation of Passivhaus principles, resulting in more energy-efficient, sustainable, and livable urban environments. (4) Analyze the Policy Framework Supporting Energy-Efficient Urban Development: Critically review existing policies, such as the UK's 2050 Net Zero target, and assess the regulatory and governance support needed to scale the adoption of Passivhaus Standards in smart city frameworks. (5) Propose Policy Recommendations and Best Practices for Sustainable Urban Development: Based on systematic analysis, provide actionable recommendations for policymakers, urban planners, and stakeholders, aimed at accelerating the integration of Passivhaus Standards into the broader context of UK smart city development. The study emphasizes the critical role of cohesive policy frameworks that actively support sustainable building practices.

This study takes a comprehensive approach, examining not only the technical intricacies of implementing Passivhaus in smart cities but also scrutinizing the associated policy implications, societal impacts, and the feasibility of scalability across diverse urban contexts in the UK (Passivhaus Trust n.d.). Furthermore, in this research, we discuss design incentives and regulations that promote the widespread adoption of Passivhaus Standards, potentially yielding significant benefits in terms of energy savings and reduced environmental impact. These policies could serve as a blueprint for other nations grappling with similar challenges in achieving urban sustainability and energy efficiency. Through this holistic exploration, the research endeavors to offer nuanced insights that extend beyond mere technical considerations, contributing to a more robust foundation for informed decision-making in the realm of sustainable urban development.

This paper begins by introducing Passivhaus Standards and explaining the basics of this energy-efficient design. It explores the impact of Information and Communication Technology (ICT) on infrastructure and sustainable development, focusing on Passivhaus Global Strategies and their UK implications. Furthermore, it explores Energy Conservation and the Passivhaus Concept in the UK, undertaking a thorough policy analysis via a systematic literature review. The conclusion interprets the literature review results, contributing insights to sustainable urban development, particularly in smart cities, and briefly discusses the management implications of Passivhaus principles.

1.2 Overview of Passivhaus Standard

Passivhaus, commonly known as the Passive House, serves as a benchmark for enhancing energy efficiency in structures, yielding highly energy-efficient buildings (Dequaire 2012) that demand minimal energy for heating or cooling (Parry 2014). Originating in Germany in the early 1990s, the concept has since transcended global borders. Passivhaus buildings typically boast exceptional thermal performance, superior insulation, and a tightly sealed construction. Fresh air is continuously supplied into the building through highly efficient mechanical ventilation systems that adeptly recover heat. Distinguished by a well-insulated enclosure encompassing roofs, external walls, floor slabs, and windows, these buildings, with airtightness principles and high thermal insulation, effectively retain winter heat and fend off summer heat (Du and Gou 2023). Passivhaus construction strives to achieve exceptional comfort levels while using minimal energy for temperature regulation, thereby significantly reducing their environmental impact as shown in Fig. 2.

Initially designed with cold climates in mind, particularly in central and northern Europe (Pitts 2017), Passivhaus architecture has demonstrated adaptability to diverse climate zones globally (Schnieders et al. 2015). In a case study conducted in Malaysia, a tropical climate region, findings revealed that a Passivhaus, through wall and roof insulation, along with efficient shading techniques, can substantially reduce cooling requirements (Tatarestaghi et al. 2018). Australian research in

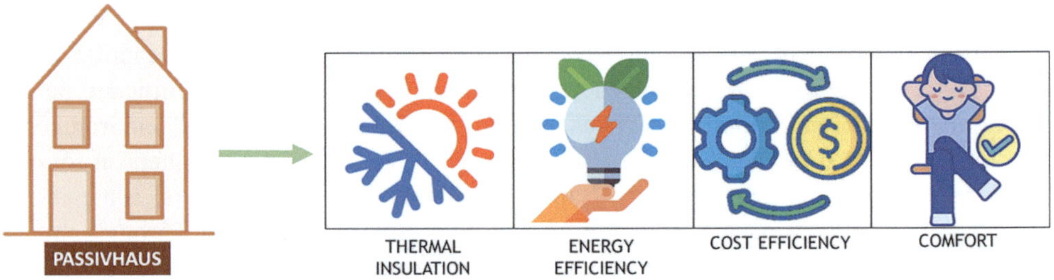

Fig. 2 Passivhaus benefit (by the authors)

Chifley reported a 64% reduction in energy usage compared to a typical dwelling in the same area, emphasizing the substantial energy-saving potential of Passivhaus principles (Truong and Garvie 2017). Research by Feist et al. (2005) indicated an 80% reduction in space heating energy usage and a 50% reduction in overall primary energy consumption (including services and appliances) in Passive houses compared to traditional dwellings.

While the Passivhaus Standard has showcased remarkable adaptability to various climates, challenges persist, particularly in urban environments with diverse architectural styles and space constraints. In the context of smart city development, especially within the UK, the integration of Passivhaus principles aligns with the broader goal of establishing sustainable, energy-efficient urban habitats. However, innovative solutions are required to address challenges such as urban density, heritage conservation, and varying regional climatic conditions. Studies have commenced exploring these complexities, uncovering both the potential and limitations of applying Passivhaus Standards in urban settings. The synergy between the stringent energy efficiency of Passivhaus and the advanced technological integration of smart cities presents a promising avenue for sustainable urban development. This convergence could usher in new paradigms in building design, where energy efficiency and intelligent infrastructure coexist seamlessly, meeting the evolving needs of urban populations while prioritizing environmental sustainability (Lynch 2014).

1.3 ICT, Infrastructure, and Sustainable Development

The utilization of contemporary Information and Communication Technology (ICT) addresses current and future challenges in innovations, competitiveness, and urban optimization. The United Nations Economic Commission of Europe (UNECE) defines a smart sustainable city as an innovative urban space that harnesses ICT to elevate the quality of life and enhance competitiveness. This transformation is designed to cater to the needs of present and future generations across economic, social, environmental, and cultural domains (ECOSOC 2015).

The CasaClima standard in Italy takes a principled approach, emphasizing energy efficiency, comfort, and environmental sustainability (Garman–Kolokotsa and Synnefa 2013). Similarly, the AECB standard (Association for Environment Conscious Building) in the UK provides guidelines for crafting low-energy and sustainable buildings (Building 2006; Zhang et al. 2017). Additionally, the Code for Sustainable Homes (CSH) presents an environmental assessment methodology for appraising and certifying the effectiveness of new residential properties in the UK (Hashemi and Hadjri 2013). These nationally recognized standards serve as catalysts for the construction of sustainable homes, fostering a continuous evolution in eco-friendly building practices. These benchmarks offer invaluable insights for seamlessly integrating the Passivhaus Standard into the fabric of smart cities in the UK.

The integration of ICT and smart technologies has played a crucial role in advancing global strategies for sustainable urban development. A critical analysis reveals that the success of these strategies often depends on local governance models, public–private partnerships, and community engagement (World Bank 2021).

In the specific context of Passivhaus integration into smart cities, the role of ICT becomes even more crucial as it facilitates the monitoring and optimization of energy performance in buildings. However, a challenge arises in seamlessly integrating these high-efficiency standards within the existing urban infrastructure and the socio-economic fabric of cities. Studies emphasize the need for tailored approaches that take into account local climate, cultural practices, and economic conditions to ensure the successful adoption of Passivhaus principles in smart city frameworks.

This underlines the importance of developing a nuanced understanding of how global strategies can be adapted and implemented effectively. Such an approach ensures that the ambition of creating smart, sustainable cities is not only technologically feasible but also socially and economically inclusive (Smart Plus Academy 2023).

1.4 Passivhaus Global Strategies and Its Impacts

Buildings globally contribute significantly to resource consumption and environmental impact, accounting for approximately 40% of energy usage, 25% of water consumption, and 40% of resource utilization, while contributing to around 33% of greenhouse gas emissions (UNEP 2009). Both residential and commercial buildings collectively represent approximately 60% of the world's electricity consumption. Despite notable investments and global efforts to reduce the energy intensity of buildings, the 2022 Buildings-GSR report (UNEP 2022) reveals a concerning trend. Despite a significant increase in investment and global success in reducing the energy intensity of buildings, the overall energy consumption and CO_2 emissions of the sector escalated in 2021, surpassing the levels observed before the pandemic. The demand for energy in buildings rose by approximately 4% from 2020, reaching 135 EJ, marking the most significant surge over the past decade. Additionally, CO_2 emissions stemming from building operations hit a record high of about 10 $GtCO_2$, reflecting a 5% increase from 2020, and a 2% surge compared to the previous peak in 2019.

The integration of the Passivhaus Standard is crucial in shaping a sustainable future for housing. In Germany, where residential buildings account for approximately 22% of total energy consumption for space heating (Schnieders and Hermelink 2006), evidence suggests that Passive Houses, with their low-energy consumption and optimal temperatures, are not only reasonable for occupants but also present a viable approach for new residential construction. Currently, 221 Passive House units are occupied across 14 building projects in Germany, Sweden, Austria, Switzerland, and France (Mihai et al. 2017).

Extending Passivhaus norms into the development of smart cities holds broader political implications. These buildings, characterized by reduced energy usage and comfortable yet cost-effective temperatures (Mihai et al. 2017), boast long lifespans and superior design, potentially leading to lower long-term maintenance and repair expenses. The implementation of Passivhaus criteria in the creation of smart cities in the UK could significantly influence energy efficiency and policy.

An economic analysis conducted by Audenaert et al. (2008) compared passive homes with low-energy buildings, standard structures, and passive houses. The study, examining approximately 11 buildings, revealed that in a scenario where the annual energy price increases by 10%, the payback period for a passive house is estimated to be 18 years. Moreover, in the case of a 15% annual energy price surge, the passive house is projected to yield significant profitability, underscoring its favorable investment potential.

1.5 Energy Conservation and Passivhaus Concept in the UK

In recent times, the global discourse on energy conservation in buildings has gained widespread attention, with a particular focus on the UK. Residential buildings, constituting the largest energy consumers in the UK, contribute to 30% of the total energy usage and associated CO_2 emissions (Palmer and Cooper 2013). This surpasses the energy consumption in commercial buildings, road transport, and industry, presenting a significant opportunity for reducing energy usage and mitigating carbon dioxide emissions. Responding to this challenge, the UK Government has established ambitious targets, aiming for an 80% reduction in energy consumption and carbon dioxide emissions by 2050 compared to 1990 levels (Giesekam et al. 2014; Xie et al. 2017). This commitment stresses the imperative of implementing effective energy-saving strategies and sustainable practices within the residential building sector.

Against this backdrop, the development of smart cities has become a decisive agenda for the UK. Driven by the urgency of the climate change crisis and the need to reduce human-caused greenhouse gas (GHG) emissions, the UK aligns its strategies with global initiatives such as the United Nations Climate Change (UNCC), Conference of the Parties (COP), Intergovernmental Panel on Climate Change (IPCC), and Sustainable Development Goals (SDGs) (Pitts 2017; Mohammadpourkarbasi et al. 2023). Smart cities aim to enhance urban life, particularly in terms of energy efficiency and sustainability, by leveraging technology and data. The integration of the Passivhaus Standard into the design of smart cities in the UK emerges as a strategic approach, promising significant energy savings, a reduction in greenhouse gas emissions, and improved indoor air quality for citizens. Emphasizing the need for context-specific research and data, McLeod (2013) draws attention to the relevance of Passivhaus principles in the unique environmental and architectural context of the UK.

In response to these initiatives, the Passivhaus Standard concept has garnered significant attention in the UK. This innovative building approach, distinguished by its minimal heating energy requirements, emphasis on maintaining a comfortable indoor environment, and commitment to superior indoor air quality, has emerged as a leading model for energy-efficient residential construction. Demonstrating its adaptability to various climates, the Passive House concept has seen substantial adoption, with over 50,000 Passive House structures built globally by 2013 (Liang et al. 2017).

For UK buildings aiming for Passivhaus certification, adherence to specific criteria is mandatory. These criteria (as shown in Fig. 3) include limitations on primary energy demand (maximum 120 kWh/(m2a)), space heating/cooling

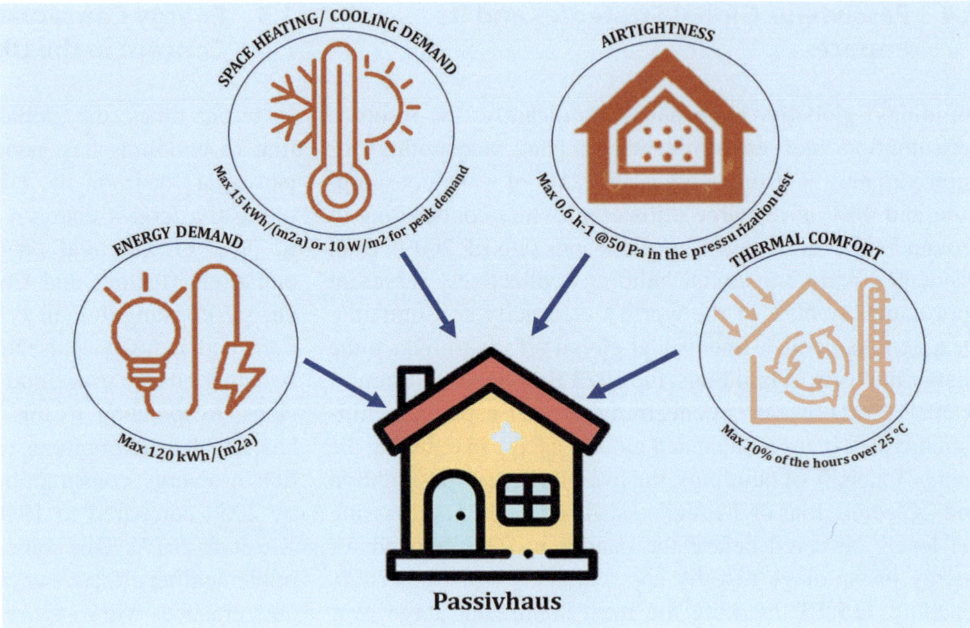

Fig. 3 Criteria for Passivhaus (by the authors)

demand (maximum 15 kWh/(m2a) or 10 W/m^2 for peak demand), airtightness (maximum 0.6 h-1 @50 Pa in the pressurization test), and thermal comfort (maximum 10% of the hours over 25 °C) (Liang et al. 2017). These stringent standards reflect the nation's commitment to promoting energy efficiency and reducing environmental impact, positioning the UK as a key player in the global movement toward sustainable and eco-friendly construction practices.

In the UK, the residential sector's journey toward energy conservation is marked by a mix of advancements and challenges. Despite significant strides in reducing energy consumption, the residential sector still represents a considerable portion of the UK's total energy use, underscoring the ongoing need for more efficient building practices (Department for Business, Energy & Industrial Strategy 2020). Passivhaus Standards, with their rigorous criteria for energy efficiency, present an intriguing contrast to the UK's traditional building practices. These standards not only align with but often exceed the requirements set out in the UK's current building regulations, offering a pathway to significantly lower energy consumption and carbon emissions in residential buildings.

However, the integration of these standards into the broader context of smart city development faces policy and regulatory challenges. While policies like the Future Homes Standard 2025 aim to encourage low-carbon heating and world-leading levels of energy efficiency, the integration of Passivhaus Standards requires a more concerted effort in policy alignment, incentives, and regulatory support. This necessitates a nuanced approach, balancing the ambitious goals of Passivhaus with the practicalities of urban development, historical preservation, and the existing housing stock. The evolution of UK policies in this area will be crucial in determining the extent to which Passivhaus Standards can be integrated into the fabric of future smart cities (HM Government 2019).

1.6 Advancing Energy Conservation and Sustainable Development—A Comprehensive Policy Analysis

The establishment of national and international policies (as shown in Fig. 4) that support sustainability and energy efficiency such as the Climate Change Act 2008[1] (Lockwood 2013), Energy Act 2011[2] (Mulliner and Kirsten 2017), Renewable Heat Incentive (RHI)[3] (Lowes et al. 2019), Feed-in Tariff Scheme (FiT)[4] (Cherrington et al. 2013),

[1] This legislation established the Committee on Climate Change to offer impartial advise on addressing climate change and sets legally binding targets for decreasing greenhouse gas emissions in the UK.

[2] The act emphasizes lowering carbon emissions and increasing energy efficiency. It contains provisions for the Green Deal, a government program created to assist individuals and companies in making improvements to their buildings' energy efficiency.

[3] The scheme offers financial incentives to encourage the use of renewable heating systems in private residences, commercial buildings, and government agencies.

[4] With the use of qualifying technologies including solar photovoltaic (PV) panels, wind turbines, and hydroelectricity, this program offers financial incentives for the production of renewable electricity.

Fig. 4 Policies fostering sustainability and promoting energy efficiency (by the authors)

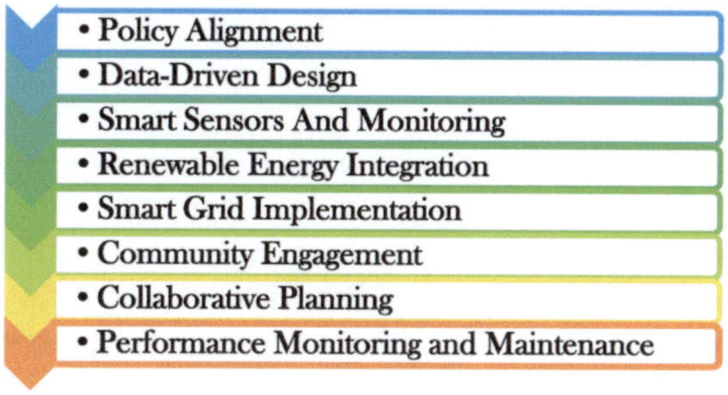

Fig. 5 Step included in the integration of the Passivhaus Standard into smart city development (by the authors)

Energy Company Obligation (ECO)[5] (Lead et al. 1708), Paris Agreement oot[6] (Schleussner et al. 2016), SDGs,[7] and the EU Energy Efficiency directive[8] (Zangheri et al. 2019) in the built environment may ultimately benefit from the same. The UK has been at the forefront of energy efficiency policy globally in various aspects. UK became the first EU nation to use the Standards of Performance model to finance energy efficiency initiatives and also the first to experiment with carbon emissions trading in 1994 (Mallaburn and Eyre 2014). The Climate Change Act was marked as one of its initial legislation internationally. The delivery models of organizations such as the Energy Saving Trust (EST) and the Carbon Trust have been emulated globally. In the past four decades, energy efficiency programs have taken almost every form available.

Policymakers in the UK must unequivocally emphasize the paramount importance of energy efficiency and sustainability in the construction industry by integrating the Passivhaus Standard into the development of smart cities. Moreover, there exists a compelling economic rationale for enhancing building efficiency, considering the escalating fuel costs and imperative climate change objectives. Passivhaus Standards harmonize with the UK's trajectory toward a low-carbon economy, as they curtail energy consumption and CO_2 emissions in buildings, substantiating a substantial contribution to energy efficiency. The Passivhaus guidelines prioritize thermal comfort and energy efficiency, rendering them exceptionally compatible for seamless integration into smart cities.

The assimilation of the Passivhaus Standard into smart city development necessitates an inclusive approach incorporating energy efficiency, sustainable building practices, and the integration of intelligent technologies. Key steps for the integration of the Passivhaus Standard into smart city development are illustrated in Fig. 5.

The potential for a transformative impact on the urban landscape in the UK, particularly through the integration of Passivhaus Standards into smart city development, arises from the adoption of the aforementioned policies. While these policies have established a foundation for energy efficiency and sustainability, their effectiveness in fostering comprehensive smart city solutions remains an area ripe for exploration. Achieving successful implementation of Passivhaus Standards in smart cities necessitates not only strict adherence to these policies but also innovative approaches that seamlessly blend energy efficiency with smart technology.

Future policy developments should prioritize creating an enabling environment that encourages such integrations, addressing both the technological and socio-economic aspects of sustainable urban development. To illustrate, enhancing the Energy Act could involve incorporating specific provisions for smart city development, thereby promoting the use of intelligent energy systems aligned with Passivhaus principles. By bridging policy with practice, the UK has the potential to lead the way in establishing smart cities that are not only technologically advanced but also sustainable and energy-efficient, thereby setting a benchmark for global urban development (Passivhaus Trust 2023).

[5] With a focus on low-income and vulnerable households, this duty mandates larger energy suppliers to offer funds for energy efficiency measures in homes.

[6] The UK is dedicated to lowering its greenhouse gas emissions and supporting the international effort to slow the rise in global temperatures as a signatory to the Paris Agreement.

[7] The SDGs, which include targets for achieving clean and affordable energy, liveable cities and communities, reducing greenhouse gas emissions, and promoting responsible consumption and production, have gained support from the UK.

[8] While the UK was a member of the EU, it adopted policies that promoted energy efficiency and the use of renewable energy sources in accordance with the EU Energy Efficiency Directive.

2 Systematic Literature Review Methodology

2.1 Introduction to the Review Process

This systematic literature review is dedicated to a detailed analysis of the integration of Passivhaus Standards in smart city development, with a specific emphasis on energy efficiency and its policy implications within the UK context. Employing a systematic approach is necessary for this review as it ensures the inclusion of a diverse range of studies, provides unbiased insights, and upholds rigorous academic standards. Through the systematic collation, review, and synthesis of pertinent literature, this study aims to offer a thorough understanding of the current state of knowledge in the field. Furthermore, it seeks to identify research gaps and propose directions for future studies.

This approach guarantees the reliability and comprehensiveness of the analysis, positioning it as a valuable resource for researchers, policymakers, and practitioners keen on advancing sustainable urban development and incorporating energy-efficient building practices. The insights derived from this review are poised to contribute significantly to the existing body of knowledge, serving as a guiding reference for those involved in shaping the future of urban environments (Turney 2022).

2.2 Literature Research Strategy

The literature research was conducted across several academic databases and search engines, including JSTOR, Scopus, PubMed, and Google Scholar, to ensure a comprehensive collection of relevant literature. Keywords and phrases such as "Passivhaus Standards," "Smart City Development," "Energy Efficiency in Urban Planning," and "Sustainable Building Practices in the UK" were used. Boolean operators like "AND" and "OR" were employed to refine the search. For instance, a typical search string was "Passivhaus Standards and Smart City Development or Energy Efficiency in Urban Planning in the UK." The step-by-step literature search is indicated in Fig. 6.

The literature search honed in on articles published between 2010 and 2023, a timeframe selected to account for the rapid advancements in this field over recent years. This specific period was chosen to encapsulate the latest practices and insights. To ensure an inclusive understanding and analysis, the search was confined to English-language publications. Moreover, emphasis was placed on peer-reviewed academic journals and reports, prioritizing sources known for their credibility and relevance. This refined approach was adopted to maintain a high standard of quality in the gathered literature and enhance the reliability of the insights obtained from the selected sources.

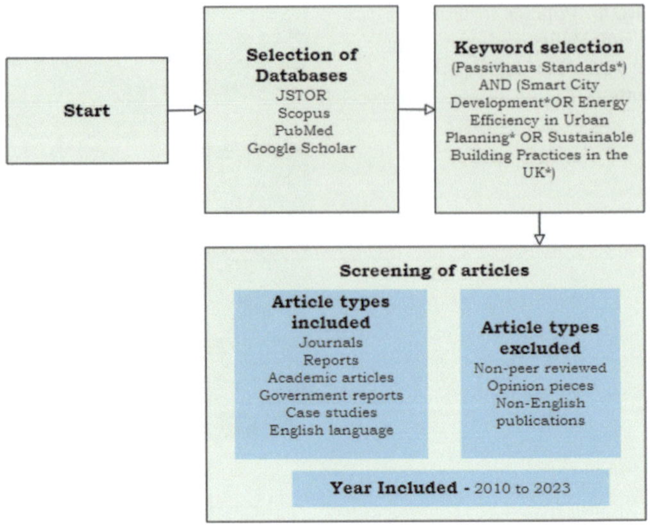

Fig. 6 Systematic literature review process flow diagram (by the authors)

2.3 Selection Criteria

The inclusion criteria for this review were thoughtfully designed to encompass a diverse array of studies directly related to the integration of Passivhaus Standards in smart city development. Encompassing peer-reviewed academic articles, governmental reports, and case studies, the focus extended to key aspects such as energy efficiency, policy analysis, and sustainable urban planning. While the primary geographical focus was the UK, international studies were also considered for comparative insights. To ensure the currency and relevance of information, studies published within the last 13 years were included.

Conversely, exclusion criteria were applied judiciously to filter out studies not directly addressing Passivhaus Standards or their application in smart city contexts. This involved excluding non-peer-reviewed articles, opinion pieces, and publications not in English. This exclusion process aimed to maintain a spotlight on high-quality, academic, and empirically grounded literature, aligning with the scholarly rigor of the review (Levett 2023).

The criteria for literature review selection were strategically devised to guarantee a thorough exploration of the most pertinent, up-to-date, and reliable research within the chosen field. This method was meticulously crafted to synthesize a comprehensive body of knowledge specifically relevant to the context of urban development and policymaking in the UK, with a primary focus on sustainable and energy-efficient building practices. By adhering to these criteria, the aim is to provide a nuanced and insightful perspective that contributes meaningfully to the discourse on sustainable urban development within the UK context.

2.4 Data Extraction Process

The data extraction process was executed to systematically procure vital information from each chosen study. Employing a standardized data extraction template ensured consistent collection of key data points across all studies. This template encompassed fields such as author(s), publication year, study location, research methodology, main findings, and conclusions. Moreover, it included specific sections tailored to the thematic focus of this review, such as energy efficiency metrics, policy implications, and practical insights within the context of smart city development.

Conducted by the authors, the data extraction involved a collaborative effort, with each study reviewed by at least two team members to uphold accuracy and uniformity. Any disparities in data extraction were judiciously addressed through thorough discussion and consensus-building. The efficiency of the process was augmented by leveraging reference management software, facilitating the organization, storage, and retrieval of information.

Throughout this phase, the emphasis was placed on extracting data directly contributing to the understanding of how Passivhaus Standards can be seamlessly integrated into smart city development, elucidating associated energy efficiency and policy implications. This rigorous and systematic approach to data extraction ensured that the review was anchored in a robust empirical foundation, thereby enabling the derivation of sound conclusions (Elamin et al. 2009).

2.5 Quality Assessment

The assessment of study quality was conducted with scrutiny, employing the Preferred Reporting Items for Systematic Reviews and Meta-Analyses (PRISMA) checklist as a robust evaluation tool. This checklist served as a standardized framework, systematically gauging the methodological robustness, transparency, and completeness of reporting in each included study. Criteria involving study design, sample size, data collection methods, analysis techniques, and the clarity of reporting were judiciously applied to assess study quality (Systematic Reviews n.d.).

Independently, two members of the research team evaluated each study, with any disagreements promptly addressed through rigorous discussion or consultation with a third researcher. This stringent quality assessment process played a crucial role in ensuring that the review was founded on reliable and valid studies, thereby bolstering the credibility of the findings and conclusions derived.

The utilization of the PRISMA checklist not only ensured a consistent and rigorous approach to quality assessment but also aligned the review with internationally recognized standards for conducting systematic literature reviews. This alignment not only fortified the academic robustness of the review but also enhanced its suitability for informing policy and practice within the field (The BMJ 2021).

2.6 Data Synthesis and Analysis

The synthesis and analysis of data from the selected studies were conducted through a rigorous thematic analysis approach. This method entailed the systematic identification, analysis, and reporting of patterns or themes inherent in the data. The resulting key themes included the effectiveness of Passivhaus Standards in diverse urban contexts, challenges in implementation, policy implications, and the role of smart technologies in enhancing energy efficiency.

The adoption of thematic analysis facilitated a holistic understanding of the subject matter by seamlessly integrating findings from a myriad of studies. This approach not only enabled the discernment of commonalities, differences, and emerging trends in the field but also provided a nuanced comprehension of the integration of Passivhaus Standards in smart city development (BMC Medical Research Methodology n.d.). This methodological depth enhances the reliability and depth of insights derived from the literature review.

2.7 Limitations of the Review Process

While this systematic literature review offers an understanding of the integration of Passivhaus Standards in smart city development, it is crucial to acknowledge certain limitations. The review's scope was confined to English-language publications, potentially leading to the exclusion of relevant studies published in other languages. Moreover, the emphasis on peer-reviewed literature and specific databases may have overlooked valuable gray literature or reports from alternative sources.

It's important to recognize that these limitations could potentially influence the breadth and depth of the findings. However, proactive measures were implemented to address these constraints. A broad spectrum of keywords and search strategies was employed, aiming to ensure a thorough search and analysis of available literature. These efforts, though, should be considered in the context of the review's constraints and their potential impact on the depth of the findings (Systematic Reviews n.d.).

3 Discussion on Passivhaus Standards Within the Landscape of Smart City Development

3.1 Interpretation of Results

This study offers an insightful exploration of integrating Passivhaus Standards within the framework of smart city development in the UK. The findings highlight both the synergies and challenges that arise when attempting to enhance energy efficiency in the multifaceted urban contexts found in the UK. The potential of Passivhaus Standards to mitigate energy consumption and reduce greenhouse gas emissions resonates with prior research. However, UK cities present distinctive architectural and urban planning challenges that make standardizing sustainable practices difficult, particularly due to heritage conservation, urban density, and the need for public acceptance (Minihane and Hashemi 2021).

Compared to regions like Northern Europe, where Passivhaus is more widely adopted, the UK demands more tailored strategies. While the nation's policies for carbon emission reduction align with the ethos of Passivhaus, effectively implementing these standards in existing infrastructure and regulatory frameworks requires a nuanced approach. The role of policy frameworks, technological innovations, and community engagement in facilitating the seamless integration of Passivhaus principles becomes essential. While the UK's commitment to carbon emission reduction and sustainable urban development aligns with the ethos of Passivhaus Standards, the practical implementation within existing infrastructures and regulatory frameworks demands a nuanced approach (Mallion 2016).

This study also sheds light on the indispensable role of community engagement and stakeholder collaboration in the successful implementation of Passivhaus Standards within the context of smart cities. The findings emphasize that the acceptance and success of these standards hinge not only on technical feasibility but also on the willingness of communities to embrace transformative changes. In the UK context, this entails an educational outreach to residents and stakeholders about the benefits of Passivhaus designs, dispelling misconceptions, and involving them in the planning process. The engagement of local communities can lead to more customized and embraced solutions, fostering a sense of ownership and commitment toward sustainable urban living. Moreover, the research reveals that successful integration often demands a multifaceted approach, amalgamating regulatory support, technological innovation, and public awareness campaigns. This approach holds the potential to significantly enhance the effectiveness of Passivhaus Standards in realizing the broader objectives of energy efficiency and sustainability in urban development (ULI United Kingdom 2022).

3.2 Implications

The findings of this research underscore several critical implications for both policymakers and urban development practitioners, emphasizing the need for comprehensive and forward-thinking approaches to embedding Passivhaus Standards into smart city frameworks. Integrating these standards not only offers immediate energy efficiency gains but also contributes to broader environmental, economic, and social objectives.

(A) *Strengthening Regulatory Frameworks and Policy Alignment*: The alignment of national and local policies with the Passivhaus Standards is vital for the realization of a sustainable built environment. By embedding Passivhaus principles into national regulations, such as building codes and planning policies, policymakers can ensure that energy efficiency becomes a non-negotiable aspect of urban development. Furthermore, targeted financial incentives, such as tax breaks, grants, or low-interest loans for Passivhaus-certified buildings, could significantly accelerate adoption.

(B) *Promoting Financial Incentives and Public–Private Partnerships*: To encourage widespread adoption, financial tools must be developed to lower the entry barrier for developers. A mix of subsidies, grants, and long-term financing options can make Passivhaus construction economically viable. Public–private partnerships should be strengthened, allowing the private sector to collaborate with municipalities on energy-efficient housing projects. Such collaborations not only distribute costs but also leverage technological innovations and expertise that may reside within private enterprises.

(C) *Addressing Urban Density and Heritage Conservation Challenges*: Adapting Passivhaus principles to densely populated and historically significant areas requires policy innovations that offer flexibility. Tailored retrofitting guidelines should be created to address the unique challenges of heritage buildings, allowing them to maintain structural integrity while improving energy performance. Urban planners and policymakers need to prioritize solutions that do not compromise cultural preservation, such as internal insulation techniques, energy-efficient window installations, and advanced HVAC systems that conform to Passivhaus Standards.

(D) *Facilitating Knowledge Sharing and Capacity Building*: Policymakers must prioritize capacity building within the construction and urban planning sectors by funding training programs focused on the technical skills necessary for Passivhaus construction. Additionally, public awareness campaigns should be launched to educate communities about the long-term benefits of living in Passivhaus buildings, including lower energy bills, improved indoor air quality, and enhanced comfort. This dual approach of professional development and public education is crucial to fostering widespread acceptance of these standards.

(E) *Encouraging Technological Innovation and Integration*: To fully realize the potential of Passivhaus Standards within smart cities, governments must encourage the adoption of emerging technologies such as IoT sensors for energy monitoring, AI for predictive energy management, and BIM for optimized building design. These technologies enhance the real-time energy efficiency of Passivhaus buildings, ensuring that they meet the evolving demands of urban environments. Policymakers can support this by offering tax incentives or grants to developers who incorporate such technologies into their projects, thus fostering a high-tech approach to sustainable urban living.

(F) *Creating Collaborative Policy Forums and International Benchmarks*: Collaborative policy forums should be established to enable knowledge sharing between local and international stakeholders. The UK can lead by setting global benchmarks for energy-efficient construction and engaging in partnerships with countries that have successfully implemented Passivhaus Standards. These forums can act as incubators for innovative policy development, enabling the UK to refine its approach to integrating Passivhaus into smart city frameworks while contributing to global sustainability efforts.

The identified potential synergy between smart city technologies and Passivhaus principles offers a promising pathway toward more livable, efficient, and sustainable urban spaces. The integration of these two approaches could usher in a new era of urban development, where technological advancements and environmental sustainability are prioritized hand in hand (Zhao 2023). This not only presents an opportunity for local transformation but also sets a precedent for global urban development practices.

3.3 Broader Perspectives

The implications extend beyond the immediate UK context. As cities globally confront challenges like climate change, resource conservation, and enhancing quality of life, the Passivhaus model emerges as a compelling blueprint for sustainable urbanization. Lessons learned from UK cities can be adapted to urban areas around the world, particularly in countries undergoing rapid urbanization and facing significant environmental challenges. This global dialogue on sustainable urban development practices must involve diverse stakeholders—urban developers, environmental scientists, policymakers, and residents—fostering the exchange of innovative solutions that prioritize sustainability, energy efficiency, and community well-being (Tzar n.d.).

The study findings highlight the necessity of a holistic approach in urban planning, integrating modern energy-efficient solutions with the existing urban fabric to achieve environmental goals while maintaining cultural identity (Lynch 2014).

4 Technological Innovations Supporting Passivhaus Integration into Smart Cities

The integration of Passivhaus Standards into smart city development necessitates leveraging advanced technologies to address energy efficiency, monitoring, and building management. Key technological innovations that align with the principles of Passivhaus and can enhance their application in smart cities include.

4.1 Internet of Things (IoT) for Real-Time Energy Monitoring

The incorporation of IoT-enabled monitoring systems is a key technological advancement facilitating the Passivhaus model's integration into smart cities. By providing real-time data on energy performance, these systems empower building managers to make informed decisions regarding energy use, ventilation adjustments, and indoor environmental quality. This continuous monitoring aligns with Passivhaus's emphasis on minimizing energy consumption while ensuring occupant comfort.

4.2 Building Information Modeling (BIM) for Design Optimization

Building Information Modeling (BIM) enhances the design and construction of Passivhaus-compliant buildings by offering detailed simulations of energy performance. BIM allows architects to optimize designs for insulation, airtightness, and heat recovery, ensuring that Passivhaus principles are adhered to from the initial planning stages. This technology not only supports energy efficiency but also ensures that these buildings are well integrated into the smart city's overall technological ecosystem.

4.3 Artificial Intelligence (AI) and Predictive Analytics for Energy Management

Artificial intelligence (AI) and predictive analytics offer innovative solutions for managing energy consumption in Passivhaus buildings. By predicting energy usage patterns and optimizing system performance in real time, AI can significantly reduce energy waste and improve building efficiency. This proactive energy management aligns with the goals of Passivhaus while contributing to the smart city's broader sustainability objectives.

4.4 Smart Grids for Optimized Energy Distribution

Smart grids facilitate the integration of Passivhaus buildings into the energy networks of smart cities by ensuring efficient energy distribution. The flexibility of smart grids allows Passivhaus-compliant structures to draw energy from renewable sources and minimize energy consumption during peak demand periods. This technology enhances the alignment of Passivhaus with the sustainable energy frameworks of smart cities.

4.5 Energy Storage Solutions for Renewable Integration

Energy storage systems, particularly battery solutions, provide critical support for integrating renewable energy into Passivhaus buildings. By storing surplus energy generated from renewable sources, these systems allow Passivhaus structures to maintain high energy efficiency and resilience, aligning with smart cities' goals of reducing dependence on non-renewable energy sources.

4.6 Smart Ventilation and Heat Recovery Systems

The deployment of smart ventilation systems with heat recovery technologies is essential to the energy efficiency goals of Passivhaus. These systems ensure continuous air exchange, maintain optimal indoor conditions, and reduce energy demand. Integrated with smart city technologies, smart ventilation systems provide a seamless solution for sustainable urban living.

5 Conclusion

In conclusion, this research has explored the integration of Passivhaus Standards into the fabric of smart city development within the UK, shedding light on its promise in advancing urban sustainability. The findings underscore the substantial benefits offered by Passivhaus Standards in terms of energy efficiency and environmental impact. However, the study illuminates the intricate challenges inherent in their implementation within the diverse urban landscapes of the UK, intricacies stemming from architectural heritage, urban densities, and existing policy frameworks.

Delving deeper into these complexities reveals a delicate equilibrium that must be struck between technological progress, historical preservation, and environmental sustainability. The study focuses on the imperative for contextually sensitive applications of Passivhaus Standards, ensuring that sustainability measures resonate effectively with and honor the cultural and historical tapestry of the UK's urban environments.

Moreover, the research emphasizes the crucial role of policy frameworks and stakeholder engagement in successfully adopting Passivhaus Standards. It posits that for smart city initiatives to be truly efficacious and sustainable, they must not only embrace technological innovations but also consider the socio-economic and cultural dimensions of urban living. This holistic approach becomes paramount for garnering public support and securing the enduring success of sustainable urban development initiatives.

Furthermore, the study directs attention to the transformative potential of Passivhaus Standards in shaping the future landscape of urban development in the UK. Through the adoption of these standards, cities can make substantial strides in mitigating their carbon footprint, bolstering energy efficiency, and elevating the overall quality of life for residents. However, this transition demands concerted efforts across diverse sectors, necessitating collaboration among government bodies, industries, and communities. As articulated by Campbell and Graham (2022), such collaborative endeavors are essential to fully realize the profound potential of sustainable urban living.

6 Contribution

This study makes a significant contribution to the field of sustainable urban development by advancing our understanding of the application of Passivhaus Standards in the context of smart cities, with a specific focus on the UK. It addresses a critical gap in the current literature, providing a nuanced examination of both the challenges and potentials associated with implementing these standards in the diverse landscape of urban settings. The insights garnered from this research hold practical implications for policy formulation and urban development practices, offering guidance for the creation of more sustainable, energy-efficient urban environments.

Beyond its immediate impact, this study acts as a catalyst for broader discussions and explorations into the integration of sustainability practices within urban planning. It underlines the imperative for multi-disciplinary collaboration in urban development, advocating for a collective effort involving architects, urban planners, environmentalists, and policymakers. This collaborative approach aims to devise solutions that not only prioritize efficiency and sustainability but also uphold cultural and social inclusivity. In doing so, the research amplifies the call for a holistic perspective in urban development that transcends individual disciplines.

In the ongoing discourse on sustainable urbanization, this research introduces a valuable dimension by emphasizing the pivotal role of innovative building standards, particularly the Passivhaus approach, in shaping the future fabric of cities. By shedding light on the intricate interplay between sustainability and urban development, this study contributes to a more complete understanding of the factors influencing the trajectory of future cities, reinforcing the importance of forward-thinking building standards in the ongoing dialogue about the sustainable evolution of urban spaces.

7 Study Limitations, Way Forward, and Future Research

While this research provides valuable insights into the integration of Passivhaus Standards within smart city development, certain limitations arise from its exclusive reliance on literature predominantly derived from the UK context. This may restrict the generalizability of the findings to regions characterized by distinct urban development dynamics. To enhance the breadth of knowledge, future research should broaden its scope to encompass diverse geographical contexts, thereby offering a more comprehensive understanding of the global applicability of Passivhaus Standards in smart cities. Furthermore, supplementing theoretical insights with empirical studies would contribute to a more robust foundation for policy and practice recommendations, reinforcing the practical relevance of the findings.

The study unveils promising avenues for future research in several key areas. Conducting comparative studies across different geographical contexts would afford a broader perspective on the global applicability of Passivhaus Standards in smart cities. Moreover, additional research could explore the long-term impacts of integrating these standards on urban development, examining social, economic, and environmental dimensions. Empirical studies involving case analyses of specific smart city projects implementing Passivhaus Standards could offer practical insights and validate the theoretical findings of this research, thereby enriching our understanding of the practical implications and effectiveness of such standards in diverse urban settings.

8 Management Implications

This study brings to the forefront the intricate yet highly advantageous challenge of integrating Passivhaus Standards into smart city development, specifically within the UK. In addition to contributing to academic discourse, the research holds valuable management implications for stakeholders in urban development. The findings derived not only highlight the significance of embracing sustainable practices but also offer actionable guidance for management teams. This information becomes a crucial tool for fostering resilience and efficiency, highlighting the strategic importance of incorporating Passivhaus Standards in the pursuit of a sustainable and thriving urban future.

References

Al Sharif, R., Pokharel, S.: Smart city dimensions and associated risks: review of literature. Sustain. Cities Soc. **77**, 103542 (2022)

Apanaviciene, R., Vanagas, A., Fokaides, P.A.: Smart building integration into a smart city (SBISC): development of a new evaluation framework. Energies **13**(9), 2190 (2020)

Audenaert, A., De Cleyn, S., Vankerckhove, B.: Economic analysis of passive houses and low-energy houses compared with standard houses. Energy Policy **36**(1), 47–55 (2008)

BMC Medical Research Methodology.: Methods for the thematic synthesis of qualitative research in systematic reviews. https://bmcmedresmethodol.biomedcentral.com (n.d.)

Building, G.: http://www.newbuilder.co.uk/news/newsFullStory.asp?ID=1744 (2006)

Campbell, N., Graham, D.: The Challenges Facing the Uptake of Passive House in the UK. BPTW (2022)

Cherrington, R., Goodship, V., Longfield, A., Kirwan, K.: The feed-in tariff in the UK: a case study focus on domestic photovoltaic systems. Renew. Energy **50**, 421–426 (2013)

Dameri, R.P., Ricciardi, F.: Leveraging smart city projects for benefitting citizens: the role of ICTs. Smart City Netw. Internet of Things 111–128 (2017)

Dequaire, X.: Passivhaus as a low-energy building standard: contribution to a typology. Energ. Effi. **5**, 377–391 (2012)

Du, Y., Gou, Z.: Predicting Passivhaus certification of dwellings using machine learning: a comparative analysis of logistic regression and gradient boosting decision trees. J. Build. Eng. 107849 (2023)

ECOSOC, U.: The UNECE–ITU smart sustainable cities indicators. UN ECOSOC: New York City, USA (2015)

Elamin, M.B., Flynn, D.N., Bassler, D., Briel, M., Alonso-Coello, P., Karanicolas, P.J., Montori, V.M.: Choice of data extraction tools for systematic reviews depends on resources and review complexity. J. Clin. Epidemiol. **62**(5), 506–510 (2009)

Emre, E., Sarul, L.S.: Investigation of smart city components by AHP-BWM-FUCOM and DEMATEL methods. Alphanumeric J. **10**(2), 197–222 (2022)

Fayomi, G., Mini, S., Fayomi, O., Mbonu, C., Udoye, N.: Sustainable smart city and its promising urban value: an overview. IOP Conf. Ser.: Earth Environ. Sci. (2019)

Feist, W., Schnieders, J., Dorer, V., Haas, A.: Re-inventing air heating: convenient and comfortable within the frame of the Passive House concept. Energy Build. **37**(11), 1186–1203 (2005)

Fokaides, P.A., Polycarpou, K., Kalogirou, S.: The impact of the implementation of the European Energy Performance of Buildings Directive on the European building stock: the case of the Cyprus Land Development Corporation. Energy Policy **111**, 1–8 (2017)

Garman–Kolokotsa, J., Synnefa, A.: Policy aspects of cool materials. Adv. Dev. Cool. Mater. Built Environ. 273 (2013)

Giesekam, J., Barrett, J., Taylor, P., Owen, A.: The greenhouse gas emissions and mitigation options for materials used in UK construction. Energy Build. **78**, 202–214 (2014)

Gregory, J.: How Passivhaus integration is revolutionising the UK's largest Passivhaus education building. AHR. https://www.ahr.co.uk/news/how-passivhaus-integration-is-revolutionising-the-uks-largest-passivhaus-education-building (2023)

Hameed, A.A.: Smart city planning and sustainable development. IOP Conf. Ser. Mater. Sci. Eng. (2019)

Hashemi, A., Hadjri, K.: Code for sustainable homes: opportunities or threats for offsite manufacturing and mass-customization? In: ZEMCH 2013 International Conference (2013)

Hui, C.X., Dan, G., Alamri, S., Toghraie, D.: Greening smart cities: an investigation of the integration of urban natural resources and smart city technologies for promoting environmental sustainability. Sustain. Cities Soc. 104985 (2023)

Lead, C., Griffin, C., Officer, S.: Subject heading: approval to access energy efficiency funding through the Green Deal & Energy Company Obligation (ECO) via a Greater London Authority (GLA). Policy 432590 (1708)

Levett, P.: Study selection and appraisal-systematic reviews. George Washington University. https://guides.himmelfarb.gwu.edu/systematic_review/study-selection-and-appraisal (2023)

Liang, X., Wang, Y., Royapoor, M., Wu, Q., Roskilly, T.: Comparison of building performance between Conventional House and Passive House in the UK. Energy Procedia **142**, 1823–1828 (2017)

Lockwood, M.: The political sustainability of climate policy: the case of the UK Climate Change Act. Glob. Environ. Change **23**(5), 1339–1348 (2013)

Lowes, R., Woodman, B., Fitch-Roy, O.: Policy change, power and the development of Great Britain's renewable heat incentive. Energy Policy **131**, 410–421 (2019)

Lynch, H.: Passivhaus in the UK: the challenges of an emerging market. A case study of innovation using mixed methods research. Doctoral dissertation, UCL (University College London) (2014)

Mallaburn, P.S., Eyre, N.: Lessons from energy efficiency policy and programmesin the UK from 1973 to 2013. Energ. Effi. **7**, 23–41 (2014)

Mallion, P.: Breathe a little easier: Passivhaus standard comparison with traditional designs. RICS (Royal Institution of Chartered Surveyors). https://www.isurv.com/info/390/features/10237/passivhaus_standard_comparison_with_traditional_designs (2016). Last accessed 19 Nov. 2023

McLeod, R.S.: An investigation into the performance of low energy and zero carbon buildings in a changing climate: applying the Passivhaus standard to the UK context, Cardiff University (2013)

Mihai, M., Tanasiev, V., Dinca, C., Badea, A., Vidu, R.: Passive house analysis in terms of energy performance. Energy Build. **144**, 74–86 (2017)

Minihane, C., Hashemi, A.: Implementation of Passive House standards in the UK, a roadmap for the local authorities. In: ZEMCH 2021: 8th Zero Energy Mass Custom Home International Conference. ZEMCH Network, Dubai, UAE (2021)

Mishra, K.N., Chakraborty, C.: A novel approach toward enhancing the quality of life in smart cities using clouds and IoT-based technologies. Digit. Twin Technol. Smart Cities 19–35 (2020)

Mohammadpourkarbasi, H., Riddle, B., Liu, C., Sharples, S.: Life cycle carbon assessment of decarbonising UK's hard-to-treat homes: A comparative study of conventional retrofit vs EnerPHit, heat pump first vs fabric first and ecological vs petrochemical retrofit approaches. Energy Build. **296**, 113353 (2023)

Mueller, T.: Revealed: The UK's Leading Smart Cities. Bee Smart City. https://www.beesmart.city/en/strategy/revealed-the-uks-leading-smart-cities (2017)

Mulliner, E., Kirsten, L.: Preparation for the Energy Act 2011 and minimum energy efficiency standards in UK commercial property. Int. J. Strateg. Prop. Manag. **21**(2), 183–198 (2017)

Palmer, J., Cooper, I.: United Kingdom housing energy fact file 2013. In: Department of Energy & Climate Change, Prepared under contract to DECC by Cambridge Architectural Research, Eclipse Research Consultants and Cambridge Energy. The Views Expressed are not Necessarily DECC'sp, p. 172 (2013)

Parry, C.: Passivhaus in Australia. Environ. Des. Guide (79), 1–10 (2014)

Passivhaus Trust.: Councils Can: Exemplar Local Plans & Policies. https://www.passivhaustrust.org.uk/news/detail/?nId=1209 (2023)

Passivhaus Trust.: Passivhaus & Planning. https://www.passivhaustrust.org.uk/competitions_and_campaigns/passivhaus-planninng/ (n.d.)

Pitts, A.: Passive house and low energy buildings: barriers and opportunities for future development within UK practice. Sustainability **9**(2), 272 (2017)

Rotuna, C., Smada, D., GheorghiÈ, A.: Smart city applications built on Big Data technologies and secure IoT. Ecoforum J. **6**(3) (2017)

Schleussner, C.-F., Rogelj, J., Schaeffer, M., Lissner, T., Licker, R., Fischer, E.M., Knutti, R., Levermann, A., Frieler, K., Hare, W.: Science and policy characteristics of the Paris Agreement temperature goal. Nat. Clim. Chang. **6**(9), 827–835 (2016)

Schnieders, J., Hermelink, A.: CEPHEUS results: measurements and occupants' satisfaction provide evidence for Passive Houses being an option for sustainable building. Energy Policy **34**(2), 151–171 (2006)

Schnieders, J., Feist, W., Rongen, L.: Passive Houses for different climate zones. Energy and Buildings **105**, 71–87 (2015)

Smart Plus Academy.: Building a Sustainable Future: Navigating the Intersection of Technology, Construction, and the Passivhaus Standard. https://smartplusacademy.com/building-a-sustainable-future-navigating-the-intersection-of-technology-construction-and-the-passivhaus-standard (2023). Last accessed 19 Nov. 2023

Su, K., Li, J., Fu, H.: Smart city and the applications. In: 2011 International Conference on Electronics, Communications and Control (ICECC) (2011)

Systematic Reviews.: Guidance to best tools and practices for systematic reviews. https://systematicreviewsjournal.biomedcentral.com (n.d.)

Systematic Reviews.: How to Properly Use the PRISMA Statement. https://systematicreviewsjournal.biomedcentral.com (n.d.)

Tahmasseby, S.: The implementation of smart mobility for smart cities: a case study in Qatar. Civ. Eng. J. **8**(10), 2154–2171 (2022)

Tatarestaghi, F., Ismail, M.A., Ishak, N.H.: A comparative study of passive design features/elements in Malaysia and passive house criteria in the tropics. J. Des. Built Environ. **18**(2), 15–25 (2018)

The BMJ.: (2021) The PRISMA 2020 statement: an updated guideline for reporting systematic reviews. https://www.bmj.com (2021)

Truong, H., Garvie, A.M.: Chifley passive house: a case study in energy efficiency and comfort. Energy Procedia **121**, 214–221 (2017)

Turney, S.: Systematic Review | Definition, Example & Guide. Scribbr. https://www.scribbr.com/methodology/systematic-review/ (2022)

Tzar, G.: Passive House and the Sustainable Development Goals (SDGs): Connecting an international building standard with global aims. Passipedia. https://passipedia.org/basics/passive_house_and_the_sustainable_development_goals (n.d.). Last accessed 19 Nov. 2023

UK Government.: Future Homes Standard 2019 Consultation. https://assets.publishing.service.gov.uk/media/5df78ecee5274a0910cb6d4d/Future_Homes_Standard_2019_Consultation.pdf (2019)

UK Government.: 2020 Energy Consumption in the UK (ECUK). https://assets.publishing.service.gov.uk/media/614c7c1ae90e077a2fc44250/2020_Energy_Consumption_in_the_UK__ECUK_.pdf (2020)

UK Government.: Net Zero Strategy. Department for Business, Energy & Industrial Strategy. https://assets.publishing.service.gov.uk/media/6194dfa4d3bf7f0555071b1b/net-zero-strategy-beis.pdf (2021)

ULI United Kingdom.: The Business Case for Passivhaus. https://uk.uli.org/business-case-passivhaus/ (2022). Last accessed 19 Nov. 2023

UN.: United Nations Department of Economic and Social Affairs. United Nations, New York (2018)

UN.: Sustainable Cities: Why they Matter. https://www.un.org/sustainabledevelopment/wp-content/uploads/2019/07/11_Why-It-Matters-2020.pdf (2020)

UNEP.: Energy Efficiency for Buildings. https://www.euenergycentre.org/images/unep%20info%20sheet%20-%20ee%20buildings.pdf (2009)

UNEP.: 2022 Global Status Report for Buildings and Construction. https://www.unep.org/resources/publication/2022-global-status-report-buildings-and-construction (2022)

World Bank.: Harnessing Smart Technology for Sustainable Development in Developing Countries. https://www.worldbank.org/en/news/feature/2021/01/18/harnessing-smart-technology-for-sustainable-development-in-developing-countries-tdlc (2021)

Xie, Y., Gilmour, M., Yuan, Y., Jin, H., Wu, H.: A review on house design with energy saving system in the UK. Renew. Sustain. Energy Rev. **71**, 29–52 (2017)

Zangheri, P., Economidou, M., Labanca, N.: Progress in the implementation of the EU energy efficiency directive through the lens of the national annual reports. Energies **12**(6), 1107 (2019)

Zhang, Y., Wang, J., Hu, F., Wang, Y.: Comparison of evaluation standards for green building in China, Britain, United States. Renew. Sustain. Energy Rev. **68**, 262–271 (2017)

Zhao, J.: Passivhaus standard as social housing model in the UK: barriers and opportunities. In: Implementing the UN Sustainable Development Goals–Regional Perspectives (Series) SDGs in the European Region. Springer. https://eprints.lincoln.ac.uk/id/eprint/50634 (2023). Last accessed 19 Nov. 2023

Hydrological and Hydraulic Modeling for Flood Protection: Case of Aglou Center, Tiznit Province

Salih Sara, Sadiki Mohamed, Chakiri Said, Bouabdli Abdelhak and Allouza Mohamed

Abstract

A flood corresponds to the slow or rapid overflow of water from its usual location. This inundation in general affects lands adjacent to a watercourse or variable level area (for example, a middle or major bed). This requires a flood protection study to define the areas that will be attenuated by the overflowing water and to take the necessary safety measures based on the results of an engineering study (for example, the construction of supporting walls, the widening of hydraulic structures, and the construction of reinforced concrete channels). In order to control flooding, a long series of flow measurements is necessary, which unfortunately is not the case in most cases. The hydrometric station of Ifni controlling the watershed of Oued Adoudou on an area of about 662 km^2 gave us a series of maximum daily rainfall for a period of 18 years, from 1996 to 2016, which fit well to several laws such as Gumbel's law, Log normal, Person III, and Gamma. From the fitting line, the values of probable rainfall for different return periods were determined; 46.1 m^3/s for a 10-year return period, 54.85 m^3/s for 20 years, 66.1 m^3/s for 50 years, and 74.7 m^3/s for 100 years. In order to estimate the flood flows by empirical formulas based on precipitation, the rainfall-flow correlation was verified by the HEC-HMS calculation code. Hydraulic modeling was made based on the calculated flows of the Oued, the object of the study, by the hydraulic modeling software HEC-RAS. This allowed us to identify the flood-prone areas and from the latter to protect the population and their property through the development of development schemes for protection against flooding.

Keywords

Flooding · Watershed · Flow · Hydraulic study · HEC-RAS · Hydraulic simulation

1 Introduction

Flooding is a natural hazard that can cause loss of life, material damage, and environmental degradation. In terms of frequency of occurrence, flooding is the second most frequent hazard in the Mediterranean basin, after earthquakes. They are the most destructive natural phenomenon, by the total number of casualties, and value of property damage, compared to any other type of natural disaster (Albu et al. 2020). The rate at which flooding occurs in recent times has been unprecedented, with the implication that only a few coastal rural, and urban environments still have their natural states unaltered (Nkwunonwo et al. 2020). Mathematical modeling is one of some scientific tools increasingly used to simulate and understand hydrological and hydraulic phenomena. In addition, geographic information systems are now indispensable to good management in this and other fields. In recent years, flood hazard assessment has considerably improved, especially due to the use of geographic information systems in integration with hydraulic modeling. Hydraulic modeling of floods, which occur as a result of dam failures, began to be applied in the 1940s and was greatly developed after 2000 when a series of studies

S. Sara (✉) · S. Mohamed · C. Said · B. Abdelhak · A. Mohamed
Laboratory of Geoscience, Department of Geology, Faculty of Science, IBN TOFAIL University, Kenitra, Morocco
e-mail: sara.salih@uit.ac.ma

S. Mohamed
e-mail: mohamed.sadiki@uit.ac.ma

B. Abdelhak
e-mail: bouabdli@uit.ac.ma

A. Mohamed
e-mail: mohamed.allouza@uit.ac.ma

P. Mohammad et al. (eds.), *Geographic Approaches to Climate Change and Mitigation: Urban and Rural Perspectives (Volume 1)*, Advances in Science, Technology & Innovation, https://doi.org/10.1007/978-3-031-92119-3_20

were carried out comparing different types of modeling and software (Albu et al. 2020). The major advantages of this integration are the possibilities to automatically prepare the geometric data and to import it into the hydraulic model. In addition, data generated by the hydraulic model can be transferred to the GIS to produce inundation maps and to perform further analysis (Abdallah 2009). Our study aims to delimit the flood-prone areas of the Tnine Aglou center, located in the province of Tiznit and exposed to flooding from the Oued Adoudou, through a hydrological study using the HEC-HMS calculation code and using the HEC-RAS software combination with Hec-GeoRAS tool for the hydraulic study. This will enable us to propose developments and protection against this risk along the watercourses that cross the center. The study is divided into the following three parts:

1. The first part is dedicated to a general presentation of the study area, as well as a diagnosis of the existing situation.
2. A second part dedicated to the hydrological study.
3. A third part consists of a hydraulic study.

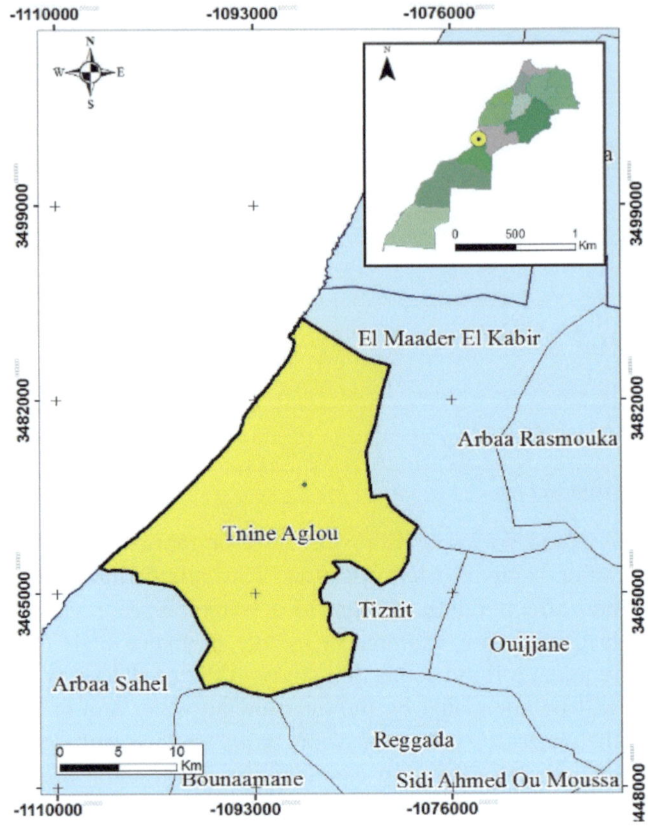

Fig. 1 Location of Aglou center (by the author)

2 Study Area

Tnine Aglou is a commune in the province of Tiznit, in the Souss-Massa region of Morocco. It has no urban center, but its administrative center is the village of Aglou. Its geographical coordinates are: 29° 46′ 30″ N, 9° 52′ 01″ W (Fig. 1).

Aglou lies at an altitude of 33 m. The prevailing climate in Tnine Aglou is steppe. At any time of year, precipitation is low in Aglou. The Köppen-Geiger climate map classifies the climate as BSh. The average temperature in Aglou Plage is 18.6 °C. Average annual rainfall is 206 mm. (https://fr.climate-data.org/). The average rainfall is lowest in July, at just 1 mm. In November, rainfall is the highest of the year with an average of 30 mm. The Tnine Aglou center is crossed by the Oued Adoudou and several Chaabats draining large watersheds. The Oued Adoudou drains a large watershed of around 668 km^2 and is characterized by a deeply incised bed, ranging in width from 20 m to around 10 m at its narrowest point just downstream of Zaouïa. Despite its large cross-section, with depths in excess of 10 m in some sections, Oued Adoudou caused the most material damage during the flood of 28/11/2016 (Fig. 2). A whole block of houses was swept away by the flooding on both banks of the wadi, as well as a major section of a tarmac road linking the two banks of the wadi. According to field surveys, the flooding problem associated with this watercourse is due to the reduced cross-section of the Oued in its downstream section. And for temperature, Aglou has a warm Mediterranean climate with dry summers according

Fig. 2 Damage on Oued Adoudou

to the Koppen-Geiger classification, with average annual temperatures ranging from 14 °C in December to around 24 °C in August (CSEC 2001).

The Aglou center belongs to the Anti-Atlasic domain, which is characterized by the extensive development of essentially calcodolomitic Infracambrian and Georgian series, locally up to 5,000 m thick (ABHSM 2003).

For basic infrastructure, the Tnine Aglou center has a 98.5% coverage of the potable water, as well as a 98.5% coverage of the electricity network (Monograph of Plan Communal). In terms of sanitation, the center does not have a liquid sewage network, but there are associations that deal with household waste, and there are also liquid sewage studies for the center. The Tnine Aglou center is accessible via the provincial road N° 1909/N°1016/N°1905 (Monograph of Plan Communal).

2.1 Existing Situation

Apart the existing situation of Adoudou Oued, many Chaabats present a risk of flooding in the study area.

Chaabat Tinmouloud

Following the flood of 28/11/2014, a detour of Chaabat Tinmouloud, by building an embankment dam, was carried out downstream of the road crossing; this development was started by the Tnine Aglou commune and completed by the Tiznit provincial planning authority. Allows the Chaabat to be diverted away from the center of Aglou via a neighboring Chaabat that crosses the provincial road 1909 via a crossing structure in the form of a simple gutter. The existing layout needs to be redesigned with good engineering practice, and the provincial road 1909 crossing needs to be hydraulically assessed to verify its capacity.

Chaabat Taghouilast

This Chaabat crosses many agricultural plots and buildings before entering the confluence with Chaabat Tinmouloud and crossing the center of Aglou. The damage caused by this Chaabat is still visible on the ground. Mostly fence walls were washed away by the floodwaters.

Confluence of Chaabat Tinmouloud and Chaabat Taghouilast

According to field surveys, the confluence of the 2 Chaabats caused significant material damage during the flood of 28/11/2016: the highest water level exceeded 2 m in places.

To protect the center, the Tiznit provincial planning authority built a gabion wall followed by a dyke along the right bank of the river. Hydraulic analysis of this structure will determine its ability to protect against exceptional rainfall events.

3 Methods

To carry out a study aimed at defining areas at risk of flooding, we will go through three stages and use several software packages to carry out hydrological and hydraulic modeling and provide a reliable result that will help land-use planning decisions.

3.1 Characteristics and Software Tools

3.1.1 Morphometric Characteristics

Before starting the study, it is essential to prepare the data needed for the modeling. To do this, we chose to work with ArcGIS to delimit the watershed (Fig. 3) and calculate its characteristics.

Seven catchment areas are the subject of this study:

- BV 1: Drained largely by Adoudou Oued.
- BV 2: Drained by Chaabat Taghouilast.
- BV 3: Drained by a small Chaabat which flows into the confluence of Chaabat Taghouilast and Tinmouloud.
- BV 4: Drained by the confluence of Chaabat.
- BV 5: Drained by Chaabat Tinmouloud.
- BV 6: Drained by a deviation of Chaabat Tinmouloud.
- BV 7: Drained by the downstream part of Adoudou Oued.

Table 1 presents the summary of the results of calculating the watershed parameters where BV: basin; S: Area; Leq: Equivalent length (Km); Hmax: maximum height (m); Hmin: minimal height (m); Ig: Overall slop index Ic: Classique slope index; Ds: the specific height difference.

3.1.2 Rainfall Characteristics

Rainfall in the study area is based on the analysis of precipitation collected at the Ifni rainfall station, which covers the watershed of the Oued concerned by this study. This main station has been in service since April 1979 and has the following Lambert coordinates (X: 37.950; Y: 273.750; Z: 20). Annual maximum daily rainfall recorded at this station over 18 years (1998/2016). The rainfall series recorded at this hydrological station was adjusted using Hyfran Plus software according to Gambel law distributions (Fig. 4).

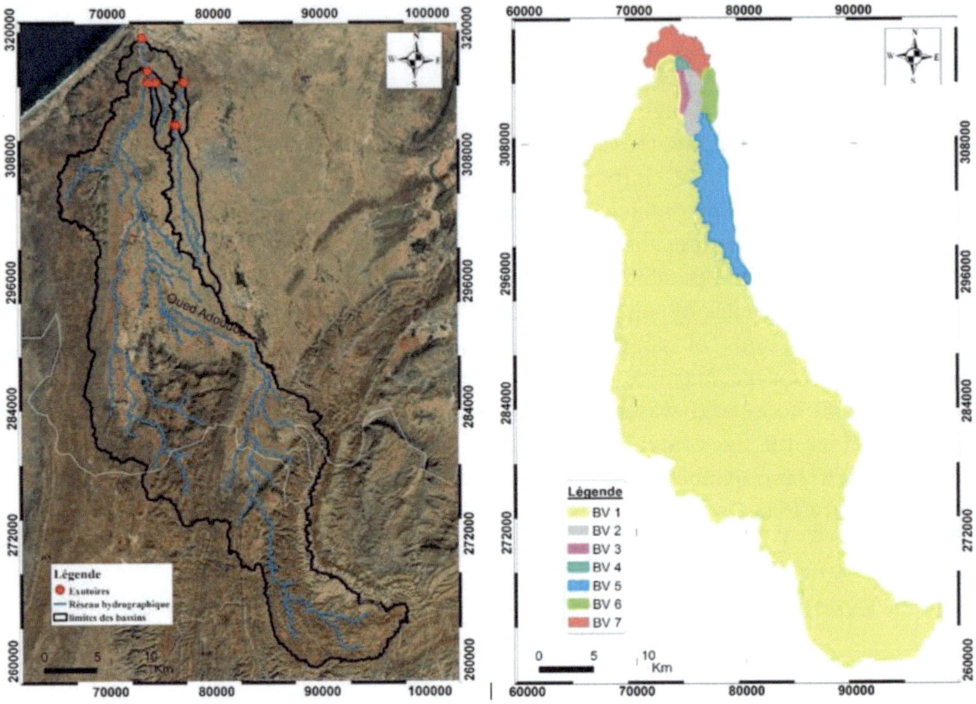

Fig. 3 Map of the hydrographic structure and delimitation basins (by the author)

Table 1 Calculation results for watershed relief characteristics

Watershed	S (Km^2)	Leq (m)	Hmax (m)	Hmin (m)	Ig (%)	Ic (m/Km)	Ds (m)
BV1	636.06	58.29	1316	111	1.32	20.67	33.96
BV2	8.50	5.2	224	134	1.24	17.31	3.62
BV3	2.54	2.91	193	130	1.41	21.65	2.25
BV4	1.08	1.51	158	120	2.38	25.17	2.47
BV5	32.00	12.28	371	186	2.83	15.07	16.01
BV6	5.73	4.25	214	140	0.50	17.41	1.20
BV7	14.59	3.61	312.00	70.00	6.56	67.04	25.06

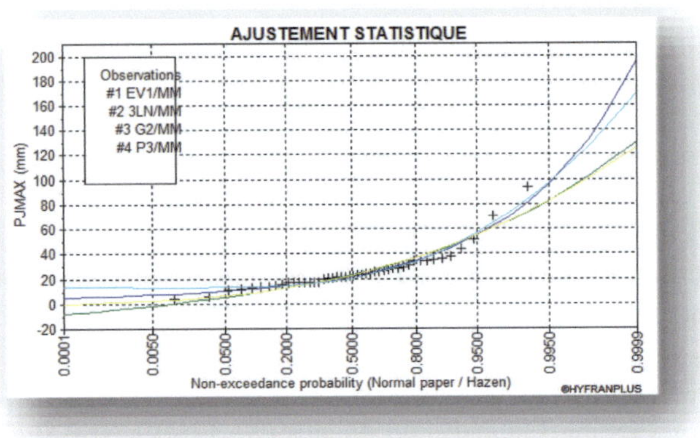

Fig. 4 Statistical fit curves with different methods

The value adopted for each return time is the average of the values adjusted and accepted by the software according to each law (Table 2).

3.1.3 Flow Characteristics

The hydrological study aims to determine the flood flows for different frequencies, for the Oued and chaabats inventoried in the diagnostic phase and which are likely to cause flooding in the Aglou center. The determination of flood flows in the study area will be based on the comparison of the results of several methods adapted to the nature of the catchment areas studied.

The runoff coefficient quantifies the flow of water to be evacuated by the network. It can be defined as the ratio of runoff to precipitation. In the present study, runoff

Table 2 Statistical adjustment of daily rainfall using several methods

Return period	Maximum daily rainfall				
T (years)	Pearson III	Lognormal	Gumbel	Gamma	Adopted
2	20	21.6	23	22.5	22.75
5	33.7	34.3	36.7	36.9	36.8
10	45	44.2	45.8	46.4	46.1
20	56.8	54.7	54.5	55.2	54.85
50	72.8	69.9	65.8	66.4	66.1
100	85.1	82.5	74.2	74.6	74.4

coefficients have been calculated based on supervised classification and image processing carried out in ArcGIS software, the latter calculating the surface area of each type occupying the ground and applying the runoff coefficient corresponding to each surface area (fig.).

$$C = \frac{\mathrm{C1A1} + C2A2 + \cdots + \mathrm{CnAn}}{\mathrm{A1} + \mathrm{A2} + \mathrm{A3}}$$

In our study, we used a Landsat satellite image to identify the various occupations. The satellite image was chosen based on an analysis of cloud cover over the study area. To create the map in ArcGIS, we used a supervised classification based on the maximum likelihood method, which is a probabilistic approach: for each pixel, we determine its probability of belonging to one class rather than another (Table 3).

3.1.4 Estimating Flood Flows

To estimate flood flows, there are several methods, using calculation with standard empirical formulas, or hydrological modeling with the SCS curve number method. In this study, we adopt the average of the values obtained between the rational method and SCS curve number (hydrological modeling).

$$Qt = \frac{\mathrm{C} * \mathrm{I(T,\ Tc)} * \mathrm{A}}{3.6}$$

Table 3 Result of runoff coefficient calculation

Basin	Runoff coefficient
B1	0.69
B2	0.78
B3	0.79
B4	0.62
B5	0.78
B6	0.75
B7	0.61

where Qt: Peak flow in m^3/s for return period T.

I: Rain intensity in mm/h.

T: Return period (years).

Tc: Time of concentration in min.

A: Catchment area in Km^2.

C: Runoff coefficient.

For a flood protection project, determining the peak flow rate is an essential step in selecting the appropriate management measures and determining the flood-prone areas.

Having delimited the catchment areas in the study zone and determined their characteristics, the focus of this step is to evaluate flood flows for return periods ranging from 10 to 100 years. The calculation methods identified, using various empirical formulas, take into account two essential factors:

- Catchment size, in terms of surface area and/or length.
- The topographical gradient of the watershed, in terms of its slope or gradient difference.

3.1.5 HEC-HMS Modeling

The HEC-HMS model is a semi-distributed hydrological model that is widely used in flood event simulations (Lin et al. 2022). The data required for watershed modeling depends on the methods used for infiltration, runoff, and groundwater flow. In the following sections, we will describe the choice of parameters used for each of these methods. We chose to model infiltration using the SCS curve number method (runoff was estimated using the SCS method). The base flow was not considered.

Each of the sub-catchments considered for this modeling was subjected to a rainfall event of varying frequency, the duration of which coincided with the time of concentration. Theoretical rainfall input to the model was calculated by statistical adjustment of the sample of maximum daily rainfall recorded at the Ifni station. Runoff volume and loss were calculated using the SCS curve number (SCS-CN) method. The SCS unit hydrograph (SCS-UH) was used to simulate direct runoff (Lin et al. 2022). The lag time is also needed for predictive dam breach analysis (Fig. 5).

In the absence of hydrometric stations in the study area, we propose to adopt the maximum values of peak flows obtained by the various methods. Flood volumes were calculated based on typical hydrographs selected for the various catchment areas (Table 4).

Some of the resulting flood hydrograms are presented here (Fig. 6).

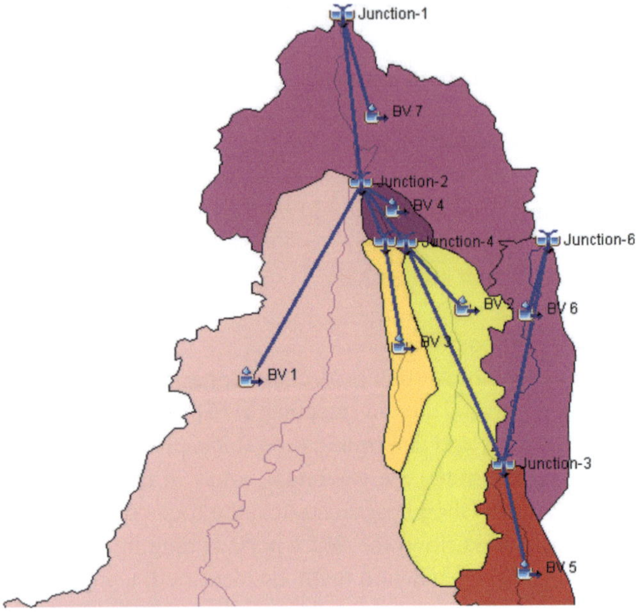

Fig. 5 Basin modeling on HEC-HMS (by the author)

Table 4 Summary table of HEC-HMS results

Watershed	Drainage area	Flow (m^3/s)
BV 1	636.06	832,28
BV 2	8.5	304,05
BV 3	2.54	86,22
BV 4	1.08	41,29
BV 5	32	460,88
BV 6	5.73	212,23
BV 7	14.59	105,01

3.1.6 Hec-GeoRAS (for ArcGIS 10.4.1)

Hec-GeoRAS is a set of procedures, tools, and utilities for processing geospatial data in ArcGIS, using a graphical user interface (Fig. 7). It is used to prepare geometric data for export to HEC-RAS and to recover simulation results from HEC-RAS. The geometric file to be exported is based on a digital terrain model, preferentially in TIN format.

To produce geometry files, we need a basic background, which can be a DEM, a topographic map, or satellite images. The more documents you have, the more likely you are to produce good results. However, the reliability of these results depends very much on the quality of the documents (scale, resolution, etc.), as well as their handling and use. We used the digital terrain model (30 m) to study the topography of the area. We created a TIN with ArcToolbox to classify the altitudes of the area and define the relief of the terrain.

The Hec-GeoRAS menu is then used to assign object identification codes (ID), nomenclatures, and other essential attributes. It also enables topography to be extracted, whether for cross-sections, watercourses, hydraulic structures, or storage surfaces. The final operation is to save the result and export it as a RASimport.sdf file that HEC-RAS recognizes.

3.1.7 HEC-RAS Modeling

Hydrologic Engineering Center-River Analysis System (HEC-RAS) version 5.0.7 is designed to model one-dimensional free-surface flows in natural and artificial channels. HEC-RAS is a hydraulic model designed by the US Corporation Engineers-Hydraulic Engineering Center to model river flow (Kashfy et al. 2020). HEC-RAS software was used to determine the flood inundation area and floodwater depth using a flood hydrograph from the HEC-HMS simulation and DEM. Again, one- and two-dimensional (1D and 2D) unsteady flow analyses were used for simulation since both are essential parts geometrically defined in setting the boundary conditions (Phyo et al. 2023). The creation of geometry files requires a basic background, which can be a digital terrain model, a topographic map, or satellite images. In reality, the more documents available, the better the results tend to be. However, the reliability of these results depends closely on the quality of the documents (scale, resolution, etc.), as well as their handling and use. The digital terrain model (30 m) was used to study the topography of the area. A TIN was created with ArcToolbox to classify the area's elevations and define the terrain's relief. The Hec-GeoRAS menu is then used to assign object identification codes (ID), nomenclatures, and other essential attributes. It also enables the extraction of topography, whether for cross-sections, watercourses, hydraulic structures, or storage surfaces. The final operation is to save the result and export it as a RASimport.sdf file, recognized by HEC-RAS (Fig. 8).

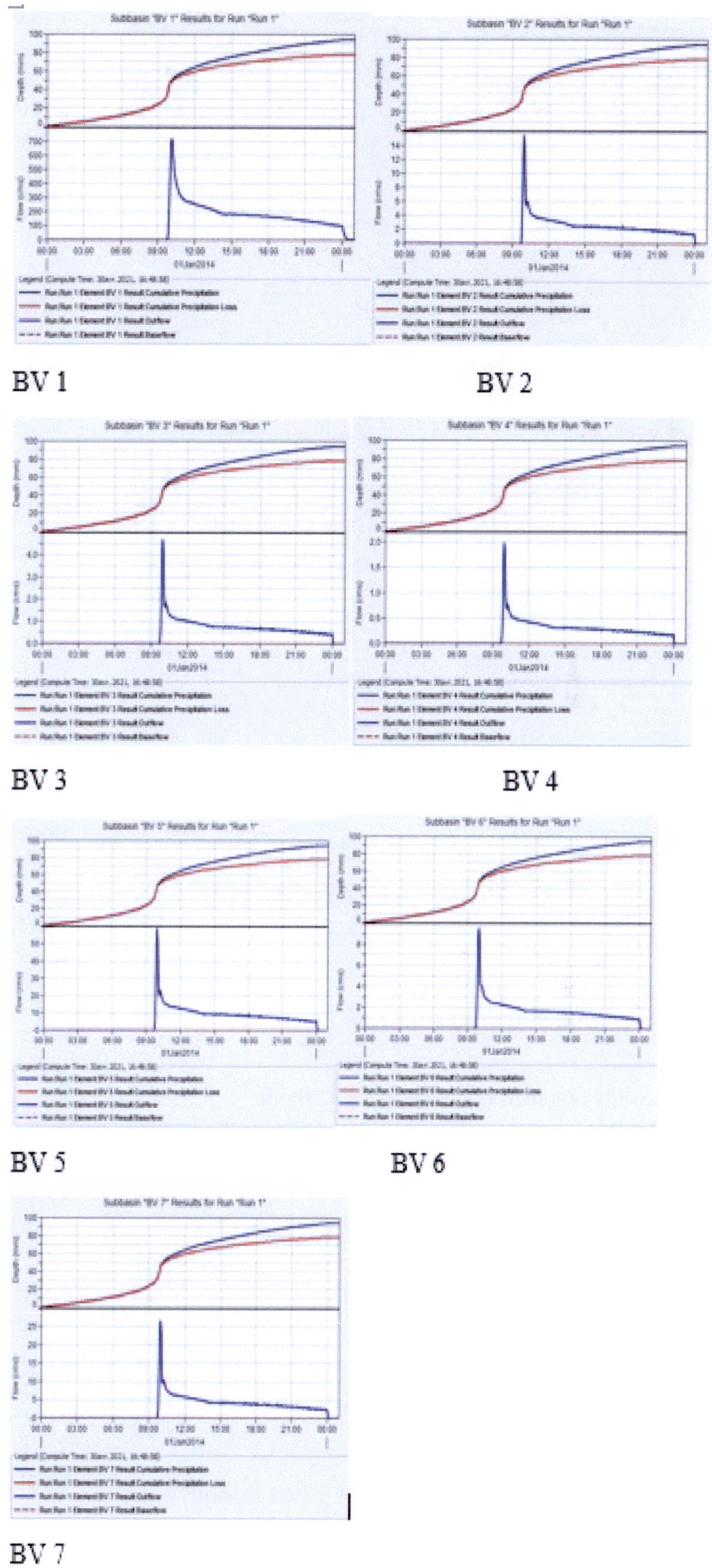

Fig. 6 Some of the resulting flood hydrograms (by the author)

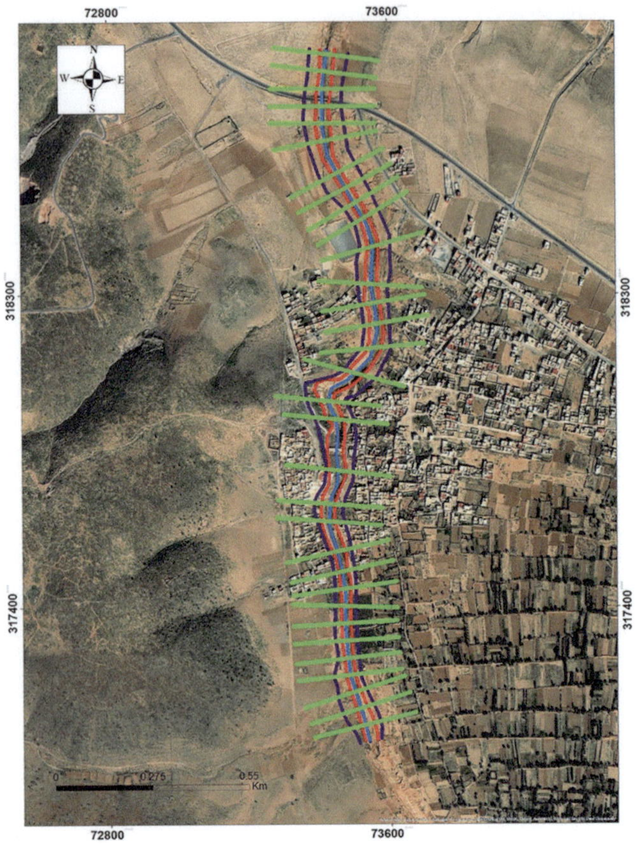

Fig. 7 Modeling the Oued Adoudou section on Hec-GeoRAS (by the author)

4 Result

4.1 Selected Flows

The results obtained by the rational method:

Basins versants	Débits de pointe en (m^3/s)			
	Q10	Q20	Q50	Q100
BV1	441.05	502.55	676.51	743.28
BV2	17.77	67.42	27.26	29.95
BV3	6.99	39.01	10.73	11.78
BV4	3.51	49.69	5.38	5.91
BV5	71.75	662.49	110.06	120.92
BV6	8.17	558.56	12.53	13.77
BV7	37.14	2539.03	56.97	62.59

It is remarked that the results obtained by software and calculation codes are much more accurate given that they take into account several data, unlike empirical formulas which are very minimalist.

The flows selected for the twenty-year, fifty-year, and hundred-year rainfall frequencies are the result of an analysis and comparison of the flows calculated by the method presented above. For this purpose, the average of the flows calculated by the empirical formula method and those selected above was used (Table 5).

The results of the hydrological study are a useful document for hydraulic modeling, enabling the calculation of flow profiles and flood propagation in rivers. The hydrological study is the foundation on which all river hydraulic modeling is built, enabling flood forecasting and hydrological risk management in flood plains.

4.2 Results of Hydraulic Modeling

4.2.1 Adoudou Oued

Simulations of the 100-year flood in its natural state have demonstrated that the flow regime of Oued Adoudou is fluvial. Velocities are high, varying in most cases from 3 to 9 m/s, but can reach maximum values of 14 m/s, particularly at profile 26, corresponding to a steepening of the Oued's riverbed. The average water level corresponding to a 100-year flood is estimated at 5 m, with a maximum height of 8 m. In the residential area, the average water level is 4 m, with a maximum of 6.6 m. The flood-prone areas include dwellings on both banks of the Oued Adoudou, as well as the cemetery on the right bank (Figs. 9 and 10).

HEC-RAS gave a representation of the variance of water head in (m) as a function of flow in m^3/s (Fig. 11).

Flow velocity (Fig. 12).

4.2.2 The Confluence of the Tinmouloud and Taghouilsat Chaabats

It should be noted that following the floods that occurred on 28/11/2014, which caused significant damage, a gabion wall was installed to protect the site, followed by a protective dike.

- The flow regime is generally fluvial, with the Froude number generally less than unity; the section of the watercourse with a torrential flow corresponds to the confluence of the two Chaabats at the Aglou center cemetery.
- The average water level corresponding to the 100-year flood is estimated at 1.5 m, with a maximum height of 3.5 m.
- The flood zones affect homes on both banks of the Oued over a distance of around 350 m on the left bank and 410 m on the right bank.

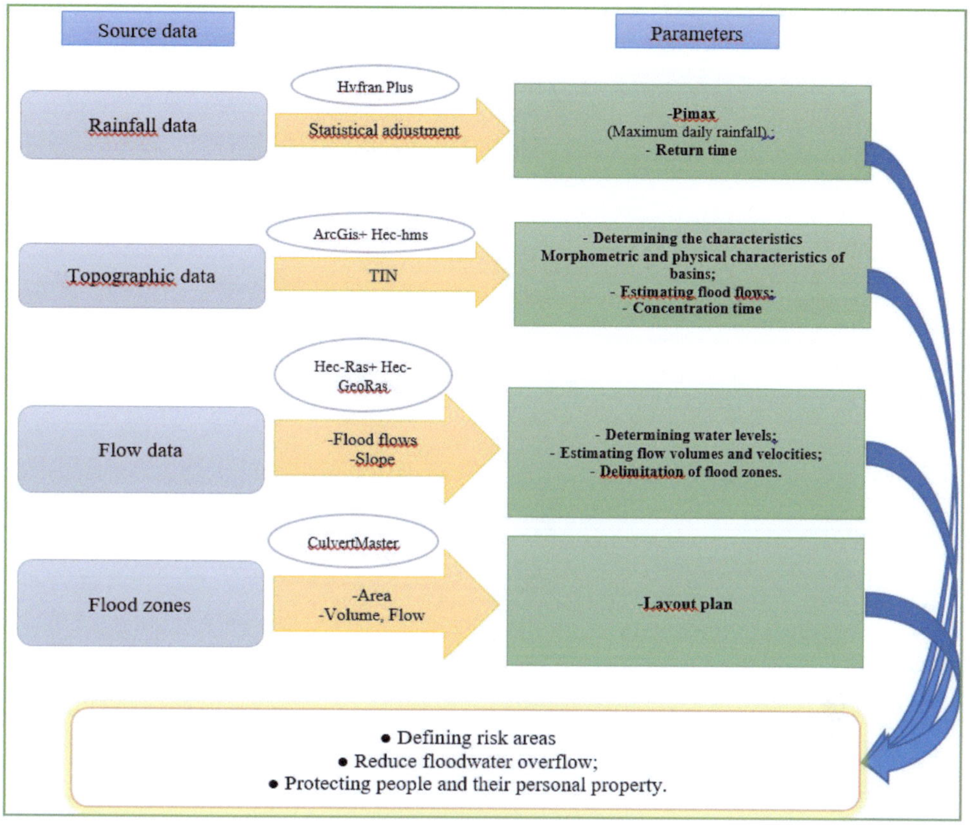

Fig. 8 Diagram summarizing the equipment used and the methodology followed (by the author)

Table 5 Flow results

Basins	Flow (m^3/s)			
	Q(T10)	Q(T20)	Q(T50)	Q(T100)
BV 1	427.4	517	730.5	787.7
BV 2	68.6	79.2	86.5	167.6
BV 3	27.4	33.6	34.4	49.5
BV 4	13	21	16.4	23.6
BV 5	217	320.8	275.3	290.9
BV 6	48.5	142.2	60.9	113
BV 7	109.5	529.4	138.9	167.6

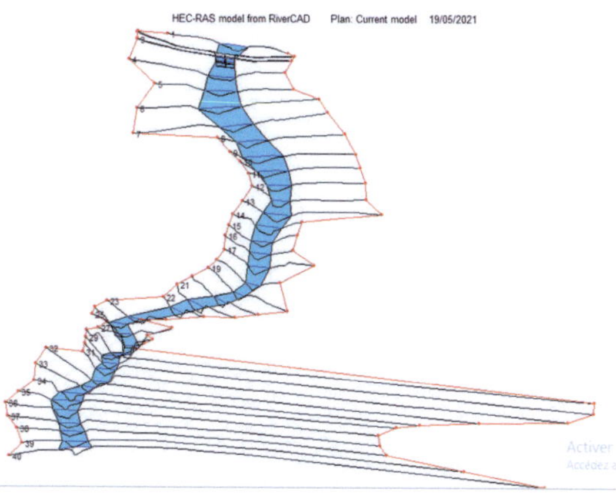

Fig. 9 3D view of Oued Adoudou on HEC-RAS

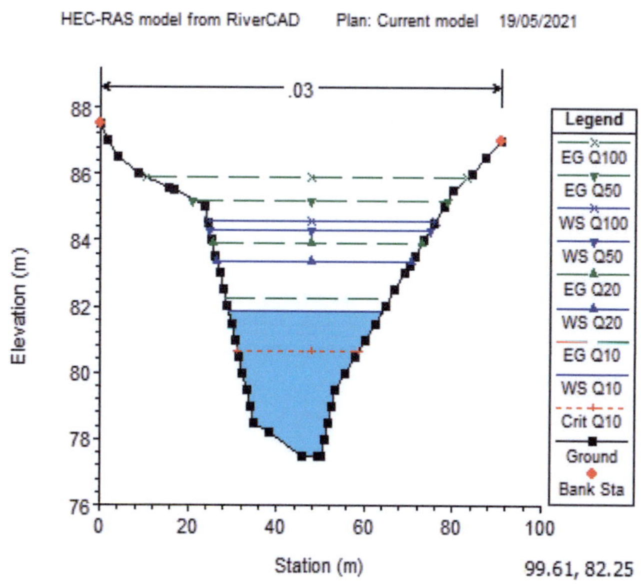

Fig. 10 Water level in some cross-sections (Oued Adoudou) during floods (by HecRAS)

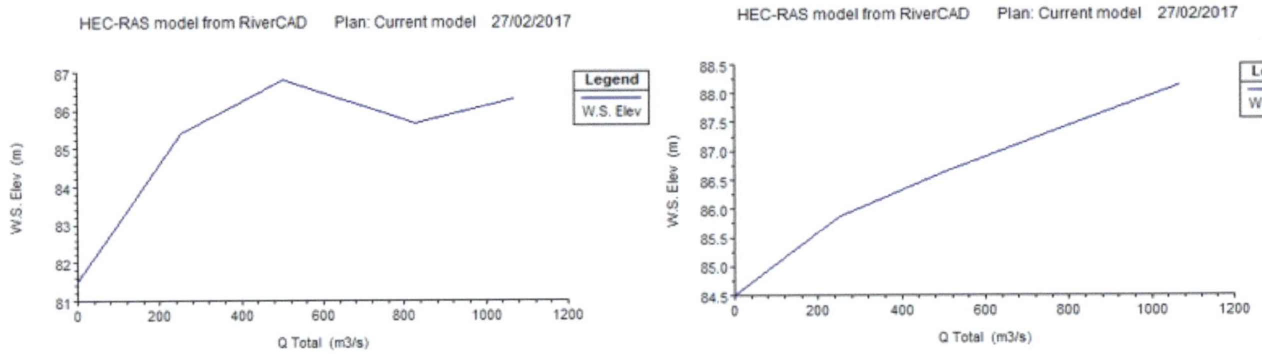

Fig. 11 Flow area curves as a function of flow rates for profiles 21 and 26

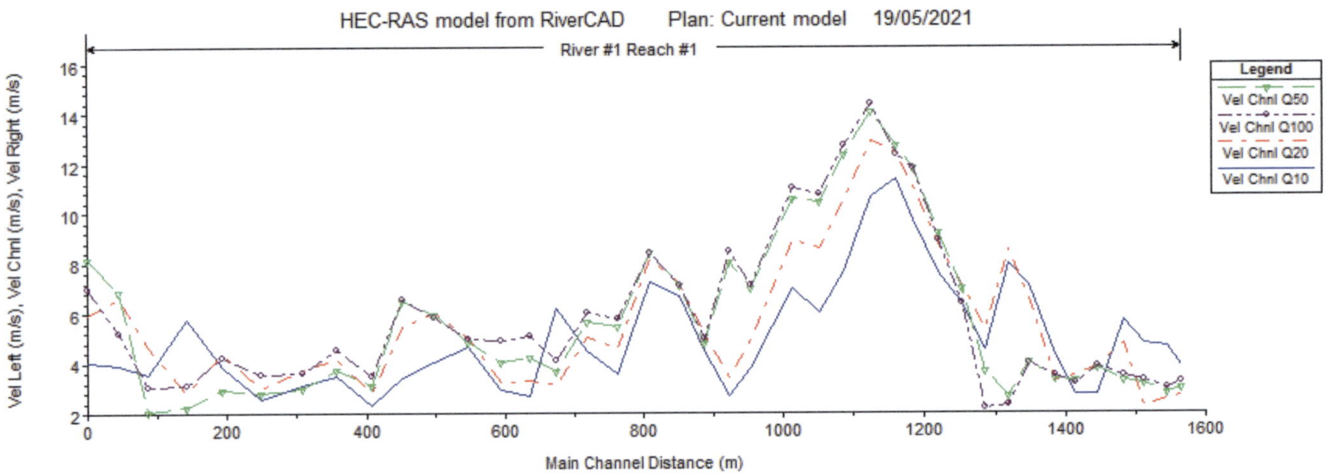

Fig. 12 Flow velocity

- Flow velocities are very high during the 100-year flood, with an average velocity of 6 m/s and a point maximum of 14 m/s.
- When the 100-year flood passes, the width of the flow from the confluence of the Chaabat Tinmouloud and Taghouilast reaches around 90 to 100 m per section (Fig. 13).

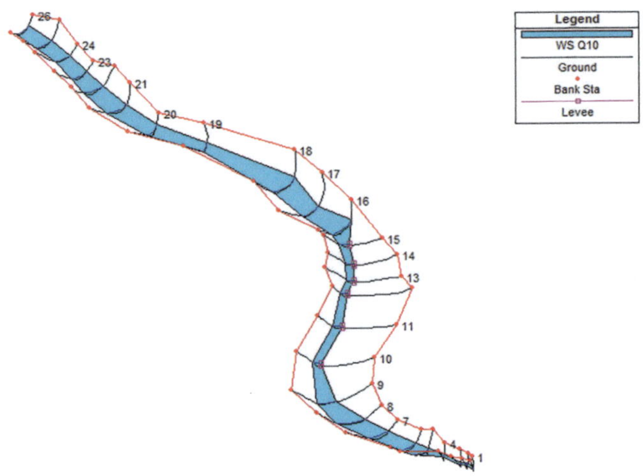

Fig. 13 3D view of the confluence

Examples of cross-sections obtained at the confluence:

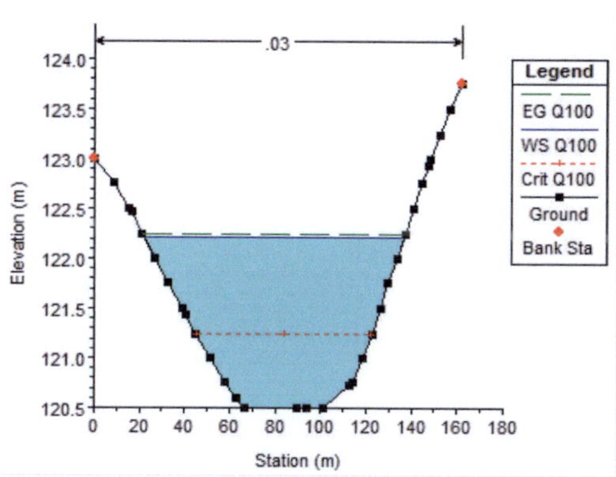

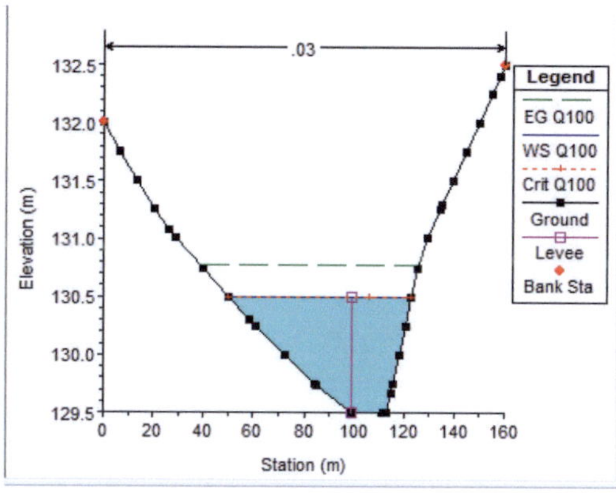

Curve of the water height as a function of the flow rates obtained at profile 11 of this Chaabat:

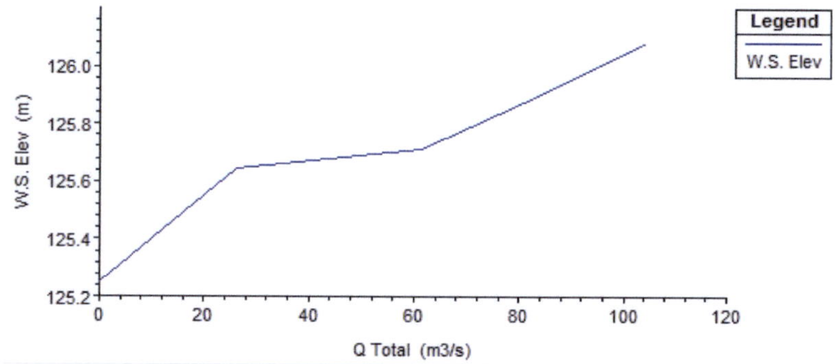

Velocity:

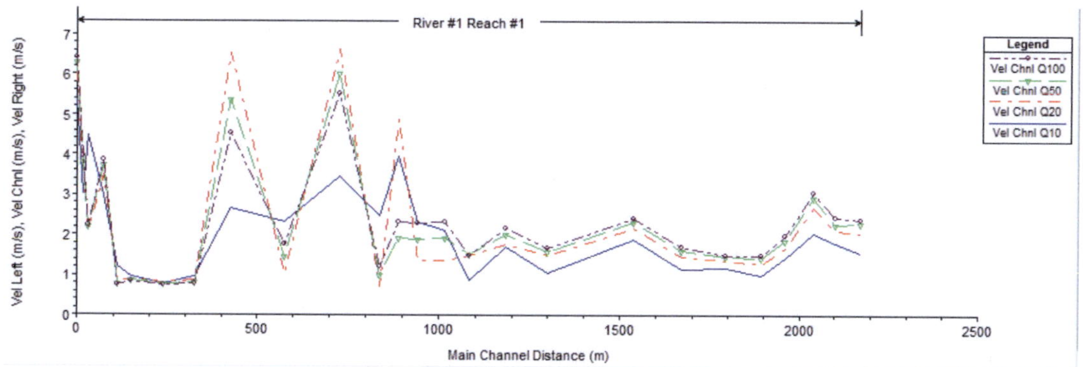

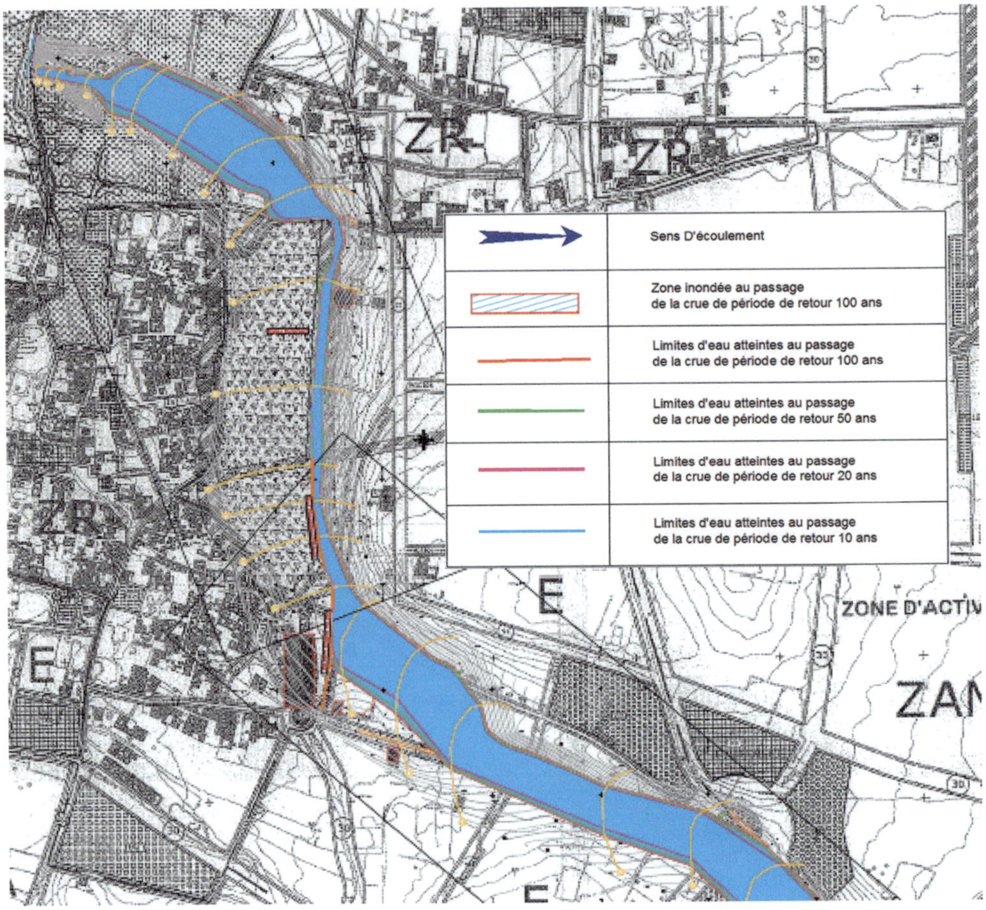

Fig. 14 Delimitation of the confluence flood zone on the development plan (by the author)

4.3 Delimitation of Flood Plains by Hec-GeoRAS

The areas affected by flooding were determined by mapping them after extrapolation using GIS software. To highlight this risk in the center of Aglou, we re-established the beds of the water sections that cross the areas based on the topographic map and rectified them using a satellite navigator that allows visualization of the affected areas. But since flows are generally low, we will map only those areas affected by the centennial flood (Fig. 14).

5 Discussion

To return to the aims of this study, it should be noted that the protection of the population and its property was the most important goal. To achieve this, the hydrological and hydraulic study uses essential physical and mathematical paths to obtain the results and scientifically interpret them to offer them to civil engineers to use as bases for their proposals and their solutions, which aim to implement works, its essential role is to react to flood disasters (Fig. 15).

Hydraulic simulation of the Oued Adoudou during historical floods has shown that its cross-section is incapable of conveying the flows generated in its catchment area without overflowing and causing relatively significant damage, due to its course through the inhabited areas of the Aglou center. Two scenarios are being considered for the development of Oued Adoudou:

- The first consists of developing the Oued without the recommended dam upstream of the Oued at the outlet of the relief.
- The second consists of developing the Oued's channel while considering the construction of a flood protection dam upstream.

6 Conclusion

This study aims to find solutions to protect the center of Tnine Aglou from flooding by establishing development schemes for the Oued Adoudou as well.

A hydrological study of the watersheds in the study area enables us to estimate the time of concentration and

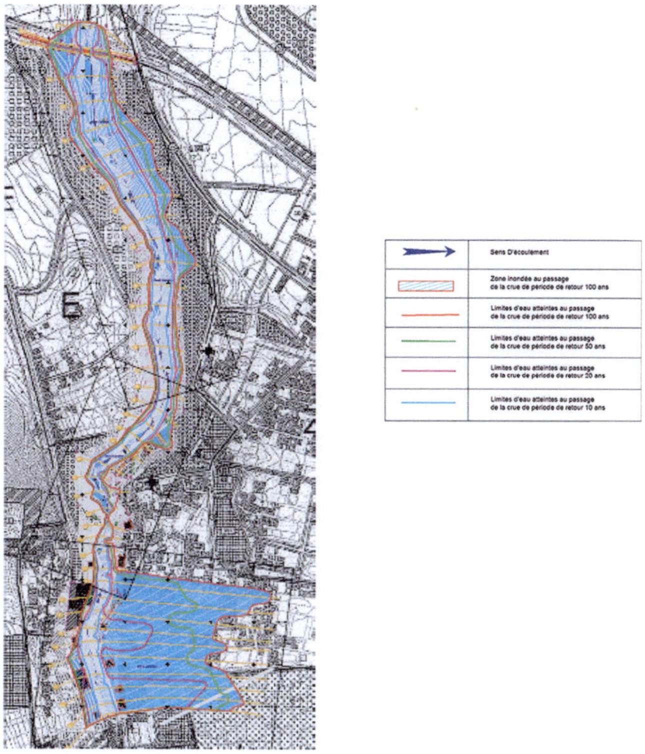

Fig. 15 Flood zone modelling on Oued Adoudou on a management plan (by the author)

flood flows for different return periods, based on rainfall data and watershed characteristics. For example, for the return period T = 100 years. The next step was to construct the geometric frameworks of the various watercourses. Hec-GeoRAS software generated cross-sections for Oued Adoudou and the Chaabats at 50m intervals. The geometric models were designed to ensure a better representation of the watercourses studied, and above all, to perceive the limits of flooding during the passage of the 100-year flood. The results of the hydraulic simulation, using HEC-RAS software, are presented in the form of summary tables containing values for flow, water level, velocity, and Froude number in each section of the three watercourses, thus providing all the parameters needed to make decisions about any future development in the area studied. These results are also presented graphically in the form of cross-sections, longitudinal profiles, and 3D views, indicating the water levels reached by the various floods. Concerning development solutions for Oued Adoudou, it was proposed to build a protective masonry barrier wall, to reinforce the hydraulic structure to protect the Tnine Aglou center from flooding, and to build sills upstream of the Oued Adoudou watershed to minimize bed load.

References

Abdallah.: Distributed GIS Approach for Flood Risk Assessment. International Journal on Advances in Security, Toronto (2009)

ABHSM.: Synthesis of Geophysical Studies Carried Out in the Souss and Chtouka Plains. ABHSM (2003)

Albu, L.M., et al.: Dam Breach Size Comparison for Flood Simulations. A HEC-RAS Based, GIS Approach for Drăcs, ani Lake, Sitna River, Romania. Iasi, Romania (2020)

CSEC.: Superior Council for Water and Climate-Master Plan for the Development of Water Resources in the Souss-Massa Basins. Agadir, 9th Session. CSEC (2001)

Kashfy, et al.: HEC-RAS One-Dimensional Hydrodynamic Modelling for Recent Major Flood Events in Pahang River. Springer Nature Switzerland AG (2020)

Lin, Q., et al.: Web-Based Prototype System for Flood Simulation and Forecasting Based On. Environmental Modelling and Software, China (2022)

Monograph of Plan Communal de Développement Commune. Monograph of Plan Communal, Aglou

Nkwunonwo, U.C., et al.: A Review of the Current Status of Flood Modelling for Urban Flood Risk Management in the Developing Countries. Scientific African (2020)

Phyo, A.P., et al.: Managing Dam Breach and Flood Inundation by HEC-RAS Modeling and GIS Mapping for Disaster Risk Management. Case Studies in Chemical and Environmental Engineering, Tsukuba (2023)

Modelling Urban Radiation Exchange Using a High-Resolution 3D Approach

Guido de Bonfioli Cavalcabo', Stavros Stagakis, Christian Feigenwinter, Markus Kalberer, Alexander Damm, Nina Buchmann, and Dominik Brunner

Abstract

Urbanization impacts Earth's systems by increasing CO_2 emissions, creating urban heat islands (UHI), and affecting climate, energy use, and public health. Understanding urban microclimates and energy and CO_2 balances is crucial for designing effective climate change adaptation and mitigation strategies. This paper presents preliminary results from the UrbaNature project, focusing on vegetation's role in urban microclimates. The goal is to develop a 3D microscale ecophysiological model for CO_2, water, and energy exchanges between plants and the urban atmosphere, including a 3D radiation exchange model for urban canyons with tree canopies. Using high-resolution digital surface models (DSMs), land cover, and leaf area index (LAI), we construct a 3D urban landscape of isometric voxels categorized as buildings, trees, terrain, or empty. This landscape supports radiation exchange simulations via a ray tracing algorithm, computing parameters like sky view factor (SVF), and light transmission coefficient, stored in look-up tables (LUTs). Meteorological data from an urban tower in Basel, Switzerland, is used to evaluate the model's accuracy by comparing simulation results with ground-level station data. Preliminary results show that the model effectively replicates observed shortwave radiation patterns, highlighting its potential to enhance urban climate research and understanding of urban energy balance.

Keywords

Modelling · Radiation exchange · Urban canyon · Tree canopies · Urban Microclimate · Climate adaptation

1 Introduction

The urbanization trend is on the rise and leads to denser and increasingly bigger cities that nowadays can have a population of more than 20 million. The process of urbanization is expected to continue in the near future, and projections show how by 2050 almost 70% of the world population will live in urban areas (UN 2019).

These areas are not only the main location where populations, resources, and infrastructure are located but they are also responsible for more than 70% of the global greenhouse gas (GHG) emissions and if we consider also commodities consumed but not produced in cities those arrive likely to 80% (Hoornweg et al. 2020). Also, their density and land

G. de Bonfioli Cavalcabo' (✉) · S. Stagakis · C. Feigenwinter · M. Kalberer
Department of Environmental Sciences, University of Basel, Basel, Switzerland
e-mail: guido.cavalcabo@unibas.ch

S. Stagakis
e-mail: stavros.stagakis@unibas.ch

C. Feigenwinter
e-mail: christian.feigenwinter@unibas.ch

M. Kalberer
e-mail: markus.kalberer@unibas.ch

A. Damm
Department of Geography, University of Zurich, Zürich, Switzerland
e-mail: alexander.damm@geo.uzh.ch

Eawag-Swiss Federal Institute of Aquatic Science and Technology, Dübendorf, Switzerland

N. Buchmann
Grassland Sciences Group, ETH Zurich, Zürich, Switzerland
e-mail: nina.buchmann@usys.ethz.ch

D. Brunner
Empa-Swiss Federal Laboratories for Materials Science and Technology, Dübendorf, Switzerland
e-mail: dominik.brunner@empa.ch

P. Mohammad et al. (eds.), *Geographic Approaches to Climate Change and Mitigation: Urban and Rural Perspectives (Volume 1)*, Advances in Science, Technology & Innovation, https://doi.org/10.1007/978-3-031-92119-3_21

consumption make them particularly vulnerable to weather extremes such as heatwaves, storms, and flooding (Handmer et al. 2012; Kundu and Pandey 2020).

Not only rapid urbanization is increasing the vulnerabilities of people residing in cities to climate change (United Nations Framework Convention on Climate Change (UNFCCC) 2017) but it also poses a growing health risk for them (Michel 2020; Lu et al. 2021). The more cities grow, the more their climate deviates from their rural counterparts. Urban heat island (UHI) is a well-known example of this deviation that depends, among other things, on urban morphology and tree canopy cover (Singh and Sharston 2022), but effects on boundary layer, wind flows, convection, and humidity are also other urban induced meteorological changes (Arnfield 2003; Karlický et al. 2020; Marelle et al. 2020). Cities do not only affect their local climate, but in fact are one of the most important anthropogenic influences on regional and global climate (Qian et al. 2022; Kalnay and Cai 2003), and as highlighted by the IPCC, they play a critical role in responding to climate change (Dodman et al. 2022). This is why climate change mitigation and adaptation strategies need to consider the transformation of cities towards resilience, sustainability, and climate neutrality (Masson et al. 2020), in fact while contributing significantly to global greenhouse gas emissions, they also present unique opportunities to sustainable strategies (Greenwalt et al. 2018).

While some aspect of how urbanization exacerbates climate change have been documented (Zhan et al. 2023), accurately quantifying its full multiscale impacts remains challenging (Kalnay and Cai 2003). This is due to several phenomenon such as the nonlinear interaction cities have with climate change (Krayenhoff et al. 2018), the intrinsic spatial and structural inhomogeneity of urban environment, and the sparse distribution of in situ weather stations combined with the low temporal resolution of the remote sensing (Kousis et al. 2021). These factors make it difficult to rely solely on measurements and cities inventories to fully understand the mechanisms driving urban microclimate, especially when a specific spatial and time resolution is required (Stagakis et al. 2023).

For a thorough understanding of the mechanisms controlling the microclimate in cities across the world, models that focus on urban environments are both crucial and more feasible than many experimental alternatives (Sharma et al. 2021). Digital simulations can be used to analyse microclimatic conditions (Ibrahim Abd El-Rahman et al. 2022) and help us understand the influence that urban areas have on climate change at local, regional, and global levels and help researchers to explore different scenarios and interventions within a controlled setting.

In this paper after an introduction on urban modelling we present the development and testing of an algorithm that assesses 3D radiation exchange and specifically the shortwave radiation interception across the urban canopy in order to estimate the radiation that reaches the leaf surface of each vegetation voxel, as our first step in the developing of a 3D urban environment model. Then in the last chapter results are discussed, and conclusions are drawn.

2 Urban Environment Modelling

UrbaNature is a multidisciplinary project that aims to investigate how urban vegetation influences local carbon, energy, and water cycles from different standpoints, and the development of a 3D ecophysiological model for urban climate analysis at the microscale is a part of it.

During the last decades, microclimate modelling has evolved substantially, with significant contributions from models like ENVI-met and SOLWEIG, which assess the impact of urban geometry on microclimatic processes and effects such as energy balance and thermal comfort within urban canyons and around buildings, incorporating vegetation primarily for its shading and evaporative cooling impacts (Bruse and Fleer 1998; Lindberg et al. 2008). For example, Eingrüber et al. (2023) recently used ENVI-met to demonstrate the utility of modelling in analysing the effectiveness of urban green infrastructures during extreme heat waves. The Urban Multi-scale Environmental Predictor (UMEP) integrates some of these aspects within a GIS framework, offering a broader and easy accessible analysis of urban climates (Lindberg et al. 2018).

The complex dynamics of plant interactions with the urban environment can be better understood with a 3D urban microscale ecophysiological model that takes into account the urban form. For instance, the MUSE model, introduced by Lee and Lee (2020), represents a significant step towards incorporating realistic urban surface processes into microscale atmospheric simulations and underscores the potential benefits of incorporating detailed vegetation processes in urban climate models. Moreover, Javanroodi and Nik (2020) explored how urban morphology affects wind speed and air temperature at the microscale significantly impacting the microclimate, suggesting that a volumetric model could enhance our understanding of these dynamics. Xu et al. (2021) review of 3D tree reconstruction for urban modelling also highlights how detailed vegetative representation that includes trees geometric features computed as voxels is required in urban microclimate simulation but rarely adopted. Also, physiological characteristics are currently obtained from empirical data instead of computed (Xu et al. 2021).

Most existing models tend to oversimplify the complex interactions between vegetation and urban atmospheres, particularly in terms of complex plant processes like radiation exchange and evapotranspiration. A comprehensive 3D urban microscale model that incorporates detailed ecophysiological processes could provide insights into how plants contribute to the urban microclimate impacting air quality, temperature, carbon dioxide, and oxygen levels, helping to explore the dynamic feedback mechanisms between urban vegetation and atmospheric conditions. Such a model would not only enhance our understanding of urban greening interventions but if combined with studies on urban plant resilience could also assist in designing better cities more suited for the future climate (Ernst et al. 2022; HosseiniHaghighi et al. 2020; Zhao et al. 2014).

3 Methodology

As shown in Fig. 1, the first action performed by the model is to combine high-resolution digital surface models (DSM) for buildings, trees, and terrain and convert those three 2D maps into a single 3D discrete array made of cubic isometric voxels. The geospatial input data have a resolution of 1 m, while each voxel is an S × S × S cube, where the size of the voxel (S) can range from 2 to 10 m depending on the run. These voxels can be only of a single urban type between tree, building, terrain or empty, and they will share all properties with voxels of the same category. The only exception is made for the leaf area index (LAI) that is assigned to each tree voxel according to a high-resolution (1 m) LAI map of the area (Mutti 2020). The LAI for each tree voxel is computed averaging all the non-zero values in the interested S × S area and then dividing the resulting value for each of the tree voxel present in the vertical stack.

In Fig. 2, we can see the study area and the respective 3D voxel grid with S = 5 m. In the process of transposing the geospatial information into a 3D grid, the predominant element within a voxel' volume dictates its assigned urban type between the four options. We acknowledge an inherent approximation error due to the discretization of continuous space that results in the merge of distinct physical features (e.g., parts of trees or buildings) into a single voxel.

For example, our analysis in the neighbourhood used as testing area, with S = 5, shows that between 10 and 13% of the voxels are generated from an area with an overlap between urban types, but also that voxels exhibiting significant overlap between two or more categories (the volume of the primary category exceeds that of the secondary by less than 25%) is around 1% of the total voxel population. This overlap indicates a noteworthy deviation from a precise correspondence between voxel representation and urban environment. While the discretization error cannot be avoided, the uniform distribution of these overlaps and of the urban type that prevail minimizes bias towards any specific category. Consequently, the overall volumetric representation of each category (trees,

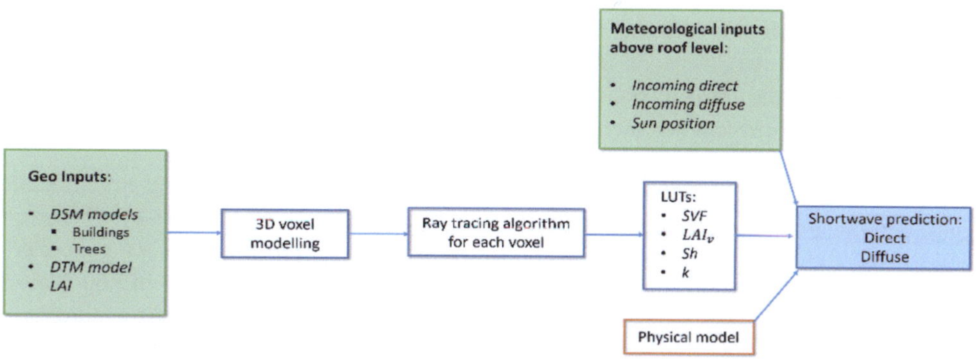

Fig. 1 Flow chart of the shortwave model. We have in green the inputs, in blue the computations, and in orange the physical model (by author)

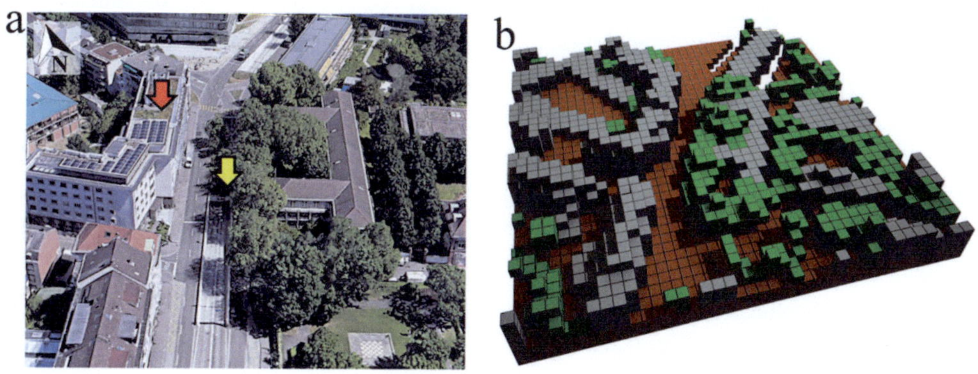

Fig. 2 Comparison between an aerial view (Google Earth) of the considered neighbourhood and the modelled version of the same area with voxels of three different types: brown voxel are terrain, grey ones are building, and green one are trees with S = 5 m. The red arrow in the left picture shows the location of the roof station, while the yellow arrow points at the street level measuring station (by author)

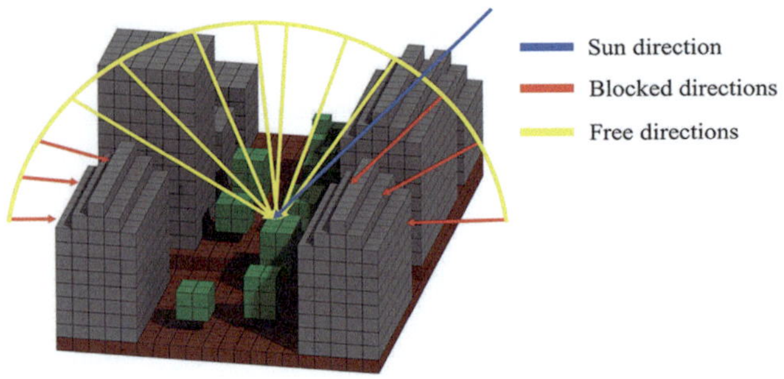

Fig. 3 3D visualization of how we determine the SVF from Eq. (1) and Sh from Eq. (2) for a tree voxel in an urban canyon with trees. The gray blocks represent buildings while the green ones trees and the brown ones the soil. The blue arrow is the direct sunlight coming from the sun at the considered hour, and since is not blocked by the building it will produce Sh = 1, the yellow arrows are a slice of the free directions (F) while the red ones are a slice of the blocked directions (by author)

buildings, terrain, and empty spaces) within the 3D grid sufficiently reflects the spatial occupancy observed in the input digital models.

Using the 3D grid, we apply a voxel traversal algorithm (Amanatides 1987; Mena-Chalco 2023) starting from each "interesting" voxel and ending when reaching a building or the horizon. We consider as interesting each tree voxel and the external voxels of buildings and terrain conformations. We chose a horizontal coordinate system precision (D) that will determine the directions discretization; for example, for D = 5°, as in all the example shown in this paper, the traversal algorithm is applied with a step of 5 degree in both azimuth and elevation. Then with the traversal algorithm we determine for each direction considered if it reaches the horizon or is blocked by a building, the amount of LAI present between a voxel and the horizon (LAI_v) and how many voxels of every type are seen by a specific voxel. All this information about the 3D spatial configuration around each "interesting" voxel is stored in lookup tables (*LUTs*) to be accessed when computing variables of the different radiation exchange processes, as for example in this paper direct and diffuse shortwave radiation reaching the street-level sensor. As shown in Fig. 3, with this information we are able to determine the sky view factor (*SVF*) using this simple equation:

$$SVF = \frac{F}{T} \tag{1}$$

where F is the number of directions that are free to the horizon, and T is the total number of the directions towards the horizon dome.

According to the location of the sun, we can estimate the amount of shortwave radiation reaching each voxel with an extinction law that takes the form of the Lambert–Beer equation (Duursma and Makela 2007; Ponce de León and Bailey 2019), for example, for direct shortwave at street level:

$$SW_{direct} = SW\downarrow_{dir} \cdot e^{(-k \cdot LAI_{dir})} \cdot Sh \tag{2}$$

where Sh is a term that accounts for the shadow (0 when the considered voxel is shadowed and 1 otherwise) based on the LUT information on building interception towards the direction of the sun, LAI_{dir} is LAI_v towards the sun direction, and $SW\downarrow_{dir}$ is the direct shortwave above the tree and roof level. k is the extinction coefficient, for which we choose a simple function of ψ, the sun zenith angle, assuming spherical leaf angle distribution (Campbell 1986; Ponce de León and Bailey 2019):

$$k = \frac{\sqrt{1 + \tan(\psi)^2}}{2} \tag{3}$$

Using similar simplification as Eq. (2) and using Eq. (3) as the extinction coefficient, we can determine the diffuse shortwave radiation at the street level:

$$SW_{diffuse} = SW\downarrow_{dif} \cdot SVF \cdot e^{(-k \cdot LAI_{dif})} \tag{4}$$

$$LAI_{dif} = \frac{1}{d}\sum LAI_v \tag{5}$$

where $SW\downarrow_{dif}$ is the incoming diffuse radiation measured on the roof, LAI_v is the amount of LAI present between the considered voxel and the horizon along all directions that are not blocked by buildings, and d is the amount of those directions for the considered voxel so that the LAI_{dif} shown in Eq. (5) is a normalized average of all the LAI blocking the atmosphere.

To evaluate this model, we run the simulation for radiation exchange between atmosphere and street level in the 3D urban canyon we generated starting from the (200 × 200)m^2 area depicted in Fig. 2, situated in the city of Basel, Switzerland. We used meteorological data from 2023 collected from the stations situated on the locations shown by the arrows in Fig. 2a. The sensors used to measure incoming radiation above tree and roof level are a 2-axis sun tracker (INTRA, BRUSAG) situated in an unobscured location on the roof of the building that thanks to a tracker can separate direct and diffuse incoming light, and a four-component CNR4 Net Radiometer (Kipp and Zonen) installed at a 2 m height over the ground at the street level. Both direct and diffuse incoming radiation measurements on the roof are used as inputs while the downward total incoming shortwave radiation from the station in the urban canyon beneath is used for comparison with the model results.

Before comparing the two set of values, we exclude from the dataset hours with rainfall exceeding 0.01 mm (recorded by the roof station) and the hour after each rain entry, because the water on the radiation sensor lenses creates highly unpredictable distortion. We also remove outlier entries, considering outliers all those observations in which the difference between the street and the roof measured shortwave is more than 700 W/m^2 where the average difference for the whole set of hourly observations is 125 W/m^2 and entries where one or both of the two measured values (street and roof) is recorded as not a number (NaN) due to instrument malfunction. Excluding all the rainy hours, the number of eliminated entries for the 2023 run are 17 outliers and 8 NaN.

4 Results and Discussion

Figure 4 shows the hourly diurnal means and standard deviations of the street level measured and simulated incoming shortwave radiation separating the periods that were not totally overcast ($SW \downarrow_{dir} > 10$ W/m^2) (Fig. 4a) and the fully overcast periods ($SW \downarrow_{dir} < 10$ W/m^2) (Fig. 4b) for the whole 2023. As we can see in Fig. 4a, while the overall values of modelled shortwave are in good accordance with the data collected at the street level, there is a time shift of approximately one hour. According to the study area characteristics (Fig. 2), during morning hours the street station area is affected by both trees and building shadowing while in afternoon hours the incoming shortwave is affected only due to building shadowing.

The model achieves better results during cloudy days (Fig. 4b), and this means that the modelling of diffuse shortwave is performing better than the direct shortwave part. This result implies that the diffuse radiation can be accurately modelled combining the contributions of the SVF and LAI_{dif} (Eq. 4).

The observed time shift of Fig. 4a reflects a discrepancy between the model and the observations that is observed only during sunny conditions. This is because volumetric models have inherent errors due to discretization, in this case of the urban geometries and of the directions taken into account, and these errors are evident only when direct sunlight is blocked by 3D objects. Therefore, taking into account this inherent effect even the not cloudy results are consistent. With the value of S in the order of meters, in the inhomogeneous urban landscape, this kind of error cannot be avoided.

The shadowing effect that produces the time shift is also clear in the scatterplot shown in Fig. 5 where we can see that afternoon hours are consistently underestimated while the morning ones are mostly overestimated. It is remarkable that there is a clear linear relationship in the middle of the day, around 14:00, but with a trend towards underestimation of the measured radiation. It is clear that errors caused by obstructions very near the street sensor, such as tree branches, street poles, and other small but relevant physical objects are not accounted for in the model. Such errors are considered minimal in the specific station set-up.

As shown in Table 1, there is quite a difference between daily and hourly error metrics. For the daily metrics, we calculate the average of simulated and measured solar radiation values for each day, considering only days where the missing daytime hours do not exceed 20%, and then we compare these daily averages over the entire dataset to assess errors. The hourly metrics instead are derived by the comparison one to one of every valid hour entry for the two datasets. On a daily basis, the model exhibits a slight overestimation that depends mainly on the discretization and on when shadow is casted on the sensor's area as shown in the shifted peak in Fig. 4. The relatively moderate RMSE (30.2 W/m^2) value suggests

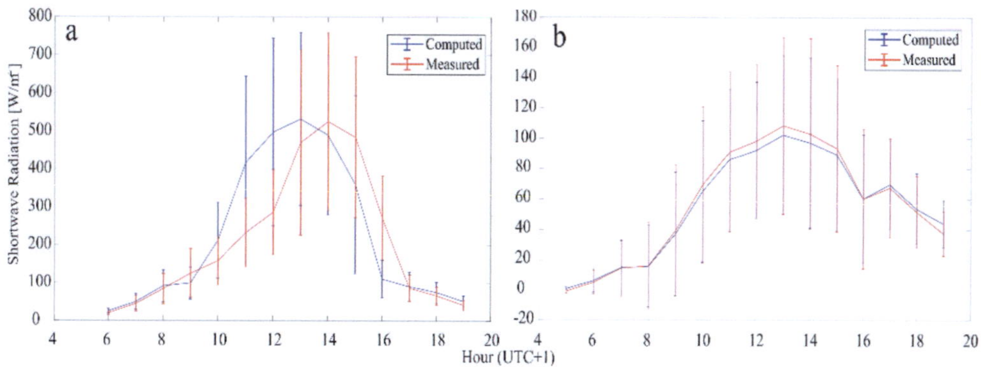

Fig. 4 Comparison between the simulated shortwave radiation (in blue) and the measured one (in red) for the whole 2023 year. The points are mean values of shortwave at the street level for the whole run, and the error bars are the standard deviation for the considered hour. Fig. a shows all the "non-cloudy" data and fig. b the "cloudy" data instead. Where we considered "cloudy" all the measurements for which the direct shortwave measured on the roof is less than 10 W/m^2 (by author)

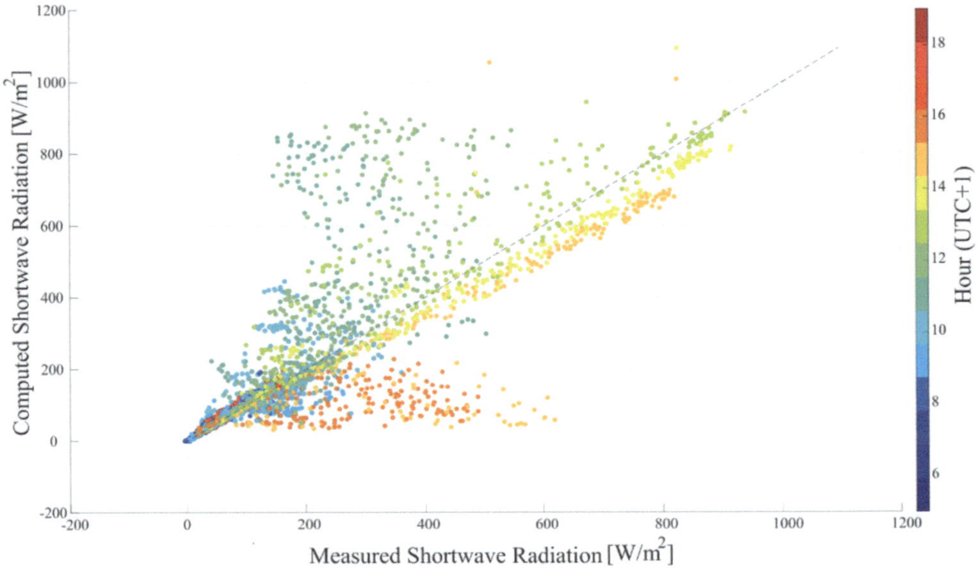

Fig. 5 Comparison of simulated street-level incoming shortwave radiation against measured data. The scatterplot is colour coded according to the hour of the day (UTC +1). (by author)

a consistent spread of daily prediction errors around a central tendency, as we can see in Fig. 4 the standard deviation is quite big instead, with an hourly average of 127 W/m^2, but it is comparable to the observed natural variations.

On the hourly analysis instead, the bias increases to reflect a marginal intensification in the model's overprediction tendency, and as expected we also have an increase in variability and errors in part due to the accumulation of the outlier deviations. As we can also see in Fig. 5, most of the data not affected by the shadowing shift aggregate around the line of equality but some outliers on the upper part of the graph greatly increase the overall error metrics. It is also clearly demonstrated in the metrics of Table 1 that during cloudy conditions the model exhibits very low errors and a slight tendency towards underestimation, while during sunny conditions the errors are much higher with a clear tendency towards overestimation. The overestimation tendency is clearly related to the discretization of the study area, but could be also a result of the modelling of the transmission of direct radiation through the tree canopy.

Nonetheless, confronting the obtained errors with other urban models, we can see that similar error metrics are found in other studies, such as the SUEWS model that accounts for an RMSE for the modelled shortwave incoming radiation between 50.6 and 110.5 W m^{-2} depending on the season (Zheng et al. 2023).

The development of this shortwave model is a first step into the ongoing effort of developing a 3D ecophysiological model that correctly simulate urban climate at the microscale. Further development on the short term are the longwave radiation model and a water balance model.

Despite the promising results, this study has some limitations as, for example, the spatial resolution of the model. In this paper, we used 5-m voxels but this resolution might be inadequate to capture some details of urban environments and especially the actual shapes of trees. This limitation at the moment contributes to the observed time-shift error but it could add other errors when other parameters and functions of the ecophysiological model will be included. And while the dimension of the voxel can be easily changed to, reach even the precision to account for small but significant physical objects, reducing the cube size by a factor of n, the number of cubes in the study area would increase by n3. Such improvement would introduce many more voxels to be traversed and of which we need to consider the interactions. To actually obtain this resolution improvement it might be necessary to also adjust the directions discretization value (D), proportionally increasing the dimension of the LUTs. All this would highly increase the computational power needed to run the simulations, and therefore, we will need to balance the potential resolution increase with computational feasibility. Furthermore, validation of the model has so far been limited to data from a single street-level station in Basel. Future validation will include additional data from another street-level station in Basel and one in Zurich, increasing the reliability of the shortwave model evaluation.

5 Conclusions

This study presents a 3D radiation exchange model as a first step of an urban ecophysiological model that will simulate the exchanges of CO_2, water, and energy between urban vegetation and the atmosphere at a microscale. We applied the model on a street canyon in Basel, Switzerland, and used measured street-level shortwave incoming radiation to evaluate its performance.

Table 1 Summary of error metrics: bias, root-mean squared error (RMSE), and mean absolute error (MAE), comparing computed and measured shortwave radiation values across two temporal scales. The "Daily" metrics represent average daily radiation values reflecting the average day-to-day model performance. The "hourly" metrics consider the entire hourly dataset, providing a broad measure of model accuracy over the year. The "C" and "NC" refer to "cloudy" and "not cloudy" conditions as explained in Fig. 4. N stands for the number of pairs used in the statistics

Time scale	Bias	RMSE	MAE	N
Daily	10.6	30.2	22.1	291
Hourly	12.5	127.6	60.7	3437
Hourly C	−1.4	7.1	4.4	1226
Hourly NC	20.2	159.0	91.9	2211

We found a consistent time-shift error of approximately one hour in the diurnal patterns of the modelled street-level direct shortwave radiation, which is attributed to the discretization of the study area in 5 m voxels. Despite the time shift, the overall daytime shortwave radiation reaching the street level is in good agreement with the observed data. A positive bias was demonstrated in the order of 10.6 and 12.5 W m^{-2} considering daytime average and hourly shortwave radiation values, respectively, indicating a slight tendency towards overestimation of the radiation reaching the street level. During cloudy conditions the model exhibited very low errors, indicating that the diffuse radiation is not affected by the discretization-related errors. The overestimation tendency can be attributed to the effects of discretization on the modelling of the direct radiation but could be also related to the modelling of the direct radiation transmission through the tree canopies.

Future developments in this research include the use of data from two new street-level stations (located in Basel and Zurich) that will help with the evaluation and refinement of the shortwave radiation model. Furthermore, the 3D voxel approach introduced here will be used to develop a microscale ecophysiological model with comprehensive plant-urban atmosphere interactions. Our objective is to create a microscale model that is precise enough to give insights about the street-level effects of plants but simple enough that, with the right adjustment, could be applied to many continental cities without the need of extraordinary computational power. Such tool can be helpful for climate scientists, urban planners, and other stakeholders to understand the physics of urban environments, predict the impacts of climate change on cities, but also to develop appropriate urban climate adaptation and mitigation strategies.

Competing Interest This study was funded by the Federal Office for Meteorology and Climatology MeteoSwiss within the framework of GAW-CH and GCOS-CH and the ICOS Switzerland (ICOS-CH) Phase 3 (Swiss National Science Foundation, grant 20F120_198227).

The authors have no conflicts of interest to declare that are relevant to the content of this paper.

References

Amanatides, J., Woo, A.: A fast voxel traversal algorithm for ray tracing. EG 1987-Technical Papers (1987)

Arnfield, A.J.: Two decades of urban climate research: a review of turbulence, exchanges of energy and water, and the urban heat island. Int. J. Climatol. **23**(1), 1–26 (2003). https://doi.org/10.1002/joc.859

Bruse, M., Fleer, H.: Simulating surface–plant–air interactions inside urban environments with a three dimensional numerical model. Environ. Model Softw. **13**(3–4), 373–384 (1998). https://doi.org/10.1016/S1364-8152(98)00042-5

Campbell, G.S.: Extinction coefficients for radiation in plant canopies calculated using an ellipsoidal inclination angle distribution. Agric. for. Meteorol. **36**(4), 317–321 (1986). https://doi.org/10.1016/0168-1923(86)90010-9

Dodman, D., Hayward, B., Pelling, M., Castán Broto, V.: Cities, settlements and key infrastructure. In: Climate Change 2022–Impacts, Adaptation and Vulnerability, pp. 907–1040. Cambridge University Press (2022). https://doi.org/10.1017/9781009325844.008

Duursma, R.A., Makela, A.: Summary models for light interception and light-use efficiency of non-homogeneous canopies. Tree Physiol. **27**(6), 859–870 (2007). https://doi.org/10.1093/treephys/27.6.859

Eingrüber, N., Schneider, K., Korres, W.: Evaluation of microclimatic variations and adaptation effects in a central European city during the most excessive heat wave in summer 2022 by ENVI-met modelling. In: EGU 2023 Proceedings (2023). https://doi.org/10.5194/egusphere-egu23-11806

Ernst, M., Le Mentec, S., Louvrier, M., Loubet, B., Personne, E., Stella, P.: Impact of urban greening on microclimate and air quality in the urban canopy layer: identification of knowledge gaps and challenges. Front. Environ. Sci. 10 (2022). https://doi.org/10.3389/fenvs.2022.924742

Greenwalt, J., Raasakka, N., Alverson, K.: Building urban resilience to address urbanization and climate change. In: Resilience, pp. 151–164. Elsevier (2018). https://doi.org/10.1016/B978-0-12-811891-7.00012-8

Handmer, J., Honda, Y., Kundzewicz, Z.W., Arnell, N., Benito, G., Hatfield, J., Mohamed, I.F., Peduzzi, P., Wu, S., Sherstyukov, B., Takahashi, K., Yan, Z., Vicuna, S., Suarez, A., Abdulla, A., Bouwer, L.M., Campbell, J., Hashizume, M., Hattermann, F., Yamano, H., et al.: Changes in impacts of climate extremes: human systems and ecosystems. In: Managing the Risks of Extreme Events and Disasters to Advance Climate Change Adaptation, pp. 231–290. Cambridge University Press (2012). https://doi.org/10.1017/CBO9781139177245.007

Hoornweg, D., Sugar, L., Gomez, C.L.T.: Cities and greenhouse gas emissions: moving forward. Urbanisation **5**(1), 43–62 (2020). https://doi.org/10.1177/2455747120923557

HosseiniHaghighi, S., Izadi, F., Padsala, R., Eicker, U.: Using climate-sensitive 3D city modeling to analyze outdoor thermal comfort in urban areas. ISPRS Int. J. Geo Inf. **9**(11), 688 (2020). https://doi.org/10.3390/ijgi9110688

Ibrahim Abd El-Rahman, S., Alaa Mandour, M., Mostafa Sobhy, T.: The effect of microclimate simulation on urban space. Eng. Res. J. **173**(0), 473–487 (2022). https://doi.org/10.21608/erj.2022.227268

Javanroodi, K., Nik, V.M.: Interactions between extreme climate and urban morphology: investigating the evolution of extreme wind speeds from mesoscale to microscale. Urban Clim. **31**, 100544 (2020). https://doi.org/10.1016/j.uclim.2019.100544

Kalnay, E., Cai, M.: Impact of urbanization and land-use change on climate. Nature **423**(6939), 528–531 (2003). https://doi.org/10.1038/nature01675

Karlický, J., Huszár, P., Nováková, T., Belda, M., Švábik, F., Ďoubalová, J., Halenka, T.: The "urban meteorology island": a multi-model ensemble analysis. Atmos. Chem. Phys. **20**(23), 15061–15077 (2020). https://doi.org/10.5194/acp-20-15061-2020

Kousis, I., Pigliautile, I., Pisello, A.L.: Intra-urban microclimate investigation in urban heat island through a novel mobile monitoring system. Sci. Rep. **11**(1), 9732 (2021). https://doi.org/10.1038/s41598-021-88344-y

Krayenhoff, E.S., Moustaoui, M., Broadbent, A.M., Gupta, V., Georgescu, M.: Diurnal interaction between urban expansion, climate change and adaptation in US cities. Nat. Clim. Chang. **8**(12), 1097–1103 (2018). https://doi.org/10.1038/s41558-018-0320-9

Kundu, D., Pandey, A.K.: World urbanisation: trends and patterns. In: Developing national urban policies, pp. 13–49. Springer Nature Singapore (2020). https://doi.org/10.1007/978-981-15-3738-7_2

Lee, D.-I., Lee, S.-H.: The microscale urban surface energy (MUSE) model for real urban application. Atmosphere **11**(12), 1347 (2020). https://doi.org/10.3390/atmos11121347

Lindberg, F., Grimmond, C.S.B., Gabey, A., Huang, B., Kent, C.W., Sun, T., Theeuwes, N.E., Järvi, L., Ward, H.C., Capel-Timms, I., Chang, Y., Jonsson, P., Krave, N., Liu, D., Meyer, D., Olofson, K.F.G., Tan, J., Wästberg, D., Xue, L., Zhang, Z.: Urban multi-scale environmental predictor (UMEP): an integrated tool for city-based climate services. Environ. Model Softw. **99**, 70–87 (2018). https://doi.org/10.1016/j.envsoft.2017.09.020

Lindberg, F., Holmer, B., Thorsson, S.: SOLWEIG 1.0 – Modelling spatial variations of 3D radiant fluxes and mean radiant temperature in complex urban settings. Int. J. Biometeorol. **52**(7), 697–713 (2008). https://doi.org/10.1007/s00484-008-0162-7

Lu, X., Yuan, D., Chen, Y., Fung, J.C.H.: Impacts of urbanization and long-term meteorological variations on global $PM_{2.5}$ and its associated health burden. Environ. Pollut. **270**, 116003 (2021). https://doi.org/10.1016/j.envpol.2020.116003

Marelle, L., Myhre, G., Steensen, B.M., Hodnebrog, Ø., Alterskjær, K., Sillmann, J.: Urbanization in megacities increases the frequency of extreme precipitation events far more than their intensity. Environ. Res. Lett. (2020). https://doi.org/10.1088/1748-9326/abcc8f

Masson, V., Lemonsu, A., Hidalgo, J., Voogt, J.: Urban climates and climate change. Annu. Rev. Environ. Resour. **45**(1), 411–444 (2020). https://doi.org/10.1146/annurev-environ-012320-083623

Mena-Chalco, J.P.: A fast voxel traversal algorithm for ray tracing. MATLAB Central File Exchange (2023)

Michel, J.-P.: Urbanization and ageing health outcomes. J. Nutr. Health Aging **24**(5), 463–465 (2020). https://doi.org/10.1007/s12603-020-1360-1

Mutti, M.: Leaf Area Index Estimation from Aerial LiDAR over Basel City Center. University of Basel (2020)

Ponce de León, M.A., Bailey, B.N.: Evaluating the use of Beer's law for estimating light interception in canopy architectures with varying heterogeneity and anisotropy. Ecol. Model. **406**, 133–143 (2019). https://doi.org/10.1016/j.ecolmodel.2019.04.010

Qian, Y., Chakraborty, T.C., Li, J., Li, D., He, C., Sarangi, C., Chen, F., Yang, X., Leung, L.R.: Urbanization impact on regional climate and extreme weather: current understanding, uncertainties, and future research directions. Adv. Atmos. Sci. **39**(6), 819–860 (2022). https://doi.org/10.1007/s00376-021-1371-9

Sharma, A., Wuebbles, D.J., Kotamarthi, R.: The need for urban-resolving climate modeling across scales. AGU Adv. **2**(1) (2021). https://doi.org/10.1029/2020AV000271

Singh, M., Sharston, R.: Evaluating the spatial variation of heatwave intensity to enhance urban resilience. IOP Conf. Ser. Earth Environ. Sci. **1101**(2), 022016 (2022). https://doi.org/10.1088/1755-1315/1101/2/022016

Stagakis, S., Feigenwinter, C., Vogt, R., Kalberer, M.: A high-resolution monitoring approach of urban CO_2 fluxes. Part 1-bottom-up model development. Sci. Total. Environ. **858**, 160216 (2023). https://doi.org/10.1016/j.scitotenv.2022.160216

UN.: World Urbanization Prospects: The 2018 Revision (2019). https://doi.org/10.18356/b9e995fe-en

United Nations Framework Convention on Climate Change (UNFCCC).: Initiatives in the Area of Human Settlements and Adaptation (2017)

Xu, H., Wang, C.C., Shen, X., Zlatanova, S.: 3D Tree reconstruction in support of urban microclimate simulation: a comprehensive literature review. Buildings **11**(9), 417 (2021). https://doi.org/10.3390/buildings11090417

Zhan, Y., Yao, Z., Groffman, P.M., Xie, J., Wang, Y., Li, G., Zheng, X., Butterbach-Bahl, K.: Urbanization can accelerate climate change by increasing soil N_2O emission while reducing CH_4 uptake. Glob. Change Biol. **29**(12), 3489–3502 (2023). https://doi.org/10.1111/gcb.16652

Zhao, L., Lee, X., Smith, R.B., Oleson, K.: Strong contributions of local background climate to urban heat islands. Nature **511**(7508), 216–219 (2014). https://doi.org/10.1038/nature13462

Zheng, Y., Havu, M., Liu, H., Cheng, X., Wen, Y., Lee, H.S., Ahongshangbam, J., Järvi, L.: Simulating heat and CO_2 fluxes in Beijing using SUEWS V2020b: sensitivity to vegetation phenology and maximum conductance. Geosci. Model. Dev. **16**(15), 4551–4579 (2023). https://doi.org/10.5194/gmd-16-4551-2023

Effects of Local Heatwave Adaptation Actions: Microclimatic Simulation of the Interventions in via Metastasio, Brescia (Italy)

Mattia Scalas, Carmela Apreda, Giuliana Barbato, Alessandro Pugliese, Alfredo Reder, Paola Mercogliano, Laura Treccani, and Graziano Lazzaroni

Abstract

In the framework of the Call for Ideas "Climate Strategy" of the Fondazione Cariplo Ambiente 2020, the partnership comprising Comune di Brescia, Parco delle Colline, Fondazione Centro euro-Mediterraneo sui Cambiamenti Climatici (CMCC), and Ambiente Parco secured resources and support for developing the climate transition strategy of the city of Brescia through the project "Un Filo Naturale". The project's ambition is to implement around thirty activities to generate and enhance adaptation, mitigation, and resilience strategies for the city, as well as pursue the well-being of citizens. These activities include the implementation of multi-benefit adaptation measures in public spaces. Among the dissemination activities, some of these measures selected to reduce the risk of heatwaves, particularly tree-planting and reflective materials pavement, have been investigated to evaluate their microclimate-scale effect in terms of variations of critical parameters for thermal comfort. This study presents the main outcomes of this evaluation. In particular, a scenario-based analysis has been performed modelling the layouts before and after the potential implementation of adaptation measures in the pilot area of Via Metastasio (Brescia), using the ENVI-met computational fluid dynamics (CFD) model to quantify the potential gains of the adaptation measures and to support the communication of these benefits to the population. The comparison of these situations (ex-ante and ex-post) shows a general positive effect of the implemented measures on thermal comfort parameters.

Keywords

Scenario-based analysis · Urban Heat Island · Urban thermal comfort · ENVI–met · Microclimate modeling · Heatwaves

1 Introduction

Rising extreme temperatures are among the most noticeable effects of climate change, which has human activity, particularly the emission of greenhouse gases, as its main driver (IPCC 2023). It is evident in Europe, particularly in southern areas and along the Mediterranean coast, where observations and climate projections indicate an increase in the intensity and frequency of extreme temperatures, especially in the summer months (European Environment Agency 2024). Prolonged periods of extreme temperatures above those typical for that time of year are known as heatwaves. Italy identifies heatwaves as one of the phenomena that can

M. Scalas (✉) · C. Apreda · G. Barbato · A. Pugliese · A. Reder · P. Mercogliano
CMCC Foundation—Euro-Mediterranean Center On Climate Change, Caserta, Italy
e-mail: mattia.scalas@cmcc.it

C. Apreda
e-mail: carmela.apreda@cmcc.it

G. Barbato
e-mail: giuliana.barbato@cmcc.it

A. Pugliese
e-mail: alessandro.pugliese@cmcc.it

A. Reder
e-mail: alfredo.reder@cmcc.it

P. Mercogliano
e-mail: paola.mercogliano@cmcc.it

L. Treccani · G. Lazzaroni
Comune di Brescia, Brescia, Italy
e-mail: LauraTreccani@comune.brescia.it

G. Lazzaroni
e-mail: GLazzaroni@comune.brescia.it

P. Mohammad et al. (eds.), *Geographic Approaches to Climate Change and Mitigation: Urban and Rural Perspectives (Volume 1)*, Advances in Science, Technology & Innovation, https://doi.org/10.1007/978-3-031-92119-3_22

amplify the natural risks the country is exposed to (Ministry of Environment and Energy Security 2023). Urban areas are particularly vulnerable to rising temperatures due to building arrangements and characteristics that make cities hotter than the surrounding areas (the so-called urban heat island, UHI, effect), as well as the presence of a high population density and human activities (Oke et al. 2017). Facing this climate threat, a high concentration of urbanized population, and a high percentage of the population vulnerable to extreme heat, especially older people, several Italian cities are including heatwaves among the targets of climate change adaptation policies.

In the framework of the project *Un Filo Naturale*, the city of Brescia is developing its climate transition strategy, identifying heatwaves among the significant climate risks and providing policies and adaptation actions, including some local interventions on public space aimed at reducing the urban heat island effect, such as the creation of oasis zones and actions such as depaving, implementing green roofs, and planting new trees (Municipality of Brescia 2021). Adaptation strategies and actions are often supported by studies analysing the climate profile of the target area and suggesting guidelines and inputs for implementing actions. Unfortunately, integrating local climate studies into performance assessment at the scale of the individual action referring to a single building, building blocks, or public space is often challenging. This lack of integration can complicate the evaluation of the adaptation measures and the ability to effectively communicate to stakeholders and citizens the relevance of these multi-benefit actions.

To address the need for scale integration and improve communication with stakeholders and citizens, we experimented with an integrated approach in the Brescia Climate Transition Strategy, coupling the elaboration of the climate profile of the city with a series of microclimate simulations of public spaces using the computational fluid dynamics (CDF) ENVI-met model (Bruse and Fleer 1998). The aim is to investigate the effectiveness of adaptation actions through scenario-based analyses modelling local microclimate layouts without and with adaptation actions, i.e., *ex-ante* and *ex-post* conditions. The overall goal is to create comparative frameworks and assess the urban microclimate dynamics significantly affecting outdoor thermal comfort (Apreda et al. 2020). This first attempt at analysis has been performed considering the design and implementation of adaptation actions in the pilot area of Via Metastasio (Brescia). The paper is structured as follows: the materials and methods section describes the geographic context and climate profile of Brescia, the pilot area and the designed interventions, the ENVI-met simulations, and the variables selected for the comparison between *ex-ante* and *ex-post* experiments; then, the results section shows and compares the results of microclimate simulations of both experiments to highlight potential gain associated with the implementation of adaptation solutions; finally, discussion and conclusion sections illustrate the microclimate improvements of the pilot area by returning helpful information to communicate the effectiveness of the actions to the local communities.

2 Materials and Methods

2.1 Context

Capital of its namesake province, Brescia is located in the central-eastern part of Lombardy, a region in northern Italy (Fig. 1). It lies in the upper Po Valley near the Trompia Valley, between the Alps to the north and the Po River basin to the south. The city experiences a humid subtropical climate (Cfa) according to Köppen's classification, with mild winters and hot, humid summers. The terrain is predominantly flat, except for the area around Mount Magdalena to the east and the northern region. Canals and rivers cross the city, the main one being the Mella River, which flows in the western area of the territory heading south. Brescia has less than 200.000 inhabitants, is located along the key A4 highway linking the entire Po Valley, and is one of Italy's leading economic and production centres in the manufacturing, industrial, and wine sectors.

2.2 Description of the Local Climate Context

The local climate context of the city of Brescia has been outlined through a series of indicators commonly used in the literature to characterise the climate and its evolution in terms of average values and extremes. The current climate has been analysed using atmospheric data derived from VHR-REA_IT (Very High-Resolution REAnalysis for ITaly; Raffa et al. 2021; Adinolfi et al. 2023), an hourly gridded simulation for the recent past over Italy, obtained by dynamically downscaling ERA5 reanalysis (Hersbach et al. 2020) from its native resolution ($\simeq$31 km) to a resolution of $\simeq$2.2 km. The future climate has been depicted in line with the National Climate Change Adaptation Plan (PNACC 2023) by processing atmospheric data from an ensemble of climate models available within the EURO-CORDEX (i.e., the European branch of the international Coordinated Downscaling Experiment; Jacob et al. 2020) initiative at the resolution of $\simeq$12 km, under three IPCC Representative Concentration Pathways (RCP) scenarios: RCP2.6 (i.e., aggressive mitigation), RCP4.5 (i.e., stabilisation scenario), and RCP8.5 (i.e., high emission) scenarios. This set of climate models allowed to assess the mean value (ensemble mean) of the expected changes against a reference period obtained from the values of the individual

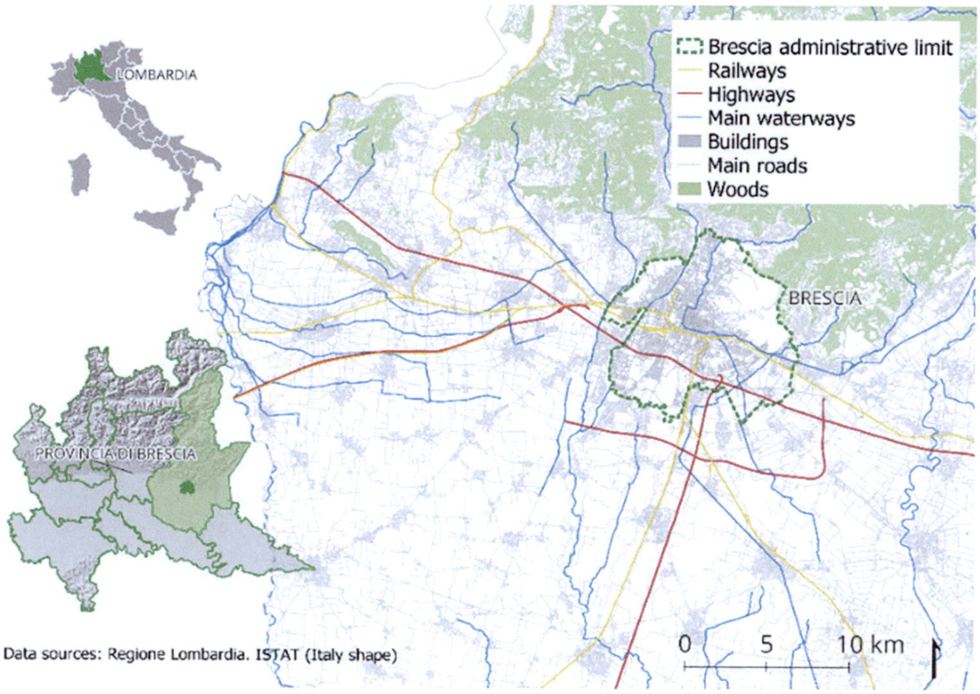

Fig. 1 Geographical framing of the Brescia area, Italy

models representing the ensemble and the dispersion of the individual models around this value (uncertainty, expressed in terms of standard deviation).

Regarding past and current climate, Brescia has an average annual temperature of approximately 15 °C (Fig. 2a), with values decreasing with altitude. The most urbanized area is characterised by temperatures reaching 16–16.5 °C, ~1 °C higher than the average, attributable to built-up areas and human activities. From a temporal perspective, the hottest months are July and August (average temperatures reaching 25 °C), while the coldest months are December and January (~5 °C). An opposite spatial trend is observed in terms of annual precipitation (Fig. 2b), with values increasing from the flatlands towards the pre-alpine area. From a temporal viewpoint (Fig. 2c), precipitation is evenly distributed across all seasons. In particular, it is concentrated mainly in May (>110 mm), with slightly lower values in April and June and from September to November (~85–100 mm).

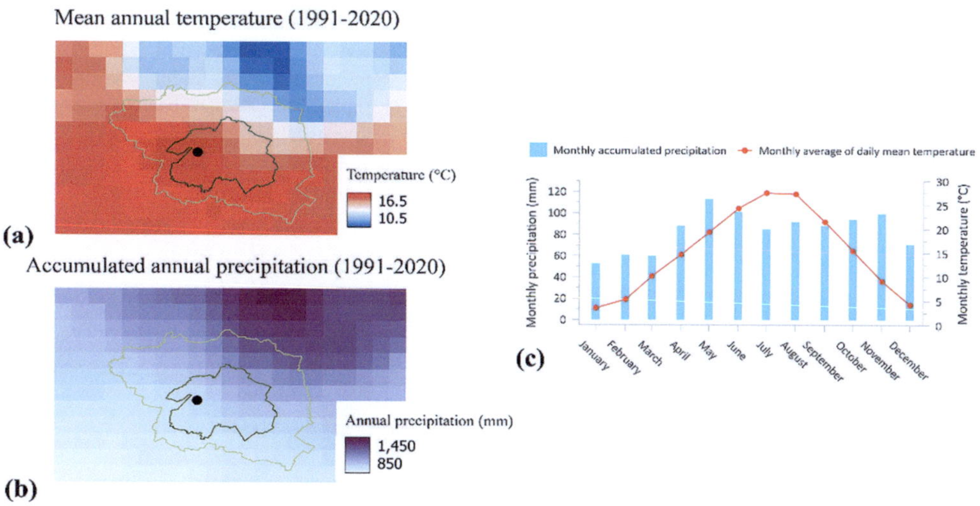

Fig. 2 Spatial distribution of **a** mean annual temperature and **b** annual precipitation calculated averaging data over 1991–2020 (the boundary of the municipality of Brescia is shown in dark green while the boundary of the municipality, including surrounding areas, is in light green); **c** multi-year cycle of monthly mean temperature and monthly accumulated precipitation over 1991–2020 calculated by averaging spatial data falling within the light green boundary. All the pictures rely on data provided by VHR-REA_IT (Raffa et al. 2021)

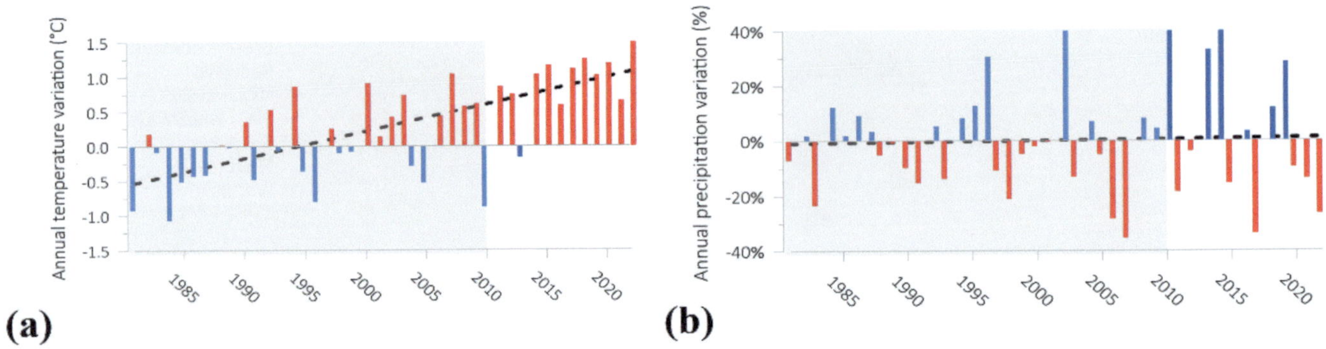

Fig. 3 Development over time for 1981–2022 of the **a** annual mean temperature and **b** annual accumulated precipitation anomalies, calculated assuming the climate normal 1981–2010 as the reference (shaded area). The dark dashed line represents the trends

The evolution of the anomalies of mean annual temperature (Fig. 3a) and mean annual precipitation (Fig. 3b), calculated against the mean value over the period 1981–2010, reveal an increase in temperature, mainly positive over the last ten years with anomalies of up to 1.5 °C, and a strong alternation of rainy and dry years for precipitation.

The expected changes of the ensemble mean (with related uncertainty) in mean temperature and precipitation (only considering days with precipitation higher than 1 mm) due to climate change are summarised in Table 1 in terms of seasonal changes for the 2050s (i.e., 2036–2065) compared to the reference period 1981–2010.

As far as the mean temperature is concerned, a general increase in magnitude is expected depending on the scenario analysed. In particular, summer projects a temperature increase (+1.9 °C under RCP4.5 and 2.2 °C under RCP8.5) twice the case of aggressive mitigation (+1.1 °C under RCP2.6); interestingly, the maximum variation is expected in autumn according to the RCP8.5 scenario (+2.4 °C), albeit this season is also one characterised by greater uncertainty than the others regardless of the scenario. Regarding accumulated precipitation on rainy days, a variation typical of the Mediterranean context is expected, with an increment in precipitation in winter and a reduction in summer; however, the climate signal is highly uncertainty, as noticeable from the fact that standard deviations are much higher than mean values regardless of seasons and scenarios.

Finally, to provide a first clue as to the expected future variation in the heat-related events, Table 1 reports the number of tropical nights and summer days, proxy indicators representing the number of days when the minimum temperature is above 20 °C, and the maximum temperature is above 29.2 °C, respectively. In line with the climatology of the Mediterranean area, these indicators are expected to vary mainly during summer, with magnitude depending on the scenario analysed. Specifically, the mean number of tropical nights is

Table 1 Changes of mean temperature, accumulated precipitation on rainy days, number of tropical nights, and number of summer days over 2036–2065 against 1981–2010 (ensemble mean and uncertainty) for autumn (SON = September–October–November), winter (DJF = December–January–February), spring (MAM = March–April–May), and summer (JJA = June–July–August)

		SON	DJF	MAM	JJA
Mean temperature change (°C)	RCP2.6	1.4 (±0.8)	1.0 (±0.2)	0.8 (±0.4)	1.1 (±0.4)
	RCP4.5	1.8 (±0.8)	1.6 (±0.5)	1.1 (±0.5)	1.9 (±0.4)
	RCP8.5	2.4 (±0.9)	2.0 (±0.4)	1.5 (±0.6)	2.2 (±0.4)
Accumulated precipitation on rainy days change (%)	RCP2.6	4 (±7)	12 (±9)	5 (±14)	4 (±15)
	RCP4.5	−1 (±13)	11 (±10)	1 (±7)	−7 (±11)
	RCP8.5	3 (±15)	4 (±14)	4 (±12)	−3 (±12)
Number of tropical nights change (days)	RCP2.6	1 (±1)	−1 (±5)	±1	11 (±10)
	RCP4.5	1 (±1)	−1 (±5)	±1	16 (±11)
	RCP8.5	2 (±3)	−1 (±5)	±1	20 (±12)
Number of summer days change (days)	RCP2.6	2 (±2)	0	0	7 (±4)
	RCP4.5	3 (±3)	0	±1	14 (±5)
	RCP8.5	4 (±4)	0	1 (±1)	15 (±6)

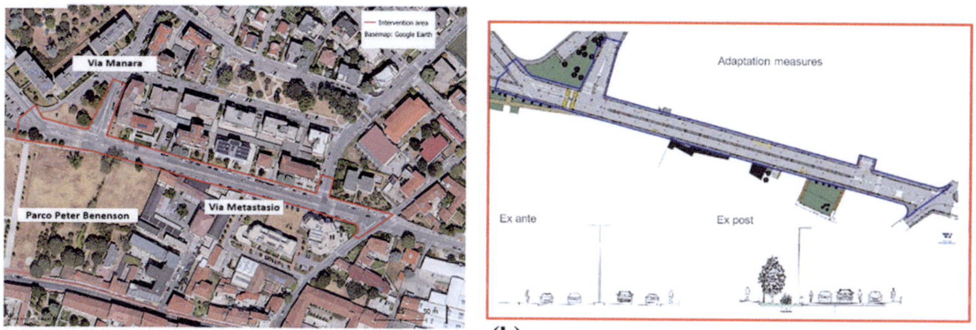

Fig. 4 **a** Google Earth image of Via Metastasio area (Brescia, Italy); **b** sketch of the adaptation measures under development in the pilot area

expected to increase by 11, 16, and 20 days under RCP2.6, RCP4.5, and RCP8.5, respectively. A similar pattern occurs for summer days (increments of 7, 14, and 15 days under RCP2.6, RCP4.5, and RCP8.5, respectively). Once again, the climate signals where variations are relevant align with the mean values.

2.3 Description of the Pilot Area and Designed Interventions

The adaptation interventions have been designed for the pilot area of Via Metastasio as part of Action 2.1 of the climate transition strategy to address the general goals of a resilient transformation of public space to counter heatwaves and floods. Via Metastasio is located in the western portion of the settlement beyond the right bank of the Mella River. Before the intervention (Fig. 4a), the street line had a full whole asphalted section consisting of two lanes separated by a curb, with a double asphalt sidewalk and parking in both directions. The buildings consist mainly of four to five constructions with low-rise structures between Metastasio and Manara streets. The area includes portions of unpaved land, including Peter Benenson Park and other smaller areas. The presence of green areas and private gardens is also detected. This pilot area has been chosen to create green areas, including a rain garden, and implement light-coloured bicycle lanes and sidewalks (Fig. 4b). The actions respond to a co-benefit approach, reducing vulnerability to both high temperatures and flooding (rain garden). Table 2 summarises some of the main design parameters.

Table 2 Project parameters of the climate adaptation actions in Via Metastasio, Brescia

Total intervention area	8180 m^2
South-est and north green area	265 m^2
High reflectance pavements	1920 m^2
Planted trees	25
Rain garden return-to-mould water	150 m^3
Drywell return-to-mould water	265 m^3

2.4 ENVI-Met Simulations

The effectiveness of the planned adaptation actions has been quantified by modelling and comparing the current situation (ex-ante) and the designed situation (ex-post) using the CFD ENVI-met v4.4 model. In brief, ENVI-met is a 3D model that simulates surface–plant–air interactions, allowing us to analyse the interaction between urban design and microclimate on a micro-scale perspective. It consists of two submodels interacting with each other: (1) a 1D boundary model, including vertical profiles of meteorological parameters; (2) a 3D core model, including atmosphere, soil, building, and vegetation processes. ENVI-met reproduces the microclimate and human thermal comfort conditions by quantifying climatic variables (such as radiation, airflow, turbulence, and humidity) and indices (such as physiological equivalent temperature, predicted mean vote, and predicted percentage of dissatisfied). Despite some limitations, ENVI-met is considered one of the most widely deployed CFD models in architectural and urban design for analysing microclimate conditions and evaluating the performance of materials and designs focused on improving thermal comfort.

For the setting of the scenario-based analysis (i.e., *ex-ante* and *ex-post* models) targeted for via Metastasio, boundary conditions (in terms of weather input), spatial and temporal resolutions, and material properties are summarised below:

- as weather input, hourly atmospheric forcing (namely air temperature at 2 m, relative humidity at 2 m, wind speed and direction at 10 m, incoming shortwave radiation—both direct and diffuse, and incoming longwave radiation) derived from the ERA5 reanalysis (Fig. 5);
- as spatial resolution, mesh element of 2 m within a computational domain size of 204 × 92 × 34 cells, except for the vertical dimension refined to 40 cm near the lowermost surface to ease numerical convergence;

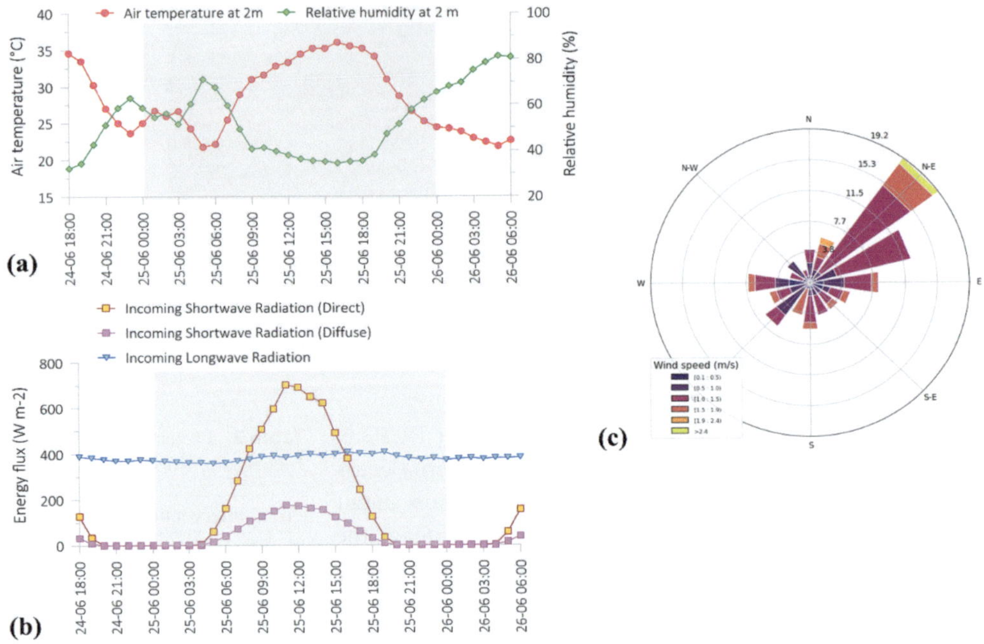

Fig. 5 Hourly weather input for ENVI-met simulations for the target day (25 June 2019, in the shaded area): **a** evolution of air temperature and relative humidity; **b** evolution of energy fluxes (i.e., incoming shortwave radiation—direct and diffuse—and longwave radiation); **c** wind speed and direction (windrose)

- as temporal resolution for outputs, 1-h;
- as experiment duration, running for 79-h (from 23 June 2019 to 26 June 2019, coinciding with a heatwave event); the interpretation of output data focuses on the period between 25 June 2019, 12 PM to 26 June 2019, 6 AM, neglecting the spin-up time to avoid the influence of initial conditions;
- as material properties for the ex-ante model: asphalt for the streets; low grass (max 25 cm) and trees for the vegetation; bare soil for other areas (Fig. 6a);
- as material properties for the ex-post model: asphalt and tactile paving (yellow concrete paving slabs) for the streets; low grass (max 25 cm), shrubs/bushes, and selected trees (*platano occidentalis* and *acer platanoides*) for the vegetation; walkways (coloured resid with high albedo); bare soil for other areas (Fig. 6b).

The simulation was carried out through a workstation with the following specifications: processor: Intel® Xeon® Silver 4210R, CPU: 2.40 GHz, RAM: 32 GB, operating system: Windows 64-bit, software version: ENVI-met v4.4. The simulation took a total calculation time of 77 h. The average calculation for each time step was 1-h circa.

According to the characteristics of the adaptation interventions, *surface temperature* (T_{surf}) and *mean radiant temperature* (T_{mrt}) at 1.60 m have been selected as relevant variables to assess the impact of the adaptation measures. Surface temperature is best suited to evaluate the effect that reflective surfaces and vegetation exert because of the lower reflectivity than asphalt or other dark-and-hard surfaces. In contrast, the mean radiant temperature at 1.60 m represents the composed radiative fluxes and air temperature for a standing person. It

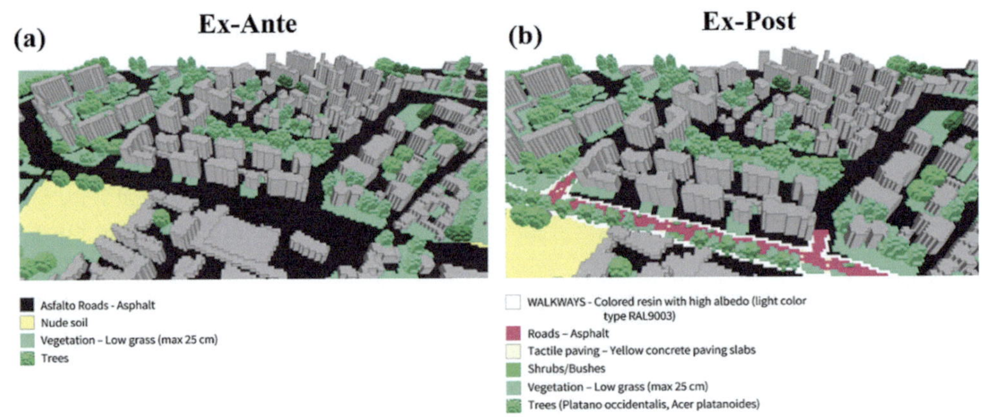

Fig. 6 **a** Ex-ante and **b** Ex-post ENVI-met models

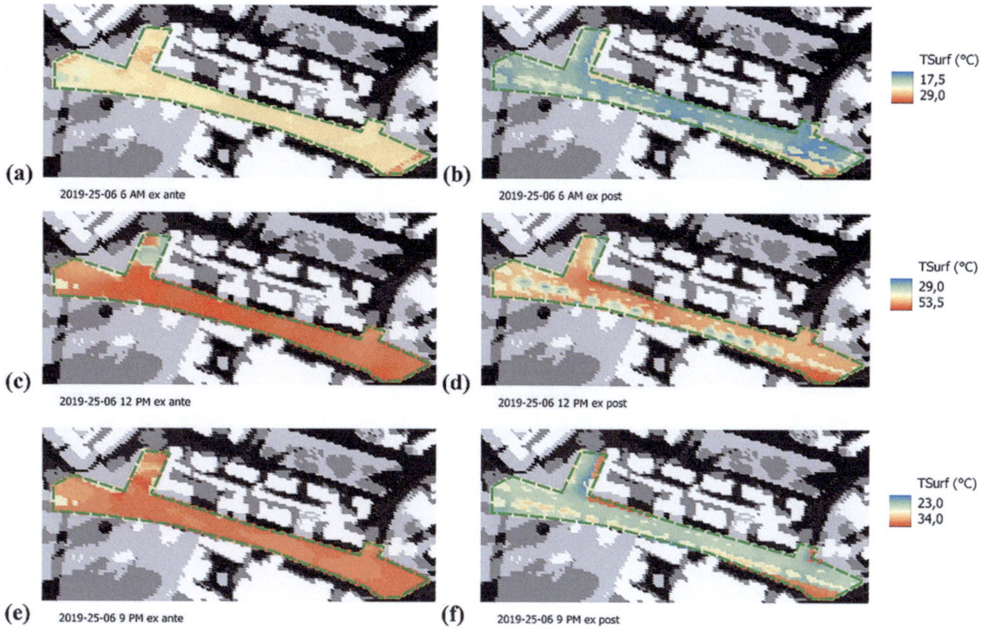

Fig. 7 Surface temperature simulated in (**a**, **c**, **e**) ex-ante and (**b**, **d**, **f**) ex-post layouts in the morning (6 AM; **a**, **b**), midday (12 PM; **c**, **d**), and evening (9 PM; **e**, **f**)

is a human biometeorology parameter that can numerically represent how human beings experience radiation based on the principle that the net exchange of radiant energy between two objects is approximately proportional to their temperature difference times their emissivity (Sinsel et al. 2022).

3 Results

Figure 7 compares the spatial distribution of surface temperature yielded by *ex-ante* and *ex-post* simulations at periods representative of the morning (6 AM; Fig. 7a and b), midday (12 PM; Fig. 7c and d), and sunset (9 PM, Fig. 7e and f). Adaptation measures generally lower surface temperatures (average 3–4 °C) comparing *ex-ante* and *ex-post* layouts. In this perspective, vegetation plays a specific role against light-coloured bicycle lanes and sidewalks. Specifically, during the morning (Fig. 7a and b) and sunset (Fig. 7e and f), the cooling effect of light-coloured bicycle lanes and sidewalks is higher than that exerted by the action of the plants near the adaptation interventions. An opposite behaviour occurs at midday (Fig. 7c and d) when the vegetation performs better than modified pavements due to the shade the plants exert on the road. The temperature increase found on Via Panigala in the *ex-post* state compared to the *ex-ante* state can be interpreted as a change in local ventilation brought about by the new vegetation.

The cooling role of vegetation and the inversion during nighttime hours, with vegetation warmer than roads, can be noted in the hourly trend of surface temperature at 1.60 m (Fig. 8). By spatially averaging vegetation and road cells for the ex-post experiment, it can be shown that in the sunless

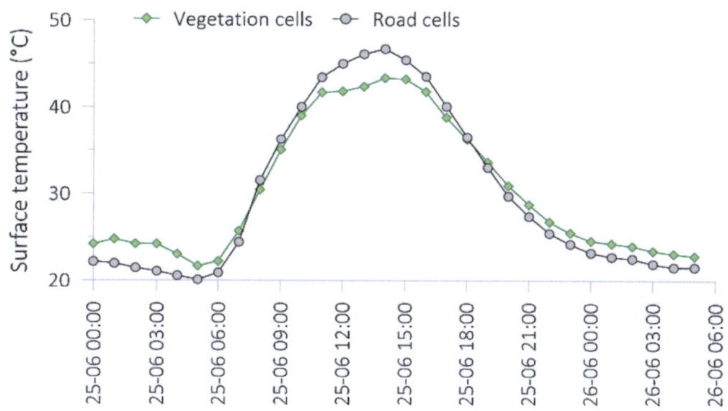

Fig. 8 Evolution of surface temperature for vegetation cells and road cells provided by ex-post simulation; data for vegetation and road have been averaged over the intervention area

hours from midnight to 7 AM and from about 8 PM on 25 June to the end of the simulation (6 AM on 26 June), the temperature of vegetation cells is higher than that of road cells. Conversely, in the daylight hours from 7 AM to 8 PM, the temperature of the road cells remains above that of the vegetation cells, with maximum peaks in the hottest hours of the day (11 AM to 5 PM).

The adaptation action in reducing the surface temperature during the hottest hours of the day can also be observed through the cross-sections shown in Fig. 9. In this representation, it is possible to appreciate the effect exerted by vegetation and reflective sidewalks locally lowering the surface temperature. In particular, lowering is particularly relevant near the new vegetation during the hottest hours, 12 and 2 PM. The reduction can reach −10 °C. Similarly, it can be observed how the temperature lowering is extremely

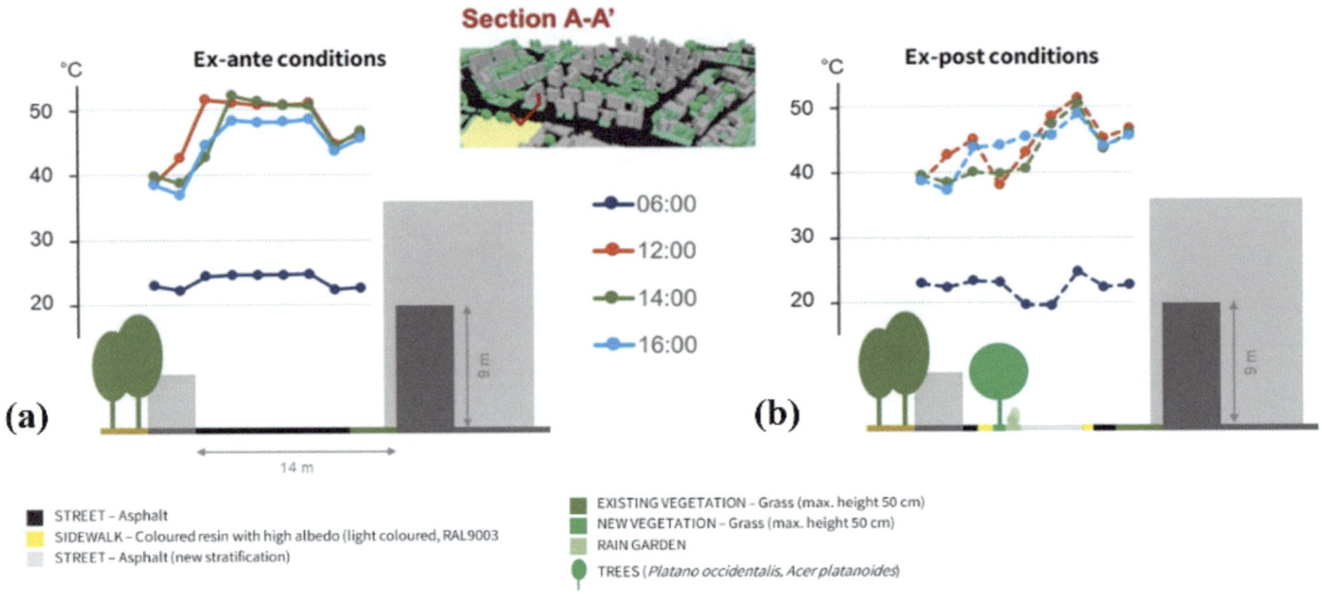

Fig. 9 Cross-section surface temperature carried out by ex-ante (**a**) and ex-post (**b**) experiments in the morning (6 AM), midday (12 PM), and afternoon (2 and 4 PM)

localised, as can be seen in Fig. 9b from the temperature values at 12 PM between the asphalt surface and the green. The values observed earlier, with the vegetation cells tending to have higher temperatures during sunless hours, can also be visualised in the section, particularly with the different temperature trends for 6 AM between the different surfaces.

Analysis of mean radiant temperature reveals an average temperature drop of 1.6 °C at 6 AM. In contrast, at 12 PM and 9 PM, there is an increase in temperature of + 0.5 °C and +0.3 °C, respectively. At 6 AM (Fig. 9a and b), it is possible to observe a reduction in the warmest area at the height of the intersection of Via Metastasio and Via Manara, along with a local reduction on Via Manara and localised warming along the new plantings consistent with the behaviour observed in the surface temperature analysis. At 12 PM, the localised cooling action exerted by the adaptation actions can be observed. Again, as in Fig. 7c and d, the disappearance of the local cooler area in Via Manara can be observed (Fig. 10c and d). At 9 PM, the localised heating phenomenon exerted by the plants is again observable (Fig. 10e and f). The attenuation of the local heating phenomenon in Via Manara has also been reported. Unfortunately, due to the large number of ground truth points required at the spatial scale involved, it was not feasible to cross-reference the simulated results with in situ measurements.

4 Discussion

The variation of surface temperature shows a general cooling trend in the *ex-post* microclimatic condition of Via Metastasio compared to the *ex-ante* condition, mainly concentrated in daylight hours. Such a reduction is consistent with the objectives of the implemented actions and coherent with the literature findings (Apreda et al. 2020) and expectations. The reduction of surface temperature seems to be related to more reflective surfaces, vegetation shading, and the combination of the two effects. The nighttime behaviour of surface temperature is also consistent with expectations. The higher heat capacity of vegetation compared with asphalt causes heat to be released slower at nighttime (Reder et al. 2018). This effect is confirmed by the moderately positive spatial correlation (0.6 and 0.5) found in cells classified as ex-post vegetation between mean radiant temperature and sensible heat flux (not shown here as maps) for 6 AM and 9 PM, which is directed from vegetation to the surrounding for these time frames. The behaviour of the mean radiant temperature is closely related to the effect exerted by vegetation, both in terms of shading and emitted radiation. In particular, there is a localised reduction in mean radiant temperature near vegetation during daylight hours due to shading. From sunset to dawn, the process reverses with plants transferring heat to the surroundings by raising the temperature. This effect is even more evident in the mean

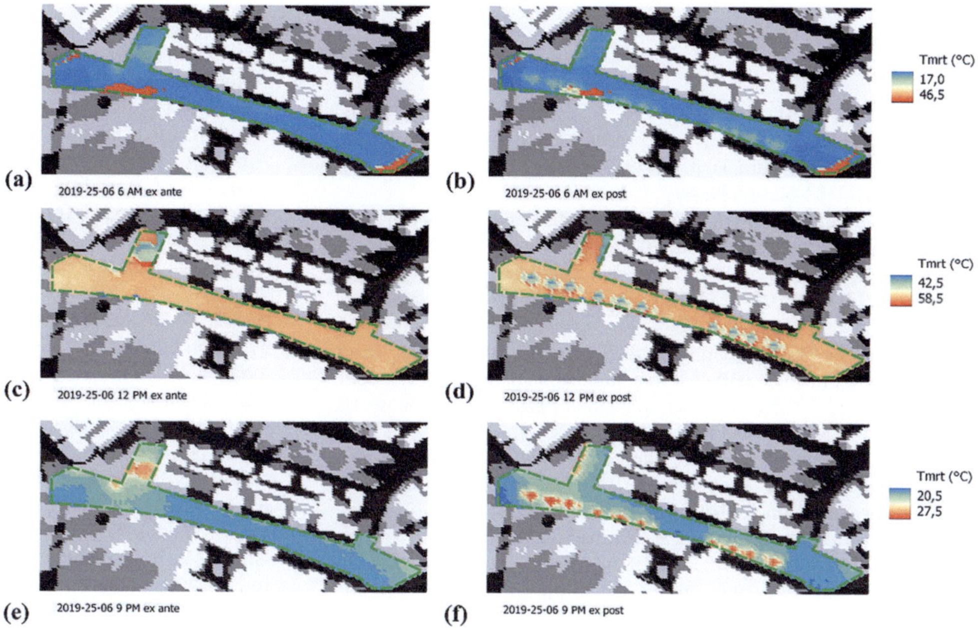

Fig. 10 Mean radiant temperature simulated in (**a**, **c**, **e**) ex-ante and (**b**, **d**, **f**) ex-post layouts in the morning (6 AM; **a**, **b**), midday (12 PM; **c**, **d**), and evening (9 PM; **e**, **f**)

radiant temperature than in the surface temperature, as can be seen from the situation at 9 PM in the ex-post simulations in Fig. 7e and f, Fig. 10e and f. The study has some limitations. Firstly, the input data used for the micorclimatic simulations are derived from the ERA5 reanalysis: possible improvements may concern the use of more detailed models such as the Very High-Resolution REAnalysis for Italy (VHR-REA_IT), a reanalysis in the Italian territory of the global ERA5 reanalysis at a resolution of approximately 2 km for the period 1981–2020 (hourly resolution) (Raffa et al. 2021). Other areas of investigation may involve comparison with satellite data. Besides, it is important to note the absence of a comparison with in situ data and some simplifications in the modelling assumed to reduce calculation time. These constrainsts may be overcome in future research developments. Despite the limitations, the study confirms the effectiveness of the adopted methodology and highlights the positive impact of the implemented actions on the urban microclimate. This is particularly evident in the reduction of surface temperature and average radiant temperature during the sunniest hours of the day.

5 Conclusion

The study compared different design layouts (without and with adaptation measures) in the Via Metastasio (Brescia) pilot area through a numerical scenario-based analysis with the CFD ENVI-met model. Such a comparison highlighted the efficiency of the proposed solutions to reduce temperature locally and potentially the intensity of the urban heat island phenomenon. These solutions can be considered viable candidates among the measures aimed at decreasing surface temperature and mean radiant temperature in the climate transition strategy of the Municipality of Brescia, showing a positive effect during the hottest hours of the day and so benefiting the population in the studied areas and aligning with the climate policies outlined in the strategy. In this sense, the behaviour of vegetation during nighttime hours, which causes a local rise in ground temperature and radiant temperature at 1.60 m, is offset by the positive effects of the vegetation itself during the most critical hours and the hours of increased use of the street by users. Using ERA5 reanalysis as input for these experiments proved effective in connecting different spatial scales, ensuring an organic approach to the study and dissemination of urban climate and microclimate and the effectiveness of adaptation measures. Future research with higher resolution reanalysis might be more interesting to adopt as forcing for microclimate simulation.

Acknowledgements This study was conducted within the framework of the project "A Natural Thread for the Drafting of the Climate Transition Strategy of Brescia", supported by funding from the Cariplo Foundation and Lombardy Region. The fundamental contribution of the project partners, Municipality of Brescia, Parco delle Colline, the Euro-Mediterranean Center on Climate Change (CMCC) Foundation, and Ambiente Parco, is acknowledged for their essential role in providing expertise and collaborative support. ERA5 have been obtained using Copernicus Climate Change Service Information and downloaded from the C3S CDS. Finally, we acknowledge the World Climate Research Programme's Working Group on Regional Climate and the Working Group on Coupled Modeling, the former coordinating body of CORDEX and the responsible panel for CMIP5, for the climate data used.

References

Adinolfi, M., Raffa, M., Reder, A., et al.: Investigation on potential and limitations of ERA5 Reanalysis downscaled on Italy by a convection-permitting model. Clim. Dyn. **61**, 4319–4342 (2023). https://doi.org/10.1007/s00382-023-06803-w

Apreda, C., Reder, A., Mercogliano, P.: Urban morphology parameterization for assessing the effects of housing blocks layouts on air temperature in the Euro-Mediterranean context. Energy Build. **223**(4), 110171 (2020). https://doi.org/10.1016/j.enbuild.2020.110171

Bruse, M., Fleer, H.: Simulating surface–plant–air interactions inside urban environments with a three dimensional numerical model. Environ Model Softw. **13**(3–4), 373–384 (1998). https://doi.org/10.1016/S1364-8152(98)00042-5

European Environment Agency.: European climate risk assessment - Executive summary. EEA Report 01/2024. Publications Office of the European Union (2024). https://doi.org/10.2800/204249

Hersbach, H., Bell, B., Berrisford, P., Hirahara, S., Horanyi, A., Munoz-Sabater, J., Thepaut, J.N.: The ERA5 global reanalysis. Q. J. R. Meteorol. Soc. **146**, 1999–2049 (2020). https://doi.org/10.1002/qj.3803

IPCC.: Climate Change 2023: Synthesis Report. Contribution of Working Groups I, II and III to the Sixth Assessment Report of the Intergovernmental Panel on Climate Change, pp. 35–115. IPCC, Geneva, Switzerland (2023). https://doi.org/10.59327/IPCC/AR6-9789291691647. (Core Writing Team, H. Lee and J. Romero (eds.))

Jacob, D., Teichmann, C., Sobolowski, S., et al.: Regional climate downscaling over Europe: perspectives from the EURO-CORDEX community. Reg. Environ. Change **20**, 51 (2020). https://doi.org/10.1007/s10113-020-01606-9

Ministry of Environment and Energy Security.: National Plan for Adaptation to Climate Change (2023)

Municipality of Brescia.: Climate Transition Strategy. City Council Resolution No. 52 of 06.25.2021 (2021)

Oke, T.R., Mills, G., Christen, A., Voogt, J.A.: Urban Climates. Cambridge University Press (2017). https://doi.org/10.1017/9781139016476

Raffa, M., Reder, A., Marras, G.F., Mancini, M., Scipione, G., Santini, M., Mercogliano, P.: VHR-REA_IT dataset: very high resolution dynamical downscaling of ERA5 reanalysis over Italy by COSMO-CLM. Data **6**(8), 88 (2021). https://doi.org/10.3390/data6080088

Reder, A., Rianna, G., Mercoliano, P., Castellari, S.: Parametric investigation of Urban Heat Island dynamics through TEB 1D model for a case study: assessment of adaptation measures. Sustain. Cities Soc. **39**, 662–673 (2018). https://doi.org/10.1016/j.scs.2018.03.023

Sinsel, T., Simon, H., Ouyang, W., dos Santos Gusson, C., Shinzato, P., Bruse, M.: Implementation and evaluation of mean radiant temperature schemes in the microclimate model ENVI-met. Urban Clim. **45**, 101279 (2022).https://doi.org/10.1016/j.uclim.2022.101279

Urban Planning for Net-Zero Energy Buildings in Italy Under Climate Changes

Paola Maria Albanese, Paolo Maria Congedo, and Cristina Baglivo

Abstract

Climate change poses a significant challenge to the energy efficiency of buildings in Italy. This study examines the effects of climate change on energy performance, focusing on the influence of building shape (S/V ratio) and window total solar energy transmittance ($g_{gl,n}$). Using a five-step methodology, the study explores three representative concentration pathways for the years 2030, 2050, and 2070, as shown in Fig. 1. Key findings reveal that rising temperatures will decrease heating demand, but significantly increase cooling needs, particularly in southern Italy. The study emphasizes the significance of building shape, showing that lower surface-to-volume (S/V) ratios lead to reduced energy consumption across different climates. For Lecce, buildings with an S/V ratio of 0.35 show $EP_{tot,nd}$ values ranging from 40.01 to 31.11 kWh/m^2 depending on the regulation and year, whereas an S/V ratio of 0.75 results in values between 99.12 and 81.59 kWh/m^2: compact buildings showed notable efficiency improvements compared to less efficient shapes. The effect of $g_{gl,n}$ on energy efficiency is evident: windows with lower $g_{gl,n}$ values improve cooling efficiency in warmer climates. The research focuses on updating regulations to include climate projections, promoting compact designs and optimized windows. It recommends a comprehensive approach that integrates thermal transmittance, S/V ratio, and glazing properties for climate-resilient buildings.

P. M. Albanese · P. M. Congedo · C. Baglivo (✉)
Department of Engineering for Innovation, University of Salento, 73100 Lecce, Italy
e-mail: cristina.baglivo@unisalento.it

P. M. Albanese
e-mail: paolamaria.albanese@unisalento.it

P. M. Congedo
e-mail: paolo.congedo@unisalento.it

Keywords

Climate change · Surface-to-volume ratio · Window total solar energy transmittance · Resilience · Building sector

Nomenclature

Abbreviations

BIM	Building Information Modelling
CA EPBD	Concerted Action EPBD
CTI	Italian Thermo-technical Committee
EGD	European Green Deal
EPBD	Energy Performance of Buildings Directive
EPCs	Energy Performance Certificates
GHG	Greenhouse Gas
RCP	Representative Concentration Pathway

Symbols

$EP_{c,nd}$	Thermal performance index for cooling [kWh/m^2]
$EP_{h,nd}$	Thermal performance index for heating [kWh/m^2]
$EP_{tot,nd}$	Total thermal performance index [kWh/m^2]
$g_{gl,n}$	Total solar energy transmittance for normal incidence
U	Thermal transmittance [W/m^2 K]
U_{floor}	Transmittance limit of floor [W/m^2 K]
U_{roof}	Transmittance limit of roof [W/m^2 K]
U_{wall}	Transmittance limit of wall [W/m^2 K]
U_{wind}	Transmittance limit of window [W/m^2 K]
S/V	Surface-to-volume building ratio

P. Mohammad et al. (eds.), *Geographic Approaches to Climate Change and Mitigation: Urban and Rural Perspectives (Volume 1)*, Advances in Science, Technology & Innovation, https://doi.org/10.1007/978-3-031-92119-3_23

1 Introduction

Carbon dioxide (CO_2) is a greenhouse gas (GHG) well known in the scientific literature as a major contributor to global warming. Current projections indicate that there is a significant threat from rising emissions, with potential increases of up to 28% by the year 2030 in the absence of mitigating strategies. This is coupled with the growing scarcity of energy resources and the well-established negative impacts of CO_2 emissions, which have driven the development and adoption of energy-saving technologies (Qureshi et al. 2021). Notably, the reduction of building energy consumption represents a vital strategy for mitigating CO_2 and other GHG emissions (Sher et al. 2019). A comprehensive approach to addressing climate change necessitates a multifaceted strategy, combining emissions reduction, improved energy efficiency, and the integration of renewable energy sources (D'Agostino et al. 2024). Moreover, adopting sustainable practices with a life-cycle perspective and promoting circularity in material use represent crucial components in combating climate change (Congedo et al. 2023). As scientific understanding of the spatial and temporal distribution of climate change impacts improves, the consequences of inaction become increasingly evident (Ulpiani and Vetters 2023).

It is well established that climate change represents a significant threat to the environment. However, the full extent of its impact on energy consumption in the built environment remains to be fully understood (Zhai and Helman 2019). The performance of buildings is influenced by climate factors. To ensure the creation of a sustainable and adaptable building stock, it is essential to design buildings with climate change in mind throughout their entire lifespan (Diaz-Lopez et al. 2021). With the progression of climate change, it is expected that summers will become more extreme, significantly increasing the demand for cooling systems, particularly in crowded areas (Salata et al. 2022). Despite the evidence presented in existing research on the impact of climate change on both buildings and energy systems (Jafarpur and Berardi 2021), a considerable number of building and component energy analyses still rely on current climate data. Optimizing the form and orientation of buildings can significantly enhance energy efficiency, reducing overall energy consumption and maximizing economic benefits, as highlighted by recent studies using building information modelling (BIM) to evaluate different building configurations (Haseeb et al. 2023).

The built environment plays a pivotal role in global greenhouse gas (GHG) emissions, with buildings accounting for a significant portion of these emissions. This is largely due to their energy consumption for heating, cooling, lighting, and powering appliances (International Energy Agency 2013). The residential sector contributes a notable proportion of these emissions, underscoring the importance of prioritizing energy efficiency and resource conservation in building design and operations (Koengkan et al. 2022). In recognition of the necessity to mitigate climate change, national governments are progressively aligning their policies and regulations with global efforts to reduce greenhouse gas emissions and transition towards low-carbon economies (Nejat et al. 2015). This shift towards sustainability encompasses various measures, including the promotion of energy-efficient buildings through building codes, standards, incentives, and the integration of renewable energy (Skillington et al. 2022). It is also of paramount importance to build resilience to climate change. To this end, it is necessary to adopt sustainable design principles, passive heating and cooling strategies, efficient insulation, and renewable energy sources, in order to withstand and adapt to changing environmental conditions (Baglivo et al. 2023a). There is a great deal of evidence to suggest that energy-efficient buildings offer a wide range of benefits. These include reduced energy costs and improved indoor comfort, in addition to lower environmental impact (D'Agostino et al. 2017). Integrating sustainable design principles into building practices is essential for reducing environmental impact, promoting energy efficiency, and supporting long-term ecological balance, as emphasized by recent studies (Congedo et al. 2024a).

The European Union has been a pioneer in the promotion of energy efficiency since the 1970s, with a particular focus on buildings since the 1980s (Economidou et al. 2020). Through directives, regulations, and standards, the EU seeks to enhance the energy performance of buildings, reduce greenhouse gas emissions, and create a more sustainable built environment for future generations. The European Union (EU) is committed to achieving climate neutrality, which entails balancing greenhouse gas emissions through removal or offset measures. This is aimed at stabilizing global temperatures and mitigating the impacts of climate change. Ambitious targets have been set, including the goal of a carbon–neutral Europe by 2050 (https://commission.europa.eu/strategy-and-policy/priorities-2019-2024/european-green-deal_it). This will require substantial reductions in emissions and the promotion of sustainable practices and technologies. The introduction of the European Green Deal (EGD) in 2019 signifies a significant step towards this goal, aiming for carbon neutrality by 2050 (Popielak et al. 2024). The Energy Efficiency Directive (European Commision 2023a), the Energy Performance of Buildings Directive (EPBD) (European Commission 2018a, 2018b), the Renewable Energy Directive (European Commission 2023b), and the Ecodesign Directive (European Commission 2009) are of pivotal importance in ensuring the effective implementation of energy efficiency aspects in line with the goals of the European Green Deal. The EPBD is concerned with reducing energy consumption in European buildings and achieving ambitious decarbonisation targets. It

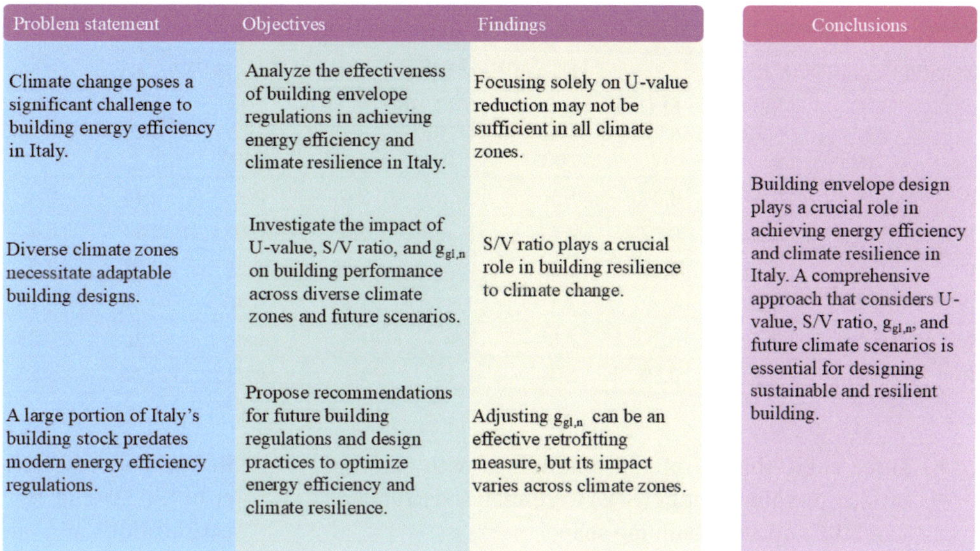

Fig. 1 Graphical abstract. (By authors)

emphasizes Energy Performance Certificates (EPCs) as crucial tools for evaluating housing energy efficiency (European Commission 2002, 2010). However, the development and implementation of EPC calculation methodologies vary across Europe, reflecting regional energy profiles and climate zones (Dascalaki et al. 2013). The Concerted Action EPBD (CA EPBD) (Concerted Actions 2020) facilitates the implementation of EPBD requirements by promoting exemplary approaches, facilitating information sharing among Member States, and standardizing approaches to improve implementation and advance progress in building energy efficiency (https://energy.ec.europa.eu/topics/energy-efficiency/financing/eu-programmes/previous-funding_en).

The Italian legislative framework, too, has undergone significant evolution in recent decades, with an increasing focus on energy efficiency. The primary national legislative document that governs obligations and measures to enhance building energy efficiency is Legislative Decree 192/2005 (L.D. 192/2005) (L.D. 2005). This document implements Directive 2002/91/EC and represents the most recent iteration in the country's legislative framework on the subject. Subsequently, the Ministerial Decree of 26 June 2015 (M.D. 26/06/2015) (M.D. 2015) was introduced to establish minimum requirements for energy efficiency and energy needs in buildings, in alignment with Directive 2010/31/EU. This decree defines and regulates minimum energy performance criteria during building design, as well as mandatory energy performance certification for new buildings and property sales and rentals. Furthermore, the Ministerial Decree of 6 August 2020 (M.D. 06/08/2020) (DM 2020) delineates the technical requirements for eligibility for tax deductions related to energy efficiency improvements in the real estate sector. It is noteworthy that the implementation of these policies has resulted in a significant reduction in thermal transmittance limits in building envelopes, particularly from L.D. 192/2005 to M.D. 26/06/2015.

Italy faces unique challenges in achieving energy-efficient and climate-resilient buildings due to its geographical diversity (Baglivo et al. 2023b). The country stretches across several climate zones, ranging from hot summers and mild winters in the south to cold winters and mild summers in the north. This variation necessitates building designs that can adapt to these contrasting needs. Additionally, Italy has a large stock of existing buildings constructed before modern energy efficiency regulations were in place. Retrofitting these buildings to meet current standards, while also considering future climate scenarios, presents a significant hurdle. Balancing the need for improved insulation in colder zones with the potential for overheating in warmer zones, all while addressing the energy deficiencies of existing structures, requires a multifaceted strategy.

2 Methodology

The methodology follows a five-step approach:

1. Selection of Italian locations: Locations representing diverse climatic zones were chosen based on national classifications and the Köppen–Geiger climate system.
2. Definition of the reference building:

    - A single building model with reinforced concrete insulated walls was used across all locations for consistency. This reflects a common and versatile building choice in Italy.

Table 1 Transmittance limit of roof (U_{roof}), window (U_{wind}), floor (U_{floor}), and wall (U_{wall}) according to Italian regulations

ZONES	U_{roof} [W/m^{2}K]			U_{wind} [W/m^{2}K]			U_{floor} [W/m^{2}K]			U_{wall} [W/m^{2}K]		
	L.D. 192/ 2005	M.D. 26/ 6/2015	M.D. 6/ 8/2020	L.D. 192/ 2005	M.D. 26/ 6/2015	M.D. 6/ 8/2020	L.D. 192/ 2005	M.D. 26/ 6/2015	M.D. 6/ 8/2020	L.D. 192/ 2005	M.D. 26/ 6/2015	M.D. 6/ 8/2020
A	0.8	0.35	0.27	5.5	3	2.6	0.8	0.44	0.4	0.85	0.43	0.38
B	0.6	0.35	0.27	4	3	2.6	0.6	0.44	0.4	0.64	0.43	0.38
C	0.55	0.33	0.27	3.3	2.2	1.75	0.55	0.38	0.3	0.57	0.34	0.3
D	0.46	0.26	0.22	3.1	1.8	1.67	0.46	0.29	0.28	0.5	0.29	0.26
E	0.43	0.22	0.2	2.8	1.4	1.3	0.43	0.26	0.25	0.46	0.26	0.23
F	0.41	0.2	0.19	2.4	1.1	1	0.41	0.24	0.23	0.44	0.24	0.22

- Three cubic-shaped reference buildings with varying surface-to-volume ratios (S/V) were analysed to understand the impact of building shape.
- Air conditioning was not considered, the analysis focused on the building envelope

3. Building envelope configuration:

- Three sets of building regulations were considered: Legislative Decree 192/2005, Ministerial Decree 26 June 2015, and Ministerial Decree 6 August 2020. These regulations specify thermal transmittance limits for each climate zone (Table 1).

4. Predictive climate analysis:

- Climate data for each location and time period were generated using Meteonorm 8 software, including outdoor temperature, wind speed, radiation, and atmospheric pressure (Meteonorm - 2012). Meteonorm was chosen as the source for climate data due to its ability to generate typical meteorological years and time series of current data for any location worldwide. This was crucial because the research needed representative climate data for the Italian locations across various time periods (2030, 2050, and 2070).

5. Building simulations:

- For the impact of climate change and building shape: Building energy simulations were conducted using Termolog software for the three reference buildings with varying S/V. In accordance with D.M. 26/6/2015, the S/V ratios employed for the building models were 0.35, 0.55, and 0.75. Due to its widespread use in building research (Congedo et al. 2021; Pastore et al. 2022) and by Italian professionals for thermal certifications and energy assessments (Baglivo 2021; Fregonara et al. 2017), Termolog was chosen as the building energy simulation software. This selection ensures that the results were relevant to the Italian context and were recognized by the official Italian Thermo-technical Committee (CTI).
- For the impact of total solar energy transmittance: Building energy simulations were conducted for a single reference building, considering two total solar energy transmittances for normal incidence ($g_{gl,n}$) values (0.55 and 0.67).
- In each case, the results focused on thermal performance indices expressed in kWh/m^2 for heating ($EP_{h,nd}$) and cooling ($EP_{c,nd}$), and the sum of these indices ($EP_{tot,nd}$). The analysis was conducted in all the locations considered under three RCP scenarios (RCP2.6, RCP4.5, and RCP8.5). Moreover, to adhere to the varying legal thermal transmittance limits established for each climatic zone, the thickness of the building envelope's insulation was strategically adjusted.

Figure 2 presents the methodological flow diagram, outlining the five-step approach used in this study. This methodology allowed for a comprehensive analysis of the combined effects of climate change, building shape, and window design on the energy performance of buildings across diverse Italian locations. In the context of this study, the results presented are limited to the case of Lecce under the RCP4.5 scenario, with a focus on $EP_{h,nd}$, $EP_{c,nd}$, and $EP_{tot,nd}$.

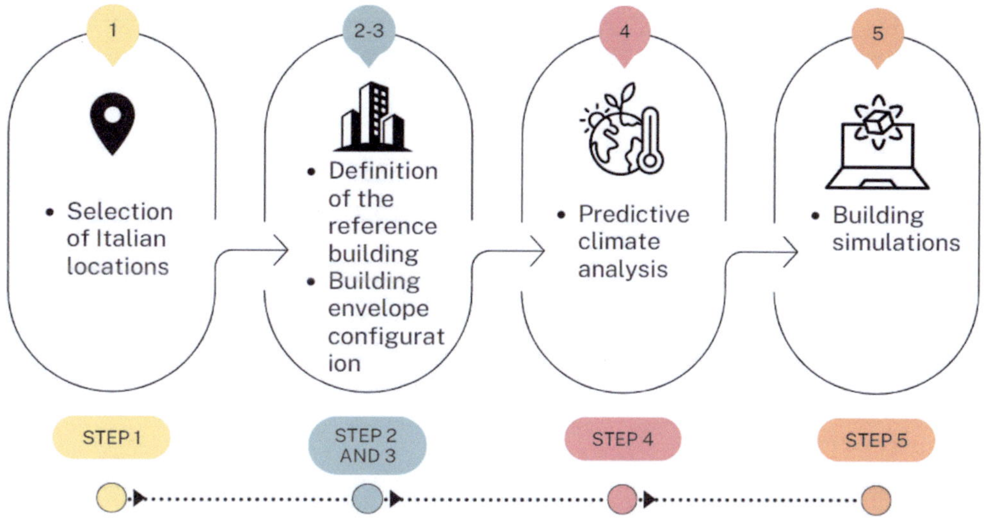

Fig. 2 Methodological flow diagram. (By authors)

3 Results

3.1 Impact of Climate Change

Considering the imperative need to adapt building designs and energy regulations to cope with climate change, it has become increasingly evident that a multifaceted approach is essential (Baglivo et al. 2023b). Beyond mere compliance with regulations, there is a pressing need for a paradigm shift towards holistic, sustainable building practices from the outset of any construction project. This shift not only mitigates the need for frequent interventions but also fosters a culture of environmental consciousness and resilience.

Moreover, the discrepancy between maintenance cycles and energy efficiency enhancements underscores the necessity for a revaluation of long-term planning and investment strategies. By aligning these cycles with the projected lifespan of buildings and their energy performance requirements, stakeholders can ensure more effective resource allocation and improve long-term sustainability outcomes. This requires a concerted effort from policymakers, industry professionals, and the public to prioritize energy efficiency measures and invest in innovative technologies and materials.

In the context of Italy regulatory framework, the evolution of heating requirements and transmittance limits highlights the ongoing effort to adapt to changing climatic conditions. However, as climate change continues to accelerate, there is a need for even more robust and dynamic regulatory mechanisms that can anticipate and respond to future challenges proactively. This might entail regular reassessment of zoning criteria, transmittance limits, and other regulatory parameters to ensure they remain effective in the face of evolving climatic conditions.

Furthermore, the proposed approaches to managing summer energy consumption through regulatory updates and flexible compliance mechanisms represent significant steps towards building resilience. However, their successful implementation hinges on effective collaboration between regulatory bodies, industry stakeholders, and researchers to develop practical solutions that balance energy efficiency with climate adaptation needs.

Overall, addressing the complex interplay between building design, energy regulations, and climate change requires a comprehensive and forward-thinking approach. By embracing sustainability as a core principle and integrating resilience into every aspect of the built environment, we can create buildings and communities that are not only more energy-efficient but also better equipped to withstand the challenges of a changing climate.

Climate change is fundamentally altering the energy landscape for buildings. To achieve a unified approach to energy transition and zero-energy buildings, it is crucial to integrate ongoing climate change dynamics into building regulations and design practices. Historically, regulations have focused primarily on enhancing thermal insulation, with successive Italian regulations like L.D. 192/2005, M.D. 26/6/2015, and M.D. 6/8/2020 progressively lowering allowable U-values (a measure of heat transfer through building components). However, a critical evaluation of these regulations is necessary to assess their effectiveness in boosting the resilience of building envelopes to climate change, particularly considering the long lifespan of buildings.

To understand the potential impact of future climate scenarios, researchers have conducted forecasting analyses for the years 2030, 2050, and 2070, considering three Representative Concentration Pathway (RCP) scenarios representing different levels of greenhouse gas emissions. The analysis focuses on trends in heating, cooling, and total thermal performance indices ($EP_{h,nd}$, $EP_{c,nd}$, and $EP_{tot,nd}$, respectively), which were calculated in a previous study

Table 2 $EP_{h,nd}$, $EP_{c,nd}$, and $EP_{tot,nd}$ of Lecce for the years 2030, 2050, and 2070, considering the three different building regulations.

Regulation	Year	RCP4,5		
		EPh,nd [kWh/m2]	EPc,nd [kWh/m2]	EPtot,nd [kWh/m2]
L.D. 192/2005	2030	34.8	28.49	63.29
L.D. 192/2005	2050	31.85	32.15	64
L.D. 192/2005	2070	28.87	35.44	64.31
M.D. 26/6/2015	2030	20.4	29.05	49.45
M.D. 26/6/2015	2050	18.31	32.03	50.34
M.D. 26/6/2015	2070	16.18	34.28	50.46
M.D. 6/8/2020	2030	15.6	30.24	45.84
M.D. 6/8/2020	2050	13.75	33.78	47.53
M.D. 6/8/2020	2070	11.61	36.77	48.38

(D'Agostino et al. 2024). The anticipated rise in outdoor temperatures, with varying intensity across different RCP scenarios, is expected to decrease winter heat demand while significantly increasing summer heat demand. Consequently, the total thermal performance index, which combines both heating and cooling needs, exhibits contrasting trends across diverse climate zones. In warmer zones (A, B, and C), the index is projected to increase, indicating a growing overall energy demand for cooling. Conversely, in colder zones (D, E, and F), the index is expected to decrease, suggesting a reduced total energy demand. This highlights the importance of considering regional climate variations when designing and regulating buildings. Table 2 reports $EP_{h,nd}$, $EP_{c,nd}$, and $EP_{tot,nd}$ values of Lecce for the years 2030, 2050, and 2070, considering the three different building regulations. All the values reported are referred to RCP4.5 scenario. For Lecce, located in climate zone C, future climate scenarios predict a significant increase in cooling demand, leading to an overall rise in the total thermal performance index ($EP_{tot,nd}$). This suggests that Lecce will face growing energy demands primarily due to higher summer temperatures. To address this, building design and regulations in the city should focus on improving cooling efficiency to manage the increased energy consumption.

Interestingly, the study revealed that while reducing the transmittance of building envelope components (lower U-values) improves the total thermal performance index, it does not necessarily enhance the envelope resilience to climate change. Despite variations in transmittance values across different regulations and scenarios, the overall trend of envelope resilience remains relatively consistent, albeit with a slight shift towards lower thermal performance over time. This suggests that focusing solely on U-values might not be sufficient for future-proofing buildings. Future research should explore additional factors that contribute to envelope resilience, such as the areal heat capacity of building components. By adopting a more holistic approach that considers both U-values and other relevant factors, building regulations and design practices can be optimized to ensure energy efficiency and resilience to climate change in the context of Italy's diverse climate zones.

3.2 Impact of Building Shape

Traditionally, building energy efficiency has focused on minimizing heat transfer through the building envelope, primarily by lowering the U-value (a measure of steady-state thermal transmittance) (Baglivo et al. 2024). However, research on surface-to-volume ratio (S/V) reveals a critical additional factor, particularly in the face of climate change. Buildings with lower S/V ratios, achieved through compact shapes and efficient space utilization, exhibit superior resilience across diverse climates.

This advantage stems from the reduced rate of heat loss or gain associated with a lower S/V. In cold climates, compact buildings retain heat more effectively, minimizing heating needs. Imagine a cube versus a long, thin rectangle. The cube has a smaller surface area relative to its volume, meaning less heat escapes compared to the rectangle. Conversely, in warm climates, compact buildings lose heat more slowly, reducing cooling requirements. This highlights the limitations of solely focusing on U-value. A holistic approach that considers both U-value and S/V ratio, alongside urban form, is essential for effective energy efficiency strategies in a changing climate.

By promoting compact building designs, policymakers and urban planners can contribute significantly to achieving better energy performance across various climate zones in Italy. This does not necessarily mean sacrificing aesthetics or functionality. Innovative architectural solutions can achieve compact layouts while incorporating courtyards, atriums, and clever use of vertical space to create a sense of openness and cater to various needs. Furthermore, urban design plays a crucial role. Densely packed buildings with narrow streets can trap heat, exacerbating cooling demands in warm climates. Conversely,

Table 3 $EP_{h,nd}$, $EP_{c,nd}$, and $EP_{tot,nd}$ of Lecce for the years 2030, 2050, and 2070, considering the three different building regulations and three different surface-to-volume ratios.

Regulations	Year	S/V 0.75			S/V 0.55			S/V 0.35		
		$EP_{h,nd}$ [kWh/m^2]	$EP_{c,nd}$ [kWh/m^2]	$EP_{tot,nd}$ [kWh/m^2]	$EP_{h,nd}$ [kWh/m^2]	$EP_{c,nd}$ [kWh/m^2]	$EP_{tot,nd}$ [kWh/m^2]	$EP_{h,nd}$ [kWh/m^2]	$EP_{c,nd}$ [kWh/m^2]	$EP_{tot,nd}$ [kWh/m^2]
L.D. 192/2005	2030	43.65	55.47	99.12	33.37	24.92	58.29	23.03	16.98	40.01
L.D. 192/2005	2050	39.59	62.81	102.4	30.67	28.12	58.79	21.21	19.23	40.44
L.D. 192/2005	2070	35.35	69.01	104.36	27.97	30.44	58.41	19.4	20.66	40.06
M.D. 26/6/2015	2030	23.08	59.23	82.31	21.15	24.87	46.02	15.5	16.47	31.97
M.D. 26/6/2015	2050	20.33	64.31	84.64	19.16	27.59	46.75	14.08	19	33.08
M.D. 26/6/2015	2070	17.59	68.25	85.84	17.18	29.59	46.77	12.68	20.42	33.1
M.D. 6/8/2020	2030	17.44	60.16	77.6	17.43	24.97	42.4	13.42	17.11	30.53
M.D. 6/8/2020	2050	15.12	64.95	80.07	15.66	27.4	43.06	12.13	18.93	31.06
M.D. 6/8/2020	2070	12.92	68.67	81.59	13.87	29.29	43.16	10.86	20.25	31.11

well-planned urban layouts with wider streets and strategically placed vegetation can promote natural ventilation and shading, further enhancing the energy efficiency benefits of compact building shapes.

Table 3 reports $EP_{h,nd}$, $EP_{c,nd}$, and $EP_{tot,nd}$ values of Lecce for the years 2030, 2050, and 2070, considering the three different building regulations and three different surface-to-volume ratios. The thermal performance indices have been calculated in a previous study (Baglivo et al. 2024). All the values reported are referred to RCP4.5 scenario. Regarding $EP_{h,nd}$, when comparing the values obtained with an S/V ratio of 0.75 to those with an S/V ratio of 0.35 under the same regulation, a reduction is observed in all cases. Similarly, for $EP_{c,nd}$, there is a consistent decrease when comparing the results for S/V ratios of 0.75 and 0.35 across all three regulations. Significant reductions in $EP_{tot,nd}$ are also observed when comparing S/V ratios of 0.75 and 0.35 across all three regulations and for each year.

3.3 Impact of Total Solar Energy Transmissivity

While building shape plays a crucial role in energy efficiency, window design adds another layer of complexity (Congedo et al. 2024b). The research delves into the influence of glazing system properties, particularly the total solar energy transmittance ($g_{gl,n}$). This value indicates how much solar energy a window transmits, impacting both heating and cooling needs.

The study emphasizes the importance of considering not just the thermal properties of building materials, but also their solar gain characteristics. In warmer zones, where cooling demands are expected to rise, employing windows with lower $g_{gl,n}$ values (achieved through more selective glazing) can significantly mitigate the need for cooling systems A window with a low $g_{gl,n}$ value functions permits to enter while blocking a significant portion of heat-producing solar radiation. Conversely, in colder zones, a balance needs to be struck. While minimizing heat loss remains important, maximizing solar heat gain for passive heating is also desirable. Here, windows with a higher $g_{gl,n}$ can be beneficial. These windows allow more solar radiation to enter, passively heating the building during colder months.

These findings underscore the need for a nuanced approach to window design, tailored to specific climate zones. Building regulations and design practices must move beyond a one-size-fits-all approach for glazing systems. By incorporating climate-specific $g_{gl,n}$ values, policymakers and designers can optimize window performance for both heating and cooling needs. This will be crucial for achieving optimal energy efficiency under the dynamic climatic conditions expected in the future. Moreover, developments in window technology present promising prospects. For instance, switchable glazing, which can adjust its $g_{gl,n}$ properties based on external conditions, could offer the best of both worlds: maximizing solar heat gain in winter and minimizing heat gain in summer.

In conclusion, optimizing building shape through a focus on S/V ratio and strategic urban design, coupled with climate-specific window design that considers total solar energy transmittance, presents a powerful approach to achieving energy-efficient and climate-resilient buildings in Italy. By embracing these strategies, Italy can create a future-proof built environment that not only minimizes energy consumption but also fosters a more sustainable and comfortable living experience for its citizens.

Table 4 reports $EP_{h,nd}$, $EP_{c,nd}$, and $EP_{tot,nd}$ values of Lecce for the years 2030, 2050, and 2070, considering the three different building regulations and two values of total solar energy transmittance. These values have been calculated in a previous study; therefore, only the results are presented here (Congedo et al. 2024b). All the values reported are referred to RCP4.5 scenario. The analysis reveals clear trends in the energy performance indices

Table 4 $EP_{h,nd}$, $EP_{c,nd}$, and $EP_{tot,nd}$ of Lecce for the years 2030, 2050, and 2070, considering the three different building regulations and two values of total solar energy transmittance.

Regulation	Year	$g_{gl,n}$ 0.50			$g_{gl,n}$ 0.67		
		$EP_{h,nd}$ [kWh/m²year]	$EP_{c,nd}$ [kWh/m²year]	$EP_{tot,nd}$ [kWh/m²year]	$EP_{h,nd}$ [kWh/m²year]	$EP_{c,nd}$ [kWh/m²year]	$EP_{tot,nd}$ [kWh/m²year]
L.D. 192/2005	2030	36.67	25.54	62.21	34.80	28.49	63.29
L.D. 192/2005	2050	33.69	28.97	62.66	31.85	32.15	64.00
L.D. 192/2005	2070	30.72	31.50	62.22	28.87	35.44	64.31
M.D. 26/6/2015	2030	22.66	25.14	47.80	20.40	29.05	49.45
M.D. 26/6/2015	2050	20.48	28.05	48.53	18.31	32.03	50.34
M.D. 26/6/2015	2070	18.33	30.19	48.52	16.18	34.28	50.46
M.D. 6/8/2020	2030	18.32	25.27	43.59	15.60	30.24	45.84
M.D. 6/8/2020	2050	16.42	27.92	44.34	13.75	33.78	47.53
M.D. 6/8/2020	2070	14.50	29.92	44.42	11.61	36.77	48.38

EPh,nd, EPc,nd, and EPtot,nd over time. EPh,nd consistently decreases, showing similar reductions for both ggl,n 0.67 and 0.50. In contrast, EPc,nd increases, with a more pronounced rise under the M.D. 06/08/2020 and M.D. 26/06/2015 regulations, especially at ggl,n 0.67. EPtot,nd also shows an upward trend, with the most significant increase observed under LD 192/2005 at ggl,n 0.67, indicating a growing impact on building resilience to climate change.

4 Conclusions

Climate change presents a significant challenge for energy use in buildings in Italy. While rising temperatures are likely to reduce heating demand, especially in the warm climate, this benefit is offset by a substantial increase in cooling needs. Current building regulations may not be sufficient to address these changing conditions. The surface-to-volume (S/V) ratio of buildings also influences energy performance, with compact structures performing better in colder climates.

Solutions include optimizing window design to balance heating and cooling requirements based on climate zones and retrofitting existing buildings to improve energy efficiency. Future climate projections should inform building regulations to avoid the risks of over-insulation in warmer areas.

A holistic approach is essential, integrating building design, urban planning, and climate projections to create sustainable and energy-efficient buildings. Collaboration among policymakers, architects, engineers, urban planners, and researchers is crucial, as is educating occupants on energy-efficient practices. Financial incentives and green building certifications can further encourage the adoption of energy-efficient construction methods.

In conclusion, to foster a more sustainable and resilient built environment in Italy, several key suggestions can be implemented:

- **Implement regulations that consider both U-value and S/V ratio**. Develop energy efficiency standards that account for the surface-to-volume ratio (S/V) of buildings alongside the U-value. This will ensure that buildings are optimized for energy performance across different climates and shapes.
- **Promote compact building designs**. Encourage compact building designs with lower S/V ratios, especially in colder climate zones. This will reduce heat loss and improve overall energy efficiency.
- **Optimize window design**. Implement regulations and guidelines for window design that consider the total solar energy transmittance ($g_{gl,n}$) appropriate for different climate zones. This will ensure that windows are optimized for both heating and cooling needs.
- **Encourage retrofitting of existing buildings**. Provide incentives and support programs for retrofitting existing buildings to improve their energy efficiency. This could include measures such as adjusting window $g_{gl,n}$, improving insulation, and upgrading cooling and heating systems.
- **Develop climate-sensitive building codes**. Update building codes to incorporate future climate projections and consider the potential impacts of climate change on energy demand. This will ensure that new buildings are designed to be resilient to future climate conditions.
- **Promote research and innovation**. Support research and development in climate-resilient building materials, technologies, and design techniques. This will pave the way for the next generation of energy-efficient and climate-adapted buildings.
- **Foster collaboration among stakeholders**. Encourage collaboration between policymakers, architects, engineers, urban planners, and researchers to develop and implement comprehensive strategies for energy-efficient and climate-resilient buildings.

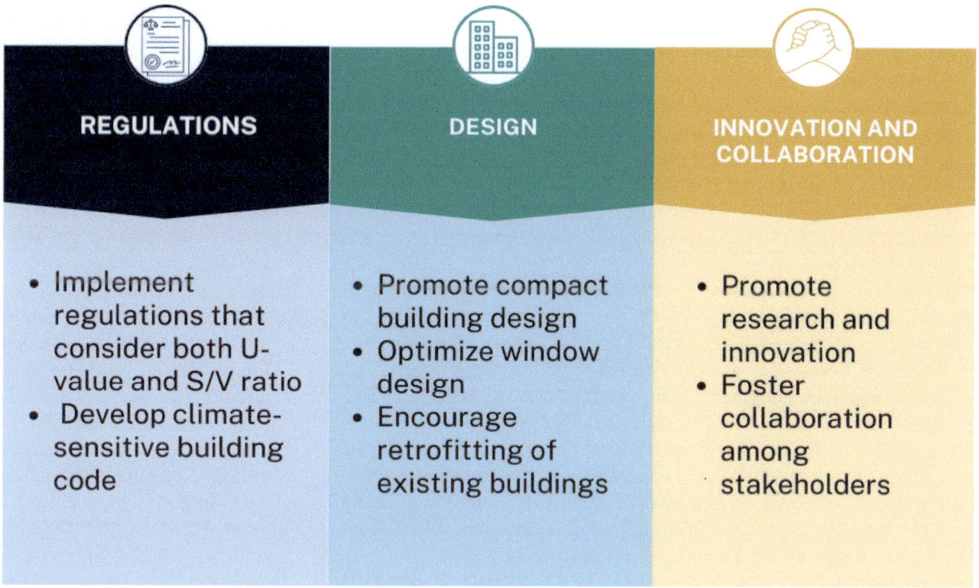

Fig. 3 Summary of key recommendations for promoting a more sustainable and resilient built environment in Italy. (By authors)

Figure 3 summarizes the conclusions of this study, highlighting key recommendations for fostering a more sustainable and resilient built environment in Italy. It illustrates the main suggestions to be implemented.

By implementing these suggestions, Italy can move towards a more sustainable and resilient built environment that is better equipped to handle the challenges of climate change.

While this study is limited to the Italian context, focusing specifically on its unique climatic challenges and building stock, it nevertheless provides an important foundation for future efforts to advance climate neutrality and enhance energy efficiency in the building sector.

Declaration of Competing Interest The authors have no conflicts of interest to declare that are relevant to the content of this chapter.

References

Baglivo, C.: Dynamic evaluation of the effects of climate change on the energy renovation of a school in a mediterranean climate. Sustainability **13**, 6375 (2021)

Baglivo, C., Congedo, P.M., Mazzeo, D.: Climate change and building performance: pervasive role of climate change on residential building behavior in different climates. In: Adapting the Built Environment for Climate Change, pp. 229–251. Woodhead Publishing (2023)

Baglivo, C., Congedo, P.M., Malatesta, N.A.: Building envelope resilience to climate change under Italian energy policies. J. Clean. Prod. **411**, 137345 (2023b)

Baglivo, C., Albanese, P.M., Congedo, P.M.: Relationship between shape and energy performance of buildings under long-term climate change. J. Build. Eng. **84**, 108544 (2024)

Concerted Actions EPBD.: Concerted Action EPBD—Enabling the implementation of the Energy Performance of Buildings Directive (2020). https://epbd-ca.eu/database-of-outputs. Accessed 25 April 2024

Congedo, P.M., Baglivo, C., Seyhan, A.K., Marchetti, R.: Worldwide dynamic predictive analysis of building performance under long-term climate change conditions. J. Build. Eng. **42**, 103057 (2021)

Congedo, P.M., Baglivo, C., D'Agostino, D., Mazzeo, D.: The impact of climate change on air source heat pumps. Energy Convers. Manage. **276**, 116554 (2023)

Congedo, P.M., Baglivo, C., D'Agostino, D., Albanese, P.M.: Overview of EU building envelope energy requirement for climate neutrality. Renew. Sustain. Energy Rev. **202**, 114712 (2024a)

Congedo, P.M., Albanese, P.M., D'Agostino, D., Baglivo, C.: The role of total solar energy transmittance for normal incidence of the glazed system in climate change adaptation under Italy energy efficiency policies. Energy Build. **307**, 113944 (2024b)

D'Agostino, D., Zacà, I., Baglivo, C., Congedo, P.M.: Economic and thermal evaluation of different uses of an existing structure in a warm climate. Energies **10**(5), 658 (2017)

D'Agostino, D., Congedo, P.M., Albanese, P.M., Rubino, A., Baglivo, C.: Impact of climate change on the energy performance of building envelopes and implications on energy regulations across Europe. Energy **288**, 129886 (2024)

Dascalaki, E.G., Kontoyiannidis, S., Balaras, C.A., Droutsa, K.G.: Energy certification of Hellenic buildings: first findings. Energy Build. **65**, 429–437 (2013)

Diaz-Lopez, C., Verichev, K., Holgado-Terriza, J.A., Zamorano, M.: Evolution of climate zones for building in Spain in the face of climate change. Sustain. Cities Soc. **74**, 103223 (2021)

DM 6/8/2020.: Italian Ministerial Decree, Requisiti tecnici per l'accesso alle detrazioni fiscali per la riqualificazione energetica degli edifici – cd Ecobonus" (In Italian) (2020)

European Commission.: Directive 2002/91/EC of the European Parliament and of the Council on the energy performance of buildings, Off. J. Eur. Commun., Brussels (2002)

European Commission.: Directive 2009/125/EC of the European Parliament and of the Council of 21 October 2009 establishing a framework for the setting of ecodesign requirements for energy-related products (recast) (2009)

European Commission.: Directive 2010/31/EU of the European Parliament and Council of 19 May 2010 on the energy performance of buildings (recast), Off. J. Eur. Union, Brussels (2010)

European Commission.: Directive 2010/31/EU of the European Parliament and of the Council of 19 May 2010 on the energy performance of buildings (recast) (2018a). https://eur-lex.europa.eu/legal-content/EN/TXT/?uri=CELEX%3A02010L0031-20210101. Accessed 25 April 2024

European Commission.: Directive 2018/844/EU of the European Parliament and of the Council of 30 May 2018 amending Directive 2010/31/EU on the energy performance of buildings and Directive 2012/27/EU on energy efficiency, Official Journal of the European Union, pp. 75–91 (2018b). https://eur-lex.europa.eu/legalcontent/EN/TXT/PDF/?uri=CELEX:32018L0844&from=EN. Accessed 25 April 2024

European Commission.: Directive 2012/27/EU of the European Parliament and of the Council of 25 October 2012 on energy efficiency, amending Directives 2009/125/EC and 2010/30/EU and repealing Directives 2004/8/EC and 2006/32/EC (2023a). https://eur-lex.europa.eu/legal-content/EN/TXT/?uri=CELEX%3A02012L0027-20230504&qid=1687869958720. Accessed 25 April 2024

European Commission.: Directive (EU) 2023/2413 of the European Parliament and of the Council of 18 October 2023 amending Directive (EU) 2018/2001, Regulation (EU) 2018/1999 and Directive 98/70/EC as regards the promotion of energy from renewable sources, and repealing Council Directive (EU) 2015/652 (2023b)

Economidou, M., Todeschi, V., Bertoldi, P., D'Agostino, D., Zangheri, P., Castellazzi, L.: Review of 50 years of EU energy efficiency policies for buildings. Energy Build. **225**, 110322 (2020)

Fregonara, E., Verso, V.R.L., Lisa, M., Callegari, G.: Retrofit scenarios and economic sustainability. A case-study in the Italian context. Energy Proc. **111**, 245–255 (2017)

Haseeb, Q.S., Yunus, S.M., Shoshan, A.A.A., Aziz, A.I.: A study of the optimal form and orientation for more energy efficiency to mass model multi-storey buildings of Kirkuk city, Iraq. Alex. Eng. J. **71**, 731–741 (2023)

https://commission.europa.eu/strategy-and-policy/priorities-2019-2024/european-green-deal_it. Accessed 24 April 2024

https://energy.ec.europa.eu/topics/energy-efficiency/financing/eu-programmes/previous-funding_en. Accessed 25 April 2024

International Energy Agency.: Technology Roadmap Energy Efficient Building Envelopes. OECD, 68 (2013)

Jafarpur, P., Berardi, U.: Effects of climate changes on building energy demand and thermal comfort in Canadian office buildings adopting different temperature setpoints. J. Build. Eng. **42**, 102725 (2021)

Koengkan, M., Fuinhas, J.A., Osmani, F., Kazemzadeh, E., Auza, A., Alavijeh, N.K., Teixeira, M.: Do financial and fiscal incentive policies increase the energy efficiency ratings in residential properties? A piece of empirical evidence from Portugal. Energy **241**, 122895 (2022)

L.D. 192/2005.: Italian Legislative Decree, Attuazione della direttiva (UE) 2018/844, che modifica la direttiva 2010/31/UE sulla prestazione energetica nell'edilizia e la direttiva 2012/27/UE sull'efficienza energetica, della direttiva 2010/31/UE, sulla prestazione energetica nell'edilizia, e della direttiva 2002/91/CE relativa al rendimento energetico nell'edilizia (In Italian)

M.D. 26/06/2015.: Italian Ministerial Decree, "Applicazione delle metodologie di calcolo delle prestazioni energetiche e definizione delle prescrizioni e dei requisiti minimi degli edifici" (In Italian) (2015)

Meteonorm—Global Meteorological Database, Meteotest (2012)

Nejat, P., Jomehzadeh, F., Taheri, M.M., Gohari, M., Majid, M.Z.A.: A global review of energy consumption, CO_2 emissions and policy in the residential sector (with an overview of the top ten CO_2 emitting countries). Renew. Sustain. Energy Rev. **43**, 843–862 (2015)

Pastore, L.M., Sforzini, M., Basso, G.L., de Santoli, L.: H2NG environmental-energy-economic effects in hybrid energy systems for building refurbishment in future National Power to Gas scenarios. Int. J. Hydrogen Energy **47**(21), 11289–11301 (2022)

Popielak, P., Majchrzak-Kucęba, I., Wawrzyńczak, D.: Climate change mitigation with CCUS-A case study with benchmarking for selected countries in adapting the European Union's Green Deal. Int. J. Greenhouse Gas Control **132**, 104057 (2024)

Qureshi, Y., Ali, U., Sher, F.: Part load operation of natural gas fired power plant with CO_2 capture system for selective exhaust gas recirculation. Appl. Therm. Eng. **190**, 116808 (2021)

Salata, F., Falasca, S., Ciancio, V., Curci, G., Grignaffini, S., de Wilde, P.: Estimating building cooling energy demand through the Cooling Degree Hours in a changing climate: a modeling study. Sustain. Cities Soc. **76**, 103518 (2022)

Sher, F., Kawai, A., Güleç, F., Sadiq, H.: Sustainable energy saving alternatives in small buildings. Sustain. Energy Technol. Assess. **32**, 92–99 (2019)

Skillington, K., Crawford, R.H., Warren-Myers, G., Davidson, K.: A review of existing policy for reducing embodied energy and greenhouse gas emissions of buildings. Energy Policy **168**, 112920 (2022)

Ulpiani, G., Vetters, N.: On the risks associated with transitioning to climate neutrality in Europe: a city perspective. Renew. Sustain. Energy Rev. **183**, 113448 (2023)

Zhai, Z.J., Helman, J.M.: Implications of climate changes to building energy and design. Sustain. Cities Soc. **44**, 511–519 (2019)

Adapting Italian School Buildings to Climate Challenges: Enhancing Energy Efficiency and Comfort Through Future-Driven Strategies

Andrea Palmieri, Cristina Baglivo, and Paolo Maria Congedo

Abstract

This study examines the challenges faced by educational institutions in the Mediterranean region in managing energy costs and thermal comfort, with a particular focus on the impact of climate change. Global temperatures are expected to increase by about 1–1.5 °C by 2040 and 1.5–3.5 °C by 2080, which could worsen environmental conditions in schools. The research examines the variation in comfort levels in schools, considering different future perspectives over a long period of time. Specifically, it focuses on the predicted thermal performance of a classroom in Lecce, a city in southern Italy. Long-term hourly dynamic thermal simulations have been performed for the year 2050. The study is structured in three, namely the analysis of the evolution of the indoor operative temperature, the effectiveness of ventilation in ensuring thermal comfort in winter, and the evaluation of indoor temperatures between current and future long-term climatic data. Ten ventilation scenarios were used for air exchange, with the aim of maintaining an operative indoor temperature between 21 °C and 23 °C during the winter period. The results highlight the critical challenge that climate change poses for schools and underline the importance of adopting integrated approaches that take into account energy efficiency, environmental impact, and the well-being of students and staff. In Mediterranean regions, one possible solution could be to discourage the installation of heating systems and promote the adoption of controlled mechanical ventilation systems.

Keywords

Schools · Indoor thermal comfort · Internal temperature operative · Mechanical ventilation · Air changes · Climate change

1 Introduction

Energy use in school buildings can be affected by climate change, as changes in temperature and weather patterns can increase or decrease the need for heating and cooling in buildings. Therefore, improving the energy efficiency of school buildings can not only help reduce operating costs and promote environmental sustainability, but also mitigate the effects of climate change on the built environment and society as a whole.

School buildings, which represent a significant portion of the building stock, are a major source of non-industrial energy consumption in a country (Barbhuiya and Barbhuiya 2013). Currently, several energy efficiency plans have been developed worldwide to improve the energy performance of schools. Some scholars (AlFaris et al. 2016) have analyzed energy management programs as sustainable strategies for schools. Energy management in schools is essential to ensure optimal thermal comfort in classrooms, considering both visual and ventilation aspects (Jovanović et al. 2014; Wang et al. 2014).

The focus on indoor air quality and energy performance in school buildings has always been high, given the critical importance of ensuring a comfortable environment for students and teachers. The high density of students in classrooms has critical aspects: an unsatisfactory thermal environment has a negative impact on student learning and performance (Barrett et al. 2015). In fact, low ventilation rates can negatively affect students' attention and concentration,

A. Palmieri · C. Baglivo (✉) · P. M. Congedo
Department of Engineering for Innovation, University of Salento, 73100 Lecce, Italy
e-mail: cristina.baglivo@unisalenot.it

P. Mohammad et al. (eds.), *Geographic Approaches to Climate Change and Mitigation: Urban and Rural Perspectives (Volume 1)*, Advances in Science, Technology & Innovation, https://doi.org/10.1007/978-3-031-92119-3_24

reducing their performance by up to 30% (Wargocki and Wyon, 2013).

Sánchez et al. (2023) propose an energy renovation strategy for Mediterranean schools, highlighting the importance of solar shading to reduce cooling requirements. Campagna and Fiorito (2023) investigate the factors influencing the heating consumption of schools in the region. A detailed analysis of the school building stock in Puglia shows that most of the schools were built before 1976 with no energy regulations. The monthly consumption profile shows a peak during the winter months, with heating dominating over electrical consumption. These results indicate that the existing building stock will be in use for at least the next 50 years.

Students spend a lot of time in school, so ensuring comfortable indoor thermal conditions is critical. High occupant densities combined with inadequate ventilation can lead to significant problems in classrooms, such as overheating and poor air quality. (Mohamed et al. 2021; Dorizas et al. 2015; Korsavi and Montazami 2020). Thermal comfort is critical to productivity, well-being, and energy conservation in schools. Zomorodian et al. (2016) reviewed thermal comfort in school buildings over the past five decades and emphasized the importance of ventilation for air quality and thermal comfort. Maintaining high indoor environmental quality in schools is essential for student health and performance. Ashrafian (2023) examined the impact of climate change on building performance, focusing on energy consumption, economic impact, and occupant comfort. In schools, balancing energy savings with comfort is critical to student performance. In addition, the projected global temperature increase, estimated to be between 1 and 1.5 °C by 2040 and between 1.5 and 3.5 °C by 2080 (Pachauri et al. 2014), could worsen environmental conditions in schools and expose them to greater risks from climate change. As a result, there is an increasing need to improve the resilience of school buildings to the effects of global warming (Dong et al. 2023).

In schools, balancing energy savings with comfort is critical to student performance. Recent studies (Congedo et al. 2016; Zacà et al. 2016) indicate that achieving an optimal balance between cost reduction and maximum comfort is critical to ensuring optimal performance for students and users. Yang et al. (2021) investigated the impact of climate change on residential buildings in Europe using comprehensive climate projections over a period of more than 90 years. The analysis considered both energy performance and thermal comfort of the buildings. The results show an increase in cooling demand, a decrease in heating demand, and significant differences in impacts between different regions and cities. Thermal discomfort may persist after renovation, leading to increased use of cooling and ventilation systems with potential energy and environmental impacts. Future projections suggest an increase in energy demand and variations in the thermal performance of buildings (Baglivo 2021). In the school environment, optimal lighting plays a critical role in enabling students and staff to perform their visual tasks without visual fatigue. A recent study has shown that replacing traditional fluorescent lighting with LED technology can reduce energy consumption by up to 33%, while the use of control systems such as dimming and modulating LED lights can lead to energy savings of up to 95% (Baglivo et al. 2021). In addition, the use of windows with double or triple low-emissivity glass can reduce energy consumption per square meter of glazing by more than 40%, depending on the materials and components used (Baglivo et al. 2017).

Optimizing thermal comfort in schools is becoming increasingly important in anticipation of future climate change. As temperature extremes become more frequent, investment in appropriate systems will be critical to ensuring student comfort and performance and mitigating negative impacts on learning and health. In addition, a comfortable school environment can help maintain high levels of productivity and performance in the face of climate change challenges.

This study aims to explore innovative solutions to improve thermal comfort in schools, considering the context of the Mediterranean climate and the impact of climate change. Through annual and hourly simulations, the thermal behavior of school buildings is studied, focusing on the monitoring of the internal operative temperature in classrooms. Assuming schools with efficient building envelopes, the heating demand can be met by internal thermal loads, mainly those generated by students and equipment. Considering the prospect of rising temperatures due to climate change, alternatives that favor the use of controlled mechanical ventilation systems over traditional heating systems are evaluated. The thermal analysis is based on a prototype school building located in the city of Lecce, known for its Mediterranean climate. Using climate data for the year 2050, the indoor operative temperature is studied through annual and hourly simulations in a dynamic regime, focusing on a single classroom in the building.

2 Methodology

In order to assess the impact of climate change, annual and hourly analyses were performed on a classroom selected as a prototype within a school characterized by a simple and regular layout. Using climate data projected for 2050 within the Termolog EpiX 15 software (Congedo et al. 2021, 2020; Baglivo et al. 2022), it was possible to study the energy behavior of the classroom over time and its response to climate change, following the current guidelines established

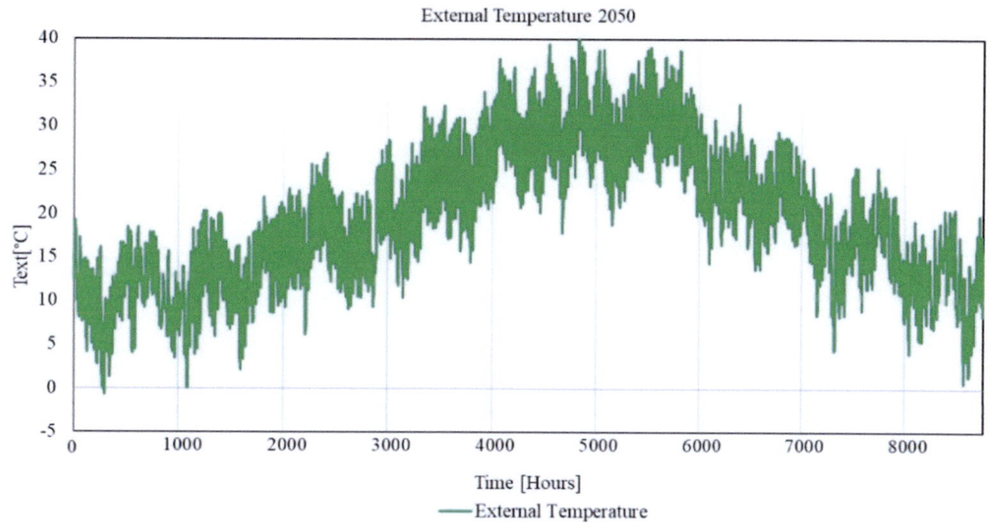

Fig. 24.1 Predicted external temperature for the city of Lecce in the year 2050 according to RCP 4.5

in the UNI EN ISO 52016:2018 standard (UNI EN ISO 2018). The climate data for 2050 are based on the Representative Concentration Pathway (RCP) 4.5 scenario and have been extrapolated from Meteonorm.

The analyzed prototype building is located in Lecce, in the southeast of Italy, specifically in the Apulia region. Lecce is classified as national climate zone C according to Italian regulations. According to Italian standards, the classification of the climate of different cities is based on heating degree days (HDD), which represent the sum of positive differences between internal and external averages for each day of the year. Lecce is characterized by a Mediterranean climate, with mild winters, warm springs, hot summers, and temperate autumns, where precipitation is usually below 500 mm. According to the Köppen climate classification (Arnfield 2023), Lecce is classified as Csa, indicating a Mediterranean climate with hot summers. The annual trend of the outdoor temperature with respect to the climate data of 2050 is shown in Fig. 24.1.

The school building under consideration is a simplified model that is insulated on all four sides to account for all possible orientations and to characterize all possible exposures. The school is designed to accommodate students from an elementary school with a short school week. The educational activities will take place from Monday to Friday from 8:00 to 16:00. The age of the students will be between 6 and 10 years old. The building design refers to the current national regulations, in particular the Ministerial Decree of December 18, 1975 (DM 1975), which states that for a primary school, the internal surface area of a classroom should be 1.8 × (m^2/student). Therefore, the building under study has a simple square floor plan with a net area of 332.70 m^2 and a net height of 3.00 m. As shown in Fig. 24.2, the area has been divided into nine classrooms of 6 × 6 m^2 each,

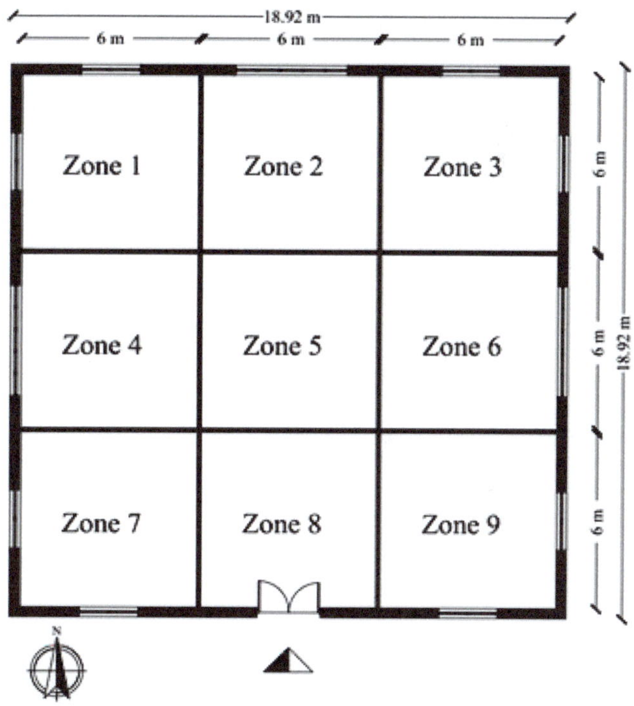

Fig. 24.2 Floor plan of the school model, designed in compliance with the Ministerial Decree of December 18, 1975, illustrating the spatial organization and functional layout required to meet regulatory standards

all of the same size, taking into account an internal load of 20.5 W/m^2 generated by the presence of 20 students.

Figure 24.2 shows the floor plan of the school building with a regular layout of the available area. The total area of 332.70 m^2 is efficiently organized into nine classrooms, each measuring 6 × 6 m^2 (36 m^2). The classrooms, all of the same size, are arranged in a regular grid. The partition walls

Table 1 Thermal transmittance limits in accordance with Ministerial Decree 26/06/2015 for Climate Zone C

Italian climate zones	Heating on		Envelope components	U (W/m^2K)
	From 1993	From 2022		
C 900 < HDD ≤ 1400	November 15–March 31 (10 h a day)	November 22–March 23 (9 h a day)	U_{roof}	0.33
			U_{wall}	0.34
			U_{floor}	0.38
			U_{window}	2.20

between the classrooms are 10 cm thick, providing adequate separation between the interior spaces without taking up too much usable space. The exterior walls are 35 cm thick to provide adequate thermal insulation. Each classroom is designed for 20 students, and the calculated internal load of 20.5 W/m^2 reflects the combined contribution of equipment, lighting, and student presence. This layout, along with the building geometric features and wall configuration, allows for a simplified thermal analysis.

In the building modeling, the thermal transmittance limits established by the Ministerial Decree 26/06/2015 (M.D. 2015) for climate zone C were taken into account, as shown in Table 1.

Table 1 shows the thermal transmittance limits (U-values) of building components in accordance with Ministerial Decree 26/06/2015 for Climate Zone C.

Figures 24.3, 24.4, 24.5 and 24.6 details the layers of the exterior wall, the floor in contact with the ground, the roof slab, and the doors and windows, respectively.

Figure 24.6 shows the thermal and geometric characteristics of the building openings. The figure shows the specifications of the French door in "Zone 7", the windows in "Zones 1, 3, 7, and 9", and the window in "Zones 2, 4, and 6". All windows have a three-chamber polyvinyl chloride (PVC) frame with a plastic spacer and are equipped with low-emissivity, 4-12-4 dual-pane glass.

Indoor operative temperature is monitored in relation to air changes, indicated by "Combo" followed by the corresponding number of air changes. For example, in "Combo 01" the indoor operative temperature values were obtained by ensuring only one air volume change per hour, and in "Combo 02" the indoor operative temperature values occur

Fig. 24.3 Ground floor stratigraphy, detailing thicknesses, and thermal conductivity values of the materials used

	Ground floor stratigraphy			
	Layers	**d [mm]**	**ρ [kg/m³]**	**λ [W/mK]**
	Internal adduntance (downward vertical flow)			5.88
1	Floor tiles	10	2300	1
2	cement mortar	10	2000	1.4
3	ordinary concrete screed	80	1700	1.1
4	Vapor barrier	0.5	360	0.4
5	EPS (expanded polyester)	20	24	0.033
6	Bituminous waterproofing membrane - Radon barrier	5	1200	0.2
7	Reinforced concrete	80	2400	1.91
8	IGLOO	200	1.2	1.39
9	Lean concrete	80	2200	1
10	Coarse gravel	150	1700	1.2
	Σ	635.5		
	U=0.775≤ 0.84 [W/m²k]			

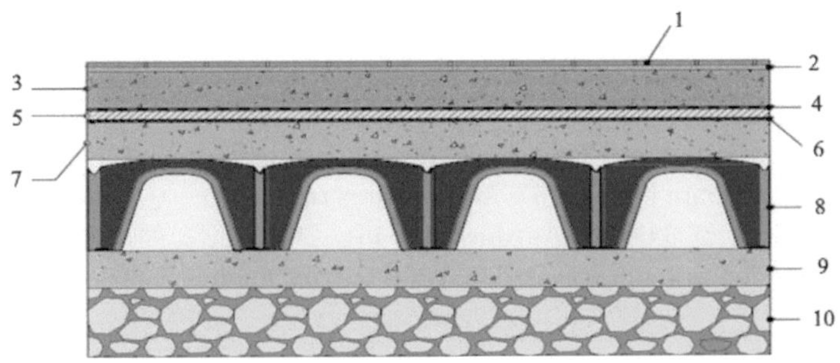

Fig. 24.4 External wall stratigraphy, detailing thicknesses, and thermal conductivity values of the materials used

External Wall				
	Layers	d [mm]	ρ [kg/m³]	λ [W/mK]
	Internal adduntance (horizontal flow)			7.69
1	Internal plaster	10	1400	0.7
2	Bricks blocks	250	1000	0.4
3	EPS (expanded polyester)	80	24	0.033
4	External plaster	10	1800	0.9
	External adductance (horizontal flow)			25
	Σ	350		
	$U=0.308 \leq 0.34$ [W/m²k]			

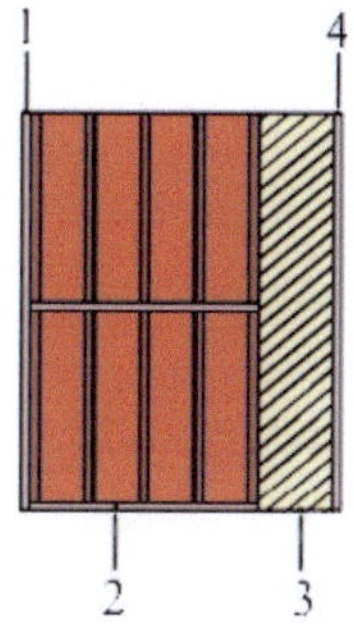

Fig. 24.5 Roof slab stratigraphy, detailing thicknesses, and thermal conductivity values of the materials used

Roof slab				
	Layers	d [mm]	ρ [kg/m³]	λ [W/mK]
	Internal adduntance (downward vertical flow)			10
1	Internal plaster	10	1400	0.7
2	Slab blocks	260	1800	0.743
3	Vapor barrier	0.5	360	0.4
4	EPS (expanded polyester)	80	24	0.033
5	Bituminous waterproofing membrane	5	1200	0.17
6	Concrete	100	1800	0.94
7	Concrete substrate - cement mortar	10	2000	1.4
8	Ceramic/porcelain tiels	10	2300	1.3
	External adduntance (upward vertical flow)			25
	Σ	475.5		
	$U=0.325 \leq 0.33$ [W/m²k]			

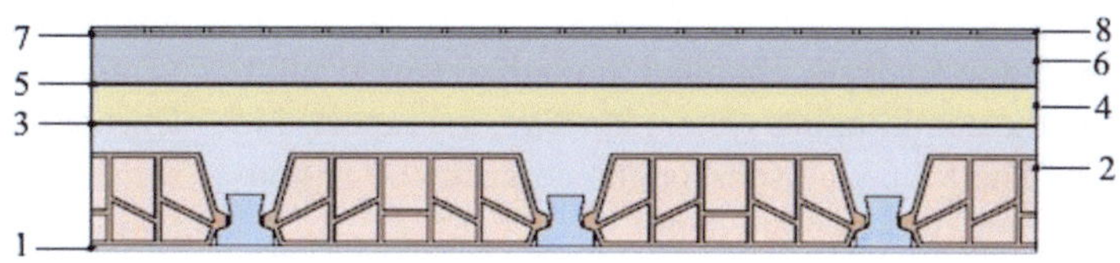

Fig. 24.6 Thermal and geometric characteristics of the building's openings, detailing the specifications for the French door in Zone 7, the windows in Zones 1, 3, 7, and 9, and the window in Zones 2, 4, and 6

Door window	Central window	Lateral window
$U_{w,set}$ = 1.926 W/m²K	$U_{w,set}$ = 2.009 W/m²K	$U_{w,set}$ = 2.004 W/m²K
$U_{w,lim}$ = 2.200 W/m²K	$U_{w,lim}$ = 2.200 W/m²K	$U_{w,lim}$ = 2.200 W/m²K

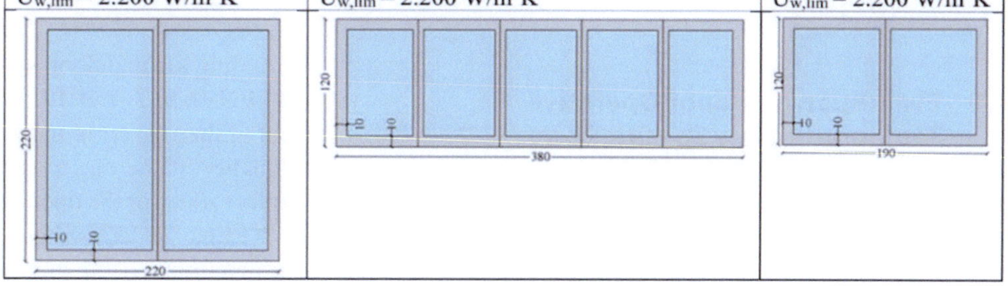

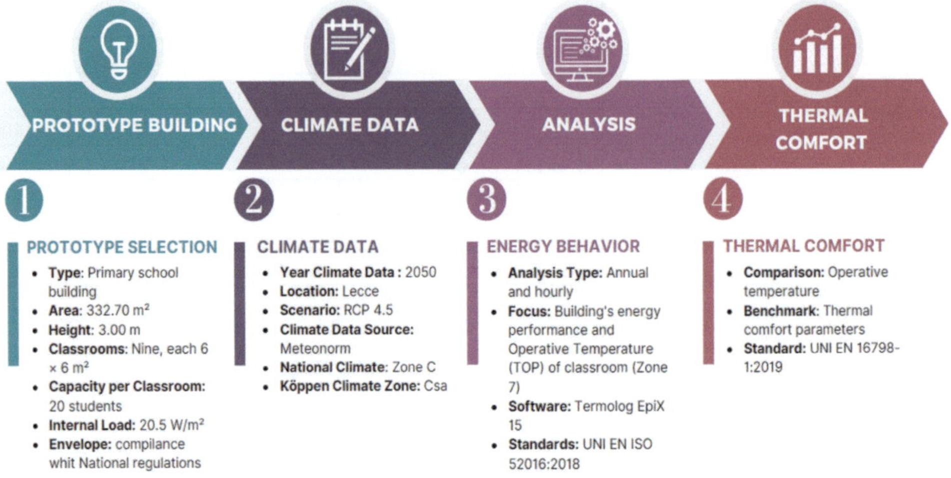

Fig. 24.7 Methodological flowchart

with two air volume changes per hour. The study combinations include a minimum of 1 to a maximum of 10 air changes. The number of students in the room was held constant.

Figure 24.6 shows the methodological flowchart, which outlines the key stages of the process, from data collection to analysis and interpretation, highlighting the logical sequence and interactions between activities that are essential to ensure the methodological coherence of the study.

3 Results and Discussions

The results of the analysis, obtained through dynamic simulations with the conditioning system inactive, are reported in this section. The following subsections describe the various aspects of the analysis in more detail:

- Evaluation of indoor operative temperature in a specific zone. (Sect. 3.1)
- Study of adaptive thermal comfort in a specific zone according to UNI EN 16,987 (Sect. 3.2)

3.1 Evaluation of Indoor Operative Temperature in a Specific Zone

This section analyzes in detail the annual operative indoor temperature trend of a specific zone within the entire building. Using future climate data for the year 2050, hourly assessments were made considering different air exchange scenarios.

The study was conducted in Zone 7, which is characterized by a southwest orientation and has two windows, one facing south and the other facing west. Characteristics of southwest facing areas include prevailing southwest sun exposure, resulting in greater sunlight exposure during the afternoon. These areas may experience high temperatures during the day due to direct sun exposure.

The indoor operative temperature in Zone 7 varies according to the different scenarios identified as "Combo". The classroom temperature analysis is shown in the graph in Fig. 24.7.

From the analysis of the different combinations from 01 to 05, it can be seen that during the coldest periods, the operative indoor temperature of the classroom ranges from a minimum of 17.3 °C with five air changes to a maximum of 35.8 °C with only one air change. This shows that it is possible to maintain adequate thermal comfort levels (between 21 °C and 23 °C) simply by adjusting the ventilation, without the need to turn on the heating system. During the intermediate periods (spring, fall) and during the summer, when only natural ventilation is used, the operative temperature of the classroom is completely uncontrollable.

3.2 Study of Adaptive Thermal Comfort in a Specific Zone According to UNI EN 16,987

According to the indoor environment quality scale defined by UNI EN 16,987–1:2019 (UNI EN ISO 2018) for the assessment of thermal comfort, the school must ensure a high level of quality in the classrooms. During the winter period, the comfort zone for the operative indoor temperature is between 21 °C and 23 °C, while during the intermediate periods (spring/autumn) and summer, the thermal comfort zone for the operative indoor temperature is determined by the two Eqs. (1) outlined in the standard.

$$\text{category I:} \begin{cases} \theta_0 = 0.33 \cdot \theta_{rm} + 18.8 + 2\, Upper\, limit \\ \theta_0 = 0.33 \cdot \theta_{rm} + 18.8 - 2\, Lower\, limit \end{cases} \tag{1}$$

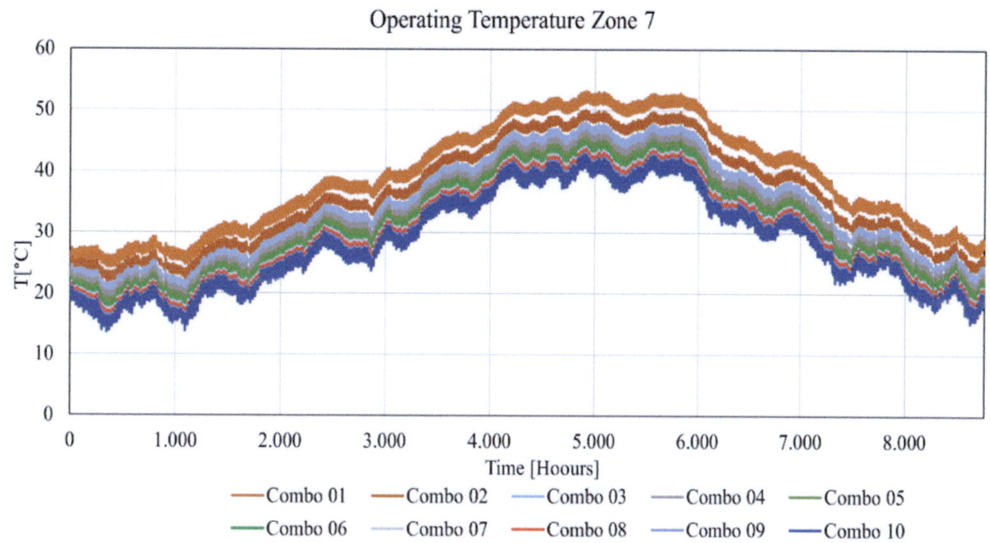

Fig. 24.8 shows the operative temperature in zone 7, with the number of combos indicating the air volume changes per hour. The relationship between air change and temperature stability is key to optimizing indoor environmental quality.

where:

$\boldsymbol{\theta}_0$ indoor operative temperature [°C].

$\boldsymbol{\theta}_{rm}$ weighted average external temperature [°C].

Figure 24.8 presents the results of the analysis of the operative temperature inside the classroom (zone 7), based on climatic data projected for 2050, compared to the thermal comfort parameters established by the UNI EN 16987-1:2019 standard [UNI EN ISO (2018)]. Figure 24.8a, b, and c shows the internal operative temperature and the adaptive thermal comfort during the coldest days of January, February, and March, for different ventilation scenarios (Combos 01–05). Figure 24.9d through 9 h shows the operative temperatures relative to the thermal comfort zone on the hottest days of April, May, June, September, and October. Finally, Fig. 24.9i and 1 shows the operative temperature versus thermal comfort on the coldest days of November and December.

The analysis of Fig. 24.8a clearly shows that Combos 01 and 02 fail to maintain the internal operative temperature within the thermal comfort zone. These scenarios show a significant deviation from the thermal comfort limits, suggesting that the strategies adopted with these combos are insufficient to ensure optimal thermal comfort during the coldest day in January. Combo 03, on the other hand, represents a significant improvement, managing to keep the internal temperature within the optimal thermal comfort limits throughout the duration of the educational activities. Although Combo 04 improves conditions compared to Combos 01 and 02, it still fails to maintain the internal operative temperature within the thermal comfort zone throughout the day.

Figure 24.8b analyzes the coldest day in February and shows that a hybrid approach is required to keep the internal operative temperature within the thermal comfort zone: starting with Combo 02 in the early hours of the day and then switching to Combo 03 for the remainder of the day. This approach highlights the importance of adapting thermal management strategies to seasonal and temporal variations. Combinations 01, 04, and 05 are found to be unsuitable because they fail to keep the internal operative temperature within the thermal comfort limits, being too far from the upper or lower limits of the defined thermal comfort zone.

Figure 24.8c examines the coldest day in March. In this case, the use of ventilation scenarios below Combo 05 results in an internal operative temperature that is constantly above the limit of the thermal comfort zone. With Combo 05, the internal operative temperature only partially remains within the thermal comfort zone: between 11:00 and 12:00 and until the end of the educational activities, the temperature values exceed the limit of the thermal comfort zone. In order to ensure an internal operative temperature within the thermal comfort zone for the entire duration of the educational activities in March, it is necessary to adopt the ventilation scenario corresponding to Combo 06. This scenario suggests that the thermal conditions in March require a more advanced control configuration compared to previous months. These results underscore the need for adaptive and scalable solutions for thermal control in classrooms, highlighting the need for different strategies to meet the challenges posed by climate change and ensure optimal thermal comfort conditions throughout the year.

Figure 24.8d analyzes the hottest day in April and shows how the internal operative temperature never falls within the defined thermal comfort zone. This phenomenon occurs both when using the minimum ventilation scenario, represented by Combo 05, and when using the maximum ventilation level, represented by Combo 10. During the in-between periods (spring and fall) and during the summer months, the thermal comfort varies daily according to Eq. (1) of the UNI EN 16,987-1:2019 standard (UNI EN ISO 2018). For example, on the hottest day in April, the thermal comfort zone oscillates between 21.47 °C and 26.47 °C. However, despite these oscillations, the internal operative temperature remains consistently outside this range, as shown in Fig. 24.8d. A similar situation is observed on the hottest day in May, shown in Fig. 24.8e, where the thermal comfort zone ranges from 23.79 °C to 28.88 °C. Again, the ventilation scenarios studied are insufficient to keep the internal operative temperature within the thermal comfort limits. Figure 24.9f, representing the hottest day in June, confirms this trend: despite the thermal comfort zone oscillating between 23.54 °C and 28.54 °C, the internal operative temperature does not fall within the defined limits. With the start of educational activities in September, the hottest day of the month, shown in Fig. 24.8 g, continues to show difficulties in maintaining the internal operative temperature within the thermal comfort zone, which in this case ranges between 23.38 °C and 28.38 °C. Even in this scenario, the ventilation strategies adopted are insufficient to ensure adequate thermal comfort conditions, suggesting the need for more effective solutions. Finally, Fig. 24.8 h, which refers to the hottest day in October, further highlights the challenges of maintaining an internal operative temperature within the thermal comfort zone, which in this month varies between 23.08 °C and 28.08 °C. Despite the adoption of higher ventilation scenarios, high external temperatures continue to prevent the achievement of optimal thermal comfort conditions.

Figure 24.8i shows the evolution of the internal operative temperature relative to the thermal comfort zone, which fluctuates between 21 °C and 23 °C during the coldest day in November. In this case, using the ventilation scenario associated with Combo 08, it is observed that the internal operative temperature remains within the thermal comfort zone until about 13:00. After that, however, the internal operative temperature exceeds the upper limit of the thermal comfort zone for the remaining hours of the educational activities. The ventilation scenario represented by Combo 09 proves to be the most effective, keeping the internal operative temperature consistently within the thermal comfort zone throughout the duration of the class. In contrast, the scenario associated with Combo 10 generates an internal operative temperature that is below the thermal comfort zone during the early hours of the school day. Only from 10:00 a.m. and until the end of the educational activities does this ventilation combination manage to ensure an internal operative temperature within the thermal comfort zone. Finally, Fig. 24.8f shows the evolution of the internal operative temperature on the coldest day of December. Also in this month, the thermal comfort zone oscillates between 21 °C and 23 °C, as in November. In this case, it can be observed that the ventilation scenarios corresponding to Combos 04 and 05 are ideal for keeping the internal operative temperature within the limits of the thermal comfort zone throughout the duration of the educational activities.

The study has several limitations. First, it relies on climate data projected to 2050, which may not accurately reflect actual future conditions. Climate projections are subject to uncertainty and can vary depending on emissions scenarios and climate policies adopted.

Another limitation is the accuracy of the simulation models used, which depends on the quality of the input parameters. Any uncertainties in the models could affect the results and their applicability to real situations. Finally, the study focuses specifically on the Italian context, with an emphasis on the climatic challenges and the school infrastructure of Mediterranean regions. While it provides an important basis, it is crucial to consider that the solutions may need to be adapted to be applicable in other contexts.

In terms of future perspectives, it would be beneficial to extend the analysis to international contexts and different climatic zones to improve the generalizability of the results. The combination of different ventilation scenarios and emerging technologies could further refine the management of thermal comfort and energy efficiency. It is also necessary to regularly update climate models and simulations to reflect the latest projections and trends to ensure that strategies are based on current and reliable data.

In addition, future studies should consider the environmental and social impacts of ventilation strategies and energy solutions adopted. Evaluating aspects such as occupant comfort, economic costs, and environmental impacts will contribute to the development of more sustainable and inclusive approaches. In summary, while this study provides a fundamental basis for the design and energy management of school buildings, further research is needed to refine and extend the proposed solutions to ensure their effectiveness in a global and dynamic context.

4 Conclusions

Many educational institutions in the Mediterranean regions of our country are facing significant challenges in managing energy costs. Reducing energy consumption while maintaining indoor air quality is a daunting task.

High student density and inadequate ventilation have a direct impact on the thermal comfort of school environments. This situation is aggravated by ongoing climate change, which has led to increasingly extreme and unpredictable thermal conditions in Mediterranean regions, further exacerbating the negative impact on the quality of the school environment.

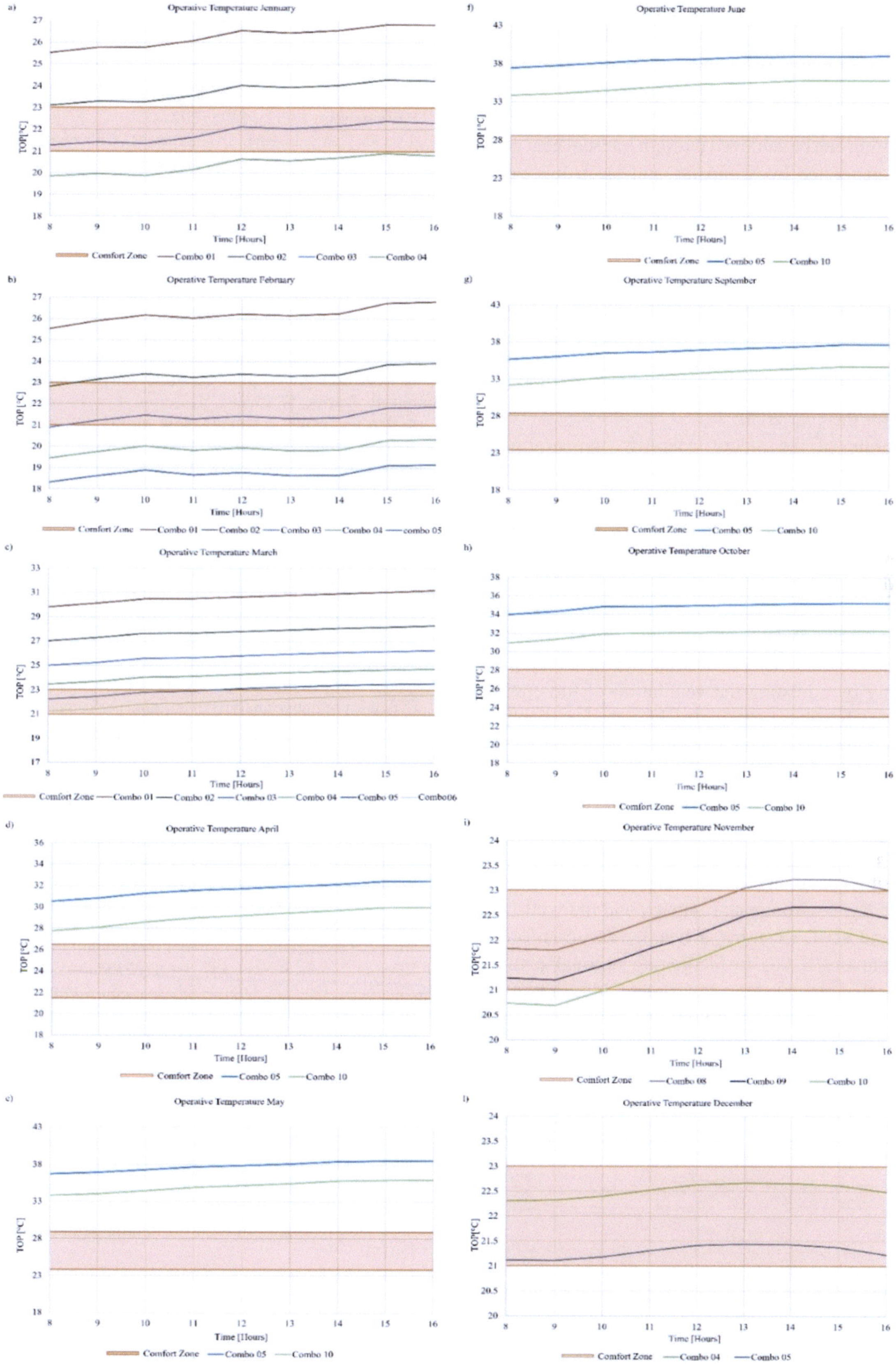

Fig. 24.9 Operative temperature in the classroom (Zone 7), based on projected climatic data for 2050, compared with the thermal comfort parameters established by the UNI EN 16,987–1:2019 standard (UNI EN ISO 2018)

The purpose of the study is to evaluate the possibility of discourage the use of heating systems in schools in anticipation of future temperatures, while still maintaining a good level of comfort during the winter months in regions with temperate climates. It is hypothesized that in efficient school buildings, heating can be achieved through internal heat gains, such as heat generated by occupants and electronic equipment. This approach involves regulating the indoor temperature by controlling ventilation, adjusting the airflow based on the outdoor temperature. The use of advanced heat exchangers provides greater flexibility. Specific strategies have been developed to maintain thermal comfort without relying on traditional heating.

This study focused on the future thermal behavior of a typical school building in the Mediterranean area, considering climate data for the year 2050.

The evolution of the indoor operative temperature of each classroom within the school building was studied. This temperature varies depending on the location and exposure of each classroom, the outside temperature, and the airflow from the ventilation system.

It was observed that during the winter months, especially in January, February, March, November, and December, it is possible to maintain a comfortable indoor operative temperature without the need to activate the air conditioning system, simply by adjusting the ventilation of the environment. However, this phase also highlighted the building difficulties in maintaining an indoor operative temperature suitable for adaptive thermal comfort during transitional periods, such as spring and autumn, and during the warmer months.

The results highlight the critical challenge that climate change poses for schools and underline the importance of adopting integrated approaches that take into account energy efficiency, environmental impact, and the well-being of students and staff. Ensuring a healthy and comfortable indoor environment is key to facilitating optimal learning, with solutions that regulate temperature, humidity, air quality, and lighting. The active involvement of stakeholders in the design and implementation of these solutions is essential, as is the promotion of environmental education and awareness of sustainable practices. In conclusion, addressing the impact of climate change on schools is essential to ensure a safe and stimulating environment for current and future students.

References

AlFaris, F., Juaidi, A., Manzano-Agugliaro, F.: Improving the efficiency of energy use in schools as a sustainable practice through an energy management program. J. Clean. Prod. **135**, 794–805 (2016)

Arnfield, A. John. "Classificazione climatica di Köppen". Enciclopedia Britannica, 16 ottobre 2023, https://www.britannica.com/science/Koppen-climate-classification. Accesso effettuato il 19 dicembre 2023

Ashrafian, T.: Enhancing school buildings energy efficiency under climate change: A comprehensive analysis of energy, cost, and comfort factors. J. Build. Eng. **80**, 107969 (2023)

Baglivo, C.: Dynamic evaluation of the effects of climate change on the energy renovation of a school in a Mediterranean climate. Sustainability **13**(11), 6375 (2021)

Baglivo, C., Congedo, P.M., Di Cataldo, M., Coluccia, L.D., D'Agostino, D.: Envelope design optimization by thermal modelling of a building in a warm climate. Energies **10**(11), 1808 (2017)

Baglivo, C., Bonomolo, M., Congedo, P.M., Beccali, M., Antonaci, S.: Technical-economic evaluation of the effectiveness of measures applied to the artificial lighting system of a school. Appl. Sci. **11**(14), 6664 (2021)

Baglivo, C., Congedo, P.M., Murrone, G., Lezzi, D.: Long-term predictive energy analysis of a high-performance building in a Mediterranean climate under climate change. Energy **238**, 121641 (2022)

Barbhuiya, S., Barbhuiya, S.: Thermal comfort and energy consumption in a UK educational building. Build. Environ. **68**, 1–11 (2013)

Barrett, P., Davies, F., Zhang, Y., Barrett, L.: The impact of classroom design on pupils' learning: Final results of a holistic, multi-level analysis. Build. Environ. **89**, 118–133 (2015)

Campagna, L.M., Fiorito, F.: On the energy performance of the Mediterranean school building stock: The case of the Apulia Region. Energy Build. **293**, 113187 (2023)

Congedo, P.M., D'Agostino, D., Baglivo, C., Tornese, G., Zacà, I.: Efficient solutions and cost-optimal analysis for existing school buildings. Energies **9**(10), 851 (2016)

Congedo, P.M., Baglivo, C., Centonze, G.: Walls comparative evaluation for the thermal performance improvement of low-rise residential buildings in warm Mediterranean climate. J. Build. Eng. **28**, 101059 (2020)

Congedo, P.M., Baglivo, C., Seyhan, A.K., Marchetti, R.: Worldwide dynamic predictive analysis of building performance under long-term climate change conditions. J. Build. Eng. **42**, 103057 (2021)

DM 18 dicembre 1975, Ministro dei lavori pubblici, "Norme tecniche aggiornate relative all'edilizia scolastica, ivi compresi gli indici di funzionalità didattica, edilizia ed urbanistica, da osservarsi nella esecuzione di opere di edilizia scolastica", in Italia, 1975

Dong, J., Schwartz, Y., Korolija, I., Mumovic, D.: The impact of climate change on cognitive performance of children in English school stock: A simulation study. Build. Environ. **243**, 110607 (2023)

Dorizas, P.V., Assimakopoulos, M.N., Santamouris, M.: A holistic approach for the assessment of the indoor environmental quality, student productivity, and energy consumption in primary schools. Environ. Monit. Assess. **187**, 1–18 (2015)

Jovanović, A., Pejić, P., Djorić-Veljković, S., Karamarković, J., Djelić, M.: Importanza dell'orientamento dell'edificio nella determinazione della qualità della luce naturale nelle stanze dei dormitori degli studenti: valori dei parametri di luce naturale fisici e simulati confrontati con i risultati soggettivi del sondaggio. Energy Build. **77**, 158–170 (2014)

Korsavi, S.S., Montazami, A.: Children's thermal comfort and adaptive behaviours; UK primary schools during non-heating and heating seasons. Energy Build. **214**, 109857 (2020)

M.D. 26/6/2015, Italian Ministerial Decree, "Applicazione delle metodologie di calcolo delle prestazioni energetiche e definizione delle prescrizioni e dei requisiti minimi degli edifici", (In Italian) (2015)

Mohamed, S., Rodrigues, L., Omer, S., Calautit, J.: Overheating and indoor air quality in primary schools in the UK. Energy Build. **250**, 111291 (2021)

Pachauri, R. K., Allen, M. R., Barros, V. R., Broome, J., Cramer, W., Christ, R., van Ypserle, J. P.: Climate change 2014: synthesis report. Contribution of Working Groups I, II and III to the fifth assessment report of the Intergovernmental Panel on Climate Change (p. 151). Ipcc, (2014)

Sánchez, E.C., Bardón, C.C., Viviana, O.P.: Architectural and environmental strategies towards a cost optimal deep energy retrofit for Mediterranean public high schools. Energy Rep. **9**, 6434–6448 (2023)

UNI EN ISO 52016–1:2018, "Prestazione energetica degli edifici - Fabbisogni energetici per riscaldamento e raffrescamento, temperature interne e carichi termici sensibili e latenti - Parte 1: Procedure di calcolo", in Italia, 1 marzo 2018.

Wang, Y., Zhao, F.Y., Kuckelkorn, J., Spliethoff, H., Rank, E.: School building energy performance and classroom air environment implemented with the heat recovery heat pump and displacement ventilation system. Appl. Energy **114**, 58–68 (2014)

Wargocki, P., Wyon, D.P.: Providing better thermal and air quality conditions in school classrooms would be cost-effective. Build. Environ. **59**, 581–589 (2013)

Yang, Y., Javanroodi, K., Nik, V.M.: Climate change and energy performance of European residential building stocks–A comprehensive impact assessment using climate big data from the coordinated regional climate downscaling experiment. Appl. Energy **298**, 117246 (2021)

Zacà, I., Tornese, G., Baglivo, C., Congedo, P.M., D'Agostino, D.: Data of cost-optimal solutions and retrofit design methods for school renovation in a warm climate. Data Brief **9**, 846–849 (2016)

Zomorodian, Z.S., Tahsildoost, M., Hafezi, M.: Thermal comfort in school buildings: a review article. Renew. Sustain. Energy Rev. **59**, 895–906 (2016)

Outdoor Thermal Comfort Assessment of Lana River Re-Qualification in Tirana, Albania

Lindita Bande, Brunilda Babameto, Anwar Ahmad, and Kim Young

Abstract

Albania is a country that had a strong transition from communism to fragile democracy. In the 30 years of this transition abusive construction happened all over the country. However, one of the most particular cases is the Lana River in the capital of Tirana. Abusive construction of commercial buildings varying from one to 5 took place along the river that crosses the capital. Several years after this invasion of public space was done, the major of the city cleared all construction along the river and returned the previously ween space to the citizens. This study aims to show the improvement of OTC in 3 different scenarios, through the use of ENVI met of one section of the river located near the area called "Blloku". The main steps of the methodology are: • Climate Analysis of Tirana; • Case Study Analysis; • Modelling of the 3 scenarios with Rhino/GH; • Simulation of the 3 scenarios with Rhino/GH; • Results. The steps of this research follow a chronological order of the changes that occurred in the capital during different regimes linked to construction along the river. The social political development of the country is connected to the public space definition in Tirana. The study analyzes 3 scenarios: the base case which is the current situation with existing trees; scenario 2 with the previous abusive construction along the river; scenario 3: with no trees and no abusive construction. The results of this research are relevant to local decision making, but also it is a proof of how decision-making can improve the life of the citizens of Tirana.

Keywords

OTC · Lana river · Tirana · Abusive construction · Green public space

1 Introduction

Tirana as the capital of Albania has gone thru drastic urban changes offer the last century. For various political and economic reasons several areas have been developed. However, the Lana River is the reference physical element that did not change over time. The river is of historical importance to the city. Therefore this analysis aims to show that re-qualifications of such areas can be beneficial for the citizens.

1.1 Historical Analysis of the Abusive Construction in Lana River

The Lana River flows east to west through the southern section of Tirana City, merging with the Tirana River in its midsection. The Tirana River then joins the Ishmi River, which subsequently empties into the Adriatic Sea (NQM 2024a).

The systemization of Lana in the bed that passes today started to be designed in the first development plans of Tirana. This area is clearly evident in the photo from 1936. (Grasshopper—algorithmic modeling for Rhino 2021) The segment for which the study is carried out has an east–west orientation and is located near the main axis of the city, the city boulevard. It extends to a length of 570 m and is the central area of the Lana's River. It is one of the first segments to be systematized. We find it presented on the map of 1959. From the beginning, the escarpments of the river bed were designed to be green, to increase the green spaces of the city. They were like that, until the change of government system in

L. Bande (✉) · B. Babameto · A. Ahmad · K. Young
Architectural Engineering Department, College of Engineering, United Arab Emirates University, AL Ain, Abu Dhabi 15551, UAE
e-mail: lindita.bande@uaeu.ac.ae; lindita.bande@uaeu.ac.ae

P. Mohammad et al. (eds.), *Geographic Approaches to Climate Change and Mitigation: Urban and Rural Perspectives (Volume 1)*, Advances in Science, Technology & Innovation, https://doi.org/10.1007/978-3-031-92119-3_25

Fig. 1 Abusive construction in Lana River (NQM 2024b)

1992. This fact is also clear in the map of 1981–1993, in which the conventional signs of greening are clearly distinguished.

From the beginning, the escarpments of the river bed were designed to be green, to increase the green spaces of the city. They were like that, until the change of government system in 1992. This fact is also clear in the map of 1981–1993, in which the conventional signs of greening are clearly distinguished. The political changes brought about an anomaly in the enforcement of law and order. The freedom to move towards Tirana created a handicap in the urban planning of the city. People who came from other cities started to build on every surface of free land, occupying public spaces day by day.

Not even Lana's escarpments escaped this fate. In 1991, the first kiosk was built on the Lana River through a permit granted by the Pluralist Executive Committee of that time. Two years later, the creation of the City Hall in Tirana would grant such one-year permits to concrete the capital's river. Many tall buildings were also built between the kiosks. (Ndërtimet anës Lanës, bashkia harton projekte për rikthimin e gjelbërimit 2002) In a short time, both sides of Lana's bed were filled with constructions up to 5 floors, most of which were for business purposes. It was a time when there was no law to prohibit these constructions. As it can be seen from the images taken by ASIG Geoportal, The river of Lana was lost among the constructions and from a very evident surface as an axis in the urban development, it became almost invisible. The pollution in the river increased extremely, as all the buildings discharged the used water into it. The situation was chaotic, as can be seen in Fig. 1. Unfortunately, the images seem to have nothing to do with the centre of a capital city.

After the passage of an extreme social and political situation, the changes that took place brought hope for the urban development of Tirana. The mayor of the time, as soon as he took office, gradually began the execution of decisions on the demolition of informal buildings, especially those on the Lana escarpment.

1.2 Current Situation of Lana River

Lana today is a very evident linear, green element in the urban silhouette of Tirana. Especially in the segment under study, a dense greenery is observed, which directly affects the urban comfort of the citizens of Tirana. Being in the central area, work began on the segment under study. From the orthophoto taken from Google, it can be seen that in 2007 the space was returned totally to the green area (Fig. 2).

The last General Local Plan of the city of Tirana approved in 2018, known as "Tirana 2030" proposes the continuation of the arrangement of the Lane riverbed, both at its eastern and western edges. Numerous interventions have been made along the length of the river, adding recreation areas. There are three bridges in the segment under study. The two bridges at the end have always been vehicle access roads, with a considerable width. The bridge in the middle, which connects the centre of Tirana with one of the most frequented areas "Blloku",

Fig. 2 Orthophotoof the Lana River, section "Blloku" (square in red) of year 2024 (https:geoportal.asig.gov.al, map, auto=true)

Fig. 3 **a** View from the old Bridge of "Blloku" 2007 (https:albanianblogger.com 2007); **b**View from the new Bridge of "Blloku" 2009 (AF1QipMjvlCe7cNlG-g5eK5pN1Cy2u_f-A3ycPN6Nekl=h1440).

until 2008 was a small pedestrian bridge. Today it is an urban element that adds value to this part of the city (Fig. 3a and b).

Based on the historical overview this study aims to quantify the outdoor thermal comfort values for a portion of the Lana River through 3 main scenarios based on the urban changes mentioned above. Afterwards, these scenarios shall be modelled in Rhino Grasshopper. The results presented in UTCI shall evaluate the impact of the urban changes into the OTC.

2 Material and Methods

The methodology is linear and focusses on the below main points:

- Climate Analysis of Tirana
- Case Study Analysis
- Modelling of the 3 scenarios with Rhino/GH
- Simulation of the 3 scenarios with Rhino/GH
- Results.

2.1 Climate Analysis

The Tirana region features a topography of gently rolling hills and spans across the boundaries of both the Fieldly Mediterranean Climate Sub-zone and the Hilly Mediterranean Climate Sub-zone. This region exhibits a characteristic Mediterranean climate, typified by humid and mild winter conditions and contrasted by hot and dry summer periods.(NQM 2024) Tirana has a typical Mediterranean weather with hot summer. It has four seasons: spring (average of March Temperature is 13.8 °C), summer (average of June temperature is 25.9 °C), autumn (average of September temperature is 25.3 °C), winter (average of December temperature is 10.9 °C). The highest rainfall is in November and the dries is in August. The sunshine varies from 6 to 12.3 h thru the year. (Yearly Monthly weather—Tirana, Albania).

2.2 Case Study Analysis

The Lana River flows through southern Tirana, joining the Tirana River, which then merges with the Ishmi River before reaching the Adriatic Sea. The river is 29 km long, with a 3.5 km concrete section in central Tirana. The river went thru changes over the last century in terms of treatment around it. During the communism era there was a limited number of trees around the river; afterwards in the fragile democracy there was abusive construction and finally a period of empty land before planting the current trees.

It's heavily polluted by untreated sewage and garbage. The densely populated central area discharges significant sewage, raising Biochemical Oxygen Demand (BOD) to 50–60 mg/l, sometimes exceeding 100 mg/l. (Lushaj 2006) However, in the "Green City Action Plan of Tirana", it is mentioned that The ongoing wastewater management project, financed by JICA, is set to upgrade one-third of the city and will be finished next year, aiming to stop the release of untreated water into the Lana River (Green City Action Plan of Tirana, 2018) This issue was addressed by starting construction on a treatment plant in 2010, but it remains incomplete to this day. The two side slopes of the Lana River now look like small urban forests. Many researches have proven that greenspaces can help cities adapt to climate change, reducing the risks of extreme heat, improving air quality, protecting health and providing connection to nature, as well as many other benefits (Fig. 4). (Power and of Urban Trees 2023).

2.3 Modelling and Simulation of the 3 Scenarios with Rhino/GH

The case study are modelled utilizing Rhinoceros software. Rhinoceros is a computer-aided design software. While the analyses conducted using Grasshopper software plug-in tools. Figure 5 shows the script utilized to analyze the case study.

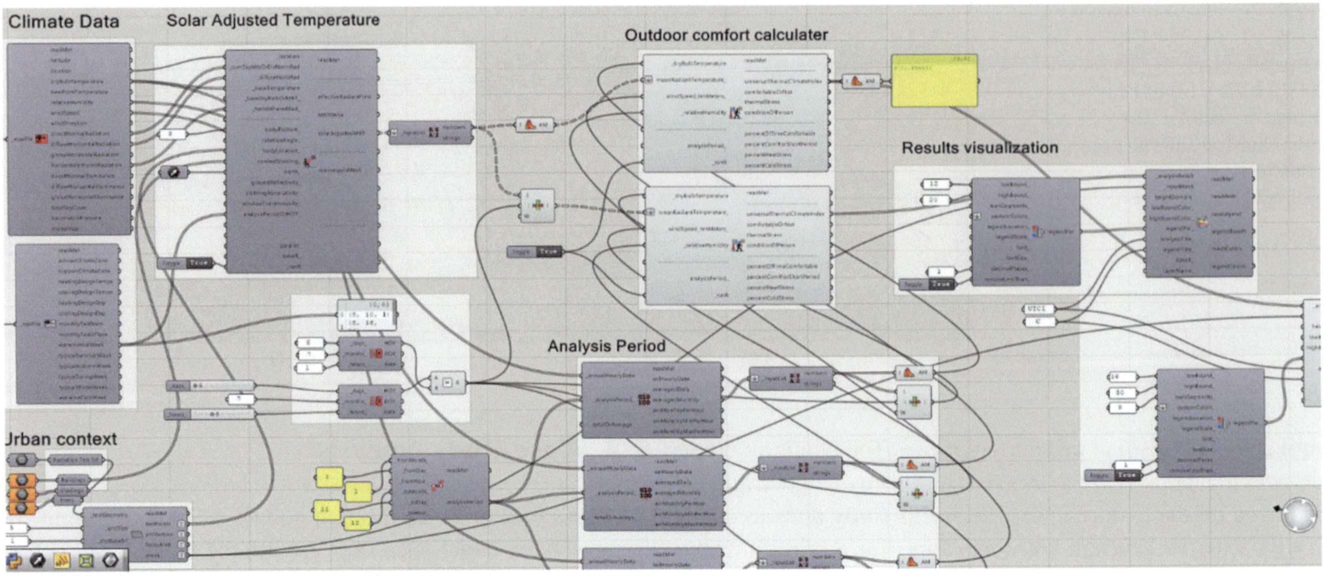

Fig. 5 UTCI analyses script in Grasshopper Software environment, Python language (Grasshopper—Algorithmic Modeling for Rhino)

Fig. 4 View of Lana River in 2023 (referring to scenario 1) (AF1QipNJgMLImxp0fs-087BMG56EqPo3klB0zZdD8Qlq=h1440)

The modelling and simulation follows the below order:

a- Modelling and simulation of the base case; in this scenario the current trees are added based on the site observation and online resources. The weather file refers to the city of Tirana.
b- Modelling and situation of scenario 2- abusive construction; the model of the abusive buildings is done based on historical references and visual material.
c- Modelling and simulation of scenario 3 where there are no trees or abusive construction.

All scenarios together with the location are shown in Figure 6.

3 Results

The aim of this study was to show the improvement of OTC in 3 different scenarios, through the use of ENVI met of one section of the river located near the area called "Blloku". Scenario 1 is the current condition of the selected area where there are trees. The second scenario has the abusive construction and the third scenario has no trees or abusive construction.

The results of the 3 simulations are shown numerically in Table 1. Even though the temperature is a yearly value (the monthly values are displayed in the table) the base scenario with the existing trees is the best performing scenario in terms of UTCI with a value of 20.04. The scenario 3 has the highest UTCI value 21.56 (where there are no trees); and scenario 2 with the abusive construction has an average UTCI value of 21.26.

4 Conclusions and Discussions

Refereeing to the history of construction and development in the city of Tirana, Lana River is one of the best case study to show how local strategies can impact the UTCI. In the case of the abusive construction the UTCI had higher value compared to the current one with the trees. Through this study, meaningful differences in the impact of vegetation on microclimate, particularly changes in temperature and humidity and solar radiation, were identified. Confirmation of changes in surrounding microclimate according to the degree of maturity of tree vegetation is also an important

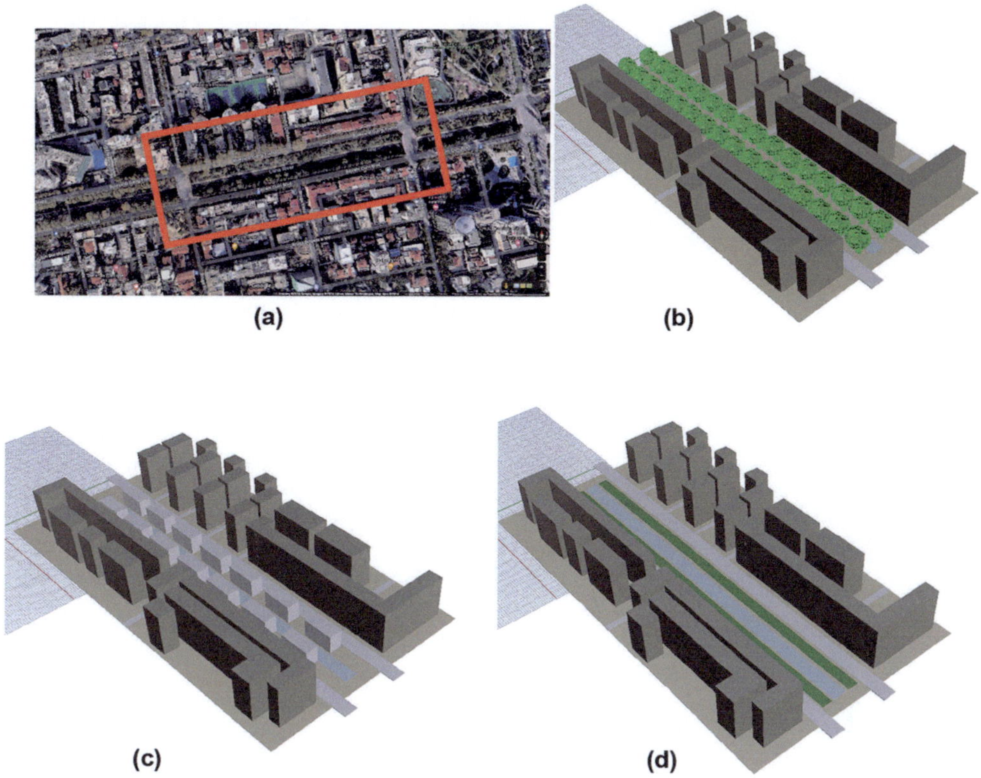

Fig. 6 **a** Location of the case study; Lana River "Blloku Zone"; **b** base case model (scenario 1) with existing trees; **c** scenario 2-abusive construction; **d** scenario 3-no trees; no abusive construction

Table 1 Results for the 3 cases analyzed

Scenario 01 UTCI analysis results												
Month	1	2	3	4	5	6	7	8	9	10	11	12
UTCI	15.72	16.19	18.45	21.17	24.15	27	28.64	26.74	23.69	21.18	18	16.20
Yearly	21.04											
Scenario 02 UTCI analysis results												
Month	1	2	3	4	5	6	7	8	9	10	11	12
UTCI	16.13	16.59	18.84	22.07	24.49	27.31	28.98	27.10	24.4	21.55	18.45	16.61
Yearly	21.26											
Scenario 03 UTCI analysis results												
Month	1	2	3	4	5	6	7	8	9	10	11	12
UTCI	16.24	16.69	18.93	20.17	21.63	27.40	29.03	27.15	24.14	21.62	18.56	16.74
Yearly	21.56											

finding. In addition, through the influence of green space in the city centre on the surrounding environment, the ratio of green space and the distance between green spaces must be considered when designing the city. Additionally, the transition period from no construction until the trees grew (3–4 years' time) the UTCHI had even a higher value (Fig. 7).

The study brings novelty to the type of analysis to the city of Tirana. Currently there are limited studies in quantifying the improvement done on various urban areas in the city. Therefore, addressing this issue with the use of the rhino/grasshopper brings new information for the particular field of study.

The study has limitations in various aspects such as: the area treated is limited to the historical zone of the city. There are other newly developed areas that can benefit from such study. Additionally, the residential blocks and the commercial blocks around the river have different treatment. Therefore, a mode detailed study might bring more knowledge in how to improve the microclimate and OTC.

There are several recommendations for policymakers and stakeholders that can be outlined in this study. The use of the local trees is very beneficial for the OTC. Empty areas around the river have a negative impact on the OTC.

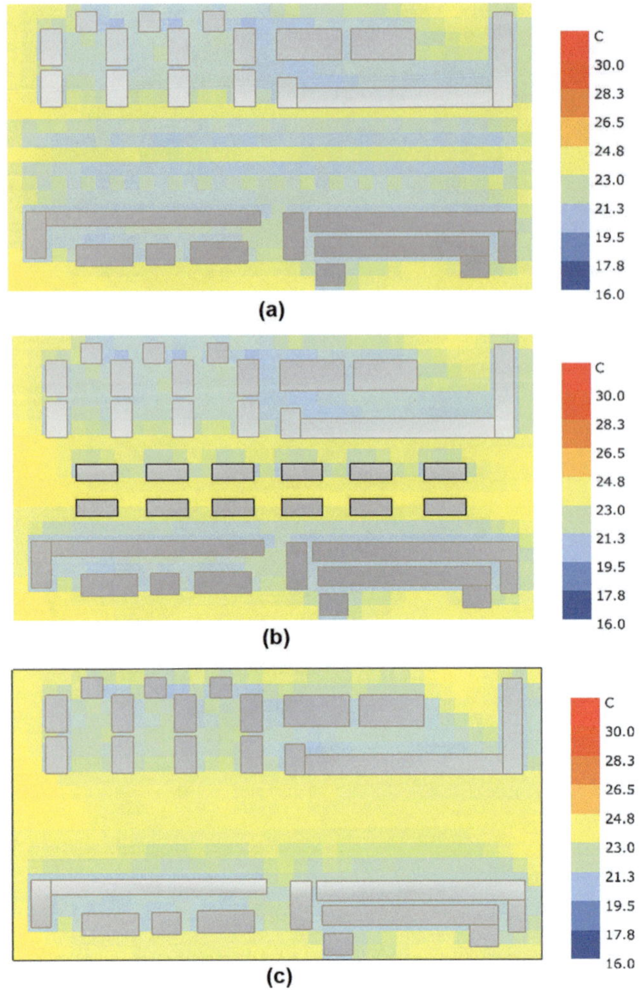

Fig. 7 Results for: **a** base case model; **b** scenario 2; **c** scenario 3

However, further studies are needed to understand the impact of the water in the UTCI considering that the construction had a blocking effect in the walkway; while the trees improve the wind flow therefore improve the UTCI.

Conflicts of Interest Declare conflicts of interest or state "The authors declare no conflict of interest".

Declaration from the Main Author The used methodology and tools (script) used in this paper have been used in the author's other work * Quantitative analysis of Outdoor Thermal Comfort in the student dorm complex of United Arab Emirates University*.

References

AF1QipMjvlCe7cNlG-g5eK5pN1Cy2u_f-A3ycPN6Nekl=h1440 (2048 × 1365) (googleusercontent.com)

AF1QipNJgMLImxp0fs-087BMG56EqPo3klB0zZdD8Qlq=h1440 (2160 × 1440) (googleusercontent.com)

Grasshopper—algorithmic modeling for Rhino. [Online]. Available: https://www.grasshopper3d.com/. Accessed: 12 Sep 2021

Grasshopper—algorithmic modeling for Rhino [Online]. Available: https://www.grasshopper3d.com/. Accessed: 12 Sep 2021

Green City Action Plan of Tirana (2018). https://www.ebrdgreencities.com/assets/Uploads/PDF/Tirana-GCAP.pdf

https://geoportal.asig.gov.al/map/?auto=true

https://albanianblogger.com/2007/10/16/the-little-pedestrian-bridge-near-blloku/

https://earth.google.com/

Lushaj. B., Japanese Team: Major International Project "The study on the development plan for sewerage system and sewage treatment plant for greater Tirana, in the Republic of Albania" (2006)

Ndërtimet anës Lanës, bashkia harton projekte për rikthimin e gjelbërimit (2 Nëntor 2002) Aviable: https://tvklan.al/ndertimet-anes-lanes-bashkia-harton-projekte-per-rikthimin-e-gjelberimit-2-nentor-2002

[NQM] The Lana River in Tirana 20 years ago vs Now : r/AskBalkans. [Online]. Available: https://www.reddit.com/r/AskBalkans/comments/191qg9d/nqm_the_lana_river_in_tirana_20_years_ago_vs_now/?rdt=38620. Accessed: 06 Sep 2024

[NQM] The Lana River in Tirana 20 years ago vs Now : r/AskBalkans. [Online]. Available: https://www.reddit.com/r/AskBalkans/comments/191qg9d/nqm_the_lana_river_in_tirana_20_years_ago_vs_now/?rdt=38620. [Accessed: 06-Sep-2024].

The Power of Urban Trees, 2023, Available: https://www.climatecentral.org/climate-matters/the-power-of-urban-trees-2023

TIRANA 1920–2020 THROUGH ARCHITECTURE. Vokshi, A (2022)

Yearly & Monthly weather - Tirana, Albania. [Online]. Available: https://www.weather-atlas.com/en/albania/tirana-climate. Accessed 14 May 2024

Sustainable Urban Development and Climate Change Challenges Within Kingdom of Saudi Arabia's Vision 2030

Waleed S. Alzamil

Abstract

Urban growth leads to the prosperity and development of communities, but it can lead to negative impacts on the environment and society if this was outside the framework of sustainability. The concept of sustainable development is one of the common concepts that seeks to achieve economic growth and prosperity in a way that preserves the environment, achieves social justice, and preserves the rights of future generations. This concept emphasizes confronting the challenges of climate change, including preserving the natural environment, effectively using non-renewable resources, and giving future generations the right to use the resources. "*Sustainable development*" provides us with a comprehensive vision of the environmental costs resulting from resource consumption and its effects on the economy and society. Building effective strategies in urban planning cannot be achieved without understanding the available resources. Urban development can lead to improving living standards and the economy. However, unbalanced development and excessive consumption of natural resources in the desire for quick economic gains can negatively impact society and the climate. As a result, development may ultimately lead us to disaster in the absence of planning for resource sustainability. This paper aims to analyze the initiatives of the National National Vision 2030 in confronting the challenges of climate change. The paper attempts to extract the most prominent principles that National National Vision 2030 focused on in the context of urban development and their role in climate change mitigation. The paper relies on the theoretical and inductive approach by reviewing the National National Vision 2030 initiatives in planning to mitigate climate change.

Keywords

Climate change · Urban development · Sustainability · National Vision 2030 · Saudi Arabia

1 Introduction

Sustainable development is one of the most important indicators set by the National Vision 2030a, b to measure the quality of life in cities. Goal 13 of the United Nations Sustainable Development Goals emphasizes "to take urgent action to combat climate change and its impacts" (United Nations 2024). In light of the major development projects that the Kingdom of Saudi Arabia has witnessed since the adoption of the National Vision in 2016, the strategic objectives emphasized the balance between the economic, environmental, and social dimensions to build a future that guarantees zero carbon neutrality for cities by 2060.

The history of urban development witnessed by the Kingdom of Saudi Arabia over the past decades confirms the efforts made by municipal agencies to keep pace with the economic changes that affected city planning. The economic boom in the early 1970s that the Kingdom of Saudi Arabia witnessed required the provision of housing, services, and infrastructure in response to population growth, and accordingly, urban development accelerated to fill the gap in demand (Al-Hathloul 2017). Since the First Development Plan (1970) many goals have been set to ensure the equitable distribution of economic services and resources. Health and educational programs expanded in response to the settlement of the desert, and that stage required a focus on curative health services and an increase in the proportion of workers in the health sector (Alzamil 2016). The plan also targeted "expanding the electricity network and lighting most cities

W. S. Alzamil (✉)
Department of Urban Planning, King Saud University, Riyadh, Saudi Arabia
e-mail: waalzamil@ksu.edu.sa

P. Mohammad et al. (eds.), *Geographic Approaches to Climate Change and Mitigation: Urban and Rural Perspectives (Volume 1)*,
Advances in Science, Technology & Innovation, https://doi.org/10.1007/978-3-031-92119-3_26

and villages with a population of more than 2,000 people, and investing the maximum possible amount of capital to establish rural lighting projects". As for water networks, delivering them to all cities and villages was a major challenge, as the first development plan set a strategic goal of completing the extension of pipes carrying water to major cities, developing and renovating 550 wells in small villages, and completing the establishment of water desalination plants (Ministry of Finance 1970). However, these efforts tended toward growing cities and accommodating population demand, far from the principles of sustainable development.

Subsequent development plans referred to diversifying the economic base horizontally and vertically and expanding the absorptive and productive capacities of the national economy. Despite this, the process of transition toward developing non-oil sectors and confronting the challenges of climate change has represented a major challenge over the past decades as a result of the lack of clear mechanisms and measuring tools for strategic goals (Ministry of Economy and Planning 2015). Thus, development projects were affected by the size of government revenues and the economic booms that accompanied the rise in oil prices, and it became necessary to push for a strategic direction that charts the next phase of development and responds to the requirements of the National Vision 2030a, b phase to build a vibrant society, a prosperous economy, and an ambitious nation (Mazzetto and Vanini 2023).

Since 2016, local efforts have begun to confront the challenges of climate change, through innovative solutions that include diversifying energy sources, to reach 50% renewable energy levels by 2030 (National Vision 2030 2022). His Royal Highness Prince Mohammed bin Salman bin Abdulaziz confirmed that:

> Climate action will enhance competitiveness, stimulate innovation and create millions of high-quality jobs. Young people, in the Kingdom and around the world, are demanding a cleaner, greener and more inclusive future, and we owe it to them to make that happen (National Vision 2030 2022).

National Vision 2030 has set a comprehensive vision for ecosystems through several initiatives in sustainable agriculture, protecting the natural environment, and preserving biodiversity, including sustainable tourism initiatives, the city development program, improving the quality of life, the Green Saudi Arabia initiative, and the Green Middle East (National Vision 2030 2022). Theoretically, the local strategic perspective of cities has changed, and the ambition is no longer to accommodate the demand in providing plans, services, and facilities. Rather, it is about the sustainability of cities in the context of confronting the challenges of climate change through alternative energy initiatives, reducing the proportion of carbon emissions, and increasing the proportion of green spaces (Alqahtany and Aravindakshan 2022). In light of National Vision 2030 there has become an urgent need to review the urban reality in Saudi cities to confront the challenges of climate change. Therefore, this paper analyzes the initiatives of the National Vision 2030 in confronting the challenges of climate change.

1.1 The Problem

Climate change is not just an environmental problem, but rather a problem of "justice" between industrialized and developing countries. The right to development among countries does not mean equality in polluting the environment; rather, international cooperation is one of the most important ways to protect the atmosphere as a shared resource through developing policies to reduce greenhouse gas emissions, global solidarity, and commitment to environmental treaties. The problem of climate change affects all countries of the world (Agarwal 2002). Based on Selin and VanDeveer (2013) industrialized countries are responsible for climate change with the percentage share of global emissions being 75.6. Countries that do not suffer from environmental challenges will be affected by negative factors resulting from greenhouse gases (Selin and VanDeveer 2013). The absence of effective environmental systems to deal with the sustainability of natural resources has been one of the main reasons that has led to the increase in resource consumption, environmental pollution, and greenhouse gas emissions in the world.

Locally, efforts to mitigate climate change began to become evident with the emergence of the National Vision in 2016. The National Vision 2030 emphasized three important pillars that constitute a framework for sustainable development: a vibrant society, a prosperous economy, and an ambitious nation. National Vision 2030 developed a set of initiatives in the field of environmental protection and city development in a manner consistent with sustainable development goals. However, there is a need to analyze the National Vision 2030 initiatives to confront the climate challenges and formulate them as strategic framework in sustainable development of Saudi cities.

1.2 Objectives

This paper analyzes the initiatives of the National Vision 2030 in confronting the challenges of climate change. The paper discusses how local efforts affect reaching the Sustainable Development Goal and what are their implications for urban policies. The results of this paper can help formulate a strategic framework for sustainable development of Saudi cities that mitigation climate change in urban and rural environments.

2 Literature Review

2.1 The Concept of Climate Change

Many studies point to climate change as one of the environmental problems that could threaten human existence in the world. However, concepts of climate change can be divergent in many health, environmental, social science, and urban science studies (Werndl 2016). The term "climate change" refers to any change in climate over time, whether caused by natural or human activity. Climate change leads to risks to public health and human existence because climate affects air, food, and water (WHO 2016). The Earth's climate changes permanently over the long term as a result of volcanic activity or movement of land masses. Scientists refer to this type as natural climate change (Riedy 2016). On the other hand, climate change occurs due to human activity, including the use of traditional methods of waste disposal, the burning of fossil fuels to produce electricity, and urban encroachment on forests. These human activities lead to the production of greenhouse gases in the atmosphere. Accumulation of greenhouse gases such as carbon dioxide, methane, halocarbons, and nitrous oxide block some of the radiant heat from the ground, which negatively affects plants and increases pollution rates (Riedy 2016).

2.2 Climate Change Mitigation in Urban and Rural Environments

Natural disasters in the world have been increasing steadily since 1960 as a result of the increasing pace of industrialization, environmental degradation, consumption of natural resources, and unpredictability of future disasters. Accordingly, the most important challenges facing disaster management are the lack of understanding of the relationship between disasters and development, such as the changes that accompanied industrial technology and their impact on the surrounding environment, or the consumption of resources in a way that could contribute to the destruction of the natural environment and climate change (Pelling 2003). Over the next hundred years, climate change will affect water levels in the world, especially coastal cities, which will affect the future of human settlement in cities. Future environmental threats will include roads, homes, utilities, and property values (Madigan 2013). The climate change are not limited to raising the Earth's temperature, but rather they negatively affect human life through the spread of disasters, floods, atmospheric pollution, and lack of natural resources (Shogren and Toman 2001). In addition, climate change also affects the economy where studies have indicated that global warming contributes to the lack of natural resources and a weak production (Porter 2013).

Climate change affects city infrastructure including water supply, energy systems, water disposal, and urban management. Climate changes lead to an increase in rainfall, floods, and the occurrence of environmental disasters. However, urban planning can manage environmental disasters resulting from climate change through disaster mitigation policies and stakeholder engagement (Shalaby and Aboelnaga 2017). Sustainable environmental planning is an organized, systematic work based on analyzing all inputs and building an implementable future vision using available resources. Dealing with the issue of climate change requires developing effective strategies based on an analysis of the urban, social and environmental situation of cities. Therefore, building a strategic framework for cities to mitigate climate change depends on analyzing a quantitative and spatial database that includes environmental pollution rates, disaster-prone areas, energy systems, and industrial activities. Urban and environmental policies to mitigate the negative effects of climate change include developing environmental control systems, analyzing potential risks, city management, and urban legislation.

Sustainable environmental planning can achieve a balance in resource consumption in a way that emphasizes the livelihood rights of future generations. These policies should promote scientific research to search for environmentally friendly alternatives. At the same time, reducing the negative impacts resulting from traditional human activity, including allocating budgets to support countries that use alternative energy. Environmental policies provide laws that govern the relationship between human and natural resources such as water, air, waste disposal, and climate change policy. The development of environmental policies in the United States came as a reaction to the problems created by the "Industrial Revolution". For example, the first environmental law established to protect air pollution was in the city of Chicago and Cincinnati in 1881 (Portney 2000). In addition, refuse Act 1899 aimed to protect water from pollution by preventing the disposal of waste through the navigable water in the United States (FreemanIII 2000).

The most prominent policies to mitigate climate change can be summarized as follows:

Long-term policies: Building a strategic plan for the city that emphasizes the principles of sustainable development that achieves a balance between economic, social, and environmental development to reduce greenhouse gas emissions. This includes urban legislation to protect the natural environment, wildlife, and vegetation. The strategic plan also includes providing a system of alternative energy sources, especially solar energy and wind energy. The increase in urban population means the need to provide an adequate supply of renewable resources and energy that does not negatively impact the environment or quality of life (Perea-Moreno 2021). Alternative energy sources can be used in productive industries and thus reduce the level of greenhouse emissions. Long-term policies encourage the transition to a clean energy economy and ensure that health is a key factor in climate change mitigation policy. Environmental policies provide systems that encourage the use of clean energy, reducing carbon rates, and using environmentally friendly transportation alternatives (WOH 2023). Moreover, these policies allow the exchange of environmental information and risk assessment between countries because the issue of climate change is not limited to a specific area but rather affects the world. Implementing long-term policies requires developing urban and environmental legislation and the structure and procedural path of national planning, including strengthening community participation, environmental protection associations, and stakeholder participation.

Short-term polices: Developing a municipal management system to enhance environmental monitoring and developing environmental models. These polices also include urgent solutions to address the negative effects of climate change, such as directing valley paths, removing squatter areas that built along valley paths, treating sources of air pollution, and increasing vegetation cover. Urban policies also emphasize water storage in cities and the construction of dams or water tanks (Tabassum et al. 2020). Short-term policies help to address the negative effects of disasters and environmental changes, but they must be consistent with long-term policies within the framework of sustainability to ensure positive returns for future generations.

Overall, these strategies must be coupled with diversification of funding sources, including government, private sector participation, and civic and volunteer associations.

3 Methods

This paper relied on the theoretical approach by analyzing the National Vision 2030 initiatives in the context of confronting the challenges of climate change. The study compared theoretical concepts and policies for climate change mitigation. The study derived the climate change actions CCAs from Goal No. 13 of the Sustainable Development Goals SDGs. The study compared CCAs to the local efforts and initiatives of the National Vision 2030 in confronting the challenges of climate change. As shown in Fig. 26.1 the study extracted the most prominent initiatives and local efforts to formulate a strategic framework for sustainable development of Saudi cities that mitigation climate change in urban and rural environments.

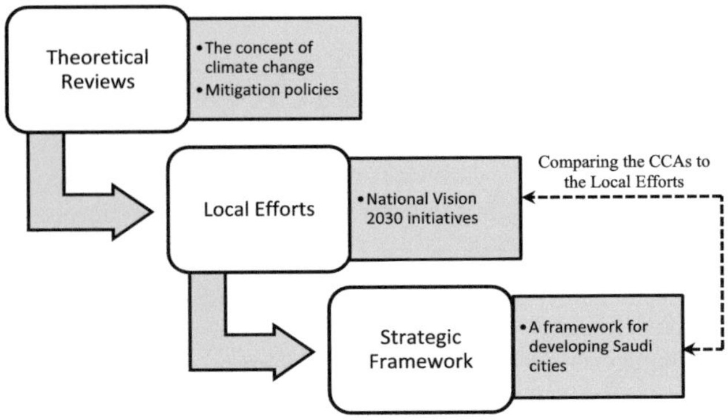

Fig. 26.1 Scheme of the research methodology

4 Results

4.1 Urban Actions of Climate Change Mitigation

The United Nations General Assembly established the Sustainable Development Goals (SDGs) to achieve a better future by 2030. The Sustainable Development Goals are a plan to achieve a better and more sustainable future for all. These goals address the global challenges we face, including challenges related to climate change, disasters, and environmental issues. Goal No. 13 focused on "to take urgent action to combat climate change and its impacts" (United Nations 2024). In light of the limited resources and excessive consumption of them, many cities around the world have become unable to cope with rapid urbanization, which has led to many urban problems such as the spread of slums, weak infrastructure, air pollution, deforestation, and problems in waste management. Achieving Goal No. 13 requires implementing actions plans for cities that take into account comprehensive urban development and give all segments of society equitable access to basic services, efficient transportation, green spaces, and environmentally friendly energy sources (United Nations 2023). Table 26.1 summarizes the most important urban actions of cities that can mitigation climate change in urban and rural environments.

4.2 Local Context

Since its launch, National Vision 2030 has sought to develop a number of programs and initiatives in the context of preserving the environment and the ecosystem in order to achieve the principles of sustainable development. The vision set goals to transform cities into economic engines that exploit existing resources efficiently without depletion or disruption. The National Vision 2030 developed the "Green Saudi Arabia" initiative with the aim of increasing green spaces and building a green economy that preserves the

Table 26.1 Urban actions of climate change mitigation

No	Goal 13 targets	Climate change actions CCAs
13.1	Strengthen resilience and adaptive capacity to climate-related hazards and natural disasters in all countries	· Environmentally friendly transportation · Safe waste disposal systems · Renewable energy alternatives · Enhancing the pedestrian environment · Increase green areas · Reducing negative environmental impact · Air quality system · Safe, comprehensive and accessible parks · Water Conservation
13.2	Integrate climate change measures into national policies, strategies and planning	· National plan · Regional plan · Local strategic plan · Governance and environmental indicators · Effective urban management · Risk Management · Disaster response policies · Developing monitoring systems · Sustainable buildings · Promoting smart cities · Flexible urban legislation
13.3	Improve education, awareness-raising and human and institutional capacity on climate change mitigation, adaptation, impact reduction and early warning	· Community participation · Raising the level of environmental awareness in education
13.4	Protect the Worlds Cultural and Natural Heritage	· Strategy to protect local heritage · Programs for the preservation of cultural heritage · Legislation to protect valleys · Legislation to protect wildlife · Register local heritage sites
13.A	Financing support for the Green Climate Fund	· Diversifying funding sources · Non-profit sector participation · Participation of civic associations · Evaluation of economic losses resulting from disasters
13.B	Promote mechanisms for raising capacity for effective climate change-related planning and management in least developed countries	· International data exchange systems · Exchange of international experiences · Policies to address social exclusion

Source United Nations (2024), The Global Goals (2024), United Nations (2023)

environment and provides tourism experiences to reach a more sustainable future (National Vision 2030 2023). The Saudi Green Initiative emphasizes global climate goals and reaching zero neutrality in 2060. There is tangible progress in increasing the percentage of green areas, reclaiming pastures, and protecting the environment, as the area of rehabilitated vegetation reached 192.4 thousand hectares by the end of 2023, exceeding the target area estimated at 69 thousand hectares. The Saudi Green Initiative adopts a sustainable strategic principle to rehabilitate green spaces through participatory efforts from the private, public and voluntary sectors. Several initiatives have been launched to build and implement parks with the aim of raising the quality of infrastructure, reducing temperatures, and improving air quality in city centers. Many cities have begun to adopt the "Green Saudi Arabia" initiative through projects to intensify green areas, such as the "Green Riyadh" initiative, the King Salman Park project, the sports track, and the New Murabba project (National Vision 2030 2023). Moreover, there are new cities and quality initiatives that seek to improve the urban environment in cities and huge projects such as NEOM, The Line, and Green Riyadh, improving the urban landscape, the sports track, Riyadh Art, the new square, King Salman Airport, the Humanization of King Salman Neighborhood project, and public transportation projects.

Saudi Arabia has developed a strategic plan to protect wildlife, through the adoption of the "Council of Royal Reserves". In the long term, this strategy preserves the natural identity, wildlife, and vegetation in a way that contributes to increasing the proportion of ecotourism. The total area of the royal reserves reached 300,000 square kilometers, equivalent to 15% of the area of Saudi Arabia. Moreover, initiatives have been developed to protect, rehabilitate and register heritage sites as world heritage (National Vision 2030 2023). However, there is a need to integrate these projects within the urban context and in line with Climate

Change Goal 13.4 "Protect the Worlds Cultural and Natural Heritage". National Vision 2030 sought to reduce dependence on fossil fuels to reach zero neutrality by 2060. Saudi Arabia has begun developing large-scale green hydrogen projects, to be one of the clean energy sources, in a manner consistent with the goals of the "Green Saudi Arabia" initiative and achieving global sustainable development goals. Oxagon's green hydrogen company NEOM obtained the first industrial license to produce hydrogen as part of local efforts to develop sustainable and innovative solutions to address the challenges of climate change. Saudi Arabia aims to be the world's largest hydrogen provider with an annual target of 4 million tons. The final operation of three solar energy projects was completed in the cities of Sudair, Jeddah and Rabigh, with a total capacity of 2.1 gigawatts, in a context that is in line with the principle of transition toward renewable energy sources (National Vision 2030 2023). Moreover, the index of the contribution of renewable energy to electricity production in Saudi Arabia rose from 0.9% in 2022 to 3.06% in 2023, a growth of more than three times. Saudi Railways (SAR) launched hydrogen train trials for the first time to contribute to the national transformation plans to a more sustainable transportation system, by improving the quality of life and protecting future generations by using clean energy sources in operational processes.

The public transportation project in the city of Riyadh is one of the largest urban transportation systems and the project aims to meeting the demands of the growing city population in urban areas, as well as reducing the severity of traffic congestion, reliance on private cars, and reducing air pollution rates. The project includes metro tracks and bus stations covering most parts of the city, including King Khalid International Airport, commercial centers and universities. In addition, Saudi Arabia has developed a road map for operating the electric vehicle system in order to fulfill its obligations toward reducing carbon dioxide emissions. The Public Investment Fund is working to diversify the economic base as an alternative to relying on oil by supporting the "Green Saudi Arabia" initiative, its commitment to green finance, the green bond standards issued by the International Capital Market Association (ICMA) in 2021, and the principles of green loans from the International Debt Market Association (LMA) for 2021. The financing aspect is important to support cities' resilience and sustainability to climate change.

5 Discussion

Saudi Arabia has made great efforts to mitigate the effects of climate change through a number of national strategic initiatives and goals. However, there is a need to include climate change goals in the local strategic plan in order to transform Saudi cities to the principles of sustainability. There is a noticeable disparity in development at the regional level, with the population concentrated in major cities. Population distribution in cities affects the quality of life, environmental resources, and pollution levels. Efforts to shift toward the use of alternative energy, smart transportation, and humanization of cities are still in their early stages. Its efficient implementation requires working on a regional plan that takes into account the balance in the distribution of population and environmental resources. The latest statistics issued by the General Authority for Statistics indicate that the population of the city of Riyadh reached 7,009,120 people in 2022, constituting 82% of the total population of the Riyadh region, which numbers 8,591,748 people. This gives an indication that urban planning efforts are focused locally on the city of Riyadh without a comprehensive understanding of the regional context. Moreover, the stages of population growth are still far from established strategies, as many new urban inputs have affected the demographics of the city. Many major cities have become attractive environments with integrated infrastructure, services and business centers. The population of the cities of Riyadh, Jeddah, Dammam, Mecca and Medina reached 16,198,139 people, constituting more than 50% of the total population of the Kingdom of Saudi Arabia (General Authority for Statistics 2022).

Urban development projects should integrate within the urban context and in line with Climate Change Goal 13.1 "to strengthen resilience and adaptive capacity to climate-related hazards and natural disasters in all countries". Therefore, urban legislation reflect the principle of "sustainable cities" as a strategic direction based on the optimal exploitation of environmental resources to improve the standard of living and create sources of income for cities. This challenge requires developing urban legislation that encourages community participation and is consistent with the characteristics of cities, demographics, and climate (Alzamil 2023). Local initiatives in transportation are still in their early stages in the context of enhancing the resilience and sustainability of cities. The number of cars in 2017 for every 1,000 people reached 310 cars (Royal Commission for Riyadh City RCRC 2018). Heavy reliance on cars has led to high levels of ownership, severe road congestion, high traffic, and pollution rates (Al-Husseini and Al-Hussein 2021). Urban transportation is one of the biggest challenges facing major cities. Public transportation projects have begun in some cities such as Riyadh, Mecca, and Medina. However, the car is a major means of transportation in these cities due to urban sprawl, the horizontal expansion of cities, and the real estate development system. Reliance on traditional vehicles causes an increase in harmful emissions, and this does not comply with Goal No. 13.1 “to strengthen resilience and adaptive capacity to climate-related hazards and natural disasters in all countries”. There is a need to activate cities as economic generators so that city revenues contribute to addressing environmental problems and achieving Goal No. 13.1 “to financing support for the Green Climate Fund”. Based on an analysis of the National Vision 2030 initiatives and local efforts, Table 26.2 summarizes the most prominent challenges facing Saudi cities to confront climate change.

Finally, Fig. 26.2 shows the strategic framework for sustainable development of Saudi cities that mitigate climate change in urban and rural environments. The proposed framework takes into account the goals of the National Vision 2030 and Goal 13 targets at the national, regional,

Table 26.2 Most prominent challenges facing Saudi cities to confront climate change

	No	Goal 13 targets	Challenges facing Saudi cities
Environmental challenges	13.1	Strengthen resilience and adaptive capacity to climate-related hazards and natural disasters in all countries	· Real estate development procedures need to be updated · Activating sustainability in real estate development projects · Implement air quality policies · The urban planning pattern does not encourage environmentally friendly transportation alternatives · The use of alternative energy is still limited · Delay in implementing waste disposal systems · Using advanced environmental monitoring and analysis systems
Urban planning and management	13.2	Integrate climate change measures into national policies, strategies and planning	· A gap between national plans and local city plans · The relationship between urban projects and the urban context · Urban legislation is not consistent with the nature and characteristics of each area · Monitoring and evaluation mechanisms · Lack of a clear plan for resource management · Applying smart growth principles to cities · The procedural aspect of urban planning
Social and cultural challenges	13.3	Improve education, awareness-raising and human and institutional capacity on climate change mitigation, adaptation, impact reduction and early warning	· There is no participation in development decision-making · Community engagement is limited to measuring satisfaction · Participation of the non-profit sector
	13.4	Protect the Worlds Cultural and Natural Heritage	· Weak tourism infrastructure in some cities · Integrating cultural elements and urban context · Monitoring and evaluation mechanisms · Realizing the economic, social and environmental returns of historical sites
Economic challenges	13.A	Financing support for the Green Climate Fund	· Exploiting city revenues · Private sector participation · Non-profit sector participation · Measuring the cost and environmental returns of projects
Education and human resources	13.B	Promote mechanisms for raising capacity for effective climate change-related planning and management in least developed countries	· Strengthening the role of scientific research · Qualification of human resources in urban planning · International partnerships

Source United Nations (2024), The Global Goals (2024), United Nations (2023), General Authority for Statistics (2022), Alzamil (2023), Al-Husseini and Al-Hussein (2021)

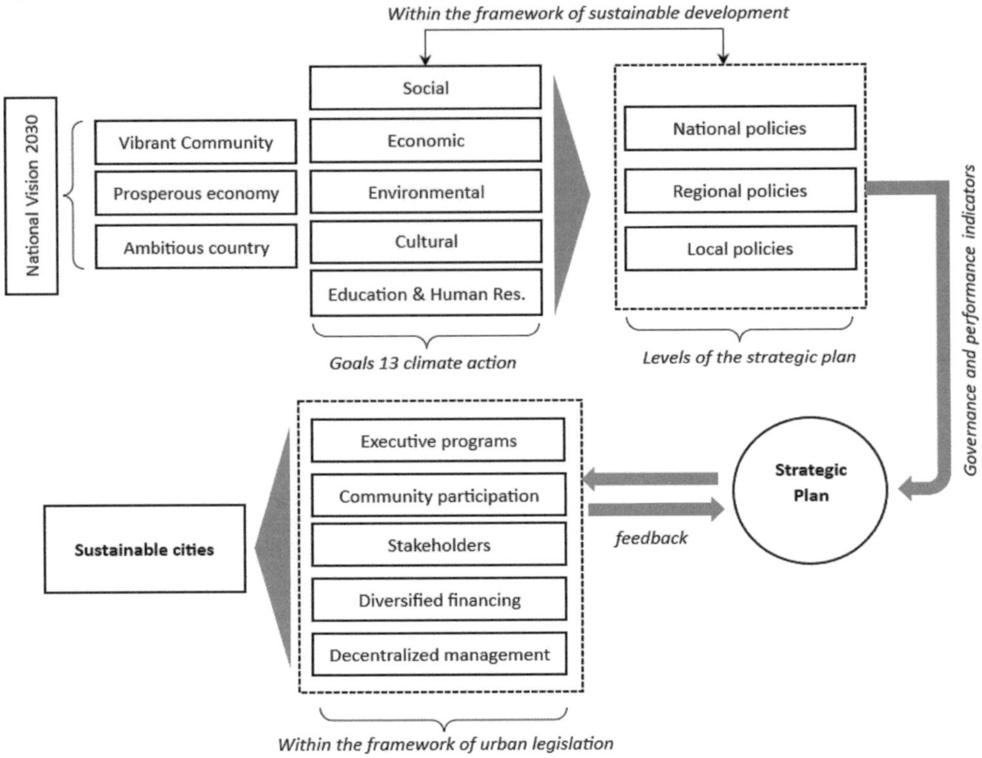

Fig. 26.2 Strategic framework for sustainable development of Saudi cities that mitigation climate change in urban and rural environments

and local planning levels. This framework also sets indicators to measure the efficiency of the strategic plan and legislation affected by stakeholder input, community participation, executive programs, and decentralized management.

6 Conclusion

The paper discussed the efforts and initiatives of the National Vision 2030 to confront the challenges of climate change within the framework of sustainable development. The study relied on an analysis of Goal No. 13 of the Sustainable Development Goals (SDGs) by referring to a wide range of relevant literature. The study discussed local efforts in the context of climate change goals, especially in transportation, energy alternatives, environment, and resources. The study concluded that there are economic challenges, urban and administrative challenges, environmental challenges, social challenges, and challenges in education and human resources development. The study developed a strategic framework for sustainable development of Saudi cities that mitigate climate change in urban and rural environments. The proposed strategic framework helps integrate efforts into national, regional and local plans. The strategic framework emphasizes upgrading urban legislation based on developing executive programs, enhancing community participation, stakeholder, diversifying financing, and decentralizing urban administration.

Based on the above, the study recommends the following:

[1] Promoting the concepts of sustainability of Saudi cities in urban planning at the local, regional, and national levels.
[2] Activating a governance framework to measure performance indicators in Saudi cities in line with Goal No. 13 of climate change.
[3] Developing urban legislation in Saudi cities in line with resources, natural and climatic characteristics.
[4] Stimulating urban growth in small cities in a way that ensures balance in the distribution of population and environmental resources.

References

Agarwal, A.: (2002) A southern perspective on curbing global climate change. In: Schneider, R., N.: (Eds.), Climate change policy: a survey (pp. 375–392). Island Press, Washington DC, USA

Al-Hathloul, S.: Riyadh development plans in the past fifty years (1967–2016). Curr. Urban Stud. **5**(1), 97–120 (2017). https://doi.org/10.4236/cus.2017.51007

Al-Husseini, A., & Al-Hussein, N.: Evaluating the accessibility of the public transportation system in the city of Riyadh using geospatial analyses. King Abdullah Center for Petroleum Studies and Research, Riyadh (2021)

Alqahtany, A., Aravindakshan, S.: Urbanization in Saudi Arabia and sustainability challenges of cities and heritage sites: heuristical insights. J. Cult. Heritage Manage. Sustain. Develop. **12**(4), 408–425 (2022). https://doi.org/10.1108/JCHMSD-07-2020-0108

Alzamil, W.: Evaluation of Affordable Housing and Subsidy Programs in Saudi Arabia. LAP LAMBERT Academic Publishing, Saarbrücken (2016)

Alzamil, W.: Priority Messages in Urban Planning and Development. Takween Publishing, Jeddah (2023)

FreemanIII, A. M.: Water pollution policy. In: Portney, P., Stavins, R.: Public policies for environmental protection (pp. 169–210). RFF Press, Washington D.C., (2000)

General Authority for Statistics. (2022). General Authority for Statistics. (Saudi Census 2022) Retrieved 3 10, 2024, from https://portal.saudicensus.sa/portal

Madigan, N. (2013). South Florida Faces Ominous Prospects From Rising Waters. Retrieved 11 12, 2013, from New York Times: http://www.nytimes.com/2013/11/11/us/south-florida-faces-ominous-prospects-from-rising-waters.html

Mazzetto, S., Vanini, F.: Urban heritage in Saudi Arabia: comparison and assessment of sustainable reuses. Sustainability 15(12), (2023). https://doi.org/10.3390/su15129819

Ministry of Economy and Planning: Summary of the tenth development plan. Ministry of Economy and Planning, Riyadh (2015)

Ministry of Finance: First development plan. Central Planning Authority, Riyadh, 1970

National Vision 2030: Annual report of the Saudi Vision. Riyadh: Saudi Vision 2030 (2022)

National Vision 2030: Annual report of the National Vision 2030. National Vision, Riyadh (2023)

Pelling, M.: Natural Disaster and Development in a Globalizing World. Routledge, Milton Park (2003)

Perea-Moreno A-J, H.-E. Q. (2021). The sustainable city: advances in renewable energy and energy saving systems. Energies 14(24). Retrieved from https://doi.org/10.3390/en14248382

Porter, E.: U.S. Revives aid program for clean energy (2013). Retrieved 9 25, 2013, from New York Times: http://economix.blogs.nytimes.com/2013/09/25/the-cost-of-climate-change/?_r=0

Portney, P.: Air pollution policy. In: P. Portney, E. Robert Stavins, Public Policies for Environmental Protection (pp. 77–122). RFF Press, Washington D.C, (2000)

Riedy, C.: Climate Change. In: Ritzer, G.: Blackwell Encyclopedia of Sociology. Blackwell (2016)

Royal Commission for Riyadh City RCRC: Riyadh City Urban Indicators. Royal Commission for Riyadh City RCRC, Riyadh (2018)

Selin, H., VanDeveer, S.: Global Climate Change. In: Vig, N.J., Kraft, M.E. (eds.) Environmental Policy: New Directions for the Twenty-First Century, pp. 278–298. Sage, Los Angeles (2013)

Shalaby, H., Aboelnaga, S.: Climate change impacts on urban planning in the cities. 1st International Conference on Towards a Better Quality of Life (2017). https://doi.org/10.2139/ssrn.3162375

Shogren, J., Toman, M. Climate change policy. In: Portney, P., Stavins, R.: Public Policies for Environmental Protection (pp. 125–160). RFF Press, Washington D.C, (2001)

Tabassum, R., Arsalan, M., Mumtaz, F., & Imam, N. (2020). Sustainable urban water conservation strategy for a planned city of a developing country: a perspective from DHA City Karachi. Arab J Geosci. Retrieved from https://doi.org/10.1007/s12517-020-06180-2

The Global Goals: The Global Goals (2024). Retrieved from Climate change: https://www.globalgoals.org/goals/13-climate-action/

United Nations. (2023). The Sustainable Development Goals Report. United Nations, New York. Retrieved from https://unstats.un.org/sdgs/report/2023/The-Sustainable-Development-Goals-Report-2023.pdf

United Nations. (2024). Goal 13. Retrieved from Sustainable Development: https://www.un.org/sustainabledevelopment/climate-change/

Vision 2030. (2024). Vision 2030. Retrieved from ENVIRONMENT & NATURE: https://www.vision2030.gov.sa/en/progress/environment-nature/

Werndl, C.: On Defining Climate and Climate Change. Br. J. Philos. **67**(2), 337–364 (2016)

WHO. (2016). Protecting health from climate change. World Health Organization

WOH. (2023). Climate change. Retrieved from World Health Organization: https://www.who.int/news-room/fact-sheets/detail/climate-change-and-health

Rural Land Use and Climate Change Mitigation

Forests Restoration in China Since 1990s: Evidence from the National Forests Inventory

Lingyue Li, Lie Wang, and Yifan Yu

Abstract

Global warming has widely been noticed among scientific researchers while forests are believed essential for stabilizing the climate. This research focuses on the forest restoration phase in China since 1990s. It builds a framework for the relationships between forest transition pathways, identifies explanatory factors through review, and examines how these factors, in association with the pathways suggested by forest transition theory, result in the restoration patterns. From the fifth to ninth National Forest Inventory (NFI), China's forest area has increased from 158 to 220 million hectare, but the increasing rate slows down in most provinces as time went by. Western provinces have faster restoration rates but the restore pace fluctuates significantly. Pathways to the restoration are multiple, in which forest scarcity pathway indicated by timber products yield contributes most and the globalization pathway measured by export volume of agricultural products impacts least. Correlations of sole factor and the restoration are mostly positive, except construction land, orchard and GDP per capita. This research identifies patterns of forest restoration and decodes how multiple pathways, indicated by explanatory factors, relate to reforestations in China. It enriches the understanding of forest transition since 1990s and generates significant insights to forestry policy makings.

Keywords

Forest transition · Reforestation · Spatiotemporal pattern · Explanatory factors · China

1 Introduction

Scientific community has reached an agreement that global warming and climate change are proceeding at an unprecedented rate since 1970s, rarely seen for the past millennia (Westerhold et al. 2020). It is believed that forests are essential for stabilizing the climate as 1/3 of the carbon dioxide released during fossil fuels combustion is absorbed by forests each year. Moreover, deforestation or degradation of forest cause greenhouse gas emission, resulting in roughly 5–10 $GtCO_2e$[1] annually, so that the loss doubled if restoration of forests failed. Forests not only play an integral part in the carbon cycle, but also regulate ecosystems, protect biodiversity, support livelihoods (Schelhas et al. 2021; Yuxing Zhang and Song 2006), and can help drive sustainable growth. Despite of the indispensable role of forest ecosystems in coping with sustainability and social-ecological challenges, loss of forests, however, remains non-negligible. From 1960 to 2019, the worldwide forest area has decreased 81.7 million ha, with forest loss (437.3 million ha) far surpassing forest gain (355.6 million ha) (Estoque et al. 2022). Halting the loss of forests and promoting their restoration, therefore, has the potential to mitigate climate change and increase the well-being of humans. Though prosperity of industrialization since mid-1800s has threatened forests on earth, leading to the decrease of forest cover, in recent years, the decreasing rates slowed down, and net reforestation was observed in many

L. Li (✉)
Department of Urban Planning, CAUP, Tongji University, Shanghai, China
e-mail: lilingyue929@gmail.com

L. Wang
Graduate Student at Department of Urban Planning, CAUP, Tongji University, Shanghai, China

Y. Yu
Department of Urban Planning, CAUP, Tongji University, Shanghai, China
e-mail: yuyifan@tongji.edu.cn

[1] $GtCO_2E$ is equivalent to 59.1 giga tons of CO_2.

P. Mohammad et al. (eds.), *Geographic Approaches to Climate Change and Mitigation: Urban and Rural Perspectives (Volume 1)*, Advances in Science, Technology & Innovation, https://doi.org/10.1007/978-3-031-92119-3_27

places (Drummond and Loveland 2010; Foster et al. 1998; Mather et al. 1999; Redo et al. 2012; Yeo and Huang 2013). Identifying how and in what ways forest restore would be useful for better forest policy making thus benefit climate change mitigation.

The loss of forests has also happened in People's Republic of China where *state forest policy* and *economic development* are considered critical to the halt of loss and the later restoration (Mather 2007; Viña et al. 2016). The immediate socialist transformation after the 1949 foundation of China has almost terminated the institution building in forestry and surge of population and economic recovery further led to extensive deforestation (Lingchao Li, Chhatre, & Liu, 2019). Forest stock volume (FSV, m^3 ha^{-1}) dropped to the historical low level of 12% in 1970s. Policy shift began after 1978 when the Chinese government has gradually implemented 11 artificial forestry ecological projects such as the "Three-North Shelterbelt Forestation Project" (Rodríguez & Pérez, 2013). Since 1990s, land degradation and flood disasters have caused substantial deforestation affecting China's social and economic development (Lu et al., 2021). Strict "No Logging (Logging Ban) Policy" was initiated and the most ambitious afforestation plan was made alongside large amount of public investment to ecological fix (Wang, Innes, Lei, Dai, & Wu, 2007). Since then, state forest policy has transited from timber production to ecological construction. Since the late 1990s, China has experienced rapid forest expansion, commercial logging in natural forests has decreased 92% by 2000 and the rate of forest cover has increased 22.96% by 2018(Mather, 2007). China's forest restoration has received wide scholarly attention. Yet extant studies mostly stress the chronic transition from net loss to net gain in a long-term span (Zhang 2000), the patterns of the restoration is less understood. A nuanced account of pathways to the transition, especially the restoration, and their specific association with explanatory factors are inadequate. This research thus decodes the spatiotemporal variation of forest restoration in China since 1990s and further explores how socioeconomic factors, in association with the five pathways, matter in the restoration.

2 Literature Review

2.1 Theory Explaining Forests Restoration

Theory of forest transition captures the rules and underlying mechanisms of forest change (MacDonald 2023; Mather and Needle 1998) that resource depletion was followed by amelioration. The five commonly referred pathways include forest scarcity, economic development, state forest policy, globalization and tree-based, smallholder land use intensification. In forest transition theory, *economic development* and *forest scarcity* are two main pathways proposed to explain forest transition, especially the increase of forest area, based on experiences from Euro-American countries (MacDonald 2023; Rudel et al. 2005). Similar pathway is also occasionally observed in less developed areas such as Latin America (Caravaggio 2022). Though discussion on forest transition is mainly rooted in Euro-American context, evidences from developing countries emerge and help complement understanding of another three pathways: the *state forest policy, globalization and tree-based, smallholder land use intensification*. The *state forest policy pathway* intervenes land use adjustment and forestry issues such as eco-tourism development and land ownership transfer for forest recovery. This was considered a key factor for forest transition in Costa Rica, China, India, Vietnam and Bangladesh etc. for the past few decades (Ahammad et al. 2023; Allen and Vásquez 2017; Mather 2007). The *globalization pathway* makes transnational trade of forest products and schemes of multilateral environmental conventions possible, thus in favor to resume forest land. This is observed in Jamaica where domestic products lose competitiveness under liberal foreign trade and left land for forest recover (Timms e al. 2013). Import of forest products further propels the process. Similar evidences are found in Honduras and the Ecuadorian Andes (Li et al. 2017). The *tree-based, smallholder land use intensification pathway* relies on innovation of agricultural system to expand forest landscape wherein high quality talents and redistribution of labor forces adapted to local conditions are critical to improve forest ecological services (Lambin and Meyfroidt 2010). Countries such as Indonesia, Costa Rica and Sri Lanka benefitted from this side (Bruggeman et al. 2016; Cerdán et al., 2012; Mohri et al. 2013).

2.2 Multiple Pathways to Forest Restoration in China

Forest resources in China are unevenly distributed and the underlying mechanisms also varied across regions and time (Rodríguez and Pérez 2013). Though this has been well-noted by researchers, limited studies have empirically examined how forest restore in Chinese context in association with multiple pathways. Indeed, any pathway alone is not sufficient to explain how forests restore in the nation. Likewise, any explanatory factor alone can hardly uncover the transition path. Moreover, in spite of the consensus on reforestation for carbon neutrality and the commitment to increase forest cover, where forestation efforts should be made is not so well understood and agreed upon (Zhang et al. 2022). Unveiling the spatiotemporal patterns of forest coverage and reforestation thus may also aid a bit in this regard.

The economic miracle since 1978 was viewed in favor of forest restoration, economic development thus represents a main pathway which is often directly measured by GDP or

GDP per capita. As Chinese economy has been driven by fast urbanization for the past few decades, the rate of urban population and area of construction land can also capture economic development. Forest scarcity often leads to increasing demand for timber products, annual average timber products yield thus is possible to represent the outcome of forests shortage. Population pressure is recognized as an important factor of deforestation (Li et al. 2017; Rudel et al. 2005). Unlike Euro-America where market force is significant for forest transition, developing countries took use of state forest policy to trigger the change. This is especially true in China where ecology is often tied to its political economic system (Zhu and Lo 2022). In China, an important time frame for policy shifts was around late 1990s when series of legislation and institutional reform were enacted, followed with massive public investment both from central and local funds (Lu et al. 2021; Rodríguez and Pérez 2013). Globalization was a non-negligible factor influencing forest transition across countries. With continuous acceleration of urbanization, series of forest-related industries such as architectural decoration grew rapidly, generating great demands on timbres (Rudel et al. 2020). Smallholder, tree-based land use intensification pathway refers to forest transition benefitted from intensive agroforestry plantation. Restoration of forests take places when agroforestry expanded (Lu et al. 2020).

3 Data Analysis

3.1 The National Forest Inventory (NFI) Data

China is one of the few countries that pioneered establishing a continuous forest inventory system. Although a national survey of forest resources was carried out between 1950 and 1962, the data can hardly unravel the reality with inconsistent methods. Since 1973, China's forestry administration has carried out the first National Forest Inventory (NFI), followed by a review every five years. Until 2018, nine forest inventory periods have been completed so far: 1973–1976 (I), 1977–1981 (II), 1984–1988 (III), 1989–1993 (IV), 1994–1998 (V), 1999–2003 (VI), 2004–2008 (VII), 2009–2013 (VIII) and 2014–2018 (IX). The NFI technical standard was revised in the fifth period, redefining "forest" as land with a forest canopy coverage over 20% (previously it was 30%) (Gu et al., 2021) (Table 1). Therefore, this study applies the fifth (1994–1998) to ninth (2014–2018) NFI data for analysis. In NFI, forest land (area of afforested land) refers to the land area where trees, bamboos, shrubs, sparse woodland, coastal mangroves and other forest trees grow. Forest area (FA) refers to the area of forest land composed of tree species with a canopy density of 0.2 or more (inclusive) or a forest belt with a crown width of more than 10 m. FA includes coniferous forest, broad-leaved forest, bamboo forest etc., but excludes shrub and sparse woodland. The total standing forest stock refers to the total volume of all trees on the land within a certain range, including forest stocking, sparse forest stocking, scattered wood stocking and surrounding tree stocking. Stock volume of forest refers to the total volume of tree trunks existing in a certain forest area. It is one of the basic indicators that reflects the total scale and level of forest resources in a country or region. Forest coverage rate refers to the percentage of forest area in a country or region's total land area. In each NFI, the final year is set to link with other socioeconomic data. The provincial data cover 31 provincial administrative regions, including 22 provinces, 5 autonomous regions, and 4 municipalities. Using NFI, this research takes FA, the area of arbor forest and bamboo forest with canopy density above 0.2, to trace how restoration varied across provinces. FA is also used as the dependent variable in follow up regression analysis.

Table 1 Nine National Forest Inventory (NFI) data

Periods	Forest land (area of afforested land) (10,000 ha)	Forest area (10,000 ha)	Plantation (man-made forest) (10,000 ha)	Forest coverage rate (%)	Total standing forest stock (10,000 m^3)	Stock volume of forest (10,000 m^3)
1973–1976	25,760	12,200	–	12.70	–	865,600
1977–1981	26,713	11,500	–	12.00	–	902,800
1984–1988	26,742	12,500	–	12.98	1,057,249	914,100
1989–1993	26,288	13,700	–	13.92	1,178,500	1,013,700
From first to fourth NFI, forest land was defined as forest canopy coverage over 30%. Since the fifth NFI, forest land had been defined as forest canopy coverage over 20%						
1994–1998	26,329	15,894	4,708	16.55	1,248,786	1,126,659
1999–2003	28,492	17,490	5,364	18.21	1,361,810	1,245,584
2004–2008	30,590	19,545	6,168	20.36	1,491,268	1,372,080
2009–2013	31,259	20,768	6,933	21.63	1,643,280	1,513,729
2014–2018	32,368	22,044	7,954	22.96	1,900,700	1,756,000

From 1994 to 2018, China's forest cover rate increased from 16.55% to 22.96%, and the forest area increased by 61.51 million hectares. The proportion of planted forests increased from 29.36% to 36.30%, increasing 33.36 million hectares. As time went by, rate of planted forests gradually decreased from 86.82% in the fifth NFI to 57.07% in the ninth NFI.

3.2 Analyzing Forest Restoration Pathways and Factors

China's forest restoration relates to diverse factors whose impacts varied across regions. In this research, 9 explanatory factors identified in the reviewed are selected as independent variables to capture how the restoration has been drive: GDP per capital, population urbanization rate, construction land area and cultivated land area are associated with economic development pathway, annual average afforestation area and annual average state forestry investment capture state forest policy, orchard area relates to smallholder intensification pathway, annual average timber products yield represents forest scarcity pathway, and export volume of agricultural products is linked to globalization pathway. These data are collected from China Statistical Yearbooks. Forest area in each province, provincial municipality and autonomous region from the fifth to ninth NFI are aggregated from China Forestry and Grassland Statistical Yearbook as dependent variables. Regression is commonly applied to identify how explanatory factors matter. Traditional regression affords to inform the significance and the positive or negative causal relation of the factors. Yet, the relative importance between these factors is hardly known. Random forest (RF), a decision tree-based ensemble learning algorithm, is proposed to supplement the traditional regression method (Liu et al. 2021).

4 Result

4.1 Characteristics of Forest Restoration in Chinese Provinces

Intensity analysis is widely used to capture forest change (Li et al. 2021). This study applies intensity analysis at each NFI interval to explore the changing characteristics of forest area in the recently past five NFIs (1994–2018) in China (Table 2). Net increase of forest area is found in most provinces during the five NFIs, but the increase rate generally slows down. Across the five NFIs, the most significant restoration was observed in the westernmost and easternmost areas, Qinghai, Xinjiang, Ningxia, Shanghai, Tibet and Jiangsu ranked top six where forest area increased 1259.29%, 349.76%, 348.09%, 308.26%, 265.30% and 237.35%. In provinces with advanced urbanization and industrialization such as Hebei, Shanghai, Tianjin and Jiangsu, deforestation occasionally happens. For instance, during the fifth NFI, forest area in Hebei and Shanghai dropped 2.17% and 13.30%; in the sixth NFI, forest area in Tianjin decreased 0.32%; in the ninth NFI, forest area in Jiangsu decreased 3.77%. In the meanwhile, spatiotemporal variations of forest area coverage are distinct among provinces (Fig. 1). Forest area coverages in Fig. 1 are

Table 2 Rate of forest area change in China from fifth to ninth NFI (%)

Provinces	1998–2003	2003–2008	2008–2013	2013–2018	Provinces	1998–2003	2003–2008	2008–2013	2013–2018
Beijing	12.27	37.41	12.99	22.12	Hubei	3.05	16.33	23.33	3.14
Tianjin	8.97	−0.32	19.74	22.22	Hunan	4.47	10.15	6.73	4.02
Hebei	−2.17	27.22	5.02	14.42	Guangdong	1.47	5.68	3.68	4.40
Shanxi	13.41	6.21	27.72	13.70	Guangxi	20.47	27.31	7.20	6.48
Inner Mongolia	39.04	15.40	5.13	5.10	Hainan	23.52	5.76	6.53	3.58
Liaoning	6.54	6.54	8.85	2.61	Chongqing	7.25	56.63	10.29	12.18
Jilin	1.86	2.28	3.71	2.75	Sichuan	26.31	13.33	2.66	7.98
Heilongjiang	2.11	7.20	1.82	1.44	Guizhou	14.47	32.45	17.31	18.01
Shanghai	−13.30	215.87	14.07	30.69	Yunnan	21.18	16.52	5.31	10.03
Jiangsu	67.41	38.88	50.78	−3.77	Tibet	240.47	5.26	0.61	1.32
Zhejiang	7.10	5.51	2.90	0.60	Shaanxi	13.24	14.49	11.16	3.94
Anhui	4.71	8.46	5.65	4.06	Gansu	37.82	56.45	8.25	0.45
Fujian	4.02	0.22	4.52	1.29	Qinghai	927.20	3.90	23.31	3.29
Jiangxi	4.68	4.54	2.89	1.92	Ningxia	175.68	26.61	20.94	6.15
Shandong	6.85	24.35	0.06	4.68	Xinjiang	171.39	36.68	5.53	14.89
Henan	29.32	24.52	6.68	12.28					

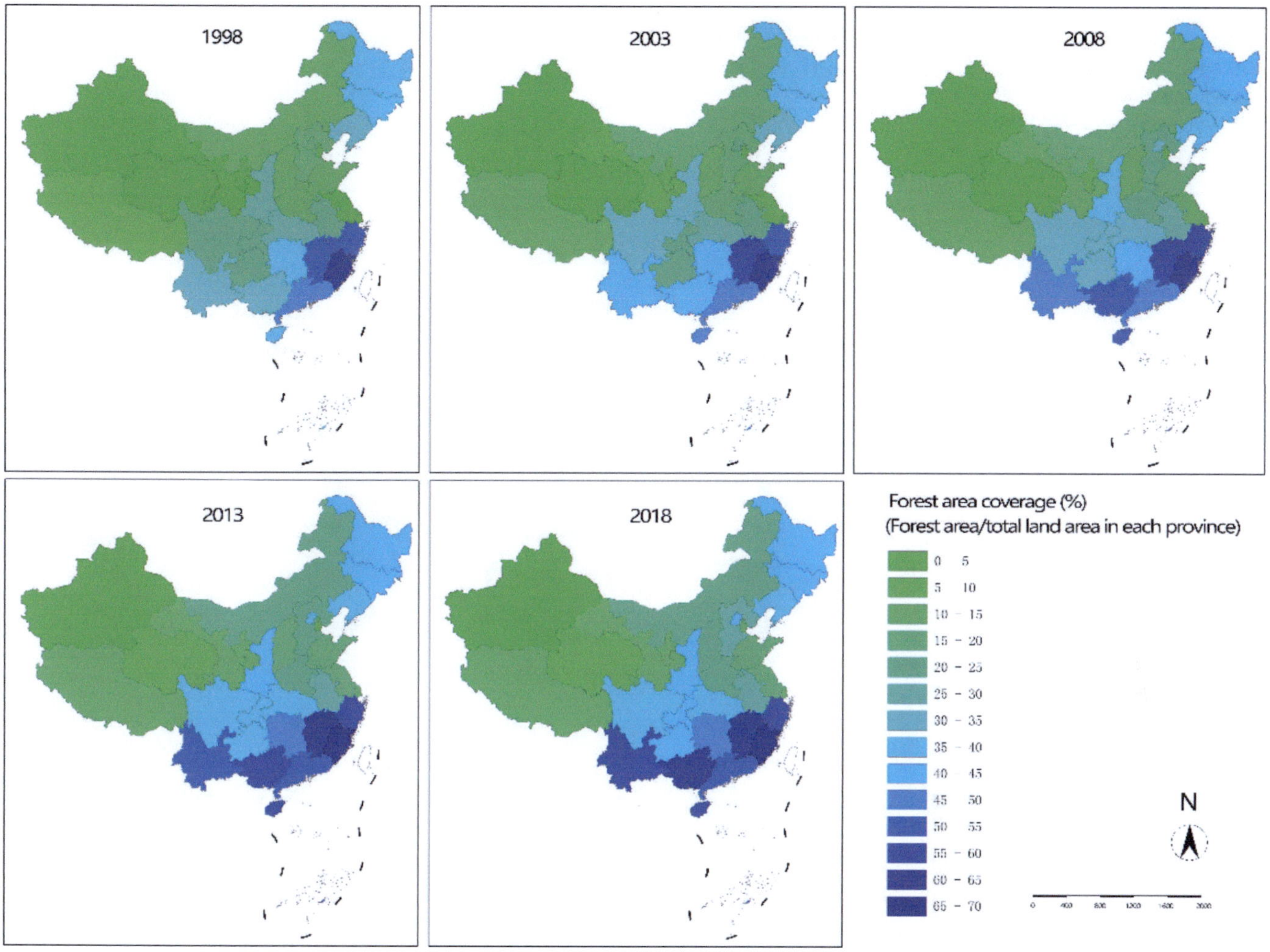

Fig. 1 Forest area coverage in China from fifth to ninth NFI. *Source* China Forestry and Grassland Statistical Yearbook (2019, 2014, 2009, 2004, 1999), edited by authors

calculated through the forest area divided by the total land area in each province, in which data of the forest area are aggregated from China Forestry and Grassland Statistical Yearbook. It can be seen from the figure the gradual increase of forest area coverage in the nation territory across time and the declining trend of this coverage from the southeast to northwest. In particular, during the fifth (1994–1998) and sixth (1999–2003) NFI, significant difference can be found between western, central and eastern parts. Forest restorations in Ningxia, Qinghai, Tibet and Xinjiang are most obvious, increasing 175.68%, 927.20%, 240.47% and 171.39% forest area during this period. The restoration of forest area in Inner Mongolia, Gansu, Sichuan, Yunnan and Guangxi are also visible, increasing 39.04%, 37.82%, 26.31%, 21.18% and 20.47% respectively from fifth to sixth NFI. The state forest policy shift from timber production to ecological construction in 1998 have facilitate massive afforestation in western China. Qinghai alone has restored forest area from 0.31 to 3.17 million hectare. In provinces other than the west, forest restoration is most significant in Jiangsu, Henan and Hainan, where forest area increased 67.41%, 29.32% and 23.52% respectively during the fifth NFI. In addition, plantation contributes greatly to forest restoration since late 1990s, in Guangdong, Guangxi, Fujian, Hunan and Sichuan, man-made forest area were 4.41, 3.71, 3.48, 3.39 and 3.24 million hectare during the fifth NFI; and across the five NFIs, plantation has been increasing in most provinces and the five provinces still have the largest increase in plantation (Fig. 2), gaining 25.58, 27.04, 18.28, 21.70 and 20.35 million hectare forest areas.

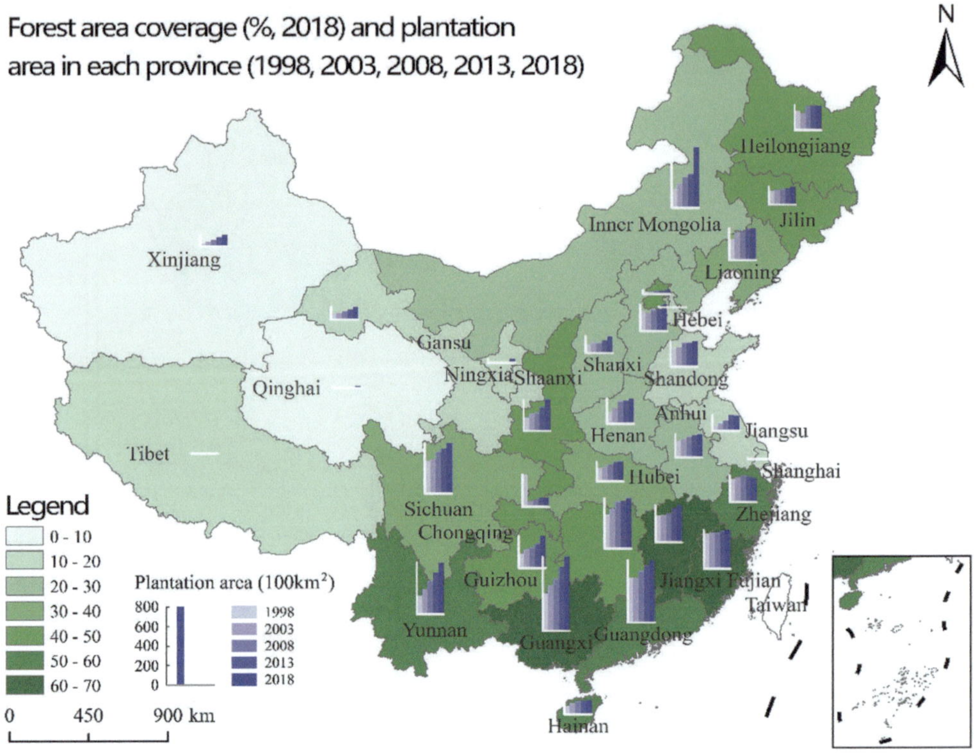

Fig. 2 Forest cover ratio and plantation area in China in the ninth NFI. *Source* China Forestry and Grassland Statistical Yearbook, edited by authors

4.2 Relative Importance of Explanatory Factors

Results of MLR indicate that annual average afforestation area, construction land area, cultivated land area and annual average timber products yield are most significantly related to forest transition, implying that state policy, economic development and forest scarcity are the main pathways to forest restoration. This adds on the recognition from Euro-American experiences that economic development and forest scarcity were two main pathways for forest transition in the initial of reforestation. Noteworthy, among the four factors representing economic development pathway, GDP per capita and population urbanization, once considered essential for forest restoration, are not significant in this research. Apart from the most notable four factors, annual average state forestry investment with p-value between 0.01 and 0.05 is also significantly related to forest restoration, again verifying the importance of state forest policy in restoring forests in China. Orchard area exemplifying smallholder intensification pathway and export volume of agricultural products exemplifying globalization pathway are not significant. In random forest (RF) model, according to the value of %IncMSE, annual average timber products yield, reflecting *forest scarcity* pathway, is the most important factor contributing to forest restoration, followed by annual average afforestation, cultivated land and orchard that report *state forest policy, economic development and tree-based, smallholder land use intensification* pathways respectively. Export volume of agricultural products reflecting *globalization* pathway is the least important factor. The study infers from the combined results of multiple linear regression and random forest that annual average timber products yield and annual average afforestation area representing outstand among all the factors, indicating the protruding importance of state forest policy pathway for reforestation in China (Table 3).

Table 3 Results by MLR and RF

Results variables	Multiple linear regression			Random forest	
	Coeff.	Std. Err	*t* Value	%IncMSE	IncNodePurity
GDP per capita	0.061	0.120	0.510	5.66	1,227,325
Annual average afforestation area	0.341***	0.078	4.389	27.04	9,646,197
Annual average state forestry investment	0.293*	0.116	2.533	17.95	5,694,705
Population urbanization rate	−0.124	0.090	−1.383	13.12	2,551,165
Construction land area	−0.383***	0.096	−3.992	20.79	4,535,525
Cultivated land area	0.521***	0.119	4.387	23.45	8,627,606
Annual average timber products yield	0.745***	0.120	6.231	31.25	8,014,732
Orchard area	−0.040	0.064	−0.624	21.22	3,974,765
Export volume of agricultural products	0.011	0.145	0.075	4.11	1,149,704
Intercept	0.002			–	
R square	0.586			–	
% Var explained	–			71.68	

Note *, **, *** denote corresponding p-value are below 0.05, 0.01 and 0.001, respectively; *Source* edited by authors

5 Conclusion

This research traces forest restoration and the possible explanatory factors in Chinese context since 1990s. Forest area in China has increased from 158 to 220 million hectare from the fifth to ninth NFI, but the increasing rates slow down as time went by and the restore pace fluctuates significantly in some provinces. Though western provinces such as Ningxia, Qinghai, Tibet and Xinjiang generally have faster restoration rates than the central and eastern provinces, net increases of forest area are almost found in all the provinces. The restoration was greatly benefitted from the following six state forestry projects initiated in late 1990s:

- Natural Forest Resources Protection Project
- Return Farmland to Forest and Grassland Project (Green for Grain Project)
- Beijing-Tianjin Sandstorm Source Control Project
- Three-North (northwest, northeast and north China plain) Shelterbelt Project and Shelter Forest System Construction Project in the Middle and Lower Reaches of the Yangtze River
- Wildlife Protection and Nature Reserve Construction Projects
- Forestry Industry Construction Projects on Fast-Growing and High-Yielding Timber Forests in Key Areas.

This can be verified in the multiple linear regression analysis which suggests that annual average afforestation area and annual average state forestry investment reflecting state forest policy pathway are significantly related to the restoration of forest area. Therefore, unlike earlier Euro-American experience that economic development pathway and forest scarcity pathway dominate in the explanation of forest transition, Chinese experience in forest restoration mixed multiple pathways and is state-led in nature. In this regard, the research echoes previous studies in favor of the importance of state forest policy in the forest recovery in developing countries (e.g., Allen & Vásquez 2017; Mather 2007). The issue is that though many restoration programs increased tree cover, ecological functionality of the system is much less addressed, which may only benefit local socioeconomic development in a short run (Ahammad et al. 2023). Assuredly, the result does not deny the significance of economic development and forest scarcity. In random forest analysis, *forest scarcity* pathway measured by timber products is the most important factor contributing to forest restoration, followed by annual average afforestation, cultivated land and orchard that report *state forest policy, economic development and tree-based, smallholder land use intensification* pathways respectively. Yet, different from some developing or underdeveloped countries where forest transition has been greatly propelled by transnational trade and the restructuring of global production chain (e.g., Jamaica), China's engagement with the liberal globalization seems less influential on its forest transition, at least in aspect of export volume of agricultural products, association between this factor and forest area restoration is weakest as indicated in the random forest analysis.

Competing Interest This study was funded by the National Key R&D Program of China [Grant No. 2023YFC3804001].

References

Ahammad, R., Hossain, M.K., Sobhan, I., Hasan, R., Biswas, S.R., Mukul, S.A.: Social-ecological and institutional factors affecting forest and landscape restoration in the Chittagong Hill Tracts of Bangladesh. Land Use Policy **125**, 106478 (2023)

Allen, K.E., Vásquez, S.P.: Forest cover, development, and sustainability in Costa Rica: can one policy fit all? Land Use Policy **67**, 212–221 (2017)

Bruggeman, D., Meyfroidt, P., Lambin, E.F.: Forest cover changes in Bhutan: Revisiting the forest transition. Appl. Geogr. **67**, 49–66 (2016)

Caravaggio, N.: Economic growth and forest transition in Latin America. Forest Policy Econ. **135**, 102667 (2022)

Cerdán, C.R., Rebolledo, M.C., Soto, G., Rapidel, B., Sinclair, F.L.: Local knowledge of impacts of tree cover on ecosystem services in smallholder coffee production systems. Agric. Syst. **110**, 119–130 (2012). https://doi.org/10.1016/j.agsy.2012.03.014

Drummond, M.A., Loveland, T.R.: Land-use pressure and a transition to forest-cover loss in the eastern United States. Bioscience **60**(4), 286–298 (2010)

Estoque, R.C., Dasgupta, R., Winkler, K., Avitabile, V., Johnson, B.A., Myint, S.W., Lasco, R.D.: Spatiotemporal pattern of global forest change over the past 60 years and the forest transition theory. Environ. Res. Lett. **17**(8), 084022 (2022)

Foster, D.R., Motzkin, G., Slater, B.: Land-use history as long-term broad-scale disturbance: regional forest dynamics in central New England. Ecosystems **1**(1), 96–119 (1998)

Gu, L., Gong, Z., Bu, Y.: Forest quality dynamic change and its driving factors accompanied by forest transition in China. Forests **12**(6), 733 (2021)

Lambin, E.F., Meyfroidt, P.: Land use transitions: Socio-ecological feedback versus socio-economic change. Land Use Policy **27**(2), 108–118 (2010)

Li, L., Chhatre, A., Liu, J.: Multiple drivers and pathways to China's forest transition. Forest Policy Econ. **106**, 101962 (2019)

Li, L., Liu, J., Long, H., de Jong, W., Youn, Y.-C.: Economic globalization, trade and forest transition-the case of nine Asian countries. Forest Policy Econ. **76**, 7–13 (2017)

Li, L., Wang, L., Qi, Z.: The spatiotemporal variation of farmland use transition and its critical influential factors in coordinated urban-rural regions: A case of Chongqing in western China. Sustain. Cities Soc. **70**, 102921 (2021)

Liu, M., Hu, S., Ge, Y., Heuvelink, G.B., Ren, Z., Huang, X.: Using multiple linear regression and random forests to identify spatial poverty determinants in rural China. Spatial Statistics **42**, 100461 (2021)

Lu, L., Huang, A., Xu, Y., Marcos-Martinez, R., Duan, Y., Ji, Z.: The influences of livelihood and land use on the variation of forest transition in a typical mountainous area of China. Sustainability **12**(22), 9359 (2020)

Lu, L., Marcos-Martinez, R., Xu, Y., Huang, A., Duan, Y., Ji, Z., Huang, L.: The spatiotemporal patterns and pathways of forest transition in China. Land Degrad. Dev. **32**(18), 5378–5392 (2021)

MacDonald, H. (2023). Envisioning better forest transitions: A review of recent forest transition scholarship. *Heliyon.*

Mather, A.S.: Recent Asian forest transitions in relation to forest transition theory. Int. for. Rev. **9**(1), 491–502 (2007)

Mather, A.S., Fairbairn, J., Needle, C.L.: The course and drivers of the forest transition: the case of France. J. Rural. Stud. **15**(1), 65–90 (1999)

Mather, A.S., Needle, C.: The forest transition: a theoretical basis. Area **30**(2), 117–124 (1998)

Mohri, H., Lahoti, S., Saito, O., Mahalingam, A., Gunatilleke, N., Irham, Herath, S.: Assessment of ecosystem services in homegarden systems in Indonesia, Sri Lanka, and Vietnam. Ecosyst. Serv. **5**, 124–136 (2013). https://doi.org/10.1016/j.ecoser.2013.07.006

Redo, D.J., Grau, H.R., Aide, T.M., Clark, M.L.: Asymmetric forest transition driven by the interaction of socioeconomic development and environmental heterogeneity in Central America. Proc. Natl. Acad. Sci. **109**(23), 8839–8844 (2012)

Rodríguez, L.G., Pérez, M.R.: Recent changes in Chinese forestry seen through the lens of Forest Transition theory. Int. for. Rev. **15**(4), 456–470 (2013)

Rudel, T.K., Coomes, O.T., Moran, E., Achard, F., Angelsen, A., Xu, J., Lambin, E.: Forest transitions: towards a global understanding of land use change. Glob. Environ. Chang. **15**(1), 23–31 (2005)

Rudel, T.K., Meyfroidt, P., Chazdon, R., Bongers, F., Sloan, S., Grau, H.R., Schneider, L.: Whither the forest transition? Climate change, policy responses, and redistributed forests in the twenty-first century. Ambio **49**, 74–84 (2020)

Schelhas, J., Brandeis, T.J., Rudel, T.K.: Planted forests and natural regeneration in forest transitions: patterns and implications from the US South. Reg. Environ. Change **21**, 1–10 (2021)

Timms, B.F., Hayes, J., McCracken, M.: From deforestation to reforestation: applying the forest transition to the Cockpit Country of Jamaica. Area **45**(1), 77–87 (2013)

Viña, A., McConnell, W.J., Yang, H., Xu, Z., Liu, J.: Effects of conservation policy on China's forest recovery. Sci. Adv. **2**(3), e1500965 (2016)

Wang, G., Innes, J.L., Lei, J., Dai, S., Wu, S.W.: China's forestry reforms. Science **318**(5856), 1556–1557 (2007). https://doi.org/10.1126/science.114724

Westerhold, T., Marwan, N., Drury, A. J., Liebrand, D., Agnini, C., Anagnostou, E., Florindo, F.: An astronomically dated record of Earth's climate and its predictability over the last 66 million years. science, 369(6509), 1383–1387 (2020)

Yeo, I.-Y., Huang, C.: Revisiting the forest transition theory with historical records and geospatial data: A case study from Mississippi (USA). Land Use Policy **32**, 1–13 (2013)

Zhang, Y. (2000). Deforestation and forest transition: theory and evidence in China. *World forests from deforestation to transition?* 41–65

Zhang, L., Sun, P., Huettmann, F., Liu, S.: Where should China practice forestry in a warming world? Glob. Change Biol. **28**(7), 2461–2475 (2022)

Zhang, Y., Song, C.: Impacts of afforestation, deforestation, and reforestation on forest cover in China from 1949 to 2003. J. Forest. **104**(7), 383–387 (2006)

Zhu, L., Lo, K.: Eco-socialism and the political ecology of forest conservation in the Greater Khingan Range, China. Pol. Geogr. **93**, 102533 (2022)

Soil Biochemical Indicators Under the Influence of Conservation and Conventional Tillage Systems for Growing Crops on Sloping Terrain

Gergana Sl. Kuncheva, Galina P. Petkova, Jonita J. Perfanova, and Tsvetelina R. Metodieva

Abstract

The decline of soil organic matter is the most common soil degradation process in arable lands. On sloping terrains, it is intensified by the flow of water erosion. A three-year experiment was conducted on Calcic chernozem on sloping terrain, with the application of conservation and traditional soil treatments. Two tillage systems on winter wheat—maize rotation where tested—conventional technology and conservation technology, including minimum tillage (no ploughing) and a cover crop between two main crops. The activity of peroxidase, polyphenol oxidase, alkaline phosphatase, protease, ß glucosidase, urease, as well as the content of labile carbon fractions, were measured twice in a year (in spring and autumn). Applied erosion control practices have a positive effect on soil organic matter, labile organic fractions and enzyme activity. The highest degree of correlation of the amount of total organic carbon and microbial biomass carbon was found by polyphenol oxidase activity ($r = +0.55$ and $r = +0.47$, $p < 0.0001$) and available potassium forms ($r = +0.52$ and $r = +0.50$, $p < 0.0001$). Minimum tillage for growing wheat and maize on sloping terrain, with the inclusion of a cover crop in the rotation, is an appropriate practice for improving soil biochemical indicators.

G. Sl. Kuncheva (✉) · G. P. Petkova · J. J. Perfanova · T. R. Metodieva
Nikola Poushkarov Institute for Soil Science, Agrotechnology and Plant Protection, Agricultural Academy, Sofia 1331, Bulgaria
e-mail: g.kuncheva@issapp-pushkarov.org

G. P. Petkova
e-mail: g.petkova@issapp-pushkarov.org

J. J. Perfanova
e-mail: jperfanova@gmail.com

T. R. Metodieva
e-mail: cv.metodieva@issapp-pushkarov.org

Keywords

Soil erosion · Cover crop · Enzyme activity · Soil organic matter · Conservation tillage

1 Introduction

The sustainable mitigation of soil degradation, the enhancement of carbon sequestration, soil structure and fertility can be achieved with the implementation of conservation practices that improve soil health and functions without causing significant yield reductions. Calcic chernozems (Kastanozems) are assessed as highly susceptible to water erosion, because of deteriorated structure and low aggregate stability (Kuncheva et al. 2023; Kercheva et al. 2011). In the Republic of Bulgaria, wheat is the most widespread crop, occupying between 1.0 and 1.3 million hectares annually, which is 45.2% of the arable land of the country and maize production occupies 400 000 hectares which is 12.6% (Dimitrov et al. 2016). A large part of these crops (about 45–50%) are located on terrains with different slopes, which are potentially exposed to the impact of water erosion processes (Dimitrov et al. 2016).

For the prevention and control of water erosion, compaction, loss of organic matter and biodiversity, as well as for degraded soils recovery the use of cover crops is recommended (García-González et al. 2018). This practice is recognized as appropriate for reducing the leaching of nitrates (Gabriel et al. 2013; Tonitto et al. 2006) and for carbon sequestration (Poeplau and Don 2015) during periods when the soils are bare and free from main crops. Carbon sequestration is a main tool for enhancing soil quality as soil organic matter affects many physical, chemical and biological processes and it is important for the formation of stable soil structure and lower susceptibility to water erosion (Fageria et al. 2005; Chahal and Eerd 2023). Cover crops can play

P. Mohammad et al. (eds.), *Geographic Approaches to Climate Change and Mitigation: Urban and Rural Perspectives (Volume 1)*, Advances in Science, Technology & Innovation, https://doi.org/10.1007/978-3-031-92119-3_28

an indirect positive role for the subsequent row crops by improving soil properties (Gabriel and Quemada 2011). The biomass of cover crops stimulates microbial activity and soil extracellular enzyme activities, involved in the biochemistry of organic matter decomposition and nutrient cycling (Sinsabaugh 2010). The enzymatic activities are indicators of microbial nutrient demands (Moorhead and Sinsabaugh 2006), key element of nutrient cycling (Burns et al. 2013; Sinsabaugh 2010), and of soil organic matter formation (Wang et al. 2022). Labile carbon fractions—permanganate oxidizable carbon (POXC) and microbial biomass carbon (Cmic) include various organic substances characterized by short turnovers, therefore they are considered more sensitive indicators for agricultural management compared to total organic carbon (Nandan et al. 2019).

The aim of the study was to determine the effect of minimum tillage and cover crops in the wheat—maize crop rotation on the enzymatic activity, and labile carbon fractions on Medium-eroded Calcic chernozem on inclined arable land. We hypothesized that by determining enzyme activity and labile organic fractions as fast, sensitive and appropriate indicators of the processes occurring in the soil we would be able to assess the improvement of soil organic matter and soil quality under this conservation practice instead of using slower ones.

2 Materials and Methods

The experiment was conducted in the experimental field of ISSAPP Nikola Pushkarov, near the village Trastenik, Ruse district, North Bulgaria on a slope of 5° (8,7%). The experimental design was performed within the crop rotation winter wheat–maize with two tillage treatments and four replicates: T1—conventional tillage; T2—minimum tillage with barley as cover crop between wheat and maize, incorporated into the soil as green manure prior to maize seeding. The conventional tillage (T1) for wheat included several (most often three) disking in different directions at a depth of 10–12 cm. The last disking was performed at a depth of 6–8 cm. On T2 variant it was carried out one disking and direct seeding (10–12 cm). Conventional main tillage for maize sowing included mouldboard plowing at 20–25 cm. For minimum tilled treatment (T2) it was implemented spraying the cover crop with glyphosate and subsequent disking with direct seeding. The cover crop was sown after shallow harrowing at 6–8 cm. The test plots were 34 m long, 22 m wide and 748 m^2 in area. Maize and wheat treatments were conducted simultaneously and each variant was grown for three years under given technologies.

The soil samples for determining the biochemical parameters were taken twice in a year, in the spring and in the autumn. Soil microbial biomass carbon (Cmic) was measured according to the method of Anderson and Domsch (1978), and permanganate oxidizable carbon (POXC) by method of Culman et al. (2012). Urease activity (URE) was determined according to the method of Hoffman and Teicher (1961); protease activity (PROT) according to Hoffman-Teicher (Hasiev 2005); alkaline phosphatase activity (PHOSPH) according to Tabatabai-Brember (Hasiev 2005); ß-glucosidase activity (ß-GLU) according to Tabatabai (Alef and Nannipieri 1995). The activity of the peroxidase (PO) and polyphenol oxidase (PPO) was assessed with the method of Galstian (1974) and expressed in relative units—mg purpurogallin (PPG) obtained per g absolutely dry soil for 30 min (mg PPG g^{-1} 30 min^{-1}). Ammonium and nitrate nitrogen were determined according to Kijeldal and the available phosphorus and potassium forms were measured with the acetate-lactate extraction method. Total organic carbon (SOC) was assessed by wet digestion (Tyurin's method). The number of the main groups of soil microorganisms (10^6colony forming units g^{-1} dry soil) were determined on agar plates: on meat peptone agar used for saprophytic bacteria, on starch ammonium agar—for actinomycetes and Chapek medium—for microscopic fungi.

Statistical analyses and ANOVA tests were performed using STATGRAPHICS software.

3 Results

The cover crop biomass incorporated into the soil affected the labile carbon fractions (Fig. 1). On average for three years the highest microbial biomass carbon was reported for two crops in T2 treatment. In the fall, Cmic enhanced compared to the spring months for all treatments and the differences between conservation and conventional practices were more pronounced (Fig. 1a).

In contrast to other studies showing an increase in POXC with reduced treatments (Shen et al. 2021), an increase in this labile organic fraction was found in wheat in the variant with conventional technology, regarding minimum treatments (Fig. 1b). In maize POXC was higher in T2 variant.

Peroxidase and polyphenol oxidase activity are related to carbon cycling (Wang et al. 2022). A trend of enhanced activity of the two enzymes was reported in T2 treatment, in wheat and in maize, and the increase in two crops was more pronounced in autumn (Fig. 2a, b). Their activity was impacted by organic matter of the cover crop, as well as plant residues under minimum tillage. The enhancement in polyphenol oxidase activity was more significant in T2 variants in two crops than peroxidase activity.

Protease activity was higher in all variants in spring and it decreased in autumn (Fig. 2c). In spring, PROT of T2 in two crops exceeded compared to variants under conventional treatments, for wheat—22,97 μmol NP g^{-1} h^{-1} and for

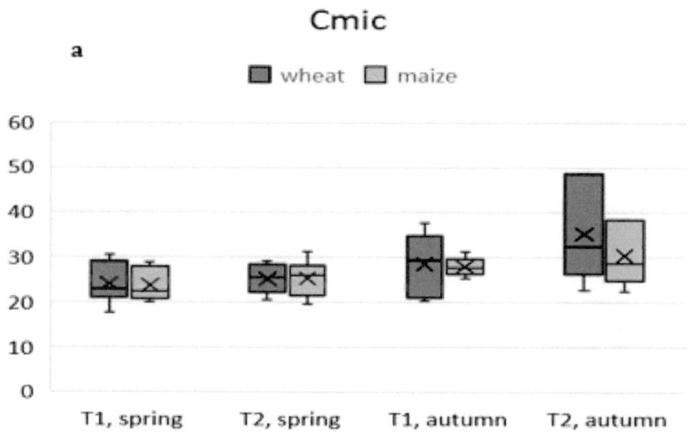

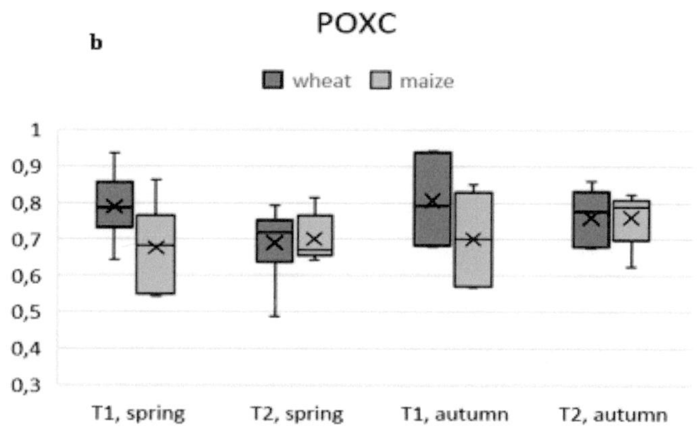

Fig. 1 Labile organic fractions; **a** Cmic (microbial biomass carbon, mg 100 g^{-1}), **b** POXC (permanganate oxidizable carbon, mg kg^{-1}), on average for period 2021–2023. T1—conventional tillage; T2—conservation tillage. ANOVA, Cmic: treatment p = 0.08; crop, NS, season p = 0.0009; POXC treatment NS; crop p = 0.06, season NS

maize—21,57 μmol NP g^{-1} h^{-1}. In autumn this enzyme activity in T2 treatments was lower than traditionally tilled variants for two crops.

The alkaline phosphatase activity in the spring was several times higher in maize variants compared to wheat variants. Treatments with conservation tillage had lower POSPH than traditionally tilled variants, in two crops (Fig. 2d). The URE increased in autumn in T2 wheat (6,15 mg amino-N 100 g^{-1}) and in T2 maize (4,55 mg amino—N 100 g^{-1}), while in the traditionally grown variants T1 URE was 3,43 and 3,12 mg amino-N 100 g^{-1} (Fig. 2e). ß glucosidase activity (Fig. 2f) in two crops, in T2 treatment was higher compared to the traditionally grown variants. The differences between variants in ß glucosidase activity were not statistically significant.

Correlation and regression analysis were performed with basic microbiological parameters, enzyme activity, labile organic fractions and soil chemical properties to determine the strength and quantitative relationships between the studied parameters. URE was moderately positively correlated with Cmic (r = +0.45) (Table 1). There was moderate correlation between phosphatase activity and mineral nitrogen (r = +0.44) and available forms of phosphorus (r = +0.47). No significant correlations were reported for any of the studied factors with ß GLU. Total organic carbon was positively correlated with PO (r = +0.41) and PPO (r = +0.55), the available forms of potassium (r = +0.52). There was a positive correlation of protease activity with the number of ammonifying bacteria, but it was negative with the amount of microscopic fungi, as well as with Cmic, which was most strongly related to the amount of fungi and actinomycetes.

The concentration of available forms of potassium had a positive correlation with many of the studied indicators, such as PO, PPO, SOC, and in the general the correlation coefficient with SOC was r = +0.52, but if the results considered by variants and crops, in T2 variant for maize it was r = +0.81, and for T2 variant wheat it was r = +0.78. PO and PPO activity were also positively correlated with microbial carbon levels. Correlation coefficients of available potassium forms with Cmic were positive. The positive relationship between available forms of potassium and Cmic and actinomycetes was also demonstrated in another study conducted on Haplic Chernozem (Kuncheva et al. 2024).

C mic had a positive correlation with the abundance of microscopic fungi (r = +0.68) and actinomycetes (r = +0.30). Between SOC, POXC and Cmic, no correlation was reported, and there was only weak positive correlation between POXC and SOC, (r = +0.26). EC was positively related with Cmic (r = +0.42) and with SOC only in T2 variants, in T2 wheat (r = +0.23) and T2 maize (r = +0.52).

Figure 3 shows the relationships between the amount of soil organic matter and polyphenol oxidase activity (a), the amount of available potassium (c) and pH (b). With increasing SOC, a decrease in pH was reported. The concentration of available forms of potassium and polyphenol oxidase activity were higher at higher amounts of SOC. There was also a positive relationship between microbial biomass carbon, the quantity of microscopic fungi and urease activity (Fig. 3d–f).

4 Discussion

Conservation tillage have a significant positive effect on the activities of 28 soil enzymes (Wen et al. 2023). Enzyme activity of the soil is an indicator highly sensitive to different agricultural practices, such as mulching, rotation, fertilization, processing, etc. (Acosta-Martinez et al. 2018) and changes very quickly as a result of the applied practices. Including a cover crop in the crop rotation affects enzyme activity (Rankoth et al. 2019). Soil organic matter is the main factor that influences the soil enzymes (Wang et al. 2022). The activity of soil enzymes increases with increasing of SOC, due to the dependence of microbiological processes on

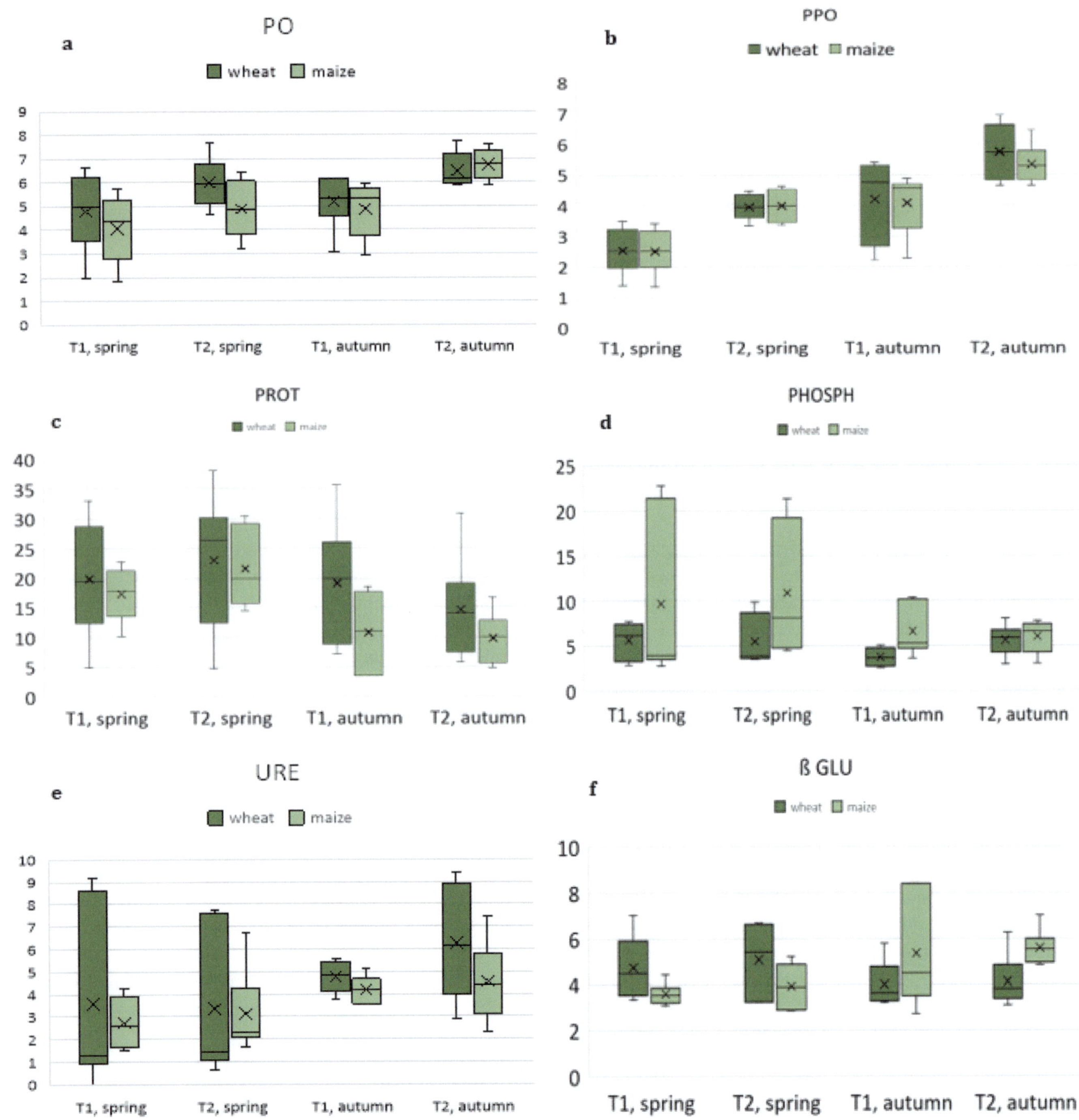

Fig. 2 Enzyme activity; **a** PO (peroxidase activity, mg PPG g^{-1} 30 min^{-1}), **b** PPO (polyphenol oxidase activity, mg PPG g^{-1} 30 min^{-1}), **c** PROT (protease, mg amino-N 100 g^{-1}), **d** POSPH (alkaline phosphatase, μmol NP g^{-1} h^{-1}), **e** URE (urease, mg NH_4-N 100 g^{-1} h^{-1}), **f** ß GLU (ß glucosidase, μmol NP g^{-1} h^{-1}). T1—conventional tillage; T2—conservation tillage. ANOVA: PO; treatment p = 0,0004; crop, NS season; p = 0,011; PPO: treatment p = 0,0000; season; p = 0,0000 crop, NS; PROT treatment p = 0,0004; crop, NS; season; p = 0,011; URE: treatment NS; crop, NS; season; p = 0,0093; ß GLU: NS; POSPH: treatment NS; crop, p = 0,018, season; p = 0,065

substrate availability (Burns et al. 2013). In our experiment, the protease, urease, ß glucosidase during certain periods increased in treatments under minimum tillage and cover crop, and the activity of peroxidase and polyphenol oxidase have been affected by conservation tillage more significantly. Peroxidase and polyphenol oxidase activities are associated with plant litter and correlate with the amount of lignin in the plant residues and with the rate of biomass decomposition (Sinsabaugh et al. 2009). This was also considered in our experiment. PO, PPO correlated strongly with Cmic and

Table 1 Correlation coefficients between enzyme activity, URE (mg NH_4-N 100 g^{-1} h^{-1}), POSPH (μmol NP g^{-1} h^{-1}), PROT (mg amino-N 100 g^{-1}), ß GLU (μmol NP g^{-1} h^{-1}), PO, PPO (mg PPG g^{-1} 30 min^{-1}), Cmic (mg 100 g^{-1}), SOC (%), POXC (mg kg^{-1}) and parameters—EC (μS cm^{-1}), pH, mineral N (mg kg^{-1}), available forms of P_2O_5, K_2O (mg 100 g^{-1}), main groups soil microorganisms—CFU (colony forming units) × 10^6 g^{-1} abs. dry soil

Parameters	PO	PPO	URE	PHOSPH	PROT	ß GLU	Cmic	POXC	SOC
Cmic	0.20*	0.47***	0.45***	−0.01	−0.44***	0.03	1.00		
POXC	0.08	0.13	−0.01	−0.25	0.11	0.13	−0.19	1.00	
SOC	0.41***	0.55***	0.14	0.06	−0.11	−0.04	0.08	0.26*	1.00
Min N	−0.04	−0.09	−0.06	0.44	0.06	−0.15	−0.02	−0.34	−0.22
Available P2O5	−0.06	0.10	−0.22	0.47	−0.01	0.08	0.18	0.13	0.21
Available K2O	0.30*	0.54***	0.23	0.03	−0.45***	−0.11	0.50***	0.09	0.52***
EC	0.08	0.18	0.11	0.07	−0.36**	−0.24*	0.42 ***	−0.23	0.11
pH	0.07	−0.06	0.21	−0.21	0.07	−0.12	0.14	−0.13	−0.48***
Saprophytic bacteria	0.31**	−0.01	0.10	−0.10	0.39***	−0.10	−0.32*	0.07	−0.12
Actinomycetes	−0.11	0.05	−0.09	0.27	−0.25	−0.06	0.30*	−0.21	0.01
Microsc. fungi	0.20	0.31**	0.31	−0.06	−0.46***	−0.07	0.68 ***	−0.14	0.05

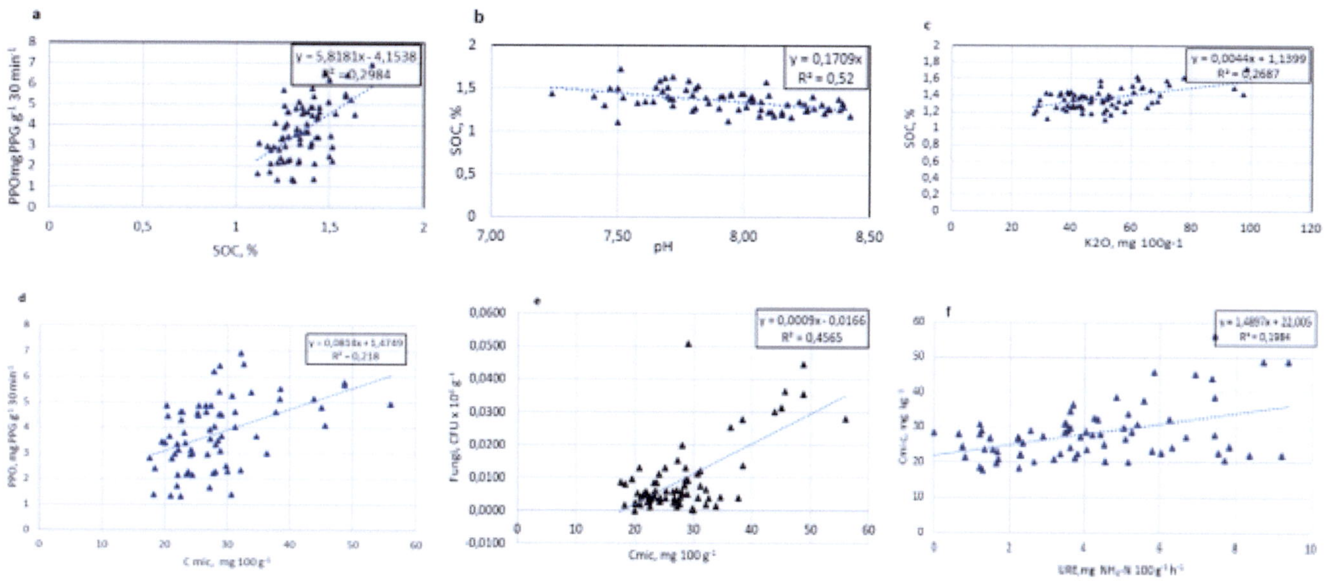

Fig. 3 Pegression dependencies: **a** of SOC (%) and PPO (mg PPG g^{-1} 30 min^{-1}) activity; **b** SOC% and pH; **c** SOC% and avalible K_2O, mg 100 g^{-1}, **d** Cmic (mg 100 g^{-1}) and PPO (mg PPG g^{-1} 30 min^{-1}) activity; **e** Cmic (mg 100 g^{-1}) and microscopic fungi (CFU × 10^6 g^{-1}; **f**) Cmic (mg 100 g^{-1}) and URE (mg NH_4-N 100 g^{-1} h^{-1})

SOC, stimulated by the greater amount of plant residues by cover crop and minimum tillage. Their activity was dependent on a season and it was higher in autumn.

It was established that β-glucosidase activity is rising at the beginning of decomposition, when the amount of labile carbon in the residues is the highest, and as labile fractions decrease, β-glucosidase activity decreases. The trend is similar for urease, which activity is higher in spring (Nevins et al. 2021). The same tendency for β-glucosidase was reported in wheat as it rose in the spring months compared to the autumn months, but in maize it was higher at the end of the growing season. In current experiment, urease activity was lower in spring and increased in autumn (opposite to cited research), but protease activity was higher in spring and decreased in autumn which is probably related to added biomass of plant residues. The higher activity of PO, PPO, URE and β GLU and the increase of labile fractions in autumn are probably related to more intense mineralization in autumn months, compared to spring. Conversely, PROT and PHOSPH are activated in the spring months.

5 Conclusion

The application of minimum tillage and inclusion of a cover crop in the wheat–maize crop rotation on sloping terrains on Calcic Chernozem soil leaded to an increase in the activity of enzymes related to the carbon cycle such as peroxidase, polyphenol oxidase, which is related to higher organic matter. A positive influence was also reported on alkaline phosphatase and β-glucosidase. Protease and urease activity were positively affected in certain phases of this conservation rotation. Applied erosion control practices have a positive impact on soil organic matter, labile organic fractions and enzyme activity. This positive effect is stronger in maize, which is the crop after the cover crop, and weaker in wheat under minimum tillage, following maize in the crop rotation.

Acknowledgements The authors acknowledge funding of research activities received from the National Science Fund under grant agreement № КП-06-Н76/2 2023 (project "Integration of satellite derived and ground-based data for soil water balance components and crop cover into models for assessment of agroecological risks and agricultural practices for their mitigation")

References

Acosta-Martinez, V., Cano, A., Johnson, J.: Simultaneous determination of multiple soil enzyme activities for soil health-biogeochemical indices. Appl Soil Ecol **126**, 0929–1393 (2018). https://doi.org/10.1016/j.apsoil.2017.11.024

Alef, K., Nannipieri, P.: Methods in Applied Soil Microbiology and Biochemistry. Academic Press, London (1995). Paperback. ISBN 9780125138406

Anderson, T., Domsch, K.: A physiological method for the quantative measurement of microbial biomass in soil. Soil Biol. Biochem. **10**, 215–221 (1978)

Burns, R.G., DeForest, J.L., Marxsen, J., Sinsabaugh, R.L., Stromberger, M.E., Wallenstein, M.D., Weintraub, M.N., Zoppini, A.: Soil enzymes in a changing environment: current knowledge and future directions. Soil Biol. Biochem. **58**, 216–234 (2013). https://doi.org/10.1016/j.soilbio.2012.11.009

Chahal, I., Van Eerd, L.L.: Do cover crops increase subsequent crop yield in temperate climates? A Meta-Analy, sis. Sustainability. **15**(8), 6517 (2023). https://doi.org/10.3390/su15086517

Culman, S.W., Snapp, S.S., Freeman, M.A., Schipanski, M.E.: Permanganate oxidizable carbon reflects a processed soil fraction that is sensitive to management. Soil Sci. Soc. Am. J. **6**, 494–504 (2012). https://doi.org/10.2136/sssaj2011.0286

Dimitrov P.D., Beloev, H., Trifonova, T., Ruseva, S., Stoyanov, K., Ilieva, D., Kuncheva, G.: Advanced soil protection technologies for minimum and unconventional soil treatment in the production of maize on sloping terrains. Nikola Poushkarov" ISSAPP—Sofia, *Publishing Center of RU "A. Kunchev "*, Rousse, 62 p (2016)

Fageria, N.K., Baligar, V.C., Bailey, B.A.: Role of cover crops in improving soil and row crop productivity. Commun. Soil Sci. Plant Anal. **36**, 2733–2757 (2005)

Gabriel, J.L., Quemada, M.: Replacing bare fallow with cover crops in a maize cropping system: yield, N uptake and fertiliser fate. Eur J Agonomy **34**, 133–143 (2011). https://doi.org/10.1016/j.eja.2010.11.006

Gabriel, J.L., Garrido, A., Quemada, M.: Cover crops effect on farm benefits and nitrate leaching: Linking economic and environmental analysis. Agric. Syst. **121**, 23–32 (2013). https://doi.org/10.1016/j.agsy.2013.06.004

Galstian, ASh.: Enzymatic activity of soils. Geoderma **12**, 43–48 (1974). https://doi.org/10.1016/0016-7061(74)90037-8

García-González, I., Hontoria, C., Gabriel, J.L., Alonso-Ayuso, M., Quemada, M.: Cover crops to mitigate soil degradation and enhance soil functionality in irrigated land. Geoderma **322**, 81–88 (2018). https://doi.org/10.1016/j.geoderma.2018.02.024

Hasiev, F.: Soil enzyme activity, p. 180. Soviet Union Academy of Sciences, Science Publishing house, Moscow (2005)

Hoffman, G., Teicher, K.: Ein metrischesverfahrenzurBestimmung de urease activitat in boden. Z. Plan. Zenetnahr Bug Bodenkunde **95**, 55–63 (1961)

Kercheva, M., Rousseva, S., Dimitrov, E., Nenov, M., Shishkov, T.: Soil aggregation estimates in CZO- Fuchsenbigl. App. Geochem. **26**(Supplement), S57-59 (2011). https://doi.org/10.1016/j.apgeochem.2011.03.028

Kuncheva, G., Kercheva, M., Enchev, Ev., Paparkova, Ts., Kolchakov, V.: Factors controlling water-stable macroaggregation of Kastanozems exposed to water erosion. Bulgarian Journal of Agricultural Sciences 29(2), 262–271 (2023). ISSN 1310–0351

Kuncheva, G., Ginchev, G., Ivanova, I.: Influence of long-term mineral fertilization on soil microbiota, organic matter content and CO_2 emissions. In: Chenchouni, H., et al. Recent advancements from aquifers to skies in hydrogeology, geoecology, and atmospheric sciences. MedGU 2022. Advances in Science, Technology & Innovation. Springer, Cham (2024). https://doi.org/10.1007/978-3-031-47079-0_31

Moorhead, D.L.S., Sinsabaugh, R.L.: A theoretical model of litter decay and microbial interaction. Ecol. Monogr. **76**(2), 151–174 (2006). https://doi.org/10.1890/0012-9615(2006)076[0151:ATMOLD]2.0.CO;2

Nandan, R., Singh, V., Singh, S.S., Kumar, V., Hazra, K.K., Nath, C.P., Poonia, S., Malik, R.K., Bhattacharyya, R., McDonald, A.: Impact of conservation tillage in rice-based cropping systems on soil aggregation, carbon pools and nutrients. Geoderma **340**, 104–114 (2019)

Nevins, C., Lacey, C., Armstrong, S.: Cover crop enzyme activities and resultant soil ammonium concentrations under different tillage systems. Eur J Agnom **126**, 126277–301 (2021). https://doi.org/10.1016/j.eja.2021.126277

Poeplau, C., Don, A.: Carbon sequestration in agricultural soils via cultivation of cover crops—a meta-analysis. Agric. Ecosyst. Environ. **200**, 33–41 (2015)

Rankoth, L.M., Udawatta, R.P., Veum, K.S., Jose, S., Alagele, S.: Cover crop influence on soil enzymes and selected chemical parameters for a Claypan Corn-soybean rotation. Agriculture **9**, 125 (2019). https://doi.org/10.3390/agriculture9060125

Shen, X., Wang, L., Yang, Q., Xiu, W., Li, G., Zhao, J., Zhang, G.: Dynamics of soil organic carbon and labile carbon fractions in soil aggregates affected by different tillage managements. Sustainability **13**, 1541 (2021). https://doi.org/10.3390/su13031541

Sinsabaugh, R.L.: Phenol oxidase, peroxidase and organic matter dynamics of soil. Soil Biol. Biochem. **42**(3), 391–404 (2010). https://doi.org/10.1016/j.soilbio.2009.10.014

Sinsabaugh R.L., Hill B.H., Follstad Shah J.J.: Ecoenzymatic stoichiometry of microbial organic nutrient acquisition in soil and sediment. Nature 10, 462(7274), 795–8 (2009). https://doi.org/10.1038/nature08632

Tonitto, C., David, M.B., Drinkwater, L.E.: Replacing bare fallows with cover crops in fertilizer-intensive cropping systems: a meta-analysis of crop yield and N dynamics. Agric. Ecosyst. Environ. **112**(1), 0167–8809 (2006). https://doi.org/10.1016/j.agee.2005.07.003

Wang, Z., Ma, S., Hu, Y., Chen, Y., Jiang, H., Duan, B., Lu, X.: Links between chemical composition of soil organic matter and soil enzyme activity in alpine grassland ecosystems of the Tibetan Plateau. CATENA **218**, 0341–8162 (2022). https://doi.org/10.1016/j.catena.2022.106565

Wen, L., Peng, Y., Zhou, Y., Cai, G., Lin, Y., Li, B.: Effects of conservation tillage on soil enzyme activities of global cultivated land: a meta-analysis. J. Environ. Manag. **345**, 118904 (2023). https://doi.org/10.1016/j.jenvman.2023.118904

Agriculture Residue as an Alternative Source of Energy Production in Togo's Coastal Region

Mawunyo Simon Pierre Kitegi, Kperkouma Walab, Madjouma Kandab, and Folega Fousseni

Abstract

The present investigation aimed to assess the viability of harnessing energy from agricultural residues across various prefectures and crops in Togo's maritime region. By employing a novel methodology termed Tolessa, data was collected from Inseed in Togo to evaluate the energy potential of five prominent crops—maize, peanuts, rice, sorghum, and soybeans. The analysis yielded several noteworthy findings. To begin with, Zio emerged as the leading contributor to the total energy potential, with 1,634,286.3 Megajoules (MJ) of energy potential generated primarily from maize and rice residues. Yoto followed closely, contributing 871,239.9 MJ of energy potential, primarily from maize residues. In total, the estimated potential energy from all prefectures was calculated to be 4,673,064.2 MJ, highlighting the substantial potential of agricultural residues as an alternative energy source in the Togo's coastal region. The results of this study demonstrate that agricultural residues can serve as a valuable alternative energy source in the region, providing a substantial amount of energy potential. This finding has significant implications for the promotion of sustainable energy development in Togo and similar regions, as it offers an alternative to fossil fuels and helps mitigate the environmental impact of energy production.

Keywords

Agriculture residue · Bioenergy · Biomass potential · Energy potential · SDG7

M. S. P. Kitegi (✉)
CERVIDA-DOUNEDON, University of Lomé, Lomé, Togo
e-mail: simonpierrekitegi@gmail.com

K. Walab · M. Kandab · F. Fousseni
Department of Botanic, Faculty of Sciences, University of Lomé, Lomé, Togo
e-mail: kperkouma.wala@gmail.com

M. Kandab
e-mail: kmadjouma@gmail.com

F. Fousseni
e-mail: ffolegamez@live.fr

1 Introduction

The availability of energy is the primary factor influencing the level of development in a society. The energy sector is central to the functioning of all other sectors, including healthcare, agriculture, commerce, and more. Energy is essential to these areas, which is why one of the United Nations' development goals (Goal 7) prioritizes access to energy. In many developing countries, access to energy is a pressing concern, hindering socio-economic development and quality of life. The lack of access to reliable and affordable energy sources can have far-reaching consequences, including limited economic opportunities, inadequate lighting and heating, and decreased access to modern technologies. This energy poverty can perpetuate cycles of poverty, limiting the ability of individuals and communities to break free from economic hardship. Moreover, the lack of access to energy can exacerbate environmental challenges, such as deforestation and air pollution, which can have severe health and environmental consequences. Therefore, addressing energy access in developing countries is crucial for promoting sustainable development and improving the quality of life for millions of people. In the maritime region, there has been a significant increase in electricity consumption, rising from approximately 20% in 1994 to over 88% in 2022. This upward trend is more pronounced in the urban area, where the access rate exceeds 90%, compared to the total region with an access rate of 65%, as reported by (Boubakar et al. 2022). This observation underscores the challenge of ensuring access to

P. Mohammad et al. (eds.), *Geographic Approaches to Climate Change and Mitigation: Urban and Rural Perspectives (Volume 1)*, Advances in Science, Technology & Innovation, https://doi.org/10.1007/978-3-031-92119-3_29

electricity in rural areas within the region. The lack of electricity in these areas poses a formidable obstacle to development across various sectors, including agriculture, industry, and social services. Recent peri-urbanization and the expansion of Lome, the capital city of the country (Guézéré 2013), have further exacerbated the energy demand in the region. The population of the region has been growing at an annual rate of 2.84%, as estimated by (Peyou 2021) which suggests that the population could potentially double by 2050. This projected increase in population will further amplify the energy demands in the region, highlighting the need for sustainable and efficient energy solutions. The increase in electricity consumption in the maritime region is a pressing concern, particularly in rural areas where access to electricity is limited. The region's rapid population growth and urbanization only serve to underscore the importance of addressing this challenge to ensure sustainable development and improved quality of life for its inhabitants. In 2020 in Togo, the energy consumption is divided into three sectors. The largest share, 76%, is in the use of biomass, followed by petroleum products 20%. Only 4% of energy is used in the form of electricity (Togo Energy Situation—Energypedia, n.d.). In Togo, biomass is used for cooking and heating purposes. However, burning biomass has several adverse effects, including impacts on the climate, air quality, and ecosystems. For instance, the radiative impact of biomass-burning aerosols can significantly affect the climate and ecosystem of the region (Calì Quaglia et al. 2022). Biomass burning can also produce black carbon, which affects the distribution of air quality and the transport of smoke over long distances, as observed during the Western US wildfire season in 2020 (Ceamanos et al. 2023). Moreover, biomass burning can influence the particle number size distribution, new particle formation, and cloud condensation nuclei concentration, all of which can impact regional and global climate (Shang et al. 2018). In addition, biomass-burning emissions in Africa have been shown to result in regional to transcontinental air pollution, with significant uncertainties in fire emission inventories, leading to challenges in estimating the spatial and temporal patterns of emissions (Heil et al. 2020). These consequences highlight the importance of understanding and addressing the adverse effects of biomass burning on the environment and human health. In the realm of biomass, our focus lies on the untapped potential of crop residues, which have traditionally been burned in the region (Agbossou et al. 2022). However, it is crucial to assess the energy potential of these biomass sources that were previously burned, as it has not been extensively explored in the existing literature. This research aims to bridge this knowledge gap and determine the feasibility of harnessing energy from these agricultural residues in the coastal areas of Togo. By doing so, we hope to contribute to the sustainable management of biomass resources and reduce the negative environmental impacts associated with burning these materials.

2 Materials and Methods

2.1 Study Area

Located at the southernmost tip of Togo, between 6°10° and 6°55° latitude north and between 0°40° and 1°45° longitude east, the Togo's coastal region call the Maritime region borders the Plateaux region to the north, the Atlantic Ocean to the south, the Republic of Benin to the east, where the Mono River serves as a natural border, and Ghana to the west. Covering an area of approximately 6,100 square kilometers, which accounts for around 11% of the country's total surface area. The climate in the area is defined by an unusual climatic abnormality specific to southern Togo, with two separate periods of rainfall: a lengthy one from March to

Fig. 1 Ombrothermic diagram for Maritime region: Monthly average rainfall and temperatures data from 1977 to 2017 (Agbewornu 2018)

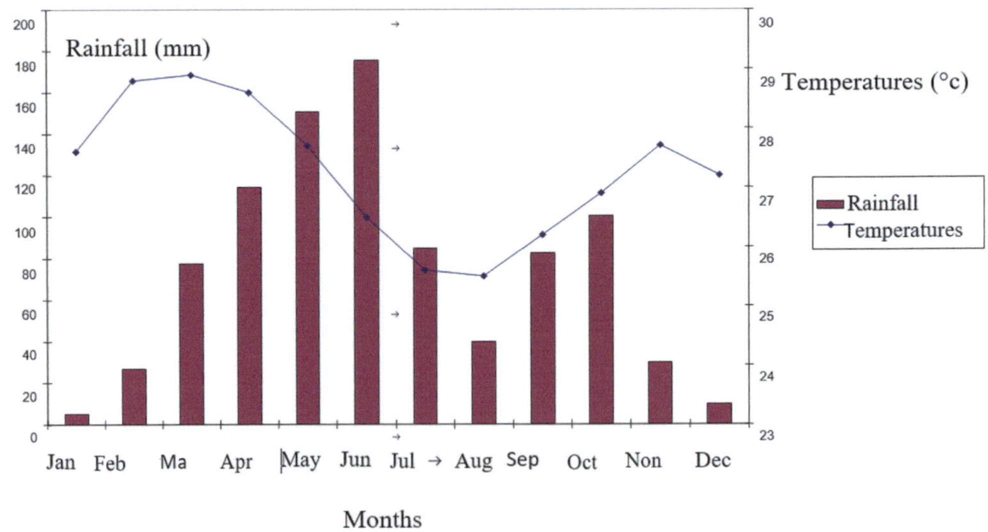

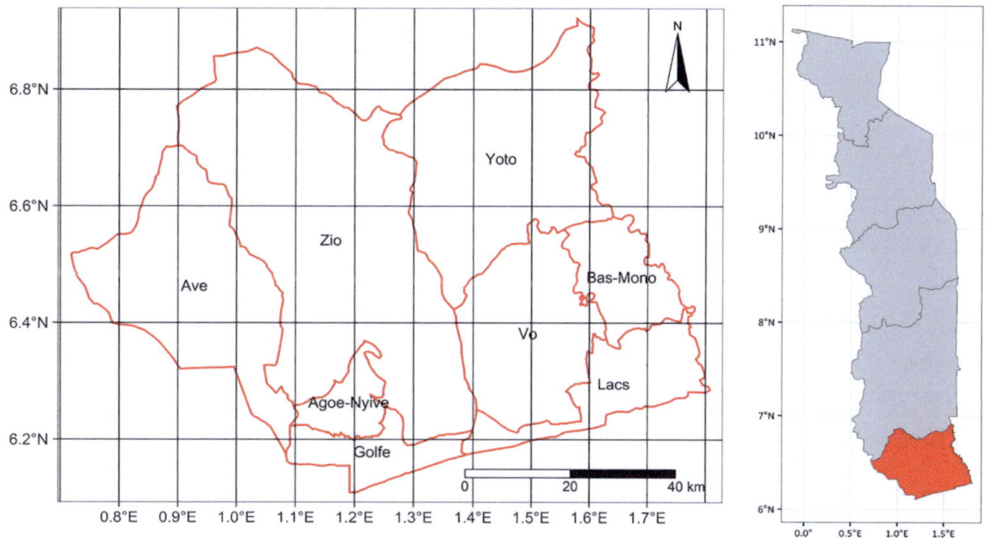

Fig. 2 Map of Togo's coastal region (Kitegi et al. 2024)

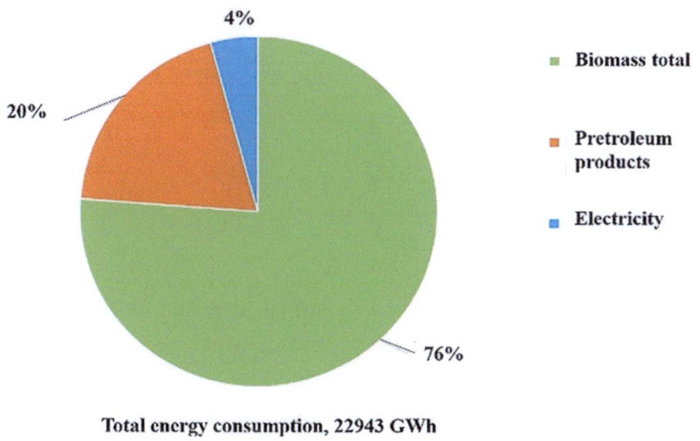

Fig. 3 Energy consumption sources in 2020 in Togo (File n.d.)

July, and a briefer one from September to December. (Agbewornu, 2018) (Figs. 1, 2 and 3).

2.2 Data

The crops statistics data (Tables 1 and 3) has been collected from the National Institute of Statistics and Economic and Demographic Studies (INSEED). It highlights the available data on crops cultivated in the region. To attempt to reach the reality on the estimation for the evaluation of the available agriculture residue biomass it is important to take in account the maximum crop data.

As the objective is to stop the release of greenhouse gases in the atmosphere through the combustion or burn of crop residues, the crop residues selected for this present study are the one who use to be used as source of energy and burnt in the Togo coastal area. They are maize (Cob, Husk and Stalk), sorghum (Straw and Stalk), rice (Straw, Husk), soy beans (Straw, pobs), peanut (shells), and yam(straw) as depicted in Table 2.

Table 1 Area cultivated per crop in Togo coastal region in 2021 (in hectares)

	Beans	Cassava	Maize	Paddy rice	Peanut	Sorghum	Soya	Sweet potato	Taro	Yam
Ave	1275	10,207	20,546	1535	1894	84	638	30	618	1461
Bas-Mono	2517	7555	16,901	1282	131	0	0	0	0	0
Lacs	1553	12,933	11,741	0	0	0	0	0	0	0
Vo	13,504	32,057	22,063	0	0	1604	0	0	0	0
Yoto	2472	35,039	46,699	963	1162	1273	338	0	615	1438
Zio	21,494	31,942	49,467	5067	904	1190	175	1190	69	1925

Source INSEED (2023)

Table 2 Crop residue. *Source* Authors

Crop	Burnt/source of energy	Feeding
Maize	Cob, Husk and Stalk	
Sorghum	Straw and Stalk	
Rice	Straw, Husk	
Soy Beans	Straw, Pobs	
Peanut	Shells	Straw
Beans	Husk	Leaves
Yam	Straw	Peeling
Cassava		Peeling, Leaves
Sweet Potato		Leaves, Peeling
Taro		Leaves, Peeling

Table 3 Production per crop in Togo coastal region in 2021 (production in Tons). *Source* INSEED (2023)

	Beans	Cassava	Maize	Paddy rice	Peanut	Sorghum	Soya	Sweet potato	Taro	Yam
Ave	919	38,542	15,056	2916	2708	57	468	77	1961	5877
Bas-Mono	1991	21,987	12,694	2370	129	0	0	0	0	0
Lacs	1385	37,873	8703	0	0	0	0	0	0	0
Vo	12,749	94,205	16,643	0	0	1132	0	0	0	0
Yoto	2456	108,937	40,766	1721	1115	929	275	0	1630	5964
Zio	19,700	104,971	48,305	10,150	1184	925	213	4379	179	8032

2.3 Methodology to Estimate the Energy Potential from Agriculture Biomass Residue

A thorough examination of the literature on energy potential from agriculture residue, as outlined by (Tolessa 2023) revealed inconsistencies in methodologies employed by various authors worldwide. The same result has been found by (Osei 2021) study in Ghana with variations in methodologies and coefficients used for bioenergy potential estimation in the country. All these results underscoring the need for a standardized approach. In response to these discrepancies, (Tolessa 2023) proposed a standardized methodology for estimating the bioenergy potential of agricultural biomass residues and used it for the study in Ethiopia. This methodology aims to establish a cohesive approach and mitigate fluctuations in the values employed for bioenergy potential estimation, particularly within a shared geographical context. The current study adopts this suggested methodology to assess the bioenergy potential of agricultural biomass residues in Ethiopia, providing a comprehensive analysis of the energy potential inherent in this valuable resource. The adoption of this standardized methodology in the current study enables a more comprehensive and accurate assessment of the bioenergy potential of agricultural biomass residues. By employing a consistent methodology, the study provides a more reliable estimate of the energy potential inherent in this valuable resource, which can inform decision-making and policy development in the agricultural sector. Furthermore, the use of a standardized methodology can facilitate comparisons across different regions and studies, enabling a more comprehensive understanding of the bioenergy potential of agricultural biomass residues globally.

The methodology is presented in three steps. It estimates the gross crop residue potential in the first step, the recoverable residue biomass potential in the second step, and the bioenergy potential in the final step.

As there are different values for each coefficient (RPR, RF, and LHV), we use the average value as used by Tolessa (Fig. 4).

2.3.1 Gross Crop Residue Potential

The gross crop residue potential of a specific crop is determined by three parameters: the area cultivated, crop produce, and the RPR (Tolessa 2023). It can be obtained by the Eq. (1).

$$GCRP = \sum_{i=1}^{t} Ca_{(i)} * Cy_{(i)} * RPR_{(i)} \qquad (1)$$

where GCRP is the potential of produced gross crop residue from “t” numbers of crops, tons (t); Cy(i) is the yield of the ith crop, ton per hectare (t.ha^{-1}); Ca(i) is crop area under the ith

Table 4 RPR, RF, and lower heating values (LHV) of crop residues. *Source* Tolessa (2023)

Crop	Residue	RPR	RF	LHV
Maize	Stalk	1.72	0.80	14.66
	Husk	0.21	1.00	14.08
	Cob	0.40	1.00	15.43
Sorghum	Straw	1.75	0.80	12.38
	Stalk	2.85	0.80	16.00
Rice	Straw	1.46	0.49	15.78
	Husk	0.27	0.81	17.44
Soybeans	Straw	2.32	0.70	15.16
	Pods	1.00	1.00	12.38
Peanut	Shells	0.40	0.93	17.56

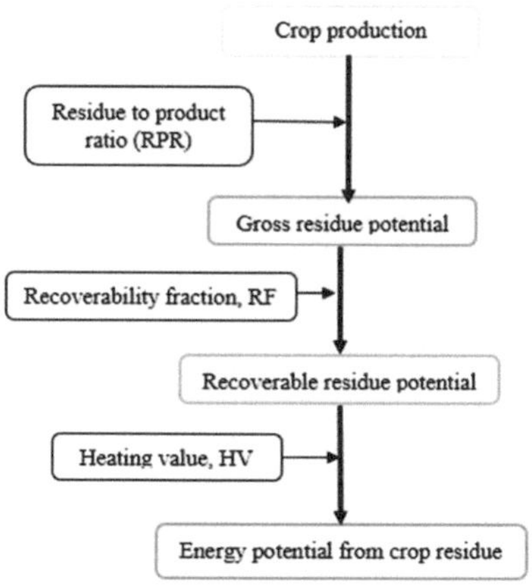

Fig. 4 Methodological flow diagram for determining the energy potential of crop residues (Tolessa 2023)

crop, hectare (ha); and RPR(i) is the residue-to-product ratio of the ith crop. The RPR used is presented in the Table 4.

2.3.2 Recoverable Residue Biomass Potential

In his study, Tolessa (2023), as not all the crop residue biomass would be available and suitable for the production of bioenergy. The author assesses the quantity of field-based residue that can practically be collected by considering the proportion of residues that can realistically be converted into bioenergy after some have been utilized for other purposes. The recoverable residue biomass potential (RCRP) can be estimated through the Eq. (2)

$$RCRP = \sum\nolimits_{i=1}^{t} GCRP_{(i)} * RF_{(i)} \quad (2)$$

where RCRP is the potential of generated recoverable crop residue from "t" number of crops, tons (t); GCRP(i) is the generated gross crop residue potential from the ith crop, tons (t); and RF(i) is recoverability factor of the ith crop. The RF used is presented in the Table 4

2.3.3 Bioenergy Potential

To estimate the bioenergy potential, (Tolessa 2023) use the following equation.

$$EP_{(j)} = \sum\nolimits_{i=1}^{t} RCRP_{(i,j)} * LHV_{(i,j)} \quad (3)$$

where EP(j) is the potential of bioenergy from t crops at the jth region, mega joule (MJ); RCRP(i,j) is the recoverable crop residue of the ith crop at the jth region, t; LHV(i,j) is lower heating value of ith crop at the jth region, MJ kg^{-1}. The lower heating values of the residues considered in this study were obtained from literature, as presented in Table 4

3 Results and Discussion

3.1 Results

3.1.1 Gross Residue Potential

The analysis of gross crop residue presented in the Table 5 depicts the gross residue potential across various prefectures and crops, revealing the quantity of residues in tons for maize, peanuts, rice, sorghum, and soybeans. Of particular note is the prefecture of Zio, which exhibits the highest total residue potential at 135,564.31 tons, primarily due to substantial amounts of maize and rice residues. Yoto follows closely, with maize residues contributing 94,984.78 tons to the total. The overall trend indicates that maize is the predominant contributor to the total residue potential across all prefectures, underscoring its pivotal role in regional agriculture. These information are crucial for devising effective agricultural waste management strategies and offers valuable insights into the regional distribution of residue production, facilitating informed decision-making by policymakers and environmentalists in the development of sustainable planning and resource allocation.

Table 5 Gross residue potential per prefectures and crop

Prefectures	Maize	Peanut	Rice	Sorghum	Soybeans	Total
Ave	35,080.48	1083.2	5044.68	262.2	1553.76	43,024.32
Bas-Mono	29,577.02	51.6	4100.1	0	0	33,728.72
Lacs	20,277.99	280.8	0	0	0	20,558.79
Vo	38,778.19	336.8	221.44	5207.2	305.44	44,849.07
Yoto	94,984.78	446	2977.33	4273.4	913	103,594.51
Zio	112,550.65	473.6	17,559.5	4273.4	707.16	135,564.31
Total	331,249.11	2672	29,903.05	14,016.2	3479.36	381,319.72

Table 6 Recoverable residue biomass potential per prefecture and crop

Prefectures	Maize	Peanut	Rice	Sorghum	Soybeans	Total
Ave	29,901.21	1007.37	2723.83	209.76	1228.03	35,070.21
Bas-Mono	25,210.28	47.98	2213.817	0	0	27,472.08
Lacs	17,284.15	261.14	0	0	0	17,545.30
Vo	33,052.99	313.22	119.56	4165.76	241.40	37,892.95
Yoto	80,961.27	414.78	1607.58	3418.72	721.6	87,123.96
Zio	95,933.73	440.44	9481.11	3418.72	558.91	109,832.92
Total	282,343.66	2484.96	16,145.91	11,212.96	2749.95	314,937.45

3.2 Recoverable Residue Biomass Potential

Table 6 exhibits the recoverable residue biomass potential per prefecture and crop, expressed in tons. It is evident that Zio emerges as the prefecture with the highest contribution to the total recoverable residue biomass, amounting to 109,832.925 tons, primarily derived from maize and rice residues. Yoto follows closely, accounting for a substantial portion of the overall total with 87,123.9621 tons, primarily sourced from maize residues. These findings underscore the prominent role of maize in the recoverable residue biomass across all prefectures. The total biomass potential is estimated to be 314,937.4525 tons, with maize again dominating, highlighting its pivotal position in sustainable biomass recovery initiatives. This data provides crucial insights for regional resource management strategies, offering valuable information on the recoverable biomass landscape and informing decision-makers in fostering environmentally conscious practices.

3.3 Energy Potential

Table 7 presents an in-depth analysis of the energy potential per prefecture and crop, quantified in units of measurement not specified. The results reveal a striking dominance of Zio, accounting for an impressive 1,634,286.325 Megajoules, primarily attributed to maize and rice residues. Yoto follows closely, contributing notably with 1,287,925.28 Mj, primarily derived from maize residues. Maize once

Table 7 Energy potential per prefecture and crop

Prefectures	Maize	Peanut	Rice	Sorghum	Soybeans	Total
Ave	441,155.25	17,689.52	44,040.75	3067.28	17,315.92	523,268.74
Bas-Mono	371,946.38	842.66	35,794.44	0	0	408,583.49
Lacs	255,006.25	4585.68	0	0	0	259,591.94
Vo	487,655.87	5500.21	1933.20	60,915.18	3403.98	559,408.46
Yoto	1,194,482.93	7283.53	25,992.50	49,991.34	10,174.95	1,287,925.28
Zio	1,415,382.87	7734.26	153,296.87	49,991.34	7880.96	1,634,286.32
Total	4,165,629.58	43,635.89	261,057.77	163,965.16	38,775.83	4,673,064.24

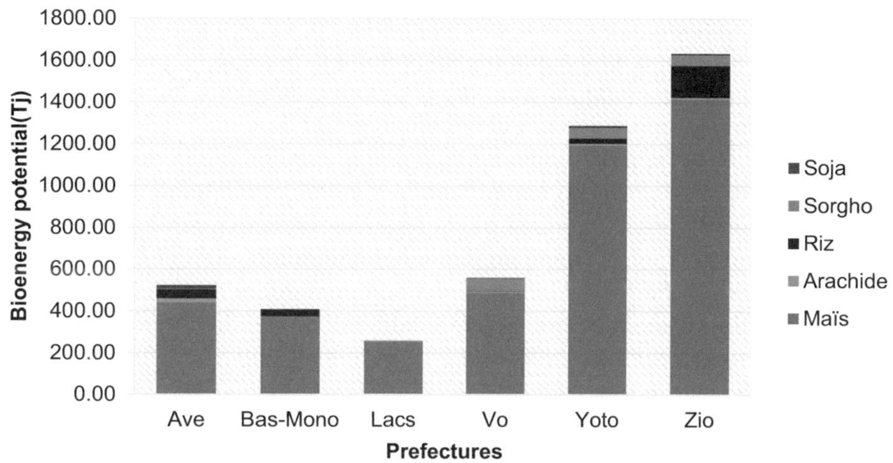

Fig. 5 Energy potential per prefecture

again emerges as the predominant source of energy potential across all prefectures, underscoring its pivotal role in shaping the regional energy landscape (Fig. 5).

The total energy potential is calculated at an impressive 4,673,064.249 Mj, highlighting the immense potential of maize as a key player in the energy potential of these prefectures. These findings are of utmost significance for energy planning and sustainable resource allocation, underscoring the crucial importance of maize in harnessing the substantial energy potential from agricultural residues in a manner that aligns with environmental and energy sustainability goals. Policymakers and energy planners can leverage these insights to develop strategies that optimize the utilization of maize-based agricultural residues for sustainable energy production.

Table 8 presents an estimation of energy potential per prefecture, categorized into three distinct types: Green Crop Residue Potential (GCRP), Recoverable Crop Residue Potential (RCRP), and Energy Potential (EP). This analysis highlights the significant contribution of both green and recoverable crop residues to the overall energy landscape, with Zio and Yoto emerging as the leading prefectures in terms of energy potential. According to the table, Zio accounts for the largest share of energy potential, amounting to 1,634,286.325 Mj, primarily derived from green crop residues. Yoto follows closely, contributing 1,287,925.28 Mj, primarily from recoverable crop residue potential. The cumulative total energy potential for all prefectures is calculated at 4,673,064.24 Mj, emphasizing the significance of both green and recoverable crop residue potentials in the energy landscape. This analysis underscores the importance of considering the diverse contributions of different residue types across various prefectures. Policymakers and energy planners can utilize this information to formulate strategies for sustainable energy generation, taking into account the unique energy potential of each prefecture. By leveraging the energy potential of green and recoverable crop residues, these strategies can help promote a more sustainable and environmentally friendly energy landscape.

Table 8 Energy potential estimation per prefecture

	GCRP	RCRP	EP
AVE	43,024.32	35,070.21	523,268.74
Bas-Mono	33,728.72	27,472.08	408,583.49
Lacs	20,558.79	17,545.30	259,591.94
Vo	44,849.07	37,892.95	559,408.46
Yoto	103,594.51	87,123.96	1,287,925.28
Zio	135,564.31	109,832.92	1,634,286.32
Total	381,319.72	314,937.45	4,673,064.24

4 Discussion

According to the study's findings, the maritime region in Togo has a significant energy potential, which has been calculated to be 4,673,064.24 Mj (or 1,298,073.40 MWh). However, with an annual consumption of 1,012,417 MWh, only approximately 65 percent of the population in this region has access to energy (INSEED 2023). Fortunately, the energy potential of agricultural residues is sufficient to cover the region's entire annual electricity consumption. This discovery emphasizes agricultural residues' enormous potential as a renewable energy source, particularly in areas with limited access to conventional energy sources. The same result was found in Ghana, a neighboring country to the study area, where the country has a total installed generation capacity of 4577 MW, which is four times less than the country's crop residue energy potential (19,781 MW) (Nelson et al. 2021).

These findings highlight the critical role of agricultural residues in a country's sustainable development strategy. Specifically, using crop residues as a source of bioenergy, such as coconut and palm tree waste and others, can significantly contribute to meeting a region's energy needs while also mitigating the negative impacts associated with burning or burning of these residues. By using crop residues as a renewable energy source, the region can not only reduce greenhouse gas emissions but also alleviate health concerns associated with crop residue burning. This emphasizes the critical need to investigate and implement sustainable practices that can contribute to a more environmentally friendly and socially responsible energy system.

It is critical to recognize that, due to a lack of relevant data, the literature coefficients used in this study may not accurately represent the specific context of the study area. Future research could benefit from looking into the RPR, EF, and LHV in our particular regional context. Furthermore, despite the potential benefits of using agricultural residues as a renewable energy source, due to a lack of comprehensive data, our study did not use all available residues. Extending research efforts or establishing a systematic database of agricultural residues in the region could help to address this limitation and contribute to the regions and beyond broader energy transition.

5 Conclusion

This study is one the first done in Togo and particularly in the maritime region to contribute to the valorization of agriculture residues. It provides a comprehensive assessment of the gross residue potential across prefectures and crops in the maritime region. This study evaluated the potential for harnessing energy from agricultural residues in Togo's maritime region, using a novel methodology called Tolessa. Data from Inseed assessed five major crops: maize, peanuts, rice, sorghum, and soybeans. Zio provided the highest energy potential at 1,634,286.3 MJ, mainly from maize and rice, while Yoto contributed 871,239.9 MJ, primarily from maize. Overall, the total energy potential from all prefectures was 4,673,064.2 MJ, highlighting the significant role of agricultural residues as an alternative energy source. This finding supports sustainable energy development in Togo and reduces reliance on fossil fuels.

6 Conflicts of Interest

The authors declare no conflict of interest.

Acknowledgements The authors are deeply grateful to the World Bank for funding this research through CERViDA-DOUNEDON (Regional Center of Sustainable Cities in Africa).

References

Agbewornu, K.: Global warming and water resources variability in the maritime region of Togo (West Africa). Environ. Nat. Res. Res. **8**, 49 (2018). https://doi.org/10.5539/enrr.v8n4p49

Agbossou, A., Fontodji, J.K., Ayassou, K., Tchegueni, S., Segla, K.N., Adjonou, K., Bokovi, Y., Ajayon, A.-L., Polo-Akpisso, A., Kuylenstierna, J.C.I., Malley, C.S., Michalopoulou, E., Slater, J.: Integrated climate change and air pollution mitigation assessment for Togo. Sci. Total Environ. **844**, 157107 (2022). https://doi.org/10.1016/j.scitotenv.2022.157107

Boubakar, Z., Williams, N., Diallo, B.A.A., Lare, Y.: Geo-temporal analysis of electricity consumption growth in Togo. IEEE PES/IAS PowerAfrica **2022**, 1–5 (2022)

Calì Quaglia, F., Meloni, D., Muscari, G., Di Iorio, T., Ciardini, V., Pace, G., Becagli, S., Di Bernardino, A., Cacciani, M., Hannigan, J.W., Ortega, I., Di Sarra, A.G.: On the radiative impact of biomass-burning aerosols in the arctic: the august 2017 case study. Remote Sens. **14**(2), 313 (2022). https://doi.org/10.3390/rs14020313

Ceamanos, X., Coopman, Q., George, M., Riedi, J., Parrington, M., Clerbaux, C.: Remote sensing and model analysis of biomass burning smoke transported across the Atlantic during the 2020 Western US wildfire season. Sci. Rep. **13**, 16014 (2023). https://doi.org/10.1038/s41598-023-39312-1

File:Figure 6 Togo's total energy consumption.png—Energypedia. (n.d.). Retrieved December 28, 2023, from https://energypedia.info/wiki/File:Figure_6_Togo%27s_total_energy_consumption.png

Guézéré, A.: Deux roues motorisées et étalement urbain à Lomé, quel lien avec la théorie des « trois âges » de la ville ? Norois **226**, 41–62 (2013). https://doi.org/10.4000/norois.4549

Heil, A., Bouarar, I., Brasseur, G.: Uncertainties in estimating biomass burning emissions for Africa: implications for atmospheric modelling (2020). https://doi.org/10.5194/egusphere-egu2020-9808

INSEED. (2023, February). *ANNUAIRE STATISTIQUE 2020–2021 DE LA REGION MARITIME*

Kitegi, M. S. P., Wala, K., Kanda, M., Fousseni, F. (2024). Understanding Farmers Utilization of Rice and Maize Residue in the Coastal Region of Togo: A Survey-Based Study

Nelson, N., Darkwa, J., Calautit, J.K., Worall, M., Mokaya, R., Adjei, E.A., Kemausuor, F., Ahiekpor, J.: Potential of bioenergy in rural Ghana. Sustainability **13**, 381 (2021). https://doi.org/10.3390/su13010381

Osei, I. (2021). Crop residues utilisation for renewable energy generation in Ghana: review of feedstocks assessment approach, conversion technologies and challenges, pp. 29–42

Peyou, S. A. N. (2021). REPUBLIQUE DU TOGO. *African Disability Rights Yearbook*, *8*. https://doi.org/10.29053/2413-7138/2020/v8a10

Shang, D., Hu, M., Zheng, J., Qin, Y., Du, Z., Li, M., Fang, J., Peng, J., Wu, Y., Lu, S., Guo, S. (2018). Particle number size distribution and new particle formation under the influence of biomass burning at a high altitude background site at Mt. Yulong (3410 m), China. Atmos. Chem. Phys. 18(21), 15687–15703. https://doi.org/10.5194/acp-18-15687-2018

Togo Energy Situation—Energypedia. (n.d.). Retrieved December 28, 2023, from https://energypedia.info/wiki/Togo_Energy_Situation#/media/File:Figure_6_Togo's_total_energy_consumption.png

Tolessa, A.: Bioenergy potential from crop residue biomass resources in Ethiopia. Heliyon **9**, 1–13 (2023). https://doi.org/10.1016/j.heliyon.2023.e13572

Climate Change and Satellite Data for Analysis of Impacts on Rural Areas

Teresa Amodio and Daniel Signorelli

Abstract

This study aims to analyze the territory of a specific municipality using Corine Land Cover data and Sentinel-2 satellite images. Corine Land Cover data allowed for a comprehensive classification of the area into categories such as urban, agricultural, and forested zones, helping to understand the spatial distribution of land use. Municipal boundaries provided by ISTAT were also used for the analysis. The surface areas of each land use category were calculated using QGIS 3.38. Additionally, Sentinel-2 satellite images were employed to calculate the Normalized Difference Vegetation Index (NDVI), which helps monitor changes in vegetation cover from 2019 to 2024. NDVI values were processed through Google Earth Engine, with data filtered for May and June each year. This analysis provided insights into vegetation health and dynamics over time. The NDVI data was then reclassified into five classes in a GIS environment, facilitating comparative analysis. Raster images were converted into polygons to quantify the areas corresponding to each NDVI class, allowing precise measurement of vegetation cover changes. The study highlights how satellite imagery and GIS tools are essential for analyzing ecological trends, particularly in tracking vegetation health and potential environmental degradation. The study highlighted a significant reduction in dense vegetation cover (NDVI class 5), decreasing from 35 to 2% between 2019 and 2022. At the same time, lower and moderate vegetation classes increased, suggesting potential ecological degradation. The results indicate that environmental events or climate changes may have impacted vegetation in the study area.

Keywords

Climate change · Remote sensing · NDVI · Change detection · Ecological degradation

1 Introduction

Climate change is a phenomenon that has intensified significantly in recent years due to growing anthropogenic pressure. Human activities have led to a considerable increase in greenhouse gases in the atmosphere, accompanied by intensive exploitation of natural resources and widespread pollution that compromises numerous ecosystems. These combined factors have destabilized natural balances and accelerated the processes of global climate change (Puppim de Oliveira and Bhuiyan 2024; Wang et al. 2024). Scientific literature offers a broad range of studies showing that lifestyles, dietary habits, and consumption patterns have already changed, or are likely to change, worldwide (Arregocés et al. 2024; Cui et al. 2024; Galveias et al. 2024) and a particular attention has been given to assessing the effects of climate change on agriculture and the development of rural areas (De Nardi et al. 2024; Luo and Dou 2024). Remote sensing technologies today represent a valuable tool for monitoring and understanding these phenomena (Yuan et al. 2020; Zhao et al. 2023). Among the available techniques, change detection analysis stands out for its ability to measure changes over time through the use of satellite images (Wang et al. 2022; Zhu et al. 2022). The combination of different satellite sensors and the use of various indices, supported by advanced analytical tools, contribute to improving the accuracy of results (Hu and Dong 2018; Roy et al. 2019). This work aims to illustrate a practical

T. Amodio · D. Signorelli (✉)
University of Salerno—Via Giovanni Paolo II, 132-84084 Fisciano, SA, Italy
e-mail: dsignorelli@unisa.it

T. Amodio
e-mail: tamodio@unisa.it

P. Mohammad et al. (eds.), *Geographic Approaches to Climate Change and Mitigation: Urban and Rural Perspectives (Volume 1)*, Advances in Science, Technology & Innovation, https://doi.org/10.1007/978-3-031-92119-3_30

case of using satellite data to analyze the impact of climate change on rural areas, focusing on change detection analysis of one of the most widely used and established indices for estimating the presence of photosynthetic activity (NDVI) and its variations over time (Arrogante-Funes et al. 2024; Gandhi et al. 2015; Yao et al. 2024).

2 Materials and Methods

The work involved a detailed analysis of the territory, based on Corine Land Cover data from the 2018 edition (https://www.isprambiente.gov.it/it/attivita/suolo-e-territorio/suolo/copertura-del-suolo/corine-land-cover) and the use of Sentinel-2 satellite images. The work was conducted in two parts: the first part focused on determining the study area, and the second on processing satellite images and interpreting land use changes from an agricultural perspective.

2.1 Selection of the Study Area

The land cover classes were grouped into main categories following the standardized Corine Land Cover legend, allowing for more efficient data interpretation and comparison. This legend divides the territory into specific classes: Class 1 identifies urban areas, including residential zones, infrastructure, and built-up areas; Class 2 represents agricultural areas, encompassing cultivated lands and spaces dedicated to food production; and Class 3 is designated for forested areas, which include forests and natural green spaces. This categorization facilitates comparative analysis and provides a clear understanding of the distribution of different land cover types. In addition to land cover data, municipal boundaries provided by ISTAT in shapefile format were used to ensure precise delineation of the study area. To identify the working area, the analysis focused on the metropolitan area of Naples to determine which municipality had the highest percentage of agricultural cover. Using map algebra methodologies and calculations performed in a GIS environment (QGIS 3.38), the classification was completed, and the specific area extents for each category were calculated, with particular emphasis on agricultural zones measured in hectares and as a percentage of municipal area (Table 1).

This analysis led to the selection of the municipality with the highest percentage of agricultural areas relative to the municipal area, namely the municipality of Acerra (Fig. 1).

Table 1 Agricultural areas for different municipalities expressed in hectares and percentage

Municipality	Area (ha)	%
Acerra	4516,42	82,9%
Afragola	1012,1	56,7%
Agerola	533,91	27,0%
Anacapri	281,65	43,7%
...	...	...

2.2 Processing of Satellite Images

Once the work area was selected, an algorithm was implemented on the Google Earth Engine (GEE) platform to allow for the download of Sentinel-2 satellite images for the chosen years, and to provide a raster calculator for calculating the vegetation index. The satellite images were used to compute the Normalized Difference Vegetation Index (NDVI) for each year in order to monitor changes in vegetation cover over time. Sentinel-2 images, with a spatial resolution of 10 m and a five-day revisit frequency, provide optimal bands for calculating NDVI, using the near-infrared band (B8) and the red band (B4). The obtained data were selected by filtering the satellite images captured in May and June of the years 2019, 2020, 2021, 2022, 2023, and 2024, allowing for the monitoring of vegetation changes over time and assessing the effects of environmental changes in the analyzed area. For the creation of the NDVI index, the GEE script included a selection filter for Sentinel-2 images from the mentioned period, with cloud coverage below 20%. The formula used to calculate the NDVI is as follows:

$$\text{NDVI} = ((\text{B8} - \text{B4})) / ((\text{B8} + \text{B4})). \quad (1)$$

This formula leverages the higher reflectance of healthy vegetation in the near-infrared band (B8) compared to the red band (B4), generating positive values that indicate the presence of active vegetation. After calculating the NDVI for each year, the resulting images were downloaded as raster files and imported into a GIS environment for further analysis.

2.3 Analysis of NDVI

Using ArcGIS PRO software, a reclassification function was applied to categorize the NDVI values into five distinct classes to facilitate the interpretation and comparison of the data. The classes were defined as follows:

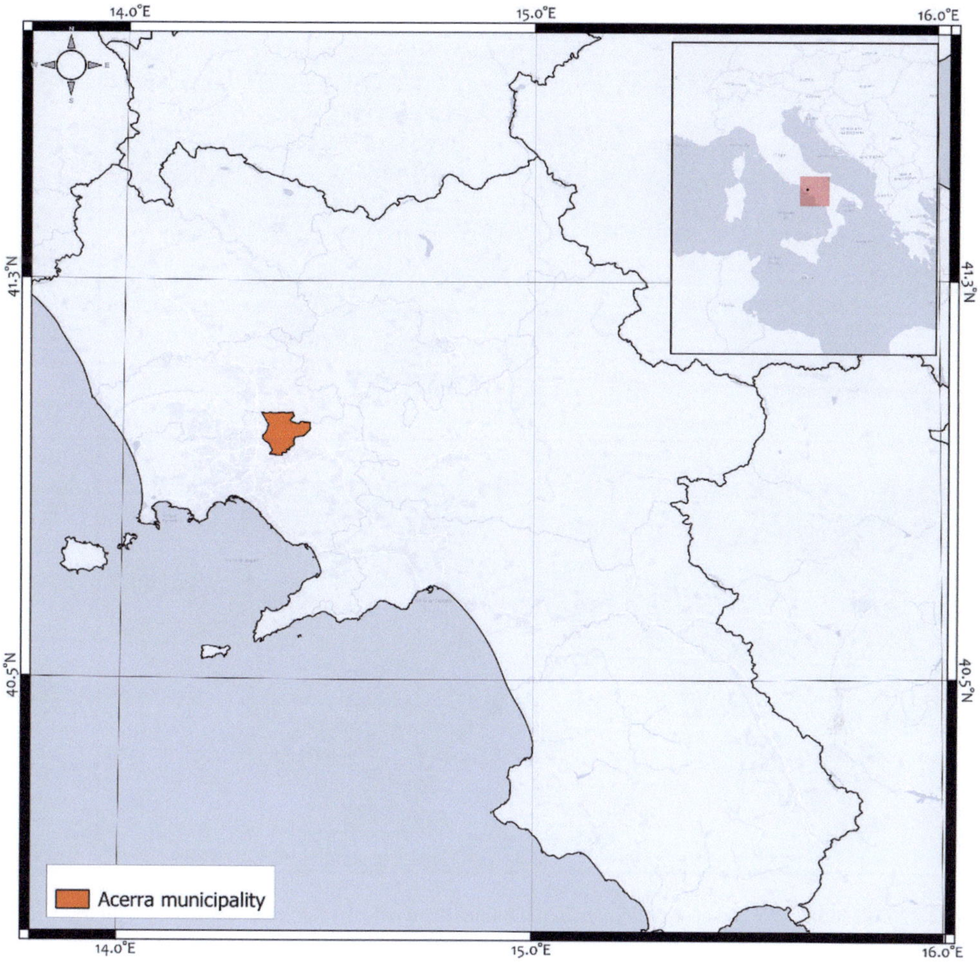

Fig. 1 Map of the municipality of Acerra located in the Campania Region. *Source* By authors

- Class 1: NDVI values below 0 (negative), indicating areas devoid of vegetation or with non-vegetated reflectance.
- Class 2: NDVI values between 0 and 0.1, associated with very sparse or degraded vegetation cover.
- Class 3: NDVI values between 0.1 and 0.25, representing sparse vegetation or early stages of growth.
- Class 4: NDVI values between 0.25 and 0.5, indicative of moderate or healthy vegetation.
- Class 5: NDVI values above 0.5, corresponding to dense and lush vegetation.

This classification allowed for highlighting changes in vegetation cover and vegetation health over the years, facilitating the study of ecological dynamics and the identification of trends and significant variations within the study area. After this step, classified rasters were obtained that represented the different NDVI classes for each year. Subsequently, a raster-to-polygon conversion function was applied, transforming the classified areas into vector geometries. This conversion facilitated the quantification of areas belonging to each class, allowing for precise measurement of the corresponding surfaces and enabling more detailed spatial analyses. The approach allowed for evaluating the extent and distribution of the different vegetation categories, providing a clear picture of vegetation cover changes for each analyzed year.

3 Results

NDVI maps were obtained for each analyzed year, classifying the index into five classes based on the measurement in order to have measurable geometries (Fig. 2).

Figure 2 presents NDVI maps for each year, depicting the vegetation trends in the municipality area of Acerra. The colors identify different NDVI classes, with darker green tones for higher classes (4 and 5), which represent healthier and denser vegetation, while lighter green tones shifting to yellow indicate sparse or absent vegetation, simply dry vegetation. Observing these maps reveals a clear decline in areas with lush vegetation (class 5), which progressively

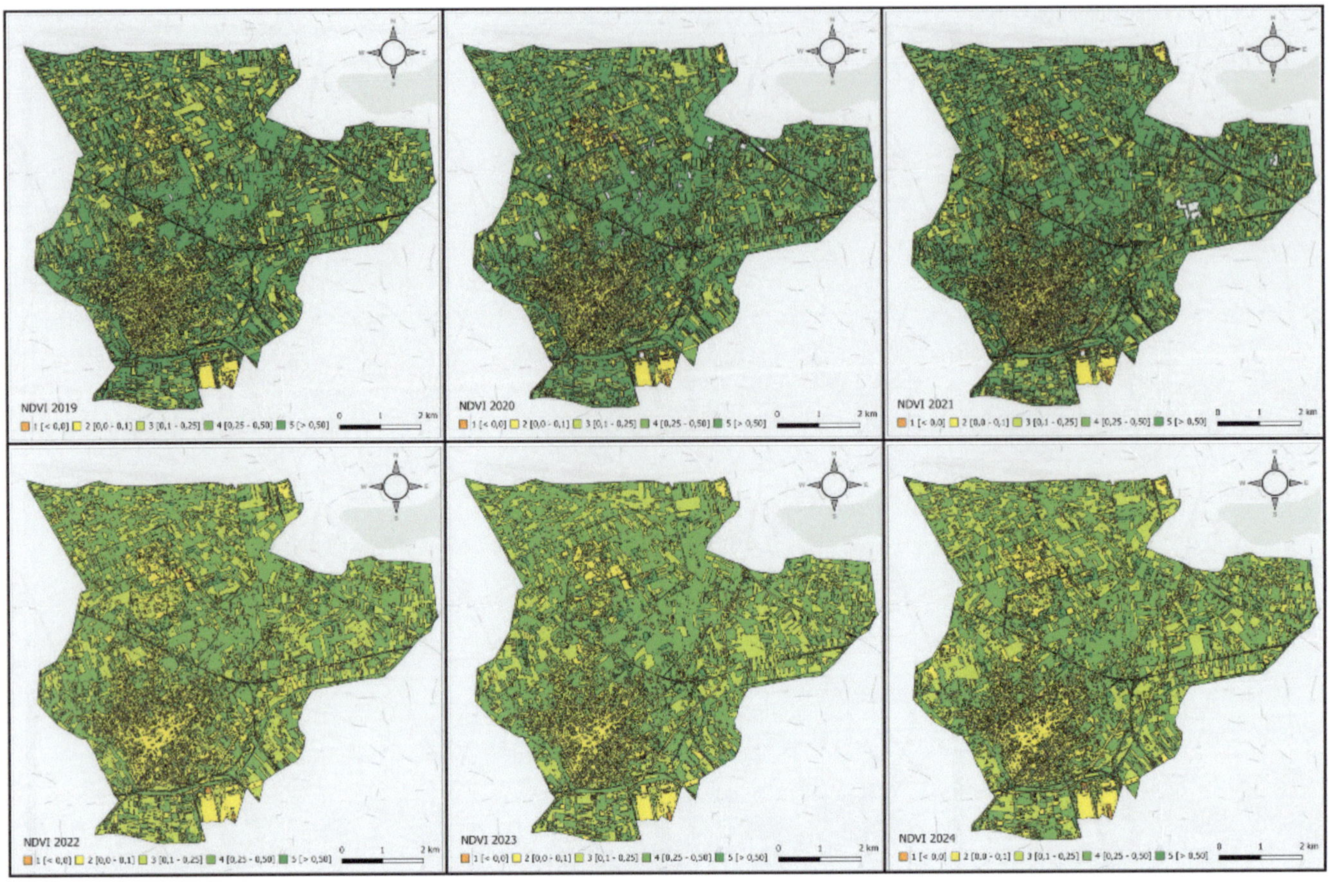

Fig. 2 NDVI indices for the analyzed years in the municipal territory of Acerra. *Source* by authors

decreases in favor of the lower classes. The spatial distribution highlights how these greener areas, initially widespread in 2019, significantly reduce over the years. This visual progression supports the hypothesis of an environmental or climatic change negatively impacting vegetation health in this region. The representation of each year facilitates comparison and allows for identifying specific areas where the most significant changes occur, also highlighting the expansion of lower NDVI classes, which may indicate more degraded or less vigorous vegetation. The analyzed areas have shown a gradual decline in classes 4 and 5, which represent, respectively, moderate vegetation (NDVI between 0.25 and 0.50) and dense and lush vegetation (NDVI > 0.50). This decline suggests a deterioration of vegetation cover and may indicate environmental degradation or changes in ecological conditions, such as drought, intensive land use, or other anthropogenic factors. The area was better quantified using bar charts to highlight the phenomenon (Fig. 3).

The analysis allowed for the precise quantification of the reduction in areas belonging to these classes, highlighting that class 5 (lush vegetation with NDVI > 0.5) decreased from around 35% in 2019, 2020, and 2022 to values around 1–2% starting from 2022. Additionally, it is evident that the other vegetation classes have increased progressively over the years. By visualizing the data according to the identified NDVI classes (Fig. 4), it is possible to better observe what has occurred over the years.

The graph shows that, over the years, classes 2 and 3 (NDVI between 0.0 and 0.1, and between 0.1 and 0.25) appear to have increased by nearly double in percentage. Class 4 (NDVI between 0.25 and 0.5) seems to have increased over the years from a minimum of 35.5% of the municipal area in 2019 to a maximum of 49.9% in 2023. Class 5 (with NDVI greater than 0.5) seems to have decreased significantly, dropping from 34.2% in 2021 to 1.2% in 2022. However, class 5 shows a slight increase in recent years, rising from 1.2% to 2.3%. This overall trend, showing a decline in higher NDVI classes and an expansion of lower ones, suggests an ecological change in the Acerra area, potentially related to both climatic and anthropogenic factors. From a climatic perspective, factors such as decreased rainfall and increased temperatures may have contributed to stressing the vegetation, reducing its ability to maintain dense and healthy cover levels. On the other hand, anthropogenic pressures, such as urban expansion and intensified agricultural activities, may have further degraded

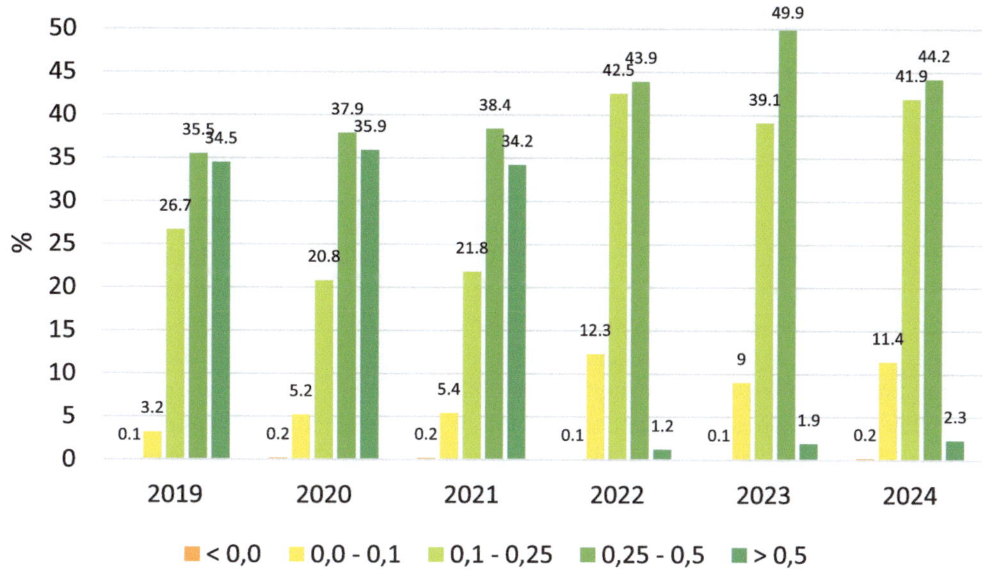

Fig. 3 Graph of the frequency distribution of NDVI classes for each individual year. *Source* by authors

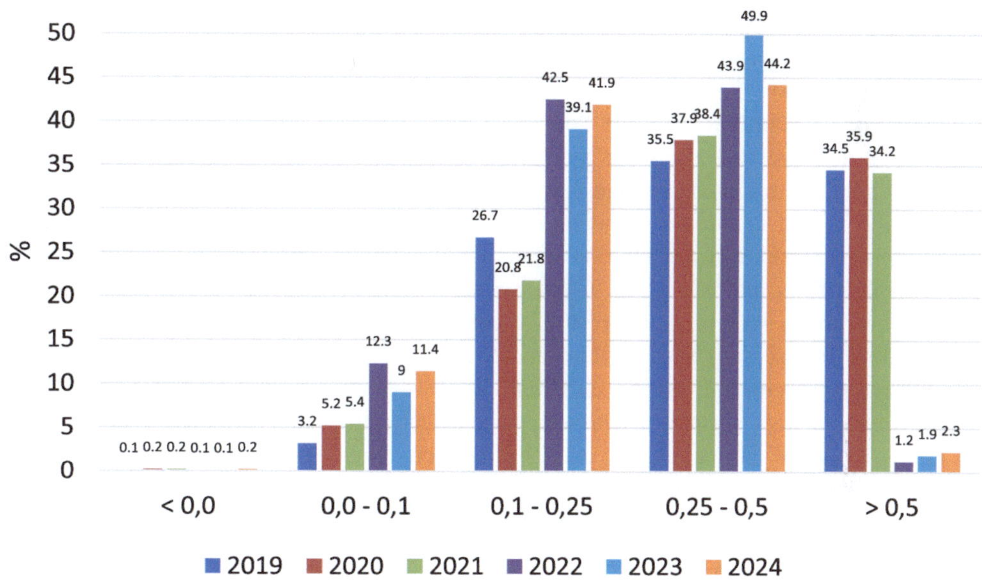

Fig. 4 Graph of the trends of the NDVI classes compared by year. *Source* by authors

the area, promoting the expansion of areas with sparser vegetation cover. These dynamics indicate a possible deterioration of the ecological health of the Acerra area, raising concerns about the area's environmental sustainability. The progressive reduction of higher NDVI classes in favor of lower ones serves as a warning signal that may reflect a degradation of the natural environment, with potential long-term implications for local biodiversity and ecosystem quality. In conclusion, the NDVI analysis conducted for the municipality of Acerra suggests a significant shift in vegetation cover toward conditions of lower density and health, with a progressive abandonment of higher NDVI classes. This change, highlighted by the maps and temporal graphs, indicates a phenomenon that requires further investigation to fully understand its causes and to plan conservation strategies. More sustainable land management and a thorough assessment of environmental pressures could be crucial in preserving local vegetation and mitigating the effects of potential environmental or anthropogenic stress factors.

4 Discussion

The results of this study, showing a significant decline in lush vegetation (NDVI class 5) and an increase in areas with less dense or degraded vegetation, align with observations from other recent studies examining the effects of climate change on vegetation cover, especially in rural and agricultural areas. Various studies have reported that rising temperatures and decreased precipitation negatively impact vegetation health, leading to biomass loss in previously vegetated areas. A similar analysis conducted in the Mediterranean Basin found a decline in dense and lush vegetation in response to prolonged drought and rising temperatures (Galveias et al. 2024). The authors observed that these climatic conditions increase the risk of desertification and soil degradation, phenomena that could also explain the changes observed in Acerra. Similarly, Yao et al. (2024), in a study on NDVI variations in mountainous areas of China, documented that climate change contributes to reducing high-density vegetative areas in favor of a sparser, less healthy cover. In both cases, the impact was attributed to water stress and temperature increases, conditions that might also be present in the study area. The expansion of lower NDVI classes (classes 2 and 3) observed in this study suggests that vegetation in many agricultural areas is becoming less lush, with reduced agricultural productivity and vegetative vigor. Similar findings have been reported in studies on agricultural areas in Southeast Asia, where intensive agriculture and excessive land exploitation practices, combined with climate change, have led to a progressive decline in soil health and vegetation (De Nardi et al. 2024). This increase in low-vegetation areas may indicate a loss of soil fertility and a buildup of environmental stress, due to phenomena such as soil erosion, compaction, and urbanization, as also suggested by Luo and Dou (2024) in their study on agricultural changes in China. The observed decline trend in lush vegetation classes is consistent with the global model of climate change and its effects on vegetation, as confirmed by other research in Europe and the Americas. Cui et al. (2024) documented that rising temperatures, altered rainfall patterns, and excessive evapotranspiration contribute to the reduction of vegetation cover, causing prolonged water stress. In contexts like the one studied in Acerra, the expansion of urban areas and intensively exploited agricultural land may further accelerate this degradation process. Furthermore, Puppim de Oliveira and Bhuiyan (2024) analyzed how growing cities and intensive agricultural practices increase pressure on local ecosystems, altering the water balance and reducing the regenerative capacity of green areas. The increase in lower NDVI classes (classes 2 and 3) is thus consistent with a trend toward environmental degradation, due to both adverse climate conditions and unsustainable land use. These results underscore the need to implement environmental management strategies to mitigate soil degradation and protect water resources. Sustainable agriculture policies, as suggested by Luo and Dou (2024), could reduce anthropogenic impact on the land and increase vegetation resilience to extreme climatic events. Continuous monitoring and sustainable planning, supported by periodic NDVI analyses, are clearly essential tools to prevent further damage to the ecosystem.

5 Conclusion

The analysis conducted highlighted a gradual decline in the most lush vegetation class (class 5) starting from 2022, suggesting a possible deterioration of the ecological conditions in the study area. This result indicates that vegetative vigor has decreased sharply, which could be caused by various factors, such as a scarcity of rainfall, or an increase in temperatures leading to a higher rate of evapotranspiration in the photosynthetic parts of plants. These findings emphasize the importance of closely monitoring the phenomenon of climate change, which can profoundly affect the health of vegetation and, consequently, the balance of local ecosystems. Climate change represents a complex challenge that requires continuous monitoring and the use of advanced techniques for spatial and temporal analysis of natural areas. However, to obtain a more comprehensive and precise picture, further studies are needed to extend the analysis at the regional and supra-regional levels, incorporating climate data on temperature and rainfall and evaluating their trends, or by filtering territorial areas, for example, by excluding urban areas. These changes suggest that the territory is vulnerable to desertification processes and biodiversity loss, phenomena that, according to the cited studies, are directly linked to climate change impacts. Future studies could expand the analysis to include data on temperatures, precipitation, and human activities to more precisely determine the specific causes of this degradation and to develop targeted conservation policies. Such research would allow for a better estimation of the impact of climate change on a larger scale and the development of effective strategies for the sustainable management of environmental resources and the mitigation of the negative effects of these changes. In conclusion, the results highlight that the use of remote sensing and NDVI is essential for monitoring vegetation trends in response to climate change. Comparison with similar studies suggests that without targeted interventions, the area could suffer further losses in biodiversity and ecological functionality.

References

Arregocés, H.A., Gómez, D., Castellanos, M.L.: Annual and monthly precipitation trends: An indicator of climate change in the Caribbean region of Colombia. Case Stud. Chem. Environ. Eng. **10**, 100834 (2024)

Arrogante-Funes, P., Osuna, D., Arrogante-Funes, F., Álvarez-Ripado, A. G., Bruzón, A.: Uncovering NDVI time trends in Spanish high mountain biosphere reserves: A detailed study. J. Environ. Manage 355, 120527 (2024)

Cui, Z., Zhang, Y., Wang, A., Wu, J.: Forest evapotranspiration trends and their driving factors under climate change. J. Hydrol. (Amst) 644, 132114 (2024)

De Nardi, S., Sangiorgi, L., Raccagni, S., Carnevale, C.: The impact of optimal climate change control on rice production in critical regions: the case of Southeast Asia. IFAC-PapersOnLine **58**, 226–231 (2024)

Galveias, A., et al.: Trends in land cover and in pollen concentration of Quercus genus in Alentejo, Portugal: Effects of climate change and health impacts. Environ. Pollut. **362**, 124996 (2024)

Gandhi, G.M., Parthiban, S., Thummalu, N., Christy, A.: Ndvi: vegetation change detection using remote sensing and GIS—a case study of Vellore district. Proc. Comput. Sci. **57**, 1199–1210 (2015)

Hu, Y., Dong, Y., Batunacun.: An automatic approach for land-change detection and land updates based on integrated NDVI timing analysis and the CVAPS method with GEE support. ISPRS J. Photogram. Remote Sensing 146, 347–359 (2018)

Luo, B., Dou, X.: Climate change, agricultural transformation and climate smart agriculture development in China. Heliyon **10**, e40008 (2024)

Puppim de Oliveira, J. A., Bhuiyan, S.: Cities and climate change in developing countries. Urban Clim 58, 102169 (2024)

Roy, D.P., et al.: Landsat-8 and Sentinel-2 burned area mapping—A combined sensor multi-temporal change detection approach. Remote Sens. Environ. **231**, 111254 (2019)

Wang, N., Li, W., Tao, R., Du, Q.: Graph-based block-level urban change detection using Sentinel-2 time series. Remote Sens. Environ. **274**, 112993 (2022)

Wang, X.L., et al.: Compound extreme inundation risk of coastal wetlands caused by climate change and anthropogenic activities in the Yellow River Delta, China. Adv. Climate Change Res. **15**, 134–147 (2024)

Yao, B., et al.: Spatiotemporal variation and GeoDetector analysis of NDVI at the northern foothills of the Yinshan Mountains in Inner Mongolia over the past 40 years. Heliyon **10**, e39309 (2024)

Yuan, H., Wu, C., Gu, C., Wang, X.: Evidence for satellite observed changes in the relative influence of climate indicators on autumn phenology over the Northern Hemisphere. Glob. Planet Change **187**, 103131 (2020)

Zhao, S., et al.: The role of satellite remote sensing in mitigating and adapting to global climate change. Sci. Total Environ. **904**, 166820 (2023)

Zhu, L., Si, R., Shen, X., Walker, J.P.: An advanced change detection method for time-series soil moisture retrieval from Sentinel-1. Remote Sens. Environ. **279**, 113137 (2022)

Geography Mapping and Innovative Systems for Rural Fragmentation Protection

Analysis of CO_2 Emissions Variations in the Region of Apulia: Assessing the Effectiveness of the Covenant of Mayors for Climate and Energy

Luca Colazzo, Cristina Baglivo, Marina Bonomolo, and Paolo Maria Congedo

Abstract

The European Commission launched the Covenant of Mayors (CoM) in 2008, an initiative that aims to coordinate, promote and incentivize thousands of local authorities to pursue the European Union's goals of reducing emissions and adapting to climate change. By voluntarily joining this movement, signatories committed to submit Sustainable Energy Action Plans (SEAPs) with the aim of reducing CO_2 emissions by at least 20% by 2020 compared to 1990 levels. In 2015, the Covenant merged with the Mayors Adapt initiative to form the new Covenant for Climate and Energy, introducing the theme of adaptation to climate change. Following the Paris Agreement and the launch of the European Green Deal, the Covenant has undergone significant adjustments. To meet the goals of these agreements, a further increase in CO_2 emission reductions of at least 55% by 2030 is required. In line with these targets, signatories have renewed their commitment by developing Sustainable Energy and Climate Action Plans (SECAPs). In Italy, the initiative has seen a growing number of participants. Despite its success, the plans face a number of problems related to the quality of emission inventories, the implementation of measures and the completion of monitoring by the signatories. The present study proposes a methodology to evaluate the effectiveness of emission reduction plans in the context of the Apulia region, statistically comparing the results achieved by the municipalities participating in the CoM initiative.

Keywords

Covenant of Mayors · Sustainable Energy and Climate Action Plans · CO_2 emissions · Climate change adaptation · Municipalities

1 Introduction

Since 1990, the European Union (EU) has taken a leading role in tackling global climate change. To this end, policies have been implemented to reduce greenhouse gas (GHG) emissions and improve the efficiency of the industrial energy sector and energy consumption. Through the implementation of targeted policies and programmes, the EU is committed to reducing energy dependency, improving the efficiency of energy consumption (Bertoldi 2018a, 2018b) and promoting the widespread use of renewable energy sources. 25 June 2009 marks an important moment for the EU with the entry into force of Directive 2009/28/EC on the promotion of the use of renewable energy. This legislative package, part of the Climate and Energy Framework, contains complementary provisions to ensure that the EU meets its ambitious climate and energy targets for 2020. Three key targets stand out: a 20% reduction in the EU's GHG emissions compared to 1990 levels, a 20% improvement in the EU's overall energy efficiency, and a 20% share of renewable energy sources such as wind, solar, biomass, etc. in the EU's energy mix. For the EU targets to be fully realized, it is essential that they are also shared and implemented at sub-national level, where local authorities can play an important role in mitigating climate change by implementing sustainable energy and climate change policies in their territories (Broto 2017).

L. Colazzo · C. Baglivo (✉) · P. M. Congedo
Department of Engineering for Innovation, University of Salento, 73100 Lecce, Italy
e-mail: cristina.baglivo@unisalento.it

L. Colazzo
e-mail: luca.colazzo@unisalento.it

P. M. Congedo
e-mail: paolo.congedo@unisalento.it

M. Bonomolo
Department of Engineering, University of Palermo, 90128 Palermo, Italy
e-mail: marina.bonomolo@unipa.it

P. Mohammad et al. (eds.), *Geographic Approaches to Climate Change and Mitigation: Urban and Rural Perspectives (Volume 1)*,
Advances in Science, Technology & Innovation, https://doi.org/10.1007/978-3-031-92119-3_31

Urban areas are particularly vulnerable to sudden and severe disasters such as climate change, global health crises and conflicts that disrupt the lives of their inhabitants. No city is completely immune to unforeseen risks; resilience refers to a city's ability to anticipate and respond to disasters and unforeseen events, allowing residents to adapt and thrive without stress (Baglivo et al. 2023). A holistic and dynamic approach can support the design of buildings and technological systems that consider the long-term impacts of climate change to create resilient and energy efficient structures (D'Agostino et al. 2024; Congedo et al. 2023).

In 2008, the European Commission launched the CoM: the first European movement to coordinate, promote and incentivize thousands of local authorities of all sizes in line with the emission reduction targets set by the EU (Gesing 2017; Jänicke and Quitzow 2017; Reckien et al. 2018). By voluntarily joining (Khan and Sovacool 2016) this bottom-up movement (Domorenok 2020; Rosenzweig et al. 2010; Schreurs 2008), signatories committed to develop SEAPs (Nuss-Girona et al. 2016) with the aim of reducing CO_2 emissions by at least 20% by 2020 compared to the reference year (ENERGY xxxx). Currently, the initiative has 11,962 signatories, 8,244 submitted action plans, and 20,672 actions.

In 2014, the European Commission integrated adaptation to climate change into the Union's strategies and set ambitious targets, including a 40% reduction in greenhouse gas emissions compared to 1990 levels, an increase in renewable energy consumption of up to 32%, and an improvement in energy efficiency of at least 32.5% (deLlano-Paz et al. 2016).

In line with this, the CoM merged with the Mayors Adapt initiative in 2015 to form the new Covenant of Mayors for Climate and Energy (Bertoldi 2018a). In addition to expanding on the original goals of accelerating the decarbonization of cities, the new initiative aims to strengthen the capacity to adapt to climate change and ensure citizens' access to safe, sustainable, and accessible energy. In this context, signatories commit to reduce CO_2 emissions by 40% by 2030 and to adopt an approach to increase climate resilience in their territories.

A significant contribution to the platform has been made by the initiatives of the CoM, which merged in 2016 to form the Global Covenant of Mayors (GCoM) (Bertoldi et al. 2018).

April 22, 2016 marked an important moment in the fight against climate change with the formal accession of the EU to the Paris Agreement. The agreement called for the adoption of a plan to limit the increase in the global average temperature to well below 2 degrees Celsius above pre-industrial levels, with a commitment to limit the increase to 1.5 degrees Celsius. In line with this agreement, Europe launched the Green Deal in 2019: a package of strategic initiatives aimed at achieving climate neutrality by 2050. In addition, European legislation adopted in 2021 makes legally binding the goal of reducing emissions by at least 55% by 2030.

In light of this significant development, the CoM is adapting to the new European objectives: signatories commit to renew their commitment by developing SECAPs, with the commitment to reduce CO_2 emissions in their territory by 55% by 2030 compared to the reference year, to increase the capacity to adapt to climate change, and to fight energy poverty as a fundamental pillar to ensure a just transition.

The Joint Research Centre (JRC) of the European Commission is in charge of evaluating the SECAPs developed by the cities, monitoring the progress of the local actions towards their respective targets and evaluating their final results. The monitoring system involves a comprehensive exercise carried out by the cities every four years (Bertoldi 2018c), which includes an update of the emissions inventory, known as the Monitoring Emissions Inventory (MEI). However, there is no final monitoring report at the end of the initiative.

Therefore, in addition to commitment and planning, monitoring plays a critical role in the CoM methodology, as without a monitoring system it would not be possible to accurately assess the effectiveness of the actions taken. Through this system, signatories can monitor the implementation of policies and measures set out in action plans, as well as progress towards emission reduction targets, and identify any shortcomings that may hinder the achievement of these targets. The introduction of the CoM can be seen as a significant evolution in the way European policies are implemented, as municipalities take a more active role in achieving European energy and climate goals (Pablo-Romero et al. 2016, 2015).

In Italy, the initiative is seeing an increasing number of participants. Despite its success, the action plans face a number of problems related to the quality of emission inventories, the implementation of measures, and the completion of monitoring by signatories. This research focuses on the analysis of energy consumption and resulting carbon dioxide emissions in the Italian region of Apulia, examining patterns and shifts at the provincial level. The study evaluates the effectiveness of the CoM in mitigating CO_2 emissions within Apulia, with a particular focus on the extent to which municipalities adhere to the Covenant at the provincial level. Particular emphasis will be placed on demonstrating the tangible benefits of participating in this initiative in terms of reducing carbon emissions. It also seeks to assess whether adherence to the covenant has facilitated the achievement of the targets set out in European climate change directives. The assessment seeks to verify whether policies and measures implemented at the local and regional level have actually contributed to the achievement of these goals.

2 Main Elements of the New Covenant of Mayors for Climate and Energy

The commitments set forth by the new Covenant of Mayors for Climate and Energy aim to reduce CO_2 emissions by at least 55% by 2030 compared to 1990 levels and integrate adaptation actions to address the inevitable impacts of climate change. These commitments are realized through the development of the SECAP for the relevant territory, following the CoM Guidelines. The SECAP encompasses two fundamental elements, alongside the cross-cutting issue of energy poverty: mitigation, which focuses on reducing CO_2 emissions (territorial decarbonization), and adaptation, a new element aimed at reducing the effects of climate change.

The SECAP drafting process is divided into three main phases: the signing of the Covenant of Mayors for Climate and Energy, the submission of the SECAP on the CoM platform within two years of joining, and the submission of an Action Monitoring Report within two years of approval, followed by a Complete Monitoring Report within four years of approval.

The SECAP includes several key components, including a Basic Emissions Inventory (BEI) that calculates emissions in the municipal territory for the reference base year, one or more Risk and Vulnerability Assessments (RVA) providing a detailed analysis of current and future scenarios, the Mitigation Action Plan, and the Adaptation Action Plan.

Within the CoM, four main sectors are identified where local administrations can influence energy consumption and reduce CO_2 emissions through targeted actions: municipal buildings, facilities/equipment; tertiary (non-municipal) buildings, facilities/equipment; residential buildings; and transportation.

The submitted SECAPs undergo quality control by the JRC, which evaluates them based on a set of eligibility criteria specified by the Covenant and the consistency of the data provided. Following the analysis, a report is drafted. Failure to meet the criteria may result in the exclusion of the SECAP.

3 Data and Descriptive Analysis

This section outlines the methodology and procedures used to evaluate the effectiveness of the CoM in the Apulia region.

Data on the percentages of compliance with the CoM by municipalities in Apulia and the percentages of variation in CO_2 emissions at the provincial level from 2005 to 2022 were analyzed.

The CoM compliance percentages were extrapolated from the latest monitoring report of the MyCovenant platform.

Electricity consumption data were extracted from the database of TERNA S.p.A., the company responsible for managing the national electricity grid. This database provides detailed information on electricity consumption by sector, broken down by region and province, from 2000 to 2022.

For data on provincial consumption related to petroleum products and natural gas, this study relies on the database provided by the Ministry of the Environment and Energy Security (MISE). To determine the carbon dioxide emissions associated with the energy consumption considered, it is necessary to multiply this consumption by the respective emission factors, according to the IPCC methodology, based on the carbon content of each energy source. This analysis estimated the percentage changes in total and per capita carbon dioxide emissions at the provincial level during the reference period 2005–2022. For this purpose, the demographic variations of the provinces of Apulia in the same years were extracted using the dataset provided by ISTAT on the resident population as of January 1. The data obtained were organized and prepared for analysis.

In the last part of this study, the results of the two analyses were compared to determine if there was a correlation between the percentages of compliance with the CoM by municipalities and the percentages of variation in CO_2 emissions at the provincial level.

The Pearson correlation coefficient was used to evaluate the relationship between the percentage of compliance with the CoM and the reduction or increase in CO_2 emissions at the provincial level. This coefficient measures the linear relationship between two continuous variables on interval or equivalent ratio scales, and provides an indication of the strength and direction of the relationship between the two variables.

3.1 Analysis of Apulian Municipalities' Adherence to the CoM

This section evaluates the percentage of compliance with the CoM by municipalities in Apulia within each province.

Since the launch of the initiative in 2008, an increasing number of local authorities have joined the GCoM initiative and developed SEAPs and/or SECAPs. The JRC provides GCoM datasets, which include a comprehensive collection of action plans and monitoring reports uploaded directly by local authorities to the MyCovenant platform. The latest monitoring report, dated March 2023, shows that Italy stands out for the number of participating municipalities: out of a total of 7,896 Italian municipalities, 4,770 have joined the CoM initiative, representing 60.4%.

The subject of this study is the region of Apulia, which, with 85.6% of municipalities joining, is one of the regions with the highest percentage in Italy. In particular, the provinces of Barletta-Andria-Trani and Brindisi have achieved 100% compliance with the CoM by municipalities. As highlighted in Fig. 1, with the exception of the provinces of

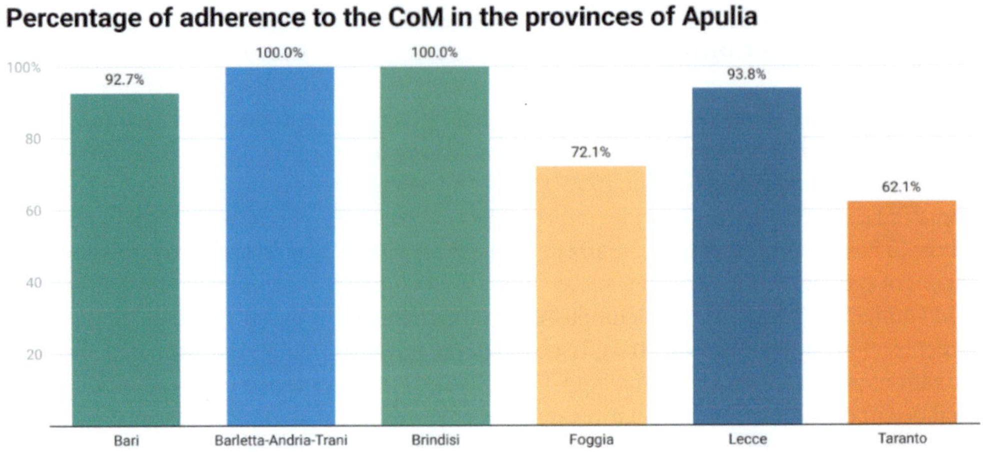

Fig. 1 Percentage of adherence to the CoM in the provinces of Apulia. *Source* (by the authors)

Taranto (62.1%) and Foggia (72.1%), the remaining provinces of Lecce and Bari have percentages above 90%. These municipalities have committed themselves over the years to develop a SEAP for the period up to 2020, a SECAP for the period up to 2030, or both.

3.2 *Analysis of CO_2 Emissions Reductions*

After evaluating the compliance percentages of the municipalities of Apulia with the CoM at the provincial level, the CO_2 emission reductions were analyzed. It is important to note that the reference years chosen by the CoM member municipalities for the BEI and MEI inventories are different. In order to partially mitigate this heterogeneity, this study considered 2022 as the monitoring year and 2005 as the reference year, as the latter is the most commonly used year for the development of SEAP and/or SECAP and the one for which more complete datasets are available. However, following the creation and activation of the new province of Barletta-Andria-Trani in June 2009, 2011 was chosen as the reference base year for the latter, as well as for the provinces of Bari and Foggia, due to the greater quantity and uniformity of available data.

3.3 Analysis of Electricity Consumption and Associated CO_2 Emissions

The data on electricity consumption comes from the TERNA database, which is managed by the company in charge of Italy's national electricity transmission grid. TERNA provides detailed information on electricity consumption by sector in different regions and provinces from 2000 to 2022. The data cover the residential, industrial, agricultural and services sectors. Specifically, the services sector includes activities such as public administration and compulsory social security; arts, sports, entertainment and recreation; accommodation and food service activities; activities of households as employers of domestic personnel and the production of goods and services for household use; financial and insurance activities; real estate activities; professional, scientific and technical activities; and other activities; professional, scientific and technical activities; wholesale and retail trade and repair of motor vehicles and motorcycles; public lighting; education; rental and leasing activities, travel agency activities and business support activities; health and social work activities; information and communication activities; road transport activities; veterinary activities; transport and storage activities and other service activities.

In this study, the electricity consumption of the residential and services sectors has been considered, since they correspond to the key sectors of the CoM: residential, municipal and tertiary. These sectors form the core of the actions implemented by the Covenant's signatory administrations to reduce carbon dioxide emissions.

For the calculation of carbon dioxide emissions, emission factors for electricity consumption at the user level in Italy, provided by Italian Institute for Environmental Protection and Research (ISPRA) for the period 1990–2022, were used.

To calculate the emission factors for consumption, ISPRA took into account the net production of electricity, which is the electricity measured at the output of the plants minus the electricity used for auxiliary production services, the share of imported electricity and grid losses (ISPRA 2019).

From Fig. 2, it is clear that the emission factor for national electricity consumption is further reduced due to the amount of energy imported from abroad whose atmospheric emission origin is outside the national territory.

Once the electricity consumption has been determined, the corresponding carbon dioxide emissions can be calculated by

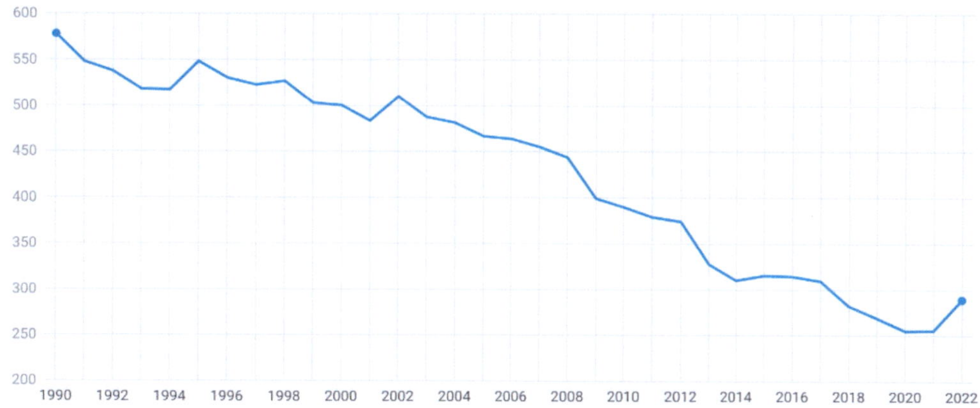

Fig. 2 National trend of CO_2 emission factor for electricity consumption (1990–2022), unit of measure: g CO_2/kWh. *Source* (by the authors)

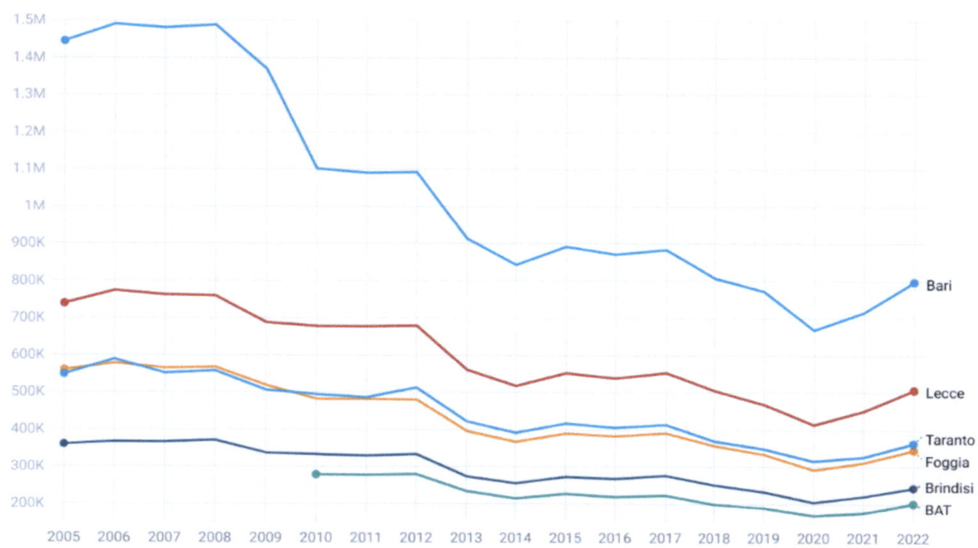

Fig. 3 CO_2 emissions from electricity in the provinces of Apulia, unit of measure: t CO_2. *Source* (by the authors)

multiplying this consumption by the electricity specific emission factor. This allows to obtain the CO_2 emissions related to the electricity consumption in the provinces of Apulia for the period 2005–2022 (see Fig. 3).

From the results obtained, it can be stated that in the considered sectors (residential and services), there is a total reduction of CO_2 emissions caused by electricity consumption of approximately 32.7% in Apulia over the period 2005–2022. This reduction seems to be evenly distributed among the territories of the considered region.

3.4 Analysis of Petroleum and Natural Gas Product Consumption and Associated CO_2 Emissions

According to the CoM guidelines, when preparing the balance of energy consumption in the residential, services, public transport and private transport sectors, sectors that are not under municipal jurisdiction, such as industries subject to the Emission Trading System (ETS) and transit roads, such as extra-urban roads and highways, are excluded.

In accordance with the CoM recommendations, this study relies on the database of the Ministry of Environment and Energy Security (MISE) for data on consumption related to petroleum products and natural gas. This database includes the Petroleum Bulletin, a report generated from data collected from 2003 to 2024 on petroleum products through the SISEN platform (Information System for National Energy Statistics). Specifically, this work used datasets on provincial consumption of gasoline, diesel, fuel oil, LPG, and lubricants from 2005 to 2022. Within these datasets, the following consumptions were considered: gasoline (regular network), diesel for automotive use (regular network), fuel oil, automotive LPG, and LPG for combustion.

At the same time, MISE provides a collection of data on imports, consumption (regional and provincial) and the balance of natural gas. In this study, only consumption related to the "distribution network" is considered. The latter refers to the use of natural gas as a fuel for domestic, commercial, or

Table 1 IPCC emission factors for local use of non-renewable energy sources (NRES)

Energy source	CO_2 (t CO_2/MWh)
Natural gas	0.202
Liquefied petroleum gases	0.227
Natural gas liquids	0.231
Gas/diesel oil	0.267
Gas/diesel oil	0.267
Motor gasoline	0.249
Lignite	0.364
Anthracite	0.354
Other bituminous coal	0.341
Sub-bituminous coal	0.346
Peat	0.382
Municipal Wastes (Non-biomass Fraction)	0.330

other purposes that require delivery through the gas distribution network. This can include residential heating, cooking, commercial use in restaurants and shops, and other similar applications.

To calculate the carbon dioxide emissions associated with the consumption of the considered petroleum products and natural gas, it is sufficient to multiply these consumptions by the respective emission factors for the combustion of fossil fuels, according to the IPCC methodology, based on the carbon content of each fuel (IPCC (Intergovernmental Panel on Climate Change) 2006). In fact, within the EU and the Global Covenant of Mayors for Climate and Energy, the JRC provides GHG emission factors related to energy (Table 1) (ENERGY, H. T. D. A. S. xxxx; Bastos et al. 2024; Koffi et al. 2017; Vullo et al. 2022). This dataset provides updated emission factors related to local energy consumption and local electricity generation according to the CoM.

This allows to obtain the CO_2 emissions associated with the consumption of gasoline (Fig. 4), diesel (Fig. 5), LPG (Fig. 6) and natural gas (Fig. 7) in the provinces of the Apulia region over the period 2005–2022.

3.5 Analysis of Total and Per Capita CO_2 Emissions

In view of the results obtained, this part of the work analyzes the total carbon dioxide emissions. In the region of Apulia, there is a significant decrease in emissions over the period from 2005 to 2022, amounting to 25.5%.

Specifically, the variations of total CO_2 emissions at provincial level are calculated for the years 2005 and 2022 (see Fig. 8) in the region of Apulia. It can be observed that the majority of the provinces show a decrease of emissions in the studied sectors during the reference period.

In this analysis, it was decided to also consider carbon dioxide emissions per capita. For this purpose, the demographic changes of the provinces in the same years were analyzed. For this purpose, ISTAT provides the dataset of the resident population as of January 1; for the years from 2001 to 2019, data from a statistical reconstruction of the resident population are used.

During the period under consideration, there is a slight decrease in the resident population of 3.1% in Apulia. Figure 9 shows the demographic evolution at the provincial level of the region under study.

Fig. 4 CO_2 emissions from gasoline in the province of Apulia, unit of measure: t CO_2. *Source* (by the authors)

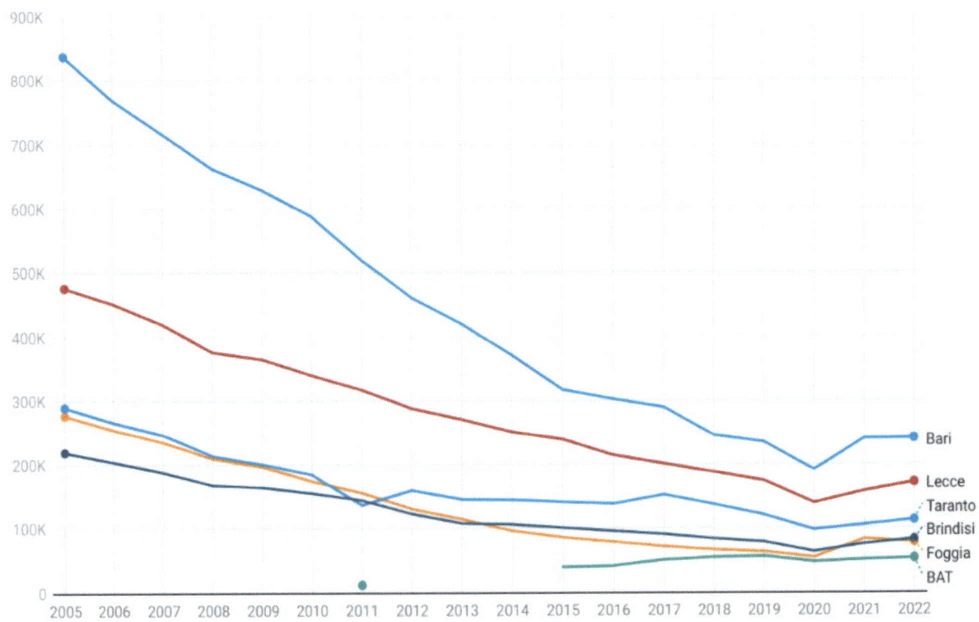

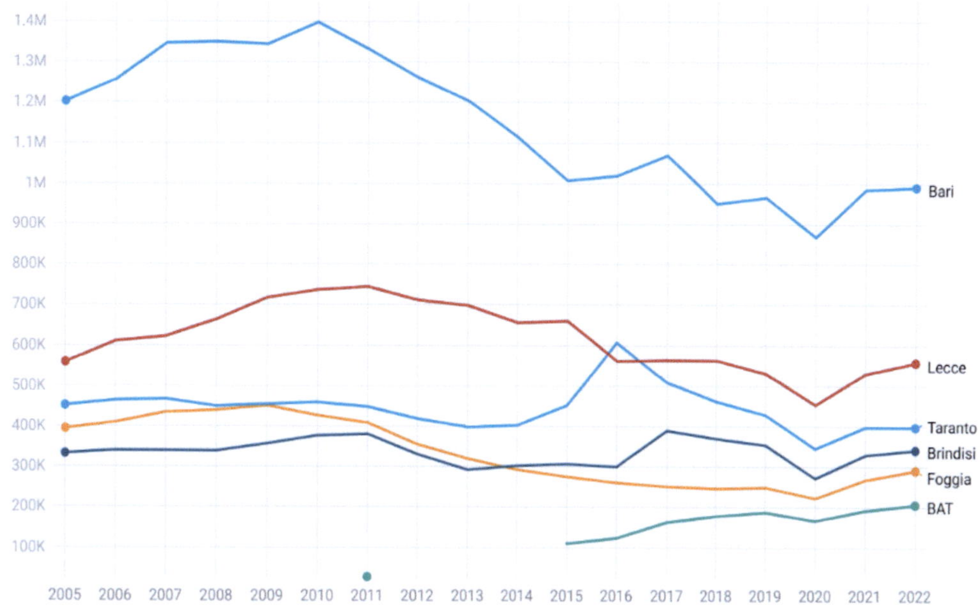

Fig. 5 CO_2 emissions from diesel in the province of Apulia, unit of measure: t CO_2. *Source* (by the authors)

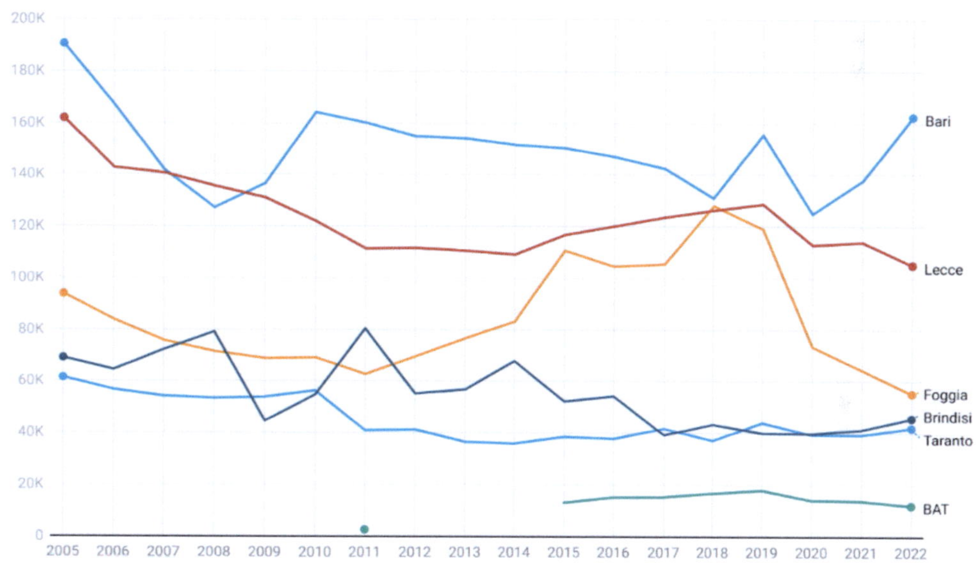

Fig. 6 CO_2 emissions from LPG in the province of Apulia, unit of measure: t CO_2. *Source* (by the authors)

The results obtained highlight the importance of accounting for demographic changes in emissions calculations, as some provinces show an increase in resident population while others show a decrease. This suggests that the impact of carbon dioxide emissions is not uniform and may be influenced by demographic factors.

Therefore, considering per capita emissions at the provincial level allows for a better understanding of the impact of CO_2 emissions in relation to the size of the local population and provides valuable insights for developing targeted policies and mitigation measures at the local level.

With these considerations in mind, per capita CO_2 emissions were calculated: in Apulia, over the period considered (2005–2022), there was a decrease from 2.88 to 2.21 tonnes of CO_2 per capita, a reduction of 23.1%. These reductions are in line with the variations in total CO_2 emissions calculated earlier.

Specifically, the variations in per capita CO_2 emissions at the provincial level were calculated for the years 2005 and 2022 (see Fig. 10). Confirming what has been observed for total emissions, it can be seen that the majority of Apulia's provinces record a decrease in per capita emissions in the studied sectors over the period considered.

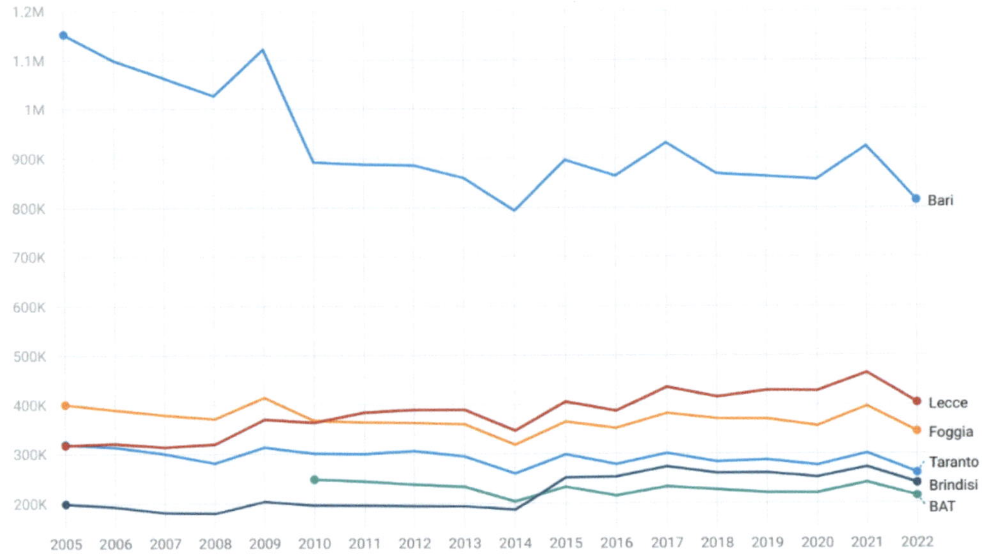

Fig. 7 CO_2 emissions from natural gas in the province of Apulia, unit of measure: t CO_2. *Source* (by the authors)

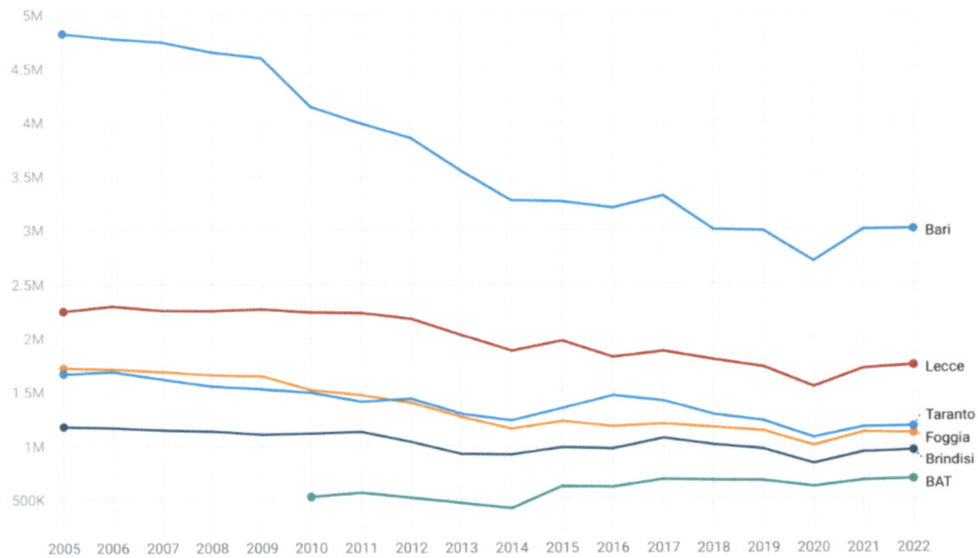

Fig. 8 CO_2 total emissions in the province of Apulia, unit of measure: t CO_2. *Source* (by the authors)

3.6 Assessment of the Effectiveness of the CoM

In a preliminary analysis, the results obtained can be summarized by Table 2.

Next, the data obtained from the two analyses on the percentage of compliance with the CoM by municipalities at the provincial level and the variations in total CO_2 emissions in the same provinces over the period considered are collected and organized.

The data are prepared for the analysis by ensuring that they are complete, consistent and free of missing or anomalous values. The variables of interest, namely the percentage of compliance with the CoM and the reduction of CO_2 emissions, are associated with the corresponding provinces.

The results of the analysis are presented in a scatter-plot (Fig. 11), where the percentage of CoM compliance is plotted on the x-axis and the changes in CO_2 emissions are plotted on the y-axis. Each point on the plot represents a province, and the coordinates of each point correspond to the percentage of compliance and the reduction in CO_2 emissions for that province over the period considered (2005–2022).

From Fig. 11, a higher concentration of points can be observed in the region corresponding to CoM compliance percentages above 90%, which represent the majority of the provinces in Apulia. In particular, only two provinces in this area have recorded a reduction in CO_2 emissions between 20 and 30%. In the left part of the scatter plot, corresponding to lower compliance percentages of the municipalities, two

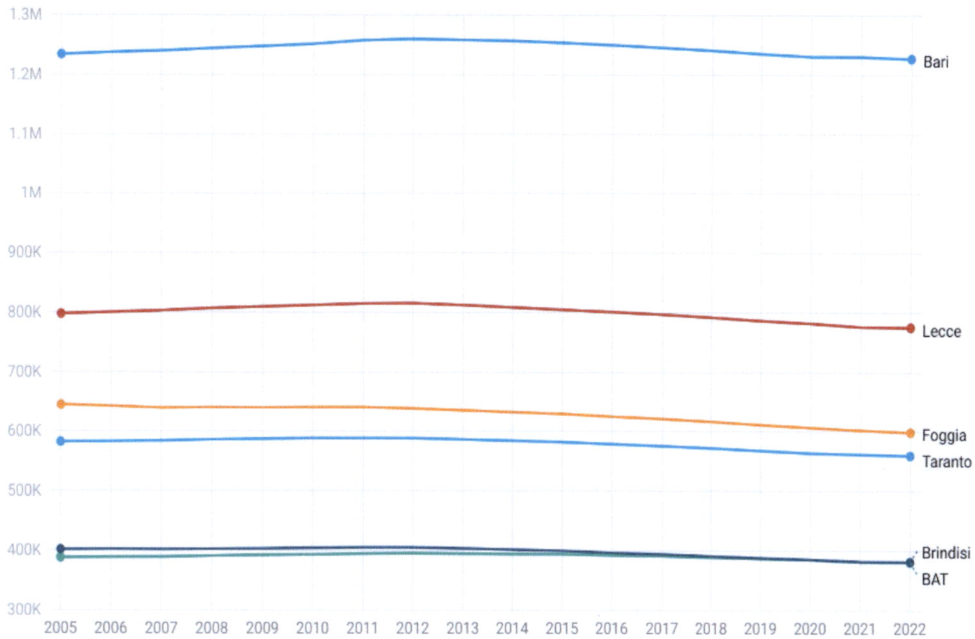

Fig. 9 Resident population on January 1st in Apulia. *Source* (by the authors)

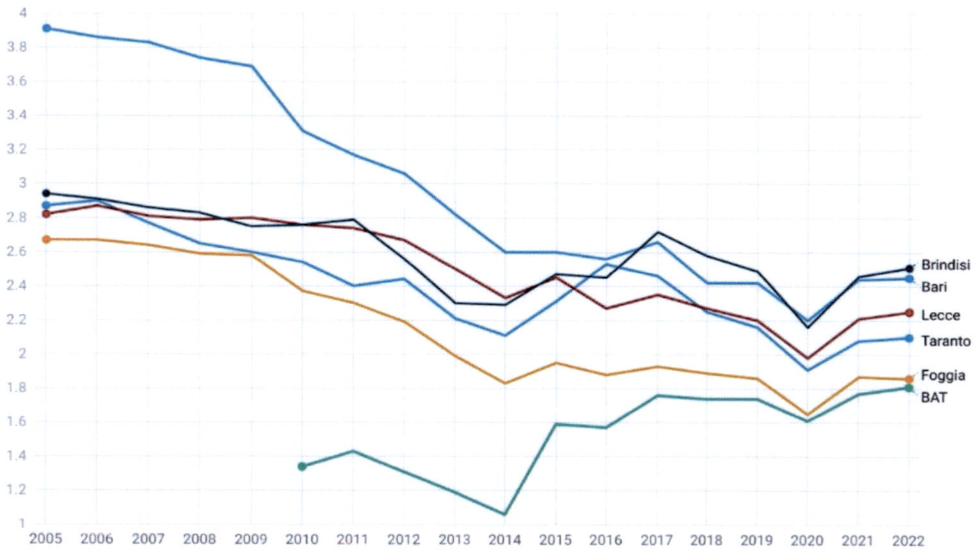

Fig. 10 CO_2 per capita emissions in the province of Apulia, unit of measure: t CO_2 per capita. *Source* (by the authors)

Table 2 Summary of results

Region	CoM adhesions	Δ CO_2 emissions (2005–2022)	Δ per capita CO_2 emissions (2005–2022)	Province	CoM adhesions (%)	Δ CO_2 emissions (2005*–2022) (%)	Δ per capita CO_2 emissions (2005*–2022) (%)
Apulia	85.6%	–25.5%	–14.7%	Bari	92.7	–24.6	–22.7
				Barletta–Andria–Trani	100.0	21.9	26.4
				Brindisi	100.0	–19.1	–14.8
				Foggia	72.1	–24.3	–30.5
				Lecce	93.8	–22.5	–20.3
				Taranto	62.1	–29.5	–26.6
				Trapani	92.0	–42.2	–40.6

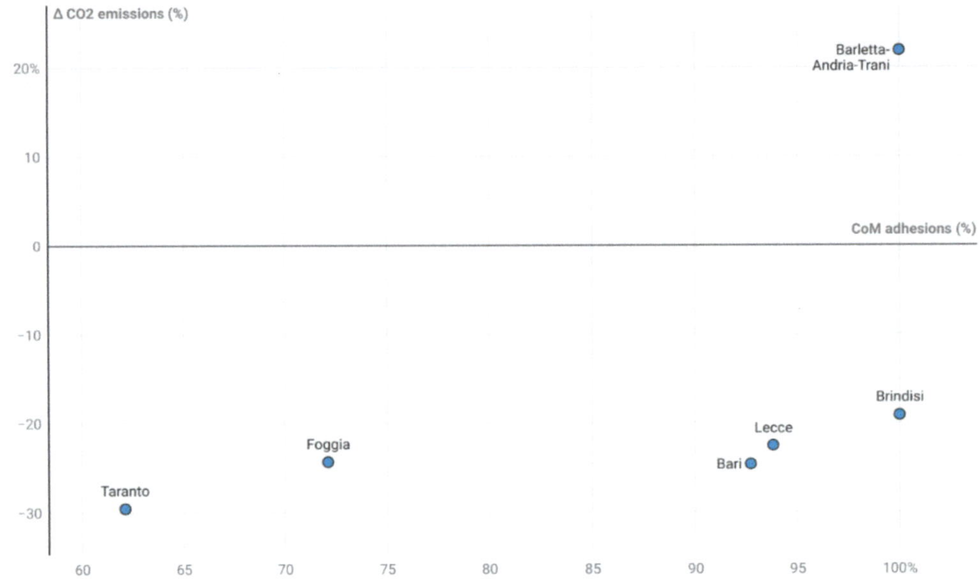

Fig. 11 Correlation between the percentages of adherence to the CoM and reductions of total CO_2 emissions at the provincial level. *Source* (by the authors)

provinces have reported a reduction in emissions ranging from 20 to 30%.

However, by looking at the scatter plot, it is possible to visually assess the presence and intensity of a relationship between the two variables. If the scatter points are evenly distributed and do not show a clear direction, this may indicate a lack of correlation. If the scatter follows an increasing or decreasing trend, it may indicate a positive or negative correlation, respectively.

In addition to the scatter plot, the Pearson correlation coefficient is calculated to numerically quantify the relationship between the two variables. This value provides a measure of the strength and direction of the correlation.

The formula for Pearson's correlation coefficient (1), denoted by r, is as follows:

$$r = \frac{\sum(x_i - \overline{x}) \bullet (y_i - \overline{y})}{\sqrt{\sum (x_i - \overline{x})^2 \bullet \sum (y_i - \overline{y})^2}}, \quad (1)$$

where

x_i and y_i are the individual observations of the two variables;

$\overline{x}$ and $\overline{y}$ are the respective means of the variables;

$\sum(x_i - \overline{x}) \bullet (y_i - \overline{y})$ is the sum of the products of the differences between the observations and the means of the two variables;

$\sqrt{\sum (x_i - \overline{x})^2 \bullet \sum (y_i - \overline{y})^2}$ is the product of the square roots of the sums of the squares of the differences between the observations and the means of the two variables.

The Pearson correlation coefficient can range from −1 to 1. A coefficient between 0 and 1 indicates a positive correlation , meaning that as one variable increases, so does the other, and as one decreases, so does the other. The closer the value is to 1, the stronger the correlation. Conversely, a coefficient between −1 and 0 indicates a negative correlation, where an increase in one variable corresponds to a decrease in the other, and vice versa. The closer the value is to −1, the stronger the negative correlation. When the Pearson correlation coefficient is exactly 0, it means that there is no linear correlation between the two variables.

In other words, the Pearson correlation coefficient measures how much variation in one variable is associated with variation in the other variable, expressing how the two variables tend to "move together" in a linear fashion.

In this study, the CoM can be said to have had a significant impact if an increase in the percentage of municipal participation at the provincial level is associated with a corresponding decrease in emissions, indicating a negative relationship.

However, if there is no clear correlation, it may be necessary to examine other factors or approaches to better understand the effectiveness of the CoM in each province.

In addition, it is important to consider other factors that may influence CO_2 emissions at the provincial level, such as national energy policies, locally adopted climate change mitigation measures, and the demographic characteristics of each province. These factors could provide important context for evaluating the impact of the CoM on provincial carbon dioxide emissions.

In this study, demographic variations at the provincial level that could influence CO_2 emission variations were considered. In this study, Pearson correlation coefficients were calculated at the provincial level. In both cases, both total emissions and per capita emissions were considered (Table 3).

In the first case, considering the variation in total CO_2 emissions and the percentage of compliance with the CoM

Table 3 Pearson correlation coefficient. *Source* (by the authors)

CO_2 total emissions	CO_2 per capita emissions
0.1534265	0.519588263

at the provincial level, the correlation coefficient is approximately 0.15: it indicates a weak positive correlation between the two variables considered. In practical terms, this could mean that an increase in the percentage of CoM compliance is associated with a modest increase in total CO_2 emissions, or vice versa, but the association is not strong enough to be considered significant or reliable for making predictions or drawing conclusions.

In the second case, considering CO_2 emissions per capita, the correlation coefficient is approximately 0.52, indicating a moderate positive correlation between the two variables considered. This suggests that there is a positive relationship between the variables, but it is not extremely strong. Specifically, this could mean that an increase in the percentage of CoM compliance is associated with a modest increase in CO_2 emissions per capita, or vice versa. However, the relationship is not strong enough to indicate a direct dependency.

4 Conclusion

The Covenant of Mayors for Climate and Energy, an initiative promoted by the European Commission, engages local governments in a voluntary agreement to achieve goals such as reducing CO_2 emissions, improving resilience to climate change, and combating energy poverty. Through this initiative, municipalities commit to develop and implement specific action plans to promote energy and environmental sustainability and play a proactive role in achieving the EU's energy and climate goals. Signatory municipalities commit to reducing pollution by increasing energy efficiency and introducing renewable energy sources in their territories.

The results obtained in the first part of this study, which focused on the analysis of CO_2 emissions in the region of Apulia at the provincial level, suggest that Community, national and regional policies have played an important role in promoting the effective reduction of emissions, even at the local level, while also recognizing the positive contribution of individual municipal initiatives. However, the analysis of the effectiveness of the CoM, developed in the second part of the study and evaluated through the Pearson correlation coefficient, used to measure the relationship between the variables studied, reveals a weak or moderate tendency towards an increase in CO_2 emissions with the increase in the adherence to the CoM at the provincial level. This result indicates that, despite the high level of formal adherence to the CoM in Apulia, there is no direct correlation between the number of affiliations and the reduction of CO_2 emissions, nor a significant acceleration towards the achievement of the ambitious emission reduction targets set by the European Green Deal.

This phenomenon can be attributed to several factors. First, adherence to the Covenant does not automatically guarantee a reduction in emissions, as the effectiveness of the policies depends on the rigorous implementation of the measures outlined in the SEAPs and/or SECAPs. In some cases, municipalities may face difficulties in implementing these measures due to financial, technical, or organizational constraints. Moreover, the effectiveness of the actions undertaken is closely linked to the implementation of an adequate monitoring system, as required by the JRC: without such a system, it is difficult to assess the impact of the policies adopted and to intervene promptly to correct any inefficiencies.

In order to overcome this problem, one of the solutions proposed is the introduction of a funding system in the form of vouchers provided by the Apulia Region to the municipalities (similar to the existing system for the preparation of SECAPs), which would help to cover the costs associated with the regular monitoring of the actions outlined in the SECAPs, thus ensuring more rigorous control and more effective implementation of the measures adopted.

Socio-economic and structural differences between municipalities may also influence the effectiveness of the measures, with municipalities with fewer resources potentially facing greater challenges in achieving the targets. A crucial aspect that emerged from the study is the need for ongoing technical and financial support to municipalities, without which implementation and monitoring of measures often prove ineffective.

In addition, external factors such as national or regional policies, economic crises, or changes in the industrial and agricultural sectors can affect CO_2 emissions independently of local initiatives, thereby reducing the positive impact of meeting the CoM. Finally, the lack of accurate and up-to-date data on emissions makes it difficult to assess the impact of the policies adopted.

This study offers significant social and political benefits by providing an in-depth analysis of the effectiveness of the CoM in reducing CO_2 emissions in the Apulia region, and by providing policy makers and local administrations with essential information to guide future strategies and improve the effectiveness of local climate policies. The study can also guide municipalities in refining their SECAPs, optimizing strategies to achieve emission reduction targets. The results could support the development of more targeted regional and national policies, identifying areas that need further intervention and promoting the creation of more effective policies tailored to local needs. The emphasis on monitoring and accountability could ultimately improve the overall effectiveness of climate initiatives.

However, the study has some limitations. The variability in reference years for emission inventories across municipalities makes it difficult to standardize comparisons and consistently assess the effectiveness of the initiative. The quality and completeness of monitoring by signatories is another challenge, as the lack of a final monitoring report limits the assessment of long-term impacts. Finally, the study does not fully account for the impact of external factors, such as economic crises, political changes, or technological advances, that may influence the effectiveness of local climate action.

References

Baglivo, C., Congedo, P. M., Mazzeo, D.: Scenarios for urban resilience—perspective on climate change resilience at the end of the 21st century of a photovoltaic-powered mixed-use energy community in two European capitals. In: Adapting the Built Environment for Climate Change (pp. 37–52). Woodhead Publishing (2023)

Bastos, J., Monforti-Ferrario, F., Melica, G.: Covenant of Mayors for Climate and Energy: Greenhouse gas emission factors for local emission inventories (2024)

Bertoldi, P.: Ten-year anniversary issue. Energ. Effi. **11**(8), 1913–1915 (2018a)

Bertoldi, P., Kona, A., Rivas, S., Dallemand, J.F.: Towards a global comprehensive and transparent framework for cities and local governments enabling an effective contribution to the Paris climate agreement. Curr. Opin. Environ. Sustain. **30**, 67–74 (2018)

Bertoldi, P.: The paris agreement 1.5 C goal: what it does mean for energy efficiency. In: Proceedings of the ACEEE Summer Study on Energy Efficiency in Buildings; ACEEE: Washington, DC, USA (2018)

Bertoldi, P. (2018). Guidebook 'how to develop a sustainable energy and climate action plan (SECAP)'. Publication Office of the European Union

Broto, V.C.: Urban governance and the politics of climate change. World Dev. **93**, 1–15 (2017)

Congedo, P.M., Baglivo, C., D'Agostino, D., Mazzeo, D.: The impact of climate change on air source heat pumps. Energy Convers. Manage. **276**, 116554 (2023)

D'Agostino, D., Congedo, P.M., Albanese, P.M., Rubino, A., Baglivo, C.: Impact of climate change on the energy performance of building envelopes and implications on energy regulations across Europe. Energy **288**, 129886 (2024)

deLlano-Paz, F., Fernandez, P.M., Soares, I.: Addressing 2030 EU policy framework for energy and climate: Cost, risk and energy security issues. Energy **115**, 1347–1360 (2016)

Domorenok, E.: Voluntary instruments for ambitious objectives? The experience of the EU Covenant of Mayors. In: The Future of European Union Environmental Politics and Policy (pp. 107–128). Routledge (2020)

ENERGY, H. T. D. A. S. How to develop a sustainable energy action plan

Gesing, F. (2017). The new Global Covenant of Mayors for Climate & Energy and the politics of municipal climate data

IPCC (Intergovernmental Panel on Climate Change): Guidelines for national greenhouse gas inventories. Chapter I. Introduction. Prepared by the National Greenhouse Gas Inventories Programme, 1–21 (2006)

ISPRA, R.: Fattori di Emissione Atmosferica di Gas a Effetto Serra Nel Settore Elettrico Nazionale e Nei Principali Paesi Europei (2019)

Jänicke, M., Quitzow, R.: Multi-level Reinforcement in European Climate and Energy Governance: Mobilizing economic interests at the sub-national levels. Environ. Policy Gov. **27**(2), 122–136 (2017)

Khan, F., Sovacool, B.K.: Testing the efficacy of voluntary urban greenhouse gas emissions inventories. Clim. Change **139**(2), 141–154 (2016)

Koffi, B., Cerutti, A., Duerr, M., Iancu, A., Kona, A., Janssens-Maenhout, G.: Covenant of Mayors for Climate and Energy: Default emission factors for local emission inventories. Joint Research Centre (JRC) (2017)

Nuss-Girona, S., Llausàs, A., Figueras, J., Morera, S.: The SEAP in the city of Girona, a crossroads between boldness and pragmatism. Local Environ. **21**(4), 476–503 (2016)

Pablo-Romero, M.D.P., Sánchez-Braza, A., Manuel González-Limón, J.: Covenant of mayors: Reasons for being an environmentally and energy friendly municipality. Rev. Policy Res. **32**(5), 576–599 (2015)

Pablo-Romero, M.D.P., Pozo-Barajas, R., Sánchez-Braza, A.: Analyzing the effects of Energy Action Plans on electricity consumption in Covenant of Mayors signatory municipalities in Andalusia. Energy Policy **99**, 12–26 (2016)

Reckien, D., Salvia, M., Heidrich, O., Church, J.M., Pietrapertosa, F., De Gregorio-Hurtado, S., Dawson, R.: How are cities planning to respond to climate change? Assessment of local climate plans from 885 cities in the EU-28. J. Clean. Prod. **191**, 207–219 (2018)

Rosenzweig, C., Solecki, W., Hammer, S.A., Mehrotra, S.: Cities lead the way in climate–change action. Nature **467**(7318), 909–911 (2010)

Schreurs, M.A.: From the bottom up: local and subnational climate change politics. J. Environ. Develop. **17**(4), 343–355 (2008)

Vullo, L., Monforti-Ferrario, F., Palermo, V., Bertoldi, P.: Greenhouse gases emission factors for local emission inventories (2022)

Farmers' Intention to Participate in Mobile Crowdsensing for Crop Pest and Disease Monitoring in Johor, Malaysia

Mohamad Jahidi Osman, Nurul Hawani Idris, Nurain Othman, Zamri Ismail, Mohamad Hafis Izran Ishak, Sheau-Ting Low, and Noor Izawati Alias

Abstract

Digitalization in the agriculture sector is a global plan to boost productivity, profitability, and resilience to climate change. Many studies have introduced digital technology such as drones and Internet of Things (IoT) under the umbrella of precision farming for crop maintenance and management, however, the acceptance of farmers to use crowdsensing technology for bottom-up reporting, specifically in crop pests and disease contexts demands further investigation. Crowdsensing relies on farmers' participation in run-on location-based technology-enabled participatory devices, which can draw collective knowledge and sensor data from local farms. Therefore, this study assesses the intention to use mobile crowdsensing technology for reporting crop pests and diseases among farmers in Johor, Malaysia based on the technology acceptance model (TAM) constructs. Using the partial least squares structural equation model (PLS-SEM), this study demonstrates that perceived usefulness (PU) did not significantly impacts farmers' attitudes toward the use (ATU) of technology. Whereas, generational cohorts (age groups) of farmers significantly affect the relationship between PU and Intention to use (ITU) on the use of the crowdsourced technology for crop pests and diseases monitoring (crowdsensing). This study will provide an understanding of how to promote the high acceptance of local farmers as human sensors in pest and disease monitoring via a mobile crowdsensing platform.

Keywords

Geographic Information System (GIS) · Mobile GIS · Crowdsensing · Technology acceptance model (TAM) · Digitalization agriculture

1 Introduction

In 2019, the world's population was expected to rise from nearly 7 billion to 8 billion by 2030 and almost 10 billion by 2050, necessitating massive agricultural demand, making food security a serious global concern (Bahar et al. 2020). Agriculture is rapidly changing to satisfy the current demands of the global population. However, the agriculture industry continues to experience new and recurring outbreaks. Globalization, climate change and diminished resilience in production ecosystems have all contributed to the emergence of pests, diseases, and epidemics. Many stakeholders have focused on farmers' resilience to globalization, trade, and climate change

M. J. Osman · N. H. Idris (✉) · N. Othman · Z. Ismail
Department of Geoinformation, Faculty of Built Environment and Surveying, Universiti Teknologi Malaysia, Johor Bahru 81310, Johor, Malaysia
e-mail: hawani@utm.my

M. J. Osman
e-mail: jahidi@graduate.utm.my

N. Othman
e-mail: nurainothman20@gmail.com

Z. Ismail
e-mail: zamriismail@utm.my

M. H. I. Ishak
Department of Control and Mechatronics, Faculty of Electrical Engineering, Universiti Teknologi Malaysia, Johor Bahru 81310, Johor, Malaysia
e-mail: hafis@utm.my

S.-T. Low
Department of Real Estate, Faculty of Built Environment and Surveying, Universiti Teknologi Malaysia, Johor Bahru 81310, Johor, Malaysia
e-mail: sheauting@utm.my

N. I. Alias
AR Mechatronics PLT, 48-01, Jalan Enau 15, Taman Teratai, Skudai 81300, Johor, Malaysia
e-mail: aliasnoorizawati@gmail.com

P. Mohammad et al. (eds.), *Geographic Approaches to Climate Change and Mitigation: Urban and Rural Perspectives (Volume 1)*, Advances in Science, Technology & Innovation, https://doi.org/10.1007/978-3-031-92119-3_32

(Thompson et al. 2023). Low-income farmers, particularly those relying on agriculture for food and income, remain the most vulnerable category. Food and Agriculture Organization (FAO, 2017) emphasized the importance of transformative processes toward 'holistic' approaches, such as agroecology, agro-forestry, climate-smart agriculture, and conservation agriculture, which rely on indigenous and traditional knowledge.

Recent technological advancements have cleared the path for innovative solutions in various industries, including agriculture. Integrating cutting-edge technology such as the Internet of Things, robots, Global Positioning System (GPS), remote sensing images through unmanned aerial vehicles (UAV) and satellites, and geographic information system (GIS) platform has transformed modern farming techniques, resulting in smart and precise farming (Chuang et al. 2020). This new paradigm incorporates the clever use of data-driven technologies to improve agricultural output, efficiency, and sustainability (Morchid et al. 2023). Location-based services (LBS) integrated into mobile apps have revolutionized the landscape of crop monitoring and smart farming by providing real-time information on pests, diseases, and site-specific farm conditions. These applications leverage mobile devices' global navigation satellite system (GNSS) capabilities to offer location data, enabling farmers to monitor their fields remotely and allow for bottom-up reporting via crowd sensing approach. One key aspect of these LBS-enabled apps is their ability to streamline pest and disease identification through image recognition and data analytics (Che'Ya et al. 2022). Farmers can capture images of affected crops using their mobile devices, gathering data from the available crowd sensors (IoT) from neighboring plots and utilizing the crowd sourced data to analyze and monitor the emergence of pests and disease at the local and regional plots. Empowering farmers with modern technology has been clearly stated in the 11th Malaysia Plan 2016–2020, 12th Malaysia Plan 2021–2025 and Shared Prosperity Vision 2030. In the 11th Malaysia Plan, modernization and transformation of the agricultural sector are required to ensure food security, increase productivity, enhance farmers' skills, strengthen agro-food supply chains, and enhance support and delivery services for all stakeholders.

The growth of mobile and communication technology has influenced the rise of crowd sensing. This bottom-level citizen engagement technique uses a human act as a sensor in data collecting. Implementing an integrated mobile platform with innovative farming technologies is thought to be capable of supporting farmers on the ground and assisting the Department of Agriculture (DOA) in data gathering related to pest and disease surveillance. However, the diversity of citizen motivation in crowdsourcing design must be custom-suited to the job and target group (Thapa et al. 2015), thus resulting in increased active citizen participation.

Therefore, this study assesses the intention to use a mobile crowdsensing to report crop pests and diseases among farmers in Johor, Malaysia, based on the technology acceptance model constructs. These are the objectives that must be achieved by the end of this study:

(1) To identify factors and indicators for measuring farmers' acceptance and engagement in using the mobile crowdsensing platform
(2) To determine user experience components of a crowdsourced platform for pest and disease monitoring

The following section describes the technology acceptance model (Sect. 2). The methodology used in this study includes a brief description of demographic profiles, sampling and a description of data collection and analysis procedures (Sect. 3). Then, Sect. 4 describes the results and discussion, followed by conclusions and recommendations (Sect. 5).

2 Literature Review

2.1 Crowdsensing in Agriculture

Ganti et al. (2011) describes two types of crowdsensing participation—personal (such as fitness tracking) and community sensing. Community sensing can be divided into 'participatory sensing' or 'opportunistic sensing.' Participatory sensing applications require the active involvement of citizens to contribute data, whereas the latter requires minimal engagement and works in an autonomous mode, such as in location tracking. Crowd data sharing plays a pivotal role in the effectiveness of location-based services for agriculture. This crowdsourced data encompass diverse geographic locations and climatic conditions, offering a comprehensive understanding of regional agricultural challenges. Farmers, researchers, and extension services can collectively contribute to a dynamic and expansive dataset by sharing their observations through these apps. Smartphone-based AI-powered tools, like Plantix (https://plantix.net/en/) and PlantVillage (https://plantvillage.psu.edu/) identify pests and diseases using image recognition and location data to suggest localized remedies. Once a disease has been identified, farmers can then seek guidance on how to respond; hence, it empowers farmers to identify and combat crop diseases. Combining location-based services and crowd data sharing (crowdsensing) creates a powerful synergy, empowering the agricultural community with actionable insights for sustainable and informed decision-making in smart farming practices. Fairuz et al. (2018) have examined local farmers' readiness and acceptance of green and sustainable farming technology; however, farmers' acceptance of crowdsensing, specifically in combating pests and diseases, demands further investigation.

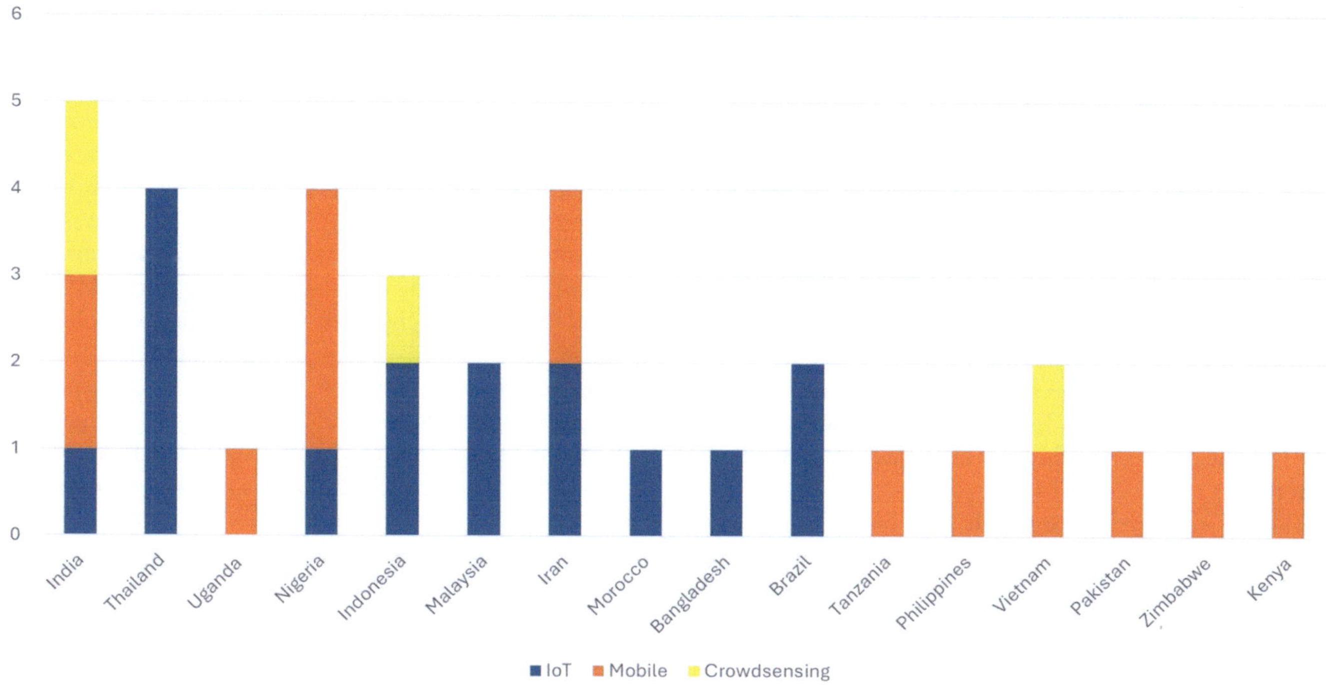

Fig. 1 Distribution of studies on TAM in agriculture across developing countries

The literature review, spanning from 2019 to 2024 in ScienceDirect and Google Scholar databases and encompassing 33 relevant studies, delved into technology adoption and acceptance in agriculture across developing countries. The used keywords (TAM + agriculture + digitization + IoT + mobile + crowdsourcing + crowdsensing) were focused on technology acceptance and adoption models toward the use of digital agricultural technologies specifically the Internet of Things (IoT), crowdsensing, and mobile applications as shown in Fig. 1. Among the countries studied, India emerged with the higher studies, covering acceptance models such as theory of planned behavior (TPB), Technology Acceptance Model (TAM), extended TAM, and TAM2. In Malaysia, two studies by Ismail et al. (2023) and Salim and Sa'don (2021), focused on IoT adoption primarily employing the TAM and TPB models. Hence, there is a lack of studies that investigate the farmers' acceptance to use and supply crowdsensing data in pests and diseases management and surveillance.

2.2 Hypotheses Based on the TAM Model

Farmers' difficulty in using technology has been identified as one of the obstacles to its application. There are several models to predict technological acceptance, and the most established is the technological acceptance model (TAM) (Davis 1989), which is based on the Theory of Planned Behavior (Azjen, 1991). The TAM assumes that the actual use or intention to use the technology is determined by two user perception processes, namely perceived usefulness (PU) and perceived ease of use (PEOU) (Mlekus et al. 2020). Durodolu (2016) states that the PU variable is associated with performance expectancy; whether they perceive the technology will improve their job performance. The PEOU variable, on the other hand, refers to effort expectancy, whereby a user believes their skills in using technology are uncomplicated. The technology referred to in this study is a crowdsensing platform for reporting crop pests and diseases, namely SmartAPT (Smart Crowdsensing Pests and Diseases Monitoring Technology). The hypotheses are as below:

H1: Perceived ease of use (PEOU) has a significant positive influence on perceived usefulness (PU) for the use of the SmartAPT application among farmers.

H2: Perceived usefulness (PU) has a significant positive influence on attitude toward use (ATU) for the use of the SmartAPT application among farmers.

H3: Perceived ease of use (PEOU) has a significant positive influence on attitude toward use (ATU) for the use of the SmartAPT application among farmers.

H4: Perceived usefulness (PU) has a significant positive influence on intention to use (ITU) for the use of the SmartAPT application among farmers.

H5: Attitude toward use (ATU) has a significant positive influence on intention to use (ITU) for the use of the SmartAPT application among farmers.

A mixed generation of farmers drawn from different age groups—young and old farmers—lifestyles and mindsets are among the demographic and individual traits challenges that need to be identified before sustainable smart farming can be successfully realized (Steinke et al. 2021). In this study, farmers' ages (acting as a moderator) were divided according to the generational cohort, namely, the Independence (Merdeka) generation (1945–1964), the Reformist generation (1965–1984), and the Internet generation (1985 onwards) considering the shared historical and social experiences that shape people's values, attitudes, and behaviors in Malaysia (see Mustafa et al. 2022). Hence, the following are the hypotheses that incorporate age as a moderator.

H6: Perceived usefulness (PU) has a significant positive influence on attitude toward use (ATU) for the use of the SmartAPT application among farmers, moderated by generational cohort.

H7: Perceived ease of use (PEOU) has a significant positive influence on attitude toward use (ATU) for the use of the SmartAPT application among farmers, moderated by generational cohort.

H8: Perceived usefulness (PU) has a significant positive influence on intention to use (ITU) for the use of the SmartAPT application among farmers, moderated by generational cohort.

3 Research Methodology

3.1 Demographic Profiles and Sampling

The study's demographic profile categorized the age of participants based on Mustafa et al.'s (2022) study following the Malaysian context, encompassing members of the Merdeka (pre- and post-Independence Day) generation (n = 33), the Reformist generation (n = 55), and the Internet generation (n = 37). The gender distribution consists of both female (36%, n = 45) and male (64%, n = 80) respondents. Education levels vary, ranging from individuals with no formal education (n = 1), primary education (n = 4), and secondary education (n = 68) to those with Bachelor's degrees and above (n = 52). This study uses purposive sampling technique which only focuses on rural farmers and community farmers who practice chili cultivation using the fertigation. Fertigation farmers uses timer for watering and fertilizing or Internet of Things (IoT) in managing their crops. This study postulates that early exposure to technology enables them to easily accept updates and improvements in technology in the field of agriculture such as this study. As many as 49 farmers from Johor Bahru and 76 from Batu Pahat district were involved in this study. The targeted respondents were farmers who have experienced using modern technology (i.e., fertigation systems) in their practices. This study assumed that farmers who have at least used smart technology might positively influence the use of crowdsensing platforms.

3.2 Data Collection

The study obtained ethical approval from the Research Ethics UTM committee. The adapted questionnaires were tested in a selected pilot group (n = 30 respondents) to assess the clarity of the questionnaires. A consent form was given during scheduled meetings or on-farm visits. The data collection was from 5 December 2022 to 9 November 2023. A demonstration of a prototype crowdsensing application, namely SmartAPT, and a video of how it might work was given before collecting responses from farmers at their sites (see Fig. 2).

The prototype combines crop pest and disease reporting and Plantix application, providing a visual walkthrough of its features and a practical example of image recognition for crop pests and diseases. The TAM section comprises four primary constructs, namely perceived usefulness (PU) with four items, perceived ease of use (PEOU) with five items, attitude toward use (ATU) with four items, and intention to use (ITU) with five items. The TAM questions were adapted from Ghani et al. (2019) and Rezaei et al. (2020).

The data collection method for distributing questionnaires to chili fertigation farmers involved several steps: firstly, the surveys were conducted either at individual farms or backyard or during chili fertigation courses organized by the Department of Agriculture, where prior permission was obtained to conduct the survey sessions. Following this, an explanation regarding the purpose of the survey was provided, and consent forms distributed to the farmers to be signed. Additionally, explanations regarding the survey questions, including sociodemographic information of the respondents, their knowledge about chili diseases and TAM, were given. Subsequently, the questionnaire was distributed to the respondents for completion. Throughout the survey process, the interviewer remained in close proximity to the respondents, providing assistance or further explanations for questions that may not be clearly understood. Upon completion of the survey, a certificate of appreciation was awarded to each respondent. The survey was conducted using two methods: paper-based and electronic mediums, specifically utilizing Google Forms. This approach accommodated varying internet connectivity in certain areas and the diverse backgrounds of respondents in terms of their proficiency in technology use. Fundamentally, respondents possessed basic literacy and technological skills to use mobile smartphones.

3.3 Data Analysis

Descriptive analysis was conducted using the Statistical Package for the Social Sciences (SPSS). Meanwhile, the measurement model underscores the relationship between latent and variable factors crucial for each construct. Ensuring a certain level of reliability and validity in the measurement

Fig. 2 Prototype of SmartAPT

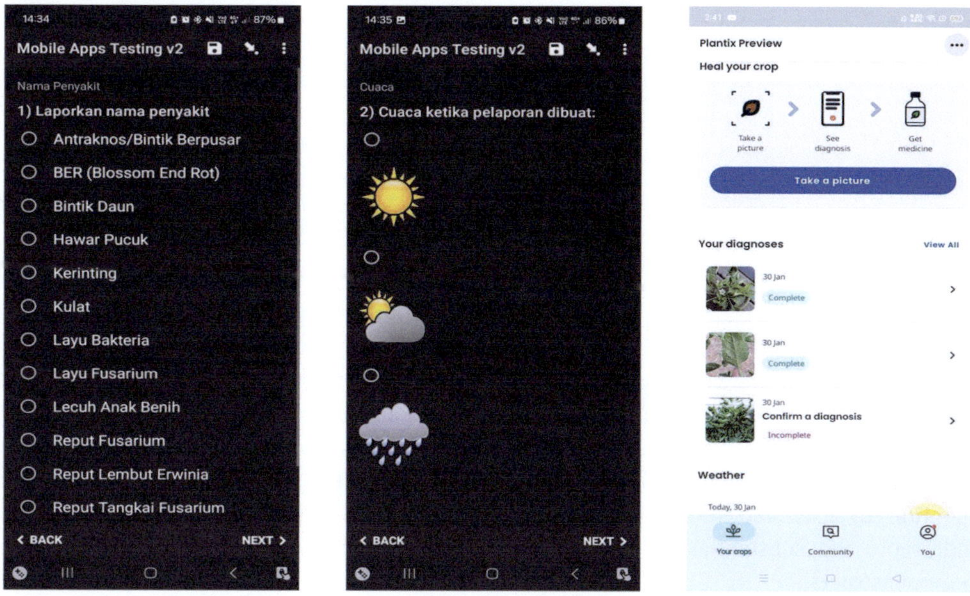

model is essential before analyzing other relationships in the structural model. Therefore, this study utilized SmartPLS 4 software to comprehensively assess the measurement model's reliability and validity through various statistical tests and indicators. First, the reliability of each construct was evaluated using Cronbach's alpha coefficient, with a threshold value of at least 0.7 to indicate acceptable reliability. Additionally, the average variance extracted (AVE) was calculated for each construct, with a threshold value of 0.5 or higher indicating convergent validity. Discriminant validity was assessed by comparing the square root of the AVE for each construct with the correlations between constructs, ensuring that the square root of the AVE for each construct was greater than its correlation with other constructs. Furthermore, correlation analysis was conducted to examine the relationships between constructs using Pearson's correlation coefficient, with a threshold value of 0.3 or higher considered indicative of a significant relationship. Finally, the structural model with moderator was tested to validate the proposed hypotheses, employing techniques such as path analysis and moderation analysis. These steps were crucial in validating the measurement model and laying the foundation for subsequent structural modeling analyses based on the TAM.

4 Results and Discussion

4.1 Descriptive Analysis

The descriptive analysis of farmers' intentions toward SmartAPT using the technology acceptance model (TAM) reveals a highly positive inclination among respondents. The perceived usefulness scores are notably high, indicating that farmers generally view the SmartAPT as valuable for their agricultural practices. Similarly, the perceived ease of use scores is strong, suggesting that farmers find the technology user-friendly. Attitudes toward use and intentions to use the SmartAPT are also positive, with mean scores reflecting a favorable disposition and a strong intention among farmers to adopt the technology. The low standard deviations across categories indicate a high level of agreement among respondents, reinforcing the overall positive trend. The findings collectively suggest a significant readiness and acceptance of smart agriculture precision technology within the surveyed farming community, showcasing a promising outlook for technology adoption in the agricultural sector. The capability for self-learning (77.60%, n = 97) to operate a smartphone contributed positively to the motivation for using smartphone applications. Additionally, from 125 respondents, it was reported that factors such as friends (31.20%) and family (43.20%) played a role in enhancing their awareness of the latest smartphone technology.

4.2 Construct Reliability and Convergent Validity

The reliability and validity metrics evaluation in Table 1 underscores the strength of our study's latent constructs. The Cronbach's alpha values, above 0.9, indicate an acceptable level of internal consistency. Additionally, composite reliability measures (rho_and rho_c) consistently surpass the recommended threshold of 0.70, further confirming the constructs' reliability. The assessment of convergent validity through average variance extracted (AVE) yields values exceeding the conventional threshold of 0.50 for all

Table 1 Reliability and validity of the four constructs in TAM

	Cronbach's alpha	Composite reliability (rho_a)	Composite reliability (rho_c)	Average variance extracted (AVE)
Attitude toward use	0.955	0.96	0.967	0.881
Intention to use	0.964	0.964	0.972	0.874
Perceived ease of use	0.893	0.946	0.923	0.752
Perceived usefulness	0.965	0.966	0.975	0.906

constructs. The R Square statistic explains the amount of variance in the dependent (ATU & ITU) variables accounted for by the exogenous (PEO and PU) variables. The R-squared values—60.5% for attitude toward use, 72.5% for intention to use, and 69.7% for perceived usefulness—underscore the substantial influence of the predictors on these constructs.

The square root of AVE for each construct (attitude toward use, intention to use, perceived ease of use, and perceived usefulness) is compared with the correlations of the constructs with all other constructs in the model. The values (square root of AVE) are consistently higher than the correlations between constructs, indicating that each construct shares more variance with its respective items than with items from other constructs. This pattern confirms discriminant validity, affirming that the latent constructs in the model effectively measure distinct aspects of farmers' perceptions of SmartAPT.

4.3 Correlations Analysis

Latent variable correlations among attitude toward use, intention to use, perceived ease of use, and perceived usefulness reveal strong positive associations (Table 2). For instance, a robust correlation of 0.834 between attitude toward use and intention to use signifies a close link between positive attitude and high intention to use SmartAPT. Similarly, positive correlations between attitude toward use and perceived ease of use (0.767) as well as perceived usefulness (0.712) indicate that a favorable attitude is associated with perceptions of ease of use and usefulness. Moreover, positive correlations between intention to use and perceived ease of use (0.797) and perceived usefulness (0.715) suggest that a strong intention to use is positively connected with perceived ease of use and usefulness. Additionally, the high correlation coefficient of 0.835 between perceived ease of use and perceived usefulness underscores their interdependence, implying that farmers perceive the technology as more useful when it is perceived to be easier to use. Overall, these correlations highlight the intricate relationships among these latent variables in shaping farmers' perceptions and intentions regarding the adoption of SmartAPT.

4.4 Structural Model with Moderator

The R Square statistic is a measure that explains the amount of variance in the endogenous (dependent) variable that is accounted for by the exogenous (independent) variable(s). The structural model depicted in Fig. 3 displays the R2 value for any given endogenous variable. Notably, PEOU (p-value < 0.0001) is not predicted by any other variables, as indicated by the absence of an arrow pointing toward it. Consequently, the R2 value reflects the extent to which the other variables in the model account for the variance in the endogenous construct. The R2 value ranges from 0 to 1, with a higher value signifying more accurate estimations in the path model.

In assessing the explained variance in endogenous constructs like attitude toward use, intention to use, and perceived usefulness, the results align with Chin's (1998)

Table 2 Latent variable correlations among four constructs

	Attitude toward use	Intention to use	Perceived ease of use	Perceived usefulness
Attitude toward use	1.00	0.834	0.767	0.712
Intention to use	0.834	1.00	0.797	0.715
Perceived ease of use	0.767	0.797	1.00	0.835
Perceived usefulness	0.712	0.715	0.835	1.00

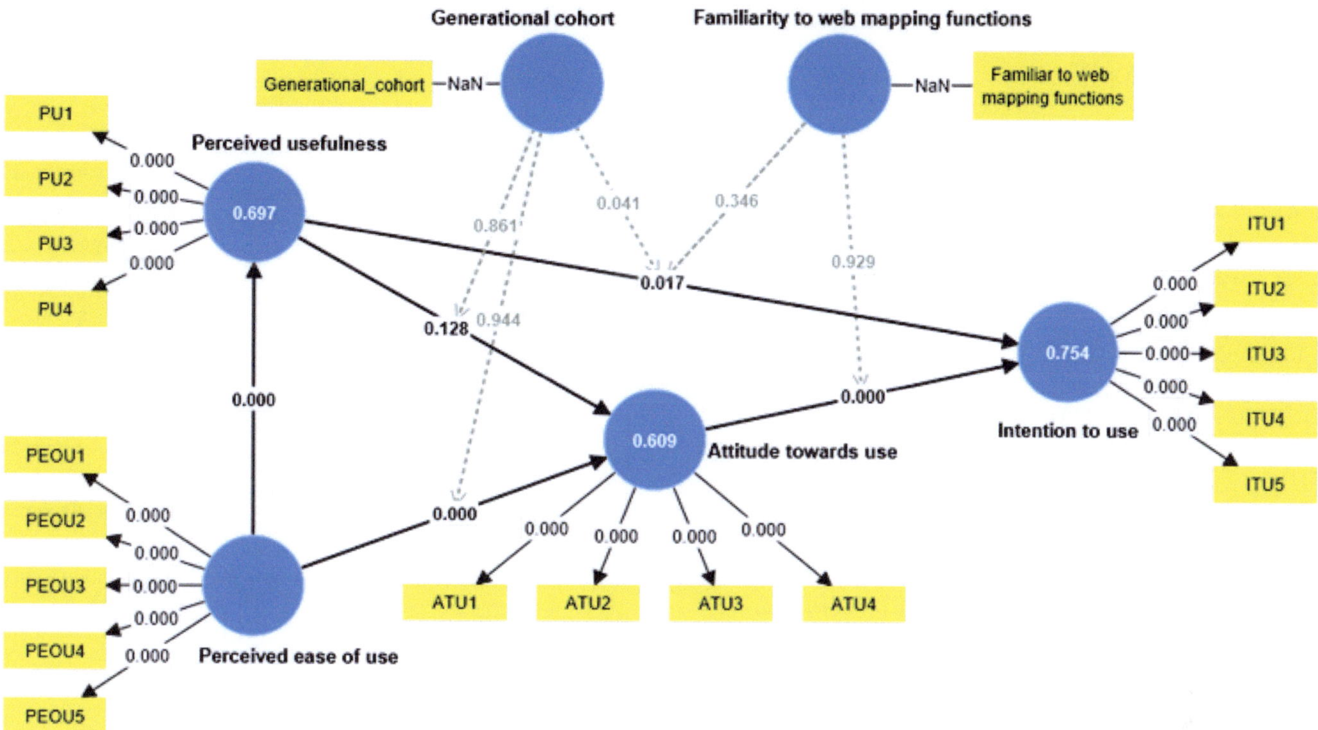

Fig. 3 Structural model with two moderators (generational cohort and familiarity to web mapping functions)

categorization of substantial impact. The R-squared values—60.9% for attitude toward use, 75.4% for intention to use, and 69.7% for perceived usefulness—underscore the substantial influence of the predictors on these constructs. This signifies a robust relationship between the independent and dependent variables, emphasizing the validity and effectiveness of the models in elucidating farmers' attitudes, intentions, and perceptions regarding the adoption of SmartAPT.

This study explores the relative importance of constructs in TAM—perceived usefulness and perceived ease of use—to predict farmers' adoption of the SmartAPT mobile application for reporting pests and diseases. Notably, perceived ease of use (PEU) strongly influences the perceived usefulness (PU) of the app (H1 Supported), highlighting the crucial role of user-friendly features in influencing the belief that the crowdsensing platform (SmartAPT) might be useful in reporting crop pests and disease, as displayed in Table 3.

On the contrary, the hypothesis that perceived usefulness (PU) significantly impacts farmers' attitudes toward the use (ATU) of the SmartAPT app is not supported (H2 Not supported). As argued by Lee (2006), PU is the extent to which individuals believe that using new technology will improve their task performance. Prastiawan et al. (2021) agreed that users' attitudes (ATU) toward using technology are influenced by the perceived benefits of its usefulness upon adopting a technology. Magsamen-Conrad and Dillon (2020) stated that it is crucial for potential users to have the opportunity to test the technology before adopting it, especially for older generations. The perceived benefits of the proposed technology might not be sufficient enough to convince farmers (respondents) as this is the first time they were introduced to this technology; hence if they are given sufficient time to try and use the technology in their daily operation, might generate positive perception toward ATU. As argued by Kim et al. (2007) knowledge gap of the quality of technology determines its usefulness. In this study, the Internet generation group, which seemed to have a positive intention to use such applications. However, their tendency to use may decrease because they do not fully grasp the real advantages and benefits of its usage.

Meanwhile, perceived ease of use is a strong predictor of farmers' attitudes toward using the SmartAPT app (H3 supported). This is in line with a study by Chen et al. (2018) on the use of supply chain systems in agriculture. In a simple analogy, ATU refers to personal opinions on the coolness of SmartAPT for reporting pests and diseases whereas ITU refers to their action on whether they will use the technology, or vice versa. Our study found a significant correlation between farmers' perception of the perceived usefulness of the SmartAPT app and their intention to use it (ITU) (H4 supported). Similarly, how people feel about using the app (ATU) considerably impacts their intention to use it (ITU) (H5 Supported). This finding aligns with Michels et al. (2019) who indicated significant positive relationships between perceived ease of use, perceived usefulness, and

Table 3 Coefficient calculation results

	Original sample (O)	Sample mean (M)	STDEV	*t*-Statistic	*p*-Value	Hypothesis
PEOU → PU	0.835	0.84	0.037	22.666	0.001	**H1 Supported**
PU → ATU	0.208	0.193	0.137	1.522	0.128	H2 Not supported
PEOU → ATU	0.593	0.606	0.121	4.886	0.001	**H3 Supported**
PU → ITU	0.226	0.211	0.095	2.379	0.017	**H4 Supported**
ATU → ITU	0.679	0.697	0.096	7.103	0.001	**H5 Supported**
GC × PU → ATU	0.025	−0.011	0.144	0.175	0.861	H6 Not supported
GC × PEOU → ATU	−0.009	0.022	0.129	0.07	0.944	H7 Not supported
GC × PU → ITU	0.115	0.112	0.056	2.047	0.041	**H8 Supported**
KOM × PU → ITU	−0.106	−0.086	0.112	0.943	0.346	H9 Not supported
KOM × ATU → ITU	−0.010	−0.031	0.111	0.089	0.929	H10 Not supported

Notes Sample size = 125 respondents; PEOU, perceived ease of use; PU, perceived usefulness; ATU, attitude toward use; ITU, intention to use; GC, generational cohort; and KOM, knowledge in online mapping

intention to use smartphone apps for herd management among German dairy farmers. Specifically, their findings suggested that farmers who perceived the apps as easy to use were more likely to find them helpful, leading to a higher intention to use them.

The study further investigated the influence of age (categorized according to which period they were born—generational cohort) as a moderator. The analysis revealed that perceived usefulness significantly impacted farmers' intentions to use the application, moderated by age group (H8 supported). In this study, the perceived usefulness of using crowdsensing technology in crop pests and disease monitoring has varied impact if viewed by generational cohorts. Chuang et al. (2020) identified young farmers might have a good perception of using digital technology in agriculture. However, the perceived usefulness of using crowdsensing to report diseases, hence being able to speed up the diagnosis and treatment processes, has varied interactions, particularly among the Internet and Reformist cohorts. Neither perceived usefulness nor perceived ease of use significantly affected attitudes toward use when moderated by the generational cohort (H6 and H7 not supported). Another moderator utilized in this study is knowledge in online mapping, which refers to respondents' understanding of three basic functions in Google Maps: search function, geolocation, and map layer. The analysis results indicate that both constructs (perceived usefulness and attitude toward use) moderated by knowledge in online mapping showed no significant effect on intention to use (H9 and H10 not supported).

Thomas et al. (2023) have highlighted the keys to farmers' acceptance of use of smart agricultural technologies are driven by 'facilitators' to use the technology. Perceived usefulness which is commonly related to the perceived benefits of using technology is influenced by facilitators such as familiarity with the technology and the views of seeing the potential to reduce farming costs and improve productivity. Sufficient training and resources to use the technology could promote farmer acceptance. Meanwhile, usability is the key driver to induce PEOU in the SmartAPT for crowdsensing-based crop pest and disease monitoring, hence the system needs to take a user-centric approach (Ouma et al. 2023) that prioritizes the demands of farmers over those of system developers or funders with a friendly and straightforward (uncomplicated) user interface (Malgaonkar et al. 2022). Offline capabilities (Mrisho et al. 2020) are essential for farmers with limited internet access, especially in rural and remote areas far from the coverage of telecommunication towers.

5 Conclusion

This study explored the adoption of a mobile crowdsensing platform, namely SmartAPT app, for pest and disease reporting among farmers. This study aims to fill the empirical gap by contributing to the knowledge gap of technology acceptance in developing countries, specifically in Malaysia, with a focus on crowdsensing applications among farmers for bottom-up reporting of pest and disease outbreaks. While ease of use significantly influenced perceived usefulness, surprisingly, usefulness did not directly impact farmers' attitudes toward the app. However, both usefulness and attitudes significantly predicted the intention to use the app. Interestingly, the influence of usefulness on intention varied by age, potentially with certain younger farmers not valuing it more, hence further studies are needed. Based on these findings, promoting SmartAPT adoption should focus on user-friendly design, targeted outreach by age group, emphasizing concrete benefits like cost reduction, and ensuring offline capabilities for limited internet areas. By addressing these factors, the Department of Agriculture via its extension agents can empower farmers to embrace this valuable tool for improved crop health management for pest and disease monitoring and surveillance.

Utilizing the TAM model, which is restricted to the perceived ease of use and, therefore, disregards the technology's feasible implementation on the ground, is a limitation of this study. Incorporating an extended version of the technology acceptance model, such as the theoretical framework of acceptability (TFA), allows for the assessment of supplementary dimensions of technology acceptability. Moreover, it facilitates the evaluation of the feasibility of implementing crowdsensing technology within their current settings.

Acknowledgements The authors would like to acknowledge the financial support from the Ministry of Higher Education Malaysia under the Fundamental Research Grant Scheme (FRGS) (FRGS/1/2021/WAB04/UTM/02/2) and Universiti Teknologi Malaysia for the funding under FRGS R.J130000.7852.5F485.

References

Ajzen, I.: The theory of planned behavior. Organ. Behav. Hum. Decis. Process. **50**(2), 179–211 (1991)

Bahar, N.H., Lo, M., Sanjaya, M., Van Vianen, J., Alexander, P., Ickowitz, A., Sunderland, T.: Meeting the food security challenge for nine billion people in 2050: What impact on forests? Glob. Environ. Chang. **62**, 102056 (2020)

Che'Ya, N. N., Mohidem, N. A., Roslin, N. A., Saberioon, M., Tarmidi, M. Z., Man, N.: Mobile computing for pest and disease management using spectral signature analysis: a review. Agronomy 12(4), 967 (2022)

Chen, X., Wu, L., Pan, Y., Siu, K. W. M., Gong, X., Zhu, D.: Consumer acceptance of an agricultural products traceability system: evidence from China. AgEcon Search (2018)

Chin, W.W.: The partial least squares approach to structural equation modeling. Modern Method. Busines. Res. **295**(2), 295–336 (1998)

Chuang, J.H., Wang, J.H., Liang, C.: Implementation of Internet of Things depends on intention: Young farmers' willingness to accept innovative technology. Int. Food Agri. Manage. Rev. **23**(2), 253–266 (2020)

Davis, F. D.: Perceived usefulness, perceived ease of use, and user acceptance of information technology. MIS Quarterly, 319–340 (1989)

Durodolu, O.: Technology acceptance model as a predictor of using information system' to acquire information literacy skills. Library Philosophy Practice (2016)

Fairuz, K., Idris, A. G., Syahrizan, S., Hatijah, K.: Assessing farmers' community readiness towards the enhancement of natural enemy population in rice fields in Malacca. In: AIP Conference Proceedings (Vol. 1940, No. 1). AIP Publishing (2018)

FAO: The future of food and agriculture—Trends and challenges. Food and Agriculture Organization of the United Nations, Rome (2017)

Ganti, R.K., Ye, F., Lei, H.: Mobile crowdsensing: current state and future challenges. IEEE Commun. Mag. **49**(11), 32–39 (2011)

Ghani, M.T.A., Hamzah, M., Ramli, S., Ab, W., Daud, A.W., Romli, T.R.M., Mokhtar, N.N.M.: A questionnaire-based approach on technology acceptance model for mobile digital game-based learning. J. Global Busin. Social Entrepren. (GBSE) **5**(14), 11–21 (2019)

Ismail, N. A., Bakar, M. A., Hilmi, M. I.: Encouraging factor for ICT usage in agriculture from supplier view: a case study in southern region, Malaysia (2023)

Kim, H.W., Chan, H.C., Gupta, S.: Value-based adoption of mobile internet: an empirical investigation. Decis. Support Syst. **43**(1), 111–126 (2007)

Lee, Y.C.: An empirical investigation into factors influencing the adoption of an e-learning system. Online Inf. Rev. **30**(5), 517–541 (2006)

Magsamen-Conrad, K., Dillon, J.M.: Mobile technology adoption across the lifespan: a mixed methods investigation to clarify adoption stages, and the influence of diffusion attributes. Comput. Hum. Behav. **112**, 106456 (2020)

Malgaonkar, S., Licorish, S.A., Savarimuthu, B.T.R.: Prioritizing user concerns in app reviews—A study of requests for new features, enhancements and bug fixes. Inf. Softw. Technol. **144**, 106798 (2022)

Michels, M., Bonke, V., Mußhoff, O.: Adoption of herd management smartphone apps in German dairy farming. In: The World's Largest Open Access Agricultural & Applied Economics Digital Library (2019)

Mlekus, L., Bentler, D., Paruzel, A., Kato-Beiderwieden, A. L., Maier, G. W.: How to raise technology acceptance: user experience characteristics as technology-inherent determinants. Gruppe. Interaktion. Organization. Zeitschrift für Angewandte Organisationspsychologie (GIO), 51(3), 273–283 (2020)

Morchid, A., El Alami, R., Raezah, A. A., Sabbar, Y.: Applications of internet of things (IoT) and sensors technology to increase food security and agricultural sustainability: Benefits and challenges. Ain Shams Eng. J. 102509 (2023)

Mrisho, L.M., Mbilinyi, N.A., Ndalahwa, M., Ramcharan, A.M., Kehs, A.K., McCloskey, P.C., Legg, J.P.: Accuracy of a smartphone-based object detection model, PlantVillage Nuru, in identifying the foliar symptoms of the viral diseases of cassava–CMD and CBSD. Front. Plant Sci. **11**, 590889 (2020)

Mustafa, H., Mukhiar, S. N. S., Jamaludin, S. S. S., Mohd Jais, N.: Malaysian generational cohorts in the new media era: Historical events and collective memory. Media Asia 49(3), 235–256 (2022)

Ouma, G.O., Awuor, F.M., Makiya, C.R., Okanda, P.: Leveraging mobile phones in crop pests and disease surveillance-perspectives of farmers in Kenya. Arch. Agricult. Res. Technol. **4**(2), 1051 (2023)

Prastiawan, D.I., Aisjah, S., Rofiaty, R.: The effect of perceived usefulness, perceived ease of use, and social influence on the use of mobile banking through the mediation of attitude toward use. APMBA (Asia Pacific Management and Business Application) **9**(3), 243–260 (2021)

Rezaei, R., Safa, L., Ganjkhanloo, M.M.: Understanding farmers' ecological conservation behavior regarding the use of integrated pest management-an application of the technology acceptance model. Global Ecol. Conserv. **22**, e00941 (2020)

Salim Sa'don, S.A.: The adoption of internet of things in urban farming. Res. Manage. Technol. Busin. **2**(2), 146–162 (2021)

Steinke, J., van Etten, J., Müller, A., Ortiz-Crespo, B., van de Gevel, J., Silvestri, S., Priebe, J.: Tapping the full potential of the digital revolution for agricultural extension: an emerging innovation agenda. Int. J. Agric. Sustain. **19**(5–6), 549–565 (2021)

Thapa, B.E., Niehaves, B., Seidel, C.E., Plattfaut, R.: Citizen involvement in public sector innovation: government and citizen perspectives. Information Polity **20**(1), 3–17 (2015)

Thomas, R.J., O'Hare, G., Coyle, D.: Understanding technology acceptance in smart agriculture: a systematic review of empirical research in crop production. Technol. Forecast. Soc. Chang. **189**, 122374 (2023)

Thompson, W.J., Varma, V., Joerin, J., Bonilla-Duarte, S., Bebber, D.P., Blaser-Hart, W., Krütli, P.: Smallholder farmer resilience to extreme weather events in a global food value chain. Clim. Change **176**(11), 152 (2023)

Climate Change versus Current Building Practices: Assessment of the Transformation of Vernacular Architecture of the Sundarbans Delta Region in Bangladesh

Gourab Kundu and Dipannita Nandi

Abstract

The Sundarbans Delta is a vast mangrove forest consisting of the lower part of the Ganga–Brahmaputra Delta which covers most of Bangladesh and West Bengal (India). This mangrove-covered coastal area has a significant historical record of human habitation spanning several centuries because of its abundant natural resources and biodiversity which, in turn, has given rise to a distinct vernacular architecture. Vernacular architecture is intentionally crafted to accommodate a particular region's specific climatic conditions and cultural context. Due to rapid urbanization and climate change, traditional architectural practices are at stake. Considering the imminent danger of climate change as a worldwide issue, it is crucial to understand the impacts of these climatic changes on the traditional building techniques and construction materials of the Sundarbans region to ensure resilience. Cyclones, tidal surges, rising sea levels, and coastal flooding have increased in frequency in the last two decades, yielding to the vulnerable living conditions of this area. To seek permanence and reduce the extent of vulnerability, the local population has started following up-to-date practices for dwelling unit construction. In fact, several shelter programs initiated by the government and non-government organizations (NGOs) in response to disasters and forest preservation emphasize modern practices in this disaster-prone region as a quick response and influenced by efficient labor force availability, mass production of construction materials, and cost-effective execution, without even considering traditional vernacular techniques. This study aims to analyze the dichotomy between climate change's influence on the Sundarbans Delta's conventional architecture and the trend of adopting new approaches to mitigate climatic hazards. A comprehensive documentation of the Sundarbans delta region's distinctive architectural alterations has been presented, contributing to its cultural identity and resistance against environmental forces. This has further unveiled the scope of incorporating traditional methods with contemporary practices to mediate ecological and climatic degradation.

Keywords

Vernacular architecture · The Sundarbans · Climate change · Building practices · Resilience and adaptation

1 Introduction

There have been substantial and unforeseen changes in global climate trends in recent decades. Consequently, severe climatic phenomena have been witnessed worldwide. Climate change is leading to more frequent and severe weather events, including storms, floods, and heatwaves (Banholzer et al. 2014). Effects of climate change has prompted a reconsideration of various aspects of human existence, including architectural methodologies (Li et al. 2021). It also threatens to erode the knowledge systems associated with vernacular architecture (Brimblecombe et al. 2008). Bangladesh is being affected by climate change in a multitude of ways. A report of World Bank in 2022, stated that Climate change would strike poor and vulnerable people the worst (The World Bank 2022). Also, various research showed that the coastal area of Bangladesh is particularly exposed to the effects of climate change, i.e., sea level rise, increasingly powerful cyclones, floods, increasing salinity, and land inundation. These climatic events are also threatening the Sundarbans along with the settlements relying on it by several means (Dasgupta et al. 2020).

G. Kundu (✉) · D. Nandi
Department of Architecture, Bangladesh University of Engineering and Technology, Dhaka, Bangladesh
e-mail: gourabkundu01@gmail.com

P. Mohammad et al. (eds.), *Geographic Approaches to Climate Change and Mitigation: Urban and Rural Perspectives (Volume 1)*,
Advances in Science, Technology & Innovation, https://doi.org/10.1007/978-3-031-92119-3_33

The Sundarbans is renowned for its unique ecological significance. The environment of the Sundarbans comprises an interconnected system of tidal rivers characterized by the presence of stable tidal flats, mangrove swamps, and dunes. It covers the south-western region of Bangladesh, consisting of three districts: Shatkhira, Khulna, and Bagerhat. The forest includes the southern portion of India's West Bengal state. Overall, this coastal area, characterized by a dominant presence of mangroves, has a significant historical background of human habitation (Eaton 1990; Sen and Ghorai 2019), which has been deeply linked with the distinctive landscape and climatic patterns. This coastal forest not only helps to sustain the biodiversity but also ensures the livelihoods of the people from the neighboring areas. Besides, it helps shield these southern coastal districts from the worst of the region's many cyclones and tidal surges (Akber et al. 2018; UNESCO n.d.). The distinctive geographical feature of this coastal area and the dependency on the mangrove forest have resulted in a unique architectural style in this region (Dutta and Rahman 2022; Bhuiyan et al. 2021).

However, these traditional construction methods are vulnerable to environmental degradation (Hossain et al. 2014), prompting local people to implement various modifications in their local architecture to combat the swift climate fluctuations (Ashrafuzzaman 2023). As the impacts of climate change intensify, the Sundarbans delta region has started witnessing a transition toward modern building practices, aimed at enhancing resilience. The incorporation of climate-resilient materials, innovative design strategies, and elevated structures has become a focal point in the efforts to adapt to the changing environment (Rahman et al. 2019; Linkon 2017). Conversely, this shift raises questions toward preserving cultural identity and potentially losing knowledge, embedded in vernacular architecture.

The study investigated the intricate correlation between climate change and modern building techniques, explicitly emphasizing the notable impacts on vernacular architecture in the Sundarbans delta region of Bangladesh. The research primarily concentrated on the south-western districts, which are in the Sundarbans Impact Zone (SIZ) of Bangladesh, i.e., areas within 20 km of Sundarbans Reserve Forest perimeter (Dasgupta et al. 2020). The study of climate change, contemporary construction techniques, and traditional architecture in the Sundarbans Delta Region highlights the necessity of comprehending these interconnections. Findings have revealed the impact of climate change on vernacular architecture and measures taken to address it, particularly in building design, construction, and material choices.

2 Aims and Objectives

The study sought to explore how climate change poses a threat to the cultural identity of a local community and how they cope with its consequences, with a specific focus on local architectural practices, which serve as a key expression of their identity. Therefore, the following objectives have been targeted to fulfill the aim of the study—

- Assess the extent of climate change in reshaping the delta landscape.
- Ascertain the evolving practices resulting from climate change and their acceptance within the local community.
- Investigate the consequences of the altered architectural landscape in the Sundarbans delta region.

3 Methodology

An empirical qualitative approach has been adopted to achieve the objectives of the study. Literature-based secondary data has been analyzed to identify the current climatic hazards in the delta region and their impact on the lifestyle of the delta-dwellers. To conduct the study, three areas of Bangladesh's Sundarbans Impact Zone (SIZ) with high climate vulnerability index (CVI) and mangrove forest vulnerability (Goosen et al. 2018) have been selected: Dacope (bordering the northern part of the Sundarbans), Mongla (covering the north-western perimeter) and Sarankhola (furthest western border) as shown in Fig. 1. Changes in building construction practices in the settlements of these areas (at least 15 homesteads from each area) have been documented from field surveys and direct observations. Resource persons from the locals were selected beforehand to validate the documentation process and focus group discussions have been conducted. Collected data has been categorized according to the hazards frequently taking a toll in the delta region and a thorough analysis has been done to discern the patterns, trends and challenges in the transformation process of the local built environment. Sketches and photographs have also been used to complement the survey findings.

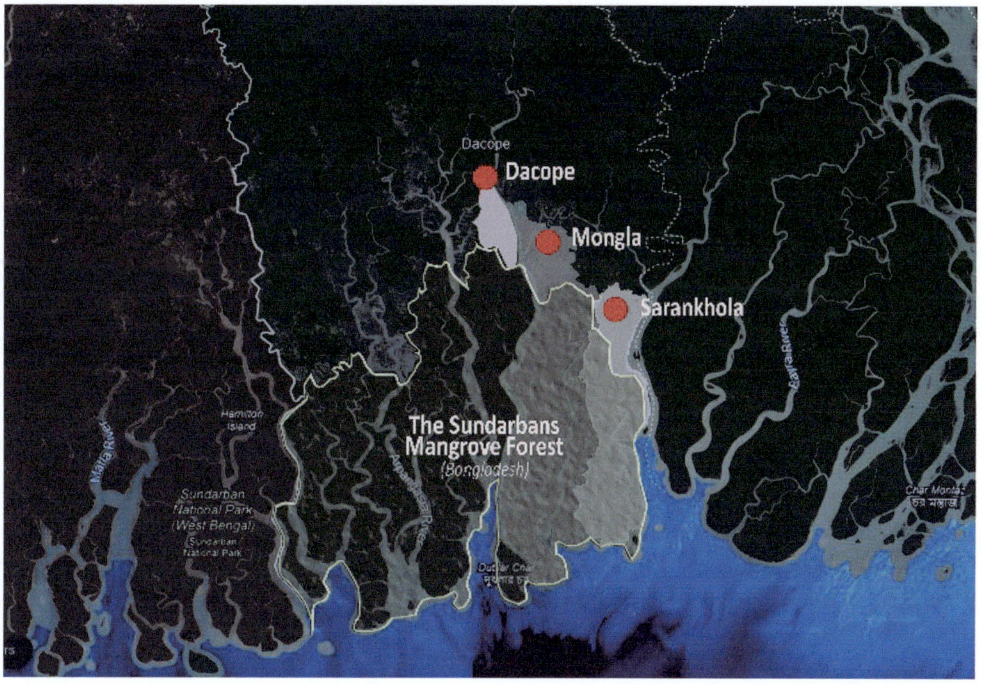

Fig. 1 Satellite map of the Sundarbans (21°32′ to 22°40′ N, and 88°05′ to 89°00′ E) with study area (by authors)

4 Literature Review

4.1 Climatic Components

The Sundarbans was formed almost 7000 years ago due to the sediments carried by the Ganges River system from the foothills of the Himalayas (Aziz and Paul 2015). Occupying the south-west part of Bangladesh, it exhibits a subtropical climate, featuring hot and humid summers and temperate winters. The average yearly temperature of the Sundarbans in Bangladesh ranges from (maximum) 30 °C to (minimum) 21 °C. The annual range of rainfall is between 1600 and 1800 mm (Ghosh et al. 2015), and the monsoon brings most rainfall in between May and October (Gosh et al. 2017). Increased water flow during the monsoon season contributes to baseline erosion in the forest area as low sediment deposition obstructs the formation of protective barriers and makes it more susceptible to erosion (Aziz and Paul 2015). The area is also frequently affected by tropical cyclones and lesser tidal events during the monsoon, resulting in significant flooding and wind damage (Mandal et al. 2019). Tropical cyclones have a great impact on this area as it has experienced more than 20 major cyclones over last two decades, leaving its impact mainly on agriculture and rural infrastructures (Dasgupta et al. 2020). Saline water intrusion is another significant occurrence in the deltaic region of the Sundarbans as the salinity of this area rises sharply from 10% (in the monsoon) to almost 40% in the dry season (Hussain et al. 2013).

4.2 Vernacular Architecture and Traditional Practices

The local inhabitants of Sundarbans have developed a distinctive vernacular architecture that reflects the specific geographic condition and possibilities posed by its unique natural environment. The monsoons, floods, and tropical cyclones have governed the lives and livelihoods of the people living in this region (Pitchaikani et al. 2015). The characteristics of this vernacular architecture are shaped by the cultural influences, availability of materials, and climatic conditions. The dwelling units, typically made from locally sourced materials, are considered as need-based, microclimate-responsive, low-cost and mostly constructed through self-help practices, utilizing knowledge accumulated over generations (Linkon 2017).

Drawing from their extensive experience with disasters, residents have developed a unique spatial layering for their homes. Their *"Ghar"* or main living area is deliberately placed surrounded by "*Pashchati*", serving as a verandah or semi covered room. This layer shields the *"Ghar"* from cyclone-induced wind and rainwater damage (Rashid 2017). Traditionally, the earthen plinth of these dwellings is raised to about 5–6 feet or sometimes built on stilts to withstand critical circumstances caused by tidal surges, flooding, and fluctuating water levels in this area (Rahman 2011).The building of the elevated floor or plinth is typically executed with meticulous care and attention, as it frequently stands as the only remaining evidence of a structure after enduring an extreme storm or flood (Rahman 2011 and Anwar 2022).

Fig. 2 Local built environment in the SIZ :Dakope, Khulna (image by authors)

Bamboo is utilized for both walls and structures due to its abundant availability and exceptional tensile strength, which effectively safeguards the units against strong wind velocities. Timber extracted from the Sundarbans (mainly from *Sundari* tree) is used as load-bearing members, walls, door and window frames, and roof truss. Generally, bamboo mats (corrugated iron sheets in some cases) are used for enclosures in semi-pucca (semi-permanent) units (Fig. 2), whereas in kutcha (temporary) houses materials like, straw, bamboo mats, jute sticks, and dry grass are used traditionally (Linkon 2017).The combination of bamboo matt and wooden frame is also used for non-load-bearing wall construction. In addition, mud and clay are used for building walls and floors. Earthen houses are built following the 'Cob technique' where the units are constructed by combining mud with water as well as additions like, straw, jute, or rice husk to enhance their strength throughout the drying and shrinking process (Gutiérrez et al. 2018). Thermal insulation to these walls is provided through applying a coating of soil and cow manure mixture. Sometime additional layer of *"Golpata"* or Nipa Palm (*Nypa fruticans)* is made to protect the mud wall from cyclone and driving rain (Hossain et al. 2008). Window shutters are mostly louvered, allowing airflow in the indoor spaces, and added perforation over the windows ensures ventilation (Linkon 2017).

Cyclones and high winds are major guiding factors for the form and shape of roofs here. Most of the houses are constructed with a hip roof and a separate lower roof (Rashid 2017; Alam et al. 2017). This preference reflects their confidence in the hip roof's resilience against cyclonic conditions, highlighting the practical wisdom accumulated over generations for living in cyclone-prone areas (Rashid 2017; Rahman 2011). In kutcha (temporary) houses, roofing systems typically incorporate organic materials like *Nipa Palm* or thatch, supported by split bamboo framing. In contrast, semi-pucca (semi-permanent) houses employ CI sheet roofing with timber or bamboo framing (Linkon 2017). In response to rising hazards, residents employ materials like polythene, cement bags, or banner cloth for enhanced rain protection. In some cases, roof tiles are also employed in this region. The pitch of a roof can range from 30 to 45 degrees to effectively mitigate significant wind pressure and minimize any damage to the roof (Hossain et al. 2008). These traditional practices of constructing houses have been proven effective against the prevalent environmental conditions in the Sundarbans deltaic region.

4.3 Threats Posed by Climate Change

The Sundarbans deltaic region experiences several consequences of climate change, such as the full impact of increasing sea levels, severe weather events, intense rainfall in the short term, extreme heat, and an influx of salinity (Mandal et al. 2019). Most of the coastal areas in Bangladesh's Sundarbans Impact Zone (SIZ), such as Sarankhola, Dacope, Shyamnagar, Mongla, and Rampal, are already facing more severe salinity than usual. Saline water penetrates up to 250 km inward during the dry season, contaminating freshwater bodies and, therefore, threatening the regeneration of tree species (Nishat and Chowdhury 2019). Alongside its extensive history of tropical cyclones (Fig. 3), this region is witnessing a heightened occurrence of intense storms because of climate change (Mandal et al. 2019). A study reported a 26% increase in the frequency of cyclones over the last 120 years in the Bay of Bengal (Rahman et al. 2021). According to a study by United Nations Development Program (UNDP) in 2004, Bangladesh was identified as the most vulnerable country to tropical cyclones and the sixth most vulnerable country to floods (Dasgupta et al. 2014).

Devastation caused by cyclone "Aila" in 2009 exemplifies a multifaceted scenario involving cyclonic winds, storm surges, flooding, water logging, and saline intrusion. These events have profoundly impacted the socio-economic conditions of coastal communities, affecting housing, assets,

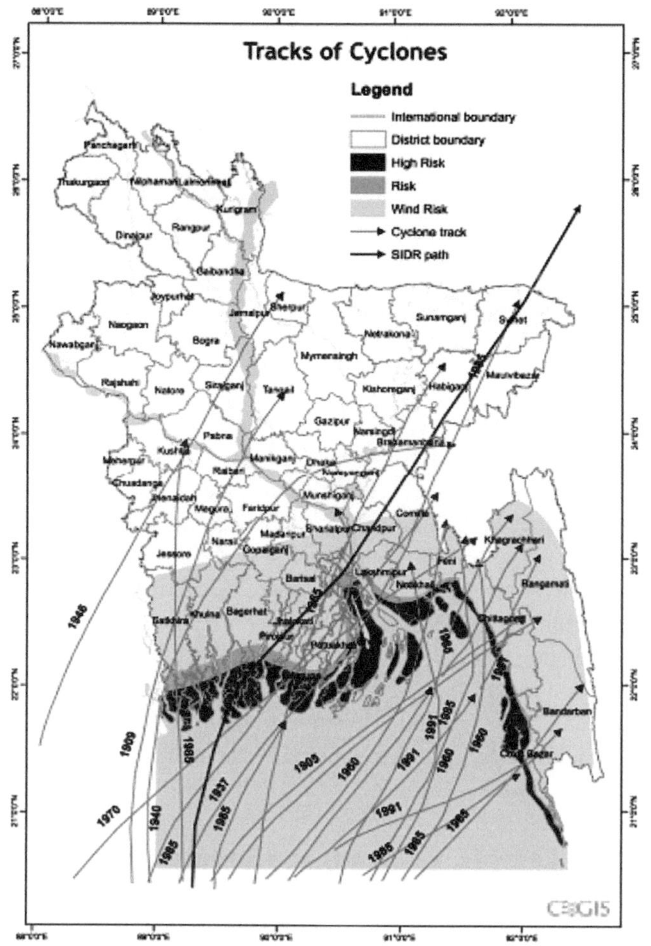

Fig. 3 Cyclone tracks of Bangladesh in last few decades (*Source* Bangladesh Climate Change Strategy and Action Plan, 2009)

income-generating activities, food consumption patterns, water supply, and sanitation (Subhani and Ahmad 2019). The escalating cyclonic activity and storm surges are not only impacting human settlements but also affecting the Sundarbans Forest area, leading to a reduction in forest cover (Sakib et al. 2015). As a result, the lives and livelihoods of the affected populations have been severely threatened. The increasing sea level leads to various issues, including heightened salinity, water logging, coastal flooding, and land erosion (Sahana and Sajjad 2019). A study pointed out that *"Hiron Point"* in the Sundarbans of Bangladesh is facing an increase of 4 mm in mean tidal level per year because of global warming, which is higher than the global rate (Islam 2019). The deltaic region is also undergoing landmass depletion and forest contraction due to the rising sea levels. According to a research article by *Rudra & Chattopadhyay*, coastal erosion impacts 90% of the population residing in the Sundarbans. Sea level rise induces saltwater intrusion, impeding mangrove growth, while increased cyclonic activity damages the forest structure (Bhargava and Friess 2022). Additionally, the changes in river dynamics affect sedimentation, further impacting mangrove habitats. To protect the Sundarbans and their biodiversity, the government has implemented regulations limiting forest resource collection. As a result, the residents of the Sundarbans Impact Zone have to limit their reliance on forest resources for livelihood, food, firewood, and building materials. A study, published in 2021, revealed that a decrease in reliance on the Sundarbans Forest resulted in a shift in dwelling patterns for 16% of the households examined (Shaikh et al. 2021).

As a whole, the region's traditional architecture is facing a substantial threat due to these severe climatic conditions. Dwellings are experiencing reduced durability due to the increased frequency of storms and rainfall, necessitating more frequent repairs and rebuilding. Rising salinity is also causing damage to the mud-built houses. As a result, efforts are underway to promote the adoption of sustainable materials and climate-resilient construction within communities (Anwar 2022). The reliance on forests for construction materials is diminishing as the mangrove forest faces depletion from frequent cyclones, rising saline levels, reduced freshwater flow from upstream, and human interventions (Mahadevia and Vikas 2012). Deforestation in this region is causing a loss of natural protection against climatic catastrophes. This phenomenon prompts individuals to embrace modern technologies and materials to adjust to emerging situations. This transition raises concerns about preserving cultural identity and highlights the potential loss of traditional expertise linked to vernacular architecture.

5 Findings

During the data collection phase of the study, four significant climatic hazards have been identified in the study area. Data gathered from surveys and interviews were meticulously documented and categorized based on the identified major climatic hazards. Later the collected data has been analyzed to discern the impacts of the hazards on vernacular architecture in the region.

5.1 Tropical Cyclones and Its Impact

5.1.1 Trend of Homogeneity

The tropical cyclones significantly impact the ever-changing practices of vernacular architecture in the Delta region of Sundarbans, which is evident from the survey findings. The southern frontier is being washed away significantly by cyclones like Sidr (2007) and Aila (2009), disrupting the settlement and livelihoods of the locals. As a result, the situation has compelled them to find alternative resources for quick recovery. Previously, the locals used to opt for various

Fig. 4 Use of CI sheet as building material in Mongla, Khulna and Sarankhola, Bagerhat (image by authors)

materials and construction techniques for their dwelling units, based on their access to resources. A survey by HIES (Household Income and Expenditure Survey) conducted in 2005 recorded a wide variation in household types based on construction materials in this region, among which almost 84% of households were semi-pucca, having CI (Corrugated Iron) sheet roof and wall constructed with materials like brick/cement (12%), CI sheet/ wood (35%), mud brick (14%) or hemp/bay/bamboo (21%), as documented in "Damage, Loss, and Needs Assessment for Recovery and Reconstruction after Cyclone Sidr" report by Government of Bangladesh in 2008 (The Government of Bangladesh, 2008). However, the study area portrays that with the increased frequency of tropical cyclones and their devastating consequences, the locals have started to opt for permanence in case of material selection and construction techniques to protect themselves during hazards and reduce damages in the post-hazard phase. As a result, a trend of homogeneity has become apparent in these coastal areas, especially in Dacope, Mongla, and Sarankhola as per the survey data. Materials like CI Sheets, cement roofing sheets, concrete blocks, and bricks are being frequently adopted over bamboo-thatch or mud-clay construction (Fig. 4).

5.1.2 Government and Non-Government Initiatives

Recovery initiatives, taken by the Government and Non-government Organizations in the post-hazard phase, have also influenced the material choices and construction practices of the locals. To provide reconstruction facilities for housing for the mass affected population, these organizations mainly source industrial materials because of their easy availability and mass supply, thereby accelerating the alteration of the local practices. This trend has further been confirmed by the present scenario of the study area. According to a report prepared by the Government of Bangladesh, with the assistance of the International Development Community in April 2008, the construction of a 15 $\times$ 15 feet house with a CI sheet roof and CI sheet fence to reduce the losses in the housing sector has been considered for easy availability of industrial material and reduced expenditure (Reza 2021). In disaster-prone rural areas, Housing and Building Research Institute has suggested to install reinforced concrete cement (RCC) or ferro-cement foundations rather than bamboo-stilt construction, as suggested by the Housing and Building Research Institute (Housing and Building Research Institute, 2017). Furthermore, multiple studies have recommended developing new design standards and guidelines to ensure infrastructure resilience in the face of climate change (Mirza et al. 2019; Flinn and Baresford 2009). These guidelines have gradually altered the landscape of rural households in the disaster-prone deltaic region.

5.1.3 Loss of Livelihoods and Reduced Income

Loss of assets and livelihood options due to frequent cyclones minimizes the scope of socio-economic recovery, which is reflected in the built forms (Subhani and Ahmad 2019). During a focus group discussion with the locals, it was recorded that households in these areas are adopting simpler forms due to insufficient income and lack of access to materials and other resources, exacerbating their capacity to resist hazards. The practice of the construction of *Pashchati* to resist wind pressure during storms and cyclones has now been almost diminished in the study areas due to reduced income and limited livelihood opportunities.

5.2 Saltwater Infiltration and Its Impact

5.2.1 Heightened Expense in Maintenance

Throughout the past twenty years, there has been a drop in run-off in the eastern rivers of the Sundarbans, an increasing trend of saltwater shrimp farming, and a rise in sea water level, leading to a continuous rise in soil salinity (Akter et al. 2020). The increased soil salinity has started to show its impact on the majority of the traditional mud buildings of the study area, leading to blistering and cracking of the mud floor and walls and raising the maintenance cost of these structures. Consequently, people opt for more

Fig. 5 Use of brick wall to protect the earthen plinth in Sarankola, Bagerhat, (image by authors)

Fig. 7 Use of plastic sheet to protect the wall, in Dacope, Khulna (image by authors)

resilient construction materials such as, *mud-concrete* over traditional ones to reduce expenses and effort. In order to safeguard the mud plinth from erosion, caused by decreased soil binding due to salinity, brick perimeter walls are now being constructed as well (Fig. 5).

5.2.2 Unavailability of Local Materials

Furthermore, the increased salinity is damaging the Sundarbans mangrove forest. Research conducted by Rahman in 2020 revealed a decline in the population of Sundari *(Heritiera fomes)* and Golpata or Nipa Palm (*Nypa fruticans)* which are the source of essential building materials in this area (Rahman M. M., 2020). Due to the lack of Golpata, individuals are driven to construct their roofs using CI sheets (Fig. 6) and utilize polythene to shield walls from cyclones and heavy rainfall (Fig. 7). The decline in paddy productivity in the Sundarbans delta region due to salinity (Miah et al. 2020) has also affected the availability of rice byproducts such as husk and straw, which are commonly used building materials in the study area. The scarcity of these building materials has led residents to explore alternative options for construction, thereby modifying the traditional vernacular architecture.

Fig. 6 Use of pre-cast concrete post in Sarankhola, Bagerhat, (image by authors)

5.3 Sea Level Rise and Land Erosion

Sea level rise is one of the significant consequences of climate change in the study area, as documented during the survey, posing significant challenges to traditional architecture. Researchers unveiled that the people of the area adjacent to the Sundarbans are losing their settlements and household properties due to the eased flow and force of water coming from the sea through rivers (Roy et al. 2022). Many traditional buildings in the study area are not designed to withstand the increased frequency and intensity of storms and tidal surges associated with sea level rise. This puts them at risk of collapse or severe damage during extreme weather events. As a result, changes in building construction methods in the study area are taking place to adapt to the increased risk of flooding and erosion.

In some cases, traditional construction techniques have also been modified and supplemented with more resilient approaches to enhance the durability and safety of buildings in this vulnerable region. The elevation of the structures is raised or sometimes constructed with stilts to set the height above the water level during the flood. Traditional building materials are being supplemented with more durable materials, i.e., concrete and steel reinforced columns, brick walls which can stand conveniently against the impacts of sea

Fig. 8 Bricks replacing traditional building materials in Mongla, Khulna (image by authors)

level rise (Fig. 8). Buildings are also being constructed or retrofitted with improved waterproofing measures to prevent water infiltration during flood. The land erosion caused by sea level rise is displacing communities living in low-lying areas of the Sundarbans. This displacement is also disrupting traditional architectural practices and may lead to the abandonment of certain settlements in the near future.

5.4 Forest Contraction and Deforestation

The decline of forested areas in the Sundarbans, resulting from a combination of factors intensified by climate change, significantly impacts traditional architecture. This primarily stems from the exhaustion of vital building materials such as timber from the "*Sundari*" tree (*Heritiera fomes*) and Golpata leaves (*Nypa fruticans*). The durability of *Sundari* tree timber, valued for its strength and resistance to decay, is highly valued in traditional construction. In addition, *Golpata* leaves are frequently utilized for roofing and wall construction due to their abundance and effectiveness in providing insulation and weatherproofing. As deforestation advances, the availability of these essential materials has evidently declined in the study area, resulting in scarcity and higher construction costs. Additionally, strict conservation policies and legal restrictions on harvesting plant-based materials from the Sundarbans make extraction challenging (BSS 2023). This scarcity not only impacts material affordability but also prompts the search for alternative materials, compromising the architectural integrity of traditional buildings.

6 Discussion

The research presented in this study sheds light on the intricate relationship between climate change and traditional vernacular architecture in the Sundarbans delta region of Bangladesh. The Sundarbans, with its unique ecological significance and historical human habitation, has fostered a distinctive architectural style that is deeply intertwined with the region's natural environment and cultural identity. However, climate change, with its various manifestations, is exerting direct and indirect influences on the region's architectural practices (Fig. 9).

The findings reveal that most districts neighboring the Sundarbans, known as the Sundarbans Impact Zone (SIZ), are experiencing significant impacts of climate change, also evidenced by the socio-economic conditions of the residents. The accelerating impacts of climate change, including rising sea levels, increased cyclonic activity, salinity intrusion, and land erosion, are posing significant challenges to the traditional building practices in the area. Changes in temperature, precipitation patterns, and the frequency of extreme weather events necessitate adaptive measures in building design and construction techniques to achieve resilience against climatic hazards. Furthermore, reduced forest areas due to climate change exacerbates the scarcity of natural building materials, traditionally sourced from the Sundarbans. This scarcity directly influences architectural practices as locals have already started to choose readily available building materials and adjust traditional techniques to accommodate modern alternatives. Hence, an urgent need to find sustainable approaches has become a significant concern in retaining cultural heterogeneity.

One of the indirect effects of climate change on vernacular architecture in the Sundarbans delta region stems from the dependence of local communities on the Sundarbans for livelihood. As environmental degradation reduces forest productivity and alters ecosystems as well as methods of survival, it prompts changes in occupation and livelihood strategies of the local inhabitants. It also causes shifts in their needs and priorities, thus impacting the architectural language. The findings of this study show that apart from disaster management initiatives, the scope of income generation opportunities of these coastal people needs to be taken into consideration as economic constraints influence construction choices and practices. Facilitating easier access to sustainable materials through economic assistance can support communities in upholding their unique architectural designs. Policy changes and interventions aimed at mitigating the impacts of climate change and promoting

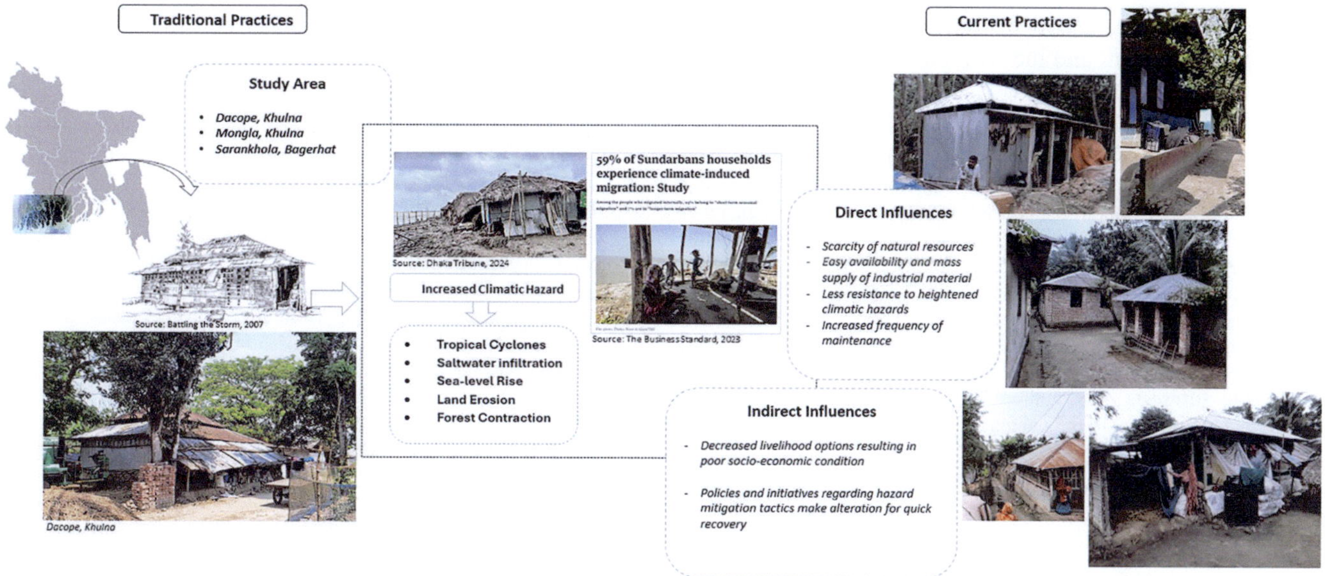

Fig. 9 Analogy of the influence of climate change on the Sundarbans delta's vernacular architecture

disaster resilience often initiate alterations in traditional practices through the adoption of modern construction materials and techniques. However, vernacular buildings are inherently adapted to their environments, offering valuable insights into resilient and energy-efficient design practices (Daoudi et al. 2019). Our research indicates that the traditional architectural language of the area under study is disappearing due to the pressure of rapid adaptation and mass execution. Nevertheless, it is possible to incorporate vernacular principles into policies and development strategies by integrating them into building codes, providing financial support and incentives to the locals, investing in research, and encouraging community participation. Additionally, our observation reveals that government and non-government organizations predominantly adopt a top-down approach in making disaster-resistant residential units. However, a comprehensive, grassroot approach can aid us with a more resilient, sustainable, and culturally enriched coastal community. Thus, co-ordination among government and non-government efforts has become mandatory to protect the Sundarbans' ecosystem and its inhabitants from natural calamities and is required for preserving cultural heritage, adapting to new climatic conditions, promoting sustainability, and ensuring economic efficiency. With local materials becoming scarcer due to rising climate risks and the over-exploitation of Sundarbans mangrove resources, educational campaigns can play a pivotal role in enhancing locals' understanding and proper utilization of these materials sourced from the mangrove ecosystem.

7 Conclusion

In conclusion, climate change is exerting significant pressure on transforming vernacular architecture in the Sundarbans Delta Region of Bangladesh. Through direct and indirect pathways, climate change impacts various aspects of the built environment, including material choices, construction methods, and architectural design. The findings of this study underscore the urgent need for a holistic approach to address the impacts of climate change on traditional vernacular architecture in the Sundarbans delta region. Efforts to ensure resilience must be informed by a deep understanding of the region's cultural heritage and ecological dynamics. While initiatives such as the Bangladesh Climate Change Strategy and Action Plan 2009, Disaster Management Act 2012, and National Plan for Disaster Management (2016–2020) have been implemented to mitigate the impacts of climate change on a broader scale, they often follow a top-down approach, excluding the local communities from the decision-making process. However, preserving traditional knowledge and architectural practices while incorporating innovative solutions is essential for ensuring communities' long-term sustainability and cultural continuity in the face of climate change. Hence, guidelines regarding the building construction process and material choice need to be derived from the combination of local practices and modern engineered techniques to fight against the vulnerabilities of climate change. Coordination between policymakers, researchers, and local communities is crucial in developing adaptive strategies that safeguard both the built environment and the intangible cultural heritage of the Sundarbans delta region.

Acknowledgements We extend our heartfelt gratitude to Mr. Mithun Nandi who has been working for several years for the economic empowerment of the inhabitants in the coastal region. His invaluable insights greatly facilitated our understanding of the local community, allowing seamless engagement with the people. Additionally, we express our sincere appreciation to Ms. Saba Islam for her assistance in documentation during the field survey of one of the study areas. Her contributions have enriched our research endeavors significantly.

Competing Interests The authors have no conflicts of interest to declare that are relevant to the content of this chapter.

References

Akber, M.A., Patwary, M.M., Islam, M.A., Rahman, M.R.: Storm protection service of the Sundarbans mangrove forest, Bangladesh. Nat. Hazards **94**, 405–418 (2018). https://doi.org/10.1007/s11069-018-3395-8

Akter, S., Ahmed, K. R., Marandi, A., Schuth, C.: Possible factors for increasing water salinity in an embanked coastal island in the southwest Bengal Delta of Bangladesh. Sci. Total Environ. 713 (2020). https://doi.org/10.1016/j.scitotenv.2020.136668

Alam, M.R., Kaish, A.B., Zain, M.F., Dev, S.K., Mahzabin, M.S.: Vulnerability assessment and construction recommendations of local houses in the cyclone prone coastal areas of Bangladesh. Int. J. Disaster Risk Reduction **21**, 118–130 (2017). https://doi.org/10.1016/j.ijdrr.2016.10.010

Anwar, S.B.: The mud houses of rural Shyamnagar battling climatic hazards. Dhaka Tribune, Dhaka (2022)

Ashrafuzzaman, M.: Local context of climate change adaptation in the south-western coastal region of Bangladesh. Sustainability 15(6664), 1–34 (2023). https://doi.org/10.3390/su15086664

Aziz, A., Paul, A. R.: Bangladesh sundarbans: present status of the environment and biota. Diversity, 7(3), 242–269 (2015). https://doi.org/10.3390/d7030242

Banholzer, S., Kossin, J., Donner, S.: The impact of climate change on natural disasters. In: Zommers, Z., Singh, A.: Reducing disaster: early warning systems for climate change (pp. 21–49). Springer, Dordrecht (2014). https://doi.org/10.1007/978-94-017-8598-3_2

Bhargava, R., Friess, D. A.: Previous shoreline dynamics determine future susceptibility to cyclone impact in the Sundarban mangrove forest. (A. G. Samaras, Ed.) Coastal Ocean Processes, 9, 1–12 (2022). https://doi.org/10.3389/fmars.2022.814577

Bhuiyan, R., Shirin, S., Paul, S.: A sustainable homestead prototype in sundarbans. In: FARU 2021—14th International Research Conference (pp. 287–298). University of Moratuwa (2021). https://doi.org/10.31705/FARU.2021.28

Brimblecombe, P., Bonazza, A., Brooks, N., Grossi, C. M., Harris, I., Sabbioni, C.: Impact of climate change on earthen buildings. In: Terra 2008: The 10th International Conference on the Study and Conservation of Earthen Architectural Heritage (pp. 278–282). The Getty Conservation Institute, Bamako, Mali (2008)

BSS (2023) Govt initiatives help improve Sundarbans ecosystem, biodiversity. Bangladesh: Dhaka Tribune. Retrieved April 04, 2024, from https://www.dhakatribune.com/bangladesh/334985/govt-initiatives-help-improve-sundarbans

Daoudi, N. S., Mestoul, D., Lamraoui, S., Boussoualim, A., Adolphe, L., Bensalem, R.: Vernacular architecture in arid climates: adaptation to climate change. In G. C. M. C. Guedes, Bioclimatic Architecture in Warm Climates (pp. 119–153). Springer Nature Switzerland AG (2019). https://doi.org/10.1007/978-3-030-12036-8_4

Dasgupta, S., Huq, M., Khan, Z.H., Ahmed, M.M., Mukherjee, N., Khan, M.F., Pandey, K.: Cyclones in a changing climate: the case of Bangladesh. Climate Dev. **6**(2), 96–110 (2014). https://doi.org/10.1080/17565529.2013.868335

Dasgupta, S., Wheeler, D., Sobhan, M.I., Bandyopadhyay, S., Paul, T.: Coping with Climate Change in the Sundarbans: Lessons from Multidisciplinary Studies. World Bank Publications, Washington (2020). https://doi.org/10.1596/978-1-4648-1587-4

Dutta, P., Rahman, D.: Learning from the root-integrating tradition into architecture towards a self-subsistent Munda community. Self-subsistent Munda community. Khulna University Studies, Special Issue (ICSTEM4IR), 878–891 (2022)

Eaton, R.M.: Human settlement and colonization in the Sundarbans, 1200–1750. Agric. Human Values **7**, 6–16 (1990). https://doi.org/10.1007/BF01530432

Flinn, B., Baresford, P.: *Post-Sidr Family Shelter Reconstruction Bangladesh.* Assesment Report (2009). Retrieved April 04, 2024, from https://www.humanitarianlibrary.org/resource/post-sidr-family-shelter-reconstruction-bangladesh-0

Ghosh, A., Schmidt, S., Fickert, T., Nusser, M.: The Indian Sundarban mangrove forests: history, utilization, conservation strategies and local perception. Diversity 149–169 (2015). https://doi.org/10.3390/d7020149

Goosen, H., Hasan, T., Saha, S. K., Rezwana, N., Rahman, M. R., Assaduzzaman, M., Scheltinga, C. T.: Nationwide Climate Vulnerability Assessment in Bangladesh. The Ministry of Environment, Forest and Climate Change, Government of the People‘s Republic of Bangladesh and GIZ (2018)

Gosh, M.K., Kumar, L., Roy, C.: Climate variability and mangrove cover dynamics at species level in the Sundarbans, Bangladesh. Sustainability **9**(5), 805 (2017). https://doi.org/10.3390/su9050805

Gutiérrez, E. S., Crété, E., Braedt, C., Morel, L. M., Molina, S.: Detailed shelter response profile: BANGLADESH | Local Building Cultures for sustainable and resilient habitat. Global shelter cluster. Psbwg (2018). Retrieved April 4, 2024, from https://craterre.hypotheses.org/2233

Hossain, R., Ahdan, M. M., Hossain, M. Z., Ashiq-ur-Rahman, M. Indigenous Knowledge And Practices For Cyclone. Solutions to Coastal Disasters, pp. 825–837 (2008). https://doi.org/10.1061/40968(312)74

Hossain, M. N., Paul, S. K., Roy, C., Hasan, M. M. (2014). Factors influencing human vulnerability to cyclones and storm surges in the coastal Bangladesh. J of Geo-Environment, 1–29 (2018)

Housing and Building Research Institute: standard guideline for rural housing in disaster prone areas of Bangladesh. Housing and Building Research Institute, Department of Disaster Management. Housing and Building Research Institute (2017). Retrieved April 04, 2024, from https://preparecenter.org/wp-content/uploads/2020/10/24BD02-Publication-housing-guidelines-1.pdf

Hussain, M. A., Islam, A. S., Hasan, M. A., Bhaskaran, B.: Changes of the seasonal salinity distribution at the Sundarbans coast due to impact of climate change. In: 4th International Conference on Water & Flood Management, pp. 565–574. ResearchGate (2013)

Islam, N. S.: Sundarbans a dynamic ecosystem: an overview of opportunities, threats and tasks. In H. S. Sen (Ed.), The sundarbans: a disaster-prone eco-region (Vols. Coastal Research Libary, Vol 30). Springer Nature (2019). https://doi.org/10.1007/978-3-030-00680-8_2

Li, S., Ding, K., Ding, A., He, L., Huang, X., Ge, Q., Fu, C.: Change of extreme snow events shaped the roof of traditional Chinese architecture in the past millennium. Sci. Adv. **7**, 1–8 (2021)

Linkon, S.B.: Autonomy in building process to adapt the climate change impacts: a study of the coastal settlements in Bangladesh. Int. J. Environ. Sustain. **6**(2), 19–39 (2017)

Mahadevia, K., Vikas, M.: 7 climate change – impact on the Sundarbans: a case study. Int. Sci. J. Environ. Sci. 7–15 (2012). Retrieved April 04, 2024, from https://www.researchgate.net/publication/311607858_Climate_Change_-_Impact_on_the_Sundarbans_A_case_study/citations

Mandal, U. K., Maji, B., Mullick, S., Nayak, D. B., Mahanta, K. K., Raut, S.: Global climate change and human interferences as risk factors, and their impacts on geomorphological features as well as on farming practices in Sundarbans eco-region. In: Sen, H. S.: (Ed.), The Sundarbans: a disaster-prone eco-region (Vols. Coastal Research Library, vol 30, pp. 405–437). Springer, Cham (2019). https://doi.org/10.1007/978-3-030-00680-8_14

Miah, M. Y., Kamal, M. Z., Salam, M. A., Islam, M. S. (2020). Impact of salinity intrusion on agriculture of Southwest Bangladesh—A review. Int. J. Agric. Policy Res., 8(2), 40–47 (2019). https://doi.org/10.15739/IJAPR.20.005

Mirza, M. M., Talpur, M. A., Nishat, A.: Mainstreaming climate change adaptation into policies in the sundarbans region: increasing livelihood security. In: H. S. Sen (Ed.), The Sundarbans: a disaster-prone eco-region increasing livelihood security, Vol. Coastal Research Library, Vol 30, pp. 555–577, 2019. https://doi.org/10.1007/978-3-030-00680-8_19

Nishat, A., Chowdhury, J. K.: Influence of climate change on biodiversity and ecosystems in the Bangladesh Sundarbans. In: Sen, H. (Ed.), The Sundarbans: a disaster-prone eco-region, Vols. Coastal Research Library, Vol 30, pp. 441–468. Springer, Cham (2019). https://doi.org/10.1007/978-3-030-00680-8_15

Pitchaikani, J. S., Sarma, K. S., Bhattacharyya, S.: First time report on the weather patterns over the Sundarban mangrove forest. Indian J. Geo-Marine Sci. 1–11 (2015)

Rahman, T.: Strengthening the coastal settlements of Bangladesh through planning. Asian J. Environ. Disaster Manage. **3**(4), 389–403 (2011). https://doi.org/10.3850/S1793924011000976

Rahman, M.M.: May 28). Impact of increased salinity on the plant community of the Sundarbans Mangrove of Bangladesh. Community Ecol. **21**, 273–284 (2020). https://doi.org/10.1007/s42974-020-00028-1

Rahman, M. R., Muhaymin, A. A., Rahman, M. M.: Adaptation to climate change through disaster-resilient housing practice in the coastal zone of Bangladesh. In: International Conference on Disaster Risk Management. Dhaka (2019)

Rahman, M. M., Hossain, R. M., Islam, M. N.: Climate change impact on Sundarbans: challenges for mitigation strategies. In: Jakariya, M., Islam, N. M. (Eds.) Climate change in Bangladesh a cross-disciplinary framework, pp. 47–64. Springer Climate (2021). https://doi.org/10.1007/978-3-030-75825-7

Rashid, M.U.: Adaptation to climate change in the built form: post sidr experience in Bangladesh. Int. J. Architect., Eng. Construct. **7**(2), 10–18 (2017). https://doi.org/10.7492/IJAEC.2018.008

Reza, A. M.: Preparation of Payra Kuakata comprehensive plan focusing on eco-tourism project examine the traditional architecture, urban form and pattern within the region. In: Urban Development Directorate (UDD, Ministry of Housing and Public Works. Government of the People's Republic of Bangladesh (2021). Retrieved April 04, 2024

Roy, B., Penha-Lopes, G. P., Uddin, M. S., Kabir, M. H., Lourenco, T. C., Torrejano, A.: Sea level rise induced impacts on coastal areas of Bangladesh. Int. J. Disaster Risk Reduct (2022). https://doi.org/10.1016/j.ijdrr.2022.102905

Rudra, A., & Chattopadhyay, A. (2019). Environmental change of coastal Sundarbans: Impact on livelihood and standard of living status of indigenous people. *Environ Qual Manage*, 77-84. https://doi.org/10.1002/tqem.21667

Sahana, M., Sajjad, H.: Vulnerability to storm surge flood using remote sensing and GIS techniques: a study on Sundarban Biosphere Reserve, India. Remote Sensing Appl.: Soc. Environ. **13**, 106–120 (2019). https://doi.org/10.1016/j.rsase.2018.10.008

Sakib, M., Nihal, F., Haque, A., Rahman, M., Ali, M.: Sundarban as a buffer against storm surge flooding. World J. Eng. Technol. **3**, 59–64 (2015). https://doi.org/10.4236/wjet.2015.33C009

Sen, H. S., Ghorai, D.: The Sundarbans: a flight into the wilderness. In: Sen, H. S.: The Sundarbans: a disaster-prone eco-region (Vol. 30). Springer Nature, Switzerland (2019). https://doi.org/10.1007/978-3-030-00680-8

Shaikh, A. A., Halder, M., Talukder, M. B., Mohibullah, S., Saha, S.: The dependency of coastal livelihood on forest resources, and alternative options in the periphery of the Sundarbans reserve forest, Patharghata, Bangladesh. Open J. Forestry 11(4) (2021). https://doi.org/10.4236/ojf.2021.114024

Subhani, R., Ahmad, M.M.: Socio-economic impacts of cyclone aila on migrant and non-migrant households in the southwestern coastal areas of Bangladesh. Geosciences **9**(11), 482 (2019). https://doi.org/10.3390/geosciences9110482

Sustainability of Climate Change Adaptation Practices in. (2019). J. Sustain. Develop. 12(5), 1–22. https://doi.org/10.5539/jsd.v12n5p1

The World Bank: Urgent Climate Action Crucial for Bangladesh to Sustain Strong Growth (2022). The World Bank. Retrieved from https://www.worldbank.org/en/news/press-release/2022/10/31/urgent-climate-action-crucial-for-bangladesh-to-sustain-strong-growth#:~:text=Climate%20change%20will%20hit%20poor,as%20much%20as%209%20percent.

The Government of Bangladesh. (2008). Cyclone Sidr in Bangladesh Damage, Loss, and Needs Assessment for Disaster Recovery and Reconstruction. Economic Relations Division. Ministry of Finance. Retrieved April 4, 2024, from https://modmr.portal.gov.bd/sites/default/files/files/modmr.portal.gov.bd/publications/83ccdd3b_1e9f_4e4 3_9583_cafeed377d73/AssessmentReport_CycloneSidr_Bangladesh_2008 (Worldbank).pdf

UNESCO. The Sundarbans (n.d.). Retrieved March 26, 2024, from UNESCO World Heritage Convention: https://whc.unesco.org/en/list/798/

Conclusion

In a time, when the world is facing rapid urbanization and global warming, the book "Geographic Approaches to Climate Change and Mitigation: Urban and Rural Perspectives" aims to be an important platform to explore the different research directions between urban and rural geography and climate change. This book volume presents different climate change issues, highlighting their profound impact on both urban and rural environments. This provides valuable information to urban planners and climate researchers through the process of understanding the geographic contexts in reshaping the urban and rural vulnerabilities and planning for effective response to the change climate by adopting suitable mitigation strategies.

It is no doubt that urban areas are at the forefront of global warming and will have to face adverse consequences due to rapid urbanization and high population density. Though cities serve as the center of innovation and development, they also significantly contribute to greenhouse gas emissions. This book provides case studies on how urban areas can harness their resources and improve the livelihood of people in the face of climate change. It highlights some important sustainable development issues and the importance of blue-green infrastructure and community engagement in creating urban resilience cities against future change climatic impacts.

On the contrary, rural areas are often overlooked in climate change discussions and talks. Although they contribute less to global warming, they still play a significant role in the narrative of climate change mitigation and adaptation. Rural communities also face the consequence of global warming by witnessing more frequent and intense extreme weather events, altered precipitation pattern, and threats to food security. This book raises the issue through discussing climate change's impact on rural communities and provides the innovation and effective strategies to adopt and thrive in the changing climate.

By integrating geographic perspectives and urban-rural environments, this book emphasizes the need for effective research collaboration and explores the interplay of urban and rural areas together in climate change context. Recognizing the urban demand for resources, the rural community's sustainability influences and impacts the rural ecological health. Researchers, practitioners, and policymakers should work together and share their wealth of knowledge across the urban-rural climate change impact by understanding rural land use and its relationship with urban environment using innovative geospatial mapping techniques.

The case studies and empirical research present in this book volume serve as the backbone for researchers, urban planners, decision-makers for effecting urban and rural planning. This book showcases successful research examples and learned lessons across diverse geographic contexts illustrating the need for improved strategies to combat climate change impact.

Ultimately, we hope this book inspires the resilient urban and rural communities to actively contribute to sustainable development by adopting effecting mitigation strategies and developing an equitable future for everyone.

P. Mohammad et al. (eds.), *Geographic Approaches to Climate Change and Mitigation: Urban and Rural Perspectives (Volume 1)*,
Advances in Science, Technology & Innovation, https://doi.org/10.1007/978-3-031-92119-3

Curriculum Vitae

Editor name	Dr. Pir Mohammad
Affiliation	Department of Land Surveying and Geo-Informatics, The Hong Kong Polytechnic University
Institution	The Hong Kong Polytechnic University
Address	ZN613, 6/F North Wing, Block Z Phase 8, 181 Chatham Road South Department of Land Surveying and Geo-Informatics The Hong Kong Polytechnic University Kowloon City, Kowloon Hong Kong
Postal Code	00000
City	Kowloon
Country	Hong Kong
Telephone number	+852-55932487
E-mail	pir.mohammad@polyu.edu.hk
Homepage URL	https://sites.google.com/view/pir-mohammad/home?authuser=0
ORCID ID	https://orcid.org/0000-0002-0075-6372

Dr. Pir Mohammad is a Research Assistant Professor in the Department of Land Surveying and Geo-Informatics at The Hong Kong Polytechnic University. He earned his PhD in Remote Sensing from the Indian Institute of Technology Roorkee, specializing in urban heat island studies in Indian cities, and was recognized with the Excellence in Doctoral Research Award for his exceptional contributions. Prior to his current role, Dr. Mohammad served as a Postdoctoral Fellow in the same department at The Hong Kong Polytechnic University. His research interests encompass remote sensing, urban climate, extreme weather events, heat mitigation, and thermal comfort. Dr. Mohammad has authored 23 peer-reviewed publications in esteemed journals and actively reviews for numerous academic publications. He has presented his research findings at various national and international conferences, further contributing to the academic community. He is also a life member of the Indian Society of Remote Sensing (ISRS), the Indian Society of Geomatics (ISG), and the Indian Science Congress Association (ISCA).

P. Mohammad et al. (eds.), *Geographic Approaches to Climate Change and Mitigation: Urban and Rural Perspectives (Volume 1)*,
Advances in Science, Technology & Innovation, https://doi.org/10.1007/978-3-031-92119-3

Editor name	Dr. Shahfahad
Affiliation	Department of Geography, Institute of Science
Institution	Banaras Hindu University
Address	Varanasi, Uttar Pradesh
Postal Code	221005
City	Varanasi
Country	India
Telephone number	(+91) 8800343539
E-mail	fahadshah921@gmail.com
Homepage URL	https://www.scopus.com/authid/detail.uri?authorId=57214364016
ORCID ID	https://orcid.org/0000-0001-5868-1062

Dr. Shahfahad is Assistant Professor of Geography at Department of Geography, Institute of Science, Banaras Hindu University (BHU). Before joining BHU, Dr. Shahfahad worked as Assistant Professor of Geography at Department of Geography, Faculty of Sciences, Jamia Millia Islamia during 2023–24. He completed his B.A. (H) Geography in 2015, M.A. Geography in 2017 and Ph.D. in 2022 from the Department of Geography, Jamia Millia Islamia, New Delhi. He has also done Post-Graduate Diploma in Geoinformatics from Centre for Distance and Online Learning, Jamia Millia Islamia during 2022–23. He was awarded prestigious National Post Doctoral Fellowship (N-PDF) by Science and Engineering Research Board (SERB) of Department of Science and Technology (DST) in 2024. He was junior research fellow (JRF) during 2017–2019 and senior research fellow (SRF) during 2019–2022 in the Department of Geography, Jamia Millia Islamia. His core areas of research are urban expansion, climate change, urban hydrology as well as applications of geospatial techniques and artificial intelligence in urban studies and natural hazards. He has extensively studied the climate pattern and change over the metropolitan cities of Indian subcontinent. He has published two edited book volume and more than 80 scientific research articles in the journals of international repute such as Habitat International, Urban Climate, Journal of Environmental Management, Scientific Reports, Ecological Indicators, Remote Sensing, Geocarto International and Ecological Informatics. Beside this, he is currently editorial board member of two internal journals. Also, he has served as the reviewer for several reputed international journals.

Editor name	Jesús Rodrigo-Comino
Affiliation	Departamento de Análisis Geográfico Regional y Geografía Física
Institution	University of Granada
Address	Facultad de Filosofía y Letras, Campus Universitario de Cartuja, Universidad de Granada
Postal Code	18071
City	Granada
Country	Spain
E-mail	jesusrc@ugr.es
Homepage URL	www.egemap.eu
ORCID ID	https://orcid.org/0000-0002-4823-0871

Ph.D. DEng Jesús Rodrigo Comino, a graduate in Geography, currently works as an Associate Professor at the University of Granada and has been included in the World's Top 2% Scientists ranking by Stanford University since 2021. He got a Master in Territorial Planning and Geographic Information Systems (2013) at the University of Málaga/Granada, whose final work was divided into 3 national publications and a monograph. During his predoc stage, he obtained three scholarships for doctoral studies: DAAD (German Academic Exchange Service), La Caixa Foundation and FPU (Ministry of Education, Spain). During this period, he completed his first doctoral thesis in Geography between the University of Trier (Germany -2 years-) and Malaga (2 years) in 2018: "Actual geomorphological processes in sloping vineyards. A comparison between Ruwer-Mosel Valley (Trier, Germany) and Montes de Málaga (Málaga, Spain)". In 2023, he defended his second doctorate in engineering in Geomatics and Topography at the Polytechnic University of Valencia. His research career consists of four complete monographs (Nova, Springer, etc.) and some edited books (Elsevier, CRC, Nova), 200 indexed publications in Scopus and WoS, leading international collaborations with research teams over the world. Regular reviewer in more than 150 indexed international journals, a member of 2 doctoral theses committees, and an evaluator of projects for the Ministries of Science of Chile, Peru, the United States, Serbia, Switzerland, Kazakhstan or Poland, and postgraduate scholarships for DAAD. He has organized several international scientific meetings and congresses (for example, COST Actions, Biohydrology, Fire in the Earth, etc.), sessions at international conferences (EGU, TerraEnvision, Conference of the Arabian Journal of Geosciences, etc.), oral presentations and conferences master classes (Germany, Bulgaria, Norway, Algeria, etc.). Editor-in-chief of the indexed journal (Scopus and ESCI; Q2) Air Soil and Water Research (SAGE) and Euro-Mediterranean Journal for Environmental Integration (Springer). In addition, he was Associate Editor at Scientific Reports (Nature) and Hydrological Science Journal (Taylor & Francis) and continues in the Arabian Journal of Geosciences (Springer), and Journal of Mountain Science (Springer). He has participated as a researcher in R + D + I projects on social issues related to housing or the census, or transfer and knowledge at a European level, such as the INTERREG Smart-Light HUB project (light pollution) or COST FIRElinks (fires). He has been invited to give lectures on agriculture, sustainable management and erosion. He has supervised 5 final degree projects and 3 master's degrees. He has taught regulated and certified teaching at the Universities of Granada, Valencia, Málaga, León, Oviedo, Trier (in German), Quito (Ecuador) or La Habana (Cuba) on geopolitics, geomorphology, Geographic Information Systems, remote sensing and statistical techniques. He acted as PI of some international (FIRElinks COST Action), national (BBVA Foundation) and regional projects and supervise several Ph.D. theses. Nowadays, he coordinate one European project H2030 (Marie Curie Staff Exchange), national project (Plan Nacional) and regional transference project (Grupos Operativos, Junta de Andalucía). Moreover, participated in other two European projects coordinating activities in different WPs (Erasmus+: Urban Inprint, and Soil Living Labs: SOILCRATES).

Editor name	Abraham R. Matamanda
Affiliation	Department of Urban and Regional Planning
Institution	University of the Free State
Address	205 Nelson Mandela Drive, Park West
Postal Code	9301
City	Bloemfontein
Country	South Africa
Telephone number	+27 639557097
E-mail	matamandaa@gmail.com
ORCID ID	https://orcid.org/0000-0001-5260-5560

ABRAHAM R MATAMANDA (Ph.D. Urban and Regional Planning) is an NRF Y2-rated Urban and Regional Planner. His lectures at the University of Free State (UFS) and is also the Deputy-Chair of the SARChI Chair on City Region Economies in the faculty of Economic and Management Sciences at UFS. Currently, he serves as the co-editor of the Town Planning Journal published by UFS and serves on the academic editorial board of Plos Water Journal. He is also the current Chairperson of the Free State Chapter of the South African Planning Institute (SAPI). He is a fellow of the Department of Higher Education and Training (DHET) Future Professorate Programme Phase 1, third Cohort. He is the South African PI for a global collaborative research project exploring how children and young people from monetary poor households have adapted to the COVID-19 pandemic focusing on food, education and play/leisure. His research focuses on urban governance and planning, climate change adaptation and resilience, informal Global South urbanism, urban land markets and housing studies. He has authored and edited several books published with Springer Nature including: Urban Geography in Postcolonial Zimbabwe: Paradigms and Perspectives for Sustainable Urban Planning and Governance; Housing and Technology: Special Focus on Zimbabwe; Urban Infrastructure in Zimbabwe: Departures, Divergences and Convergences; and Secondary Cities and Local Governance in Southern Africa.

MIX
Papier aus verantwortungsvollen Quellen
Paper from responsible sources
FSC® C105338

If you have any concerns about our products,
you can contact us on
ProductSafety@springernature.com

In case Publisher is established outside the EU,
the EU authorized representative is:
Springer Nature Customer Service Center GmbH
Europaplatz 3, 69115 Heidelberg, Germany

Printed by Libri Plureos GmbH
in Hamburg, Germany